8th Symposium on Frequency Standards and Metrology 2015

Journal of Physics: Conference Series Volume 723

Potsdam, Germany
12 – 16 October 2015

Editor:

Fritz Riehle

ISBN: 978-1-5108-2611-3
ISSN: 1742-6588

Printed from e-media with permission by:

Curran Associates, Inc.
57 Morehouse Lane
Red Hook, NY 12571

Some format issues inherent in the e-media version may also appear in this print version.

Printed by Curran Associates, Inc. (2016)

For permission requests, please contact the Institute of Physics at the address below.

Institute of Physics
Dirac House, Temple Back
Bristol BS1 6BE UK

Phone: 44 1 17 929 7481
Fax: 44 1 17 920 0979

techtracking@iop.org

Additional copies of this publication are available from:

Curran Associates, Inc.
57 Morehouse Lane
Red Hook, NY 12571 USA
Phone: 845-758-0400
Fax: 845-758-2633
Email: curran@proceedings.com
Web: www.proceedings.com

Table of contents

Volume 723

8th Symposium on Frequency Standards and Metrology 2015
12–16 October 2015, Potsdam, Germany

Accepted papers received: 1 June 2016
Published online: 4 July 2016

Preface

Papers

Keynote lecture

Microwave Clocks

Time and Frequency Transfer

Tests of Fundamental Physics with Clocks

Novel Concepts and Applications

Preface

The eighth Symposium on Frequency Standards and Metrology [1] has been organized from 12 to 16 October in the Seminaris SeeHotel near Lake Templin at Potsdam, Germany. The symposium included of a keynote lecture by David Wineland, 48 invited oral presentations and a session with 111 invited poster contributions. The symposium is the most recent one of a series that was initially organized by Jacques Vanier in 1971 in Forét Montmorency in Quebec, Canada. The next ones took place in 1976 in Copper Mountain, USA, organized by Helmut Hellwig, 1981 in Aussois, France (Claude Audoin) [2], 1988 in Ancona, Italy (Andrea De Marchi) [3], 1995 in Woods Hole, USA (James Bergquist) [4], 2001 in St Andrews, UK (Patrick Gill) [5], and 2008 Pacific Grove, USA (Lute Maleki) [6]. The symposium series serves as an international discussion forum on precision frequency standards throughout the electromagnetic spectrum and associated metrology. It focuses on the fundamental scientific aspects of the latest ideas, results and applications in relation to these frequency standards. During the seven years after the last symposium very significant progress has occurred in various associated fields. Carrying forward the historical sketch by David Wineland [7] it is interesting to wrap up the new ideas, the novel technologies, developments and inventions that have been created and implemented in the seven years since the last symposium.

In the microwave domain the accuracy of caesium fountain clocks as the primary standards of time and frequency approaches 10^{-16} [8] and small scale, miniaturized atomic clocks [9] and sensors are further developed with respect to accuracy with ample applications in science, technology [10] and life sciences [11].

Today, optical clocks are about to make the vision of Hans Dehmelt [12] a reality. Several optical clocks capable of estimated fractional uncertainties in the 10^{-18} regime, both based on neutral atoms in a lattice and trapped single ions, have been presented at the symposium. In the latter case this was only possible with newly developed schemes for the interrogation that are capable to remove the frequency shifts associated with strong light field necessary for the interrogation [13]. The achieved accuracy already asks for a new definition of the unit of time [14].

The verification of those breath-taking uncertainties by comparison of clocks at different laboratories is one of the challenges today. Conventional methods of Two-Way Satellite Time and Frequency Transfer have already been used to make intercontinental comparisons [15] with fractional uncertainty in the low 10^{-15} range. The uncertainty associated with the transfer via the microwave links may be reduced e.g. in the ACES project. Novel methods that can achieve this goal better use transportable frequency standards [16], the evaluation of direct frequency ratios between optical frequency standards [17, 18], or long haul fiber links for remote frequency comparisons that have been implemented [19].
Approaching 10^{-18} fractional uncertainty optical atomic clocks in conjunction with accurate links can serve as accurate measuring devices for the geopotential with height resolution in the cm range thereby entering the era of a relativistic geodesy [20].

Besides accuracy, long-lived optical transitions hold the promise of optical clocks with unprecedented low instability if the transitions can be interrogated by lasers of adequate stability and phase coherence. The past years have experienced tremendous progress on various fronts. The more classical approach stabilizes the interrogating laser to a high finesse optical cavity. In the past decade the different methods of vibration-insensitive mounting have made available lasers with about one Hertz linewidth available in virtually every laboratory when required. There was, however, a limit given by the Brownian motion of the spacer [21], the mirror substrates and the mirror coatings. Use of long cavities [22], crystalline spacers and substrates, cryogenic operating temperatures [23] have led to instabilities below the 10^{-16} regime in 1 to 100 seconds. The recent development of crystalline mirrors [24] and other achievements promise instabilities in the low 10^{-17} regime and below. Unconventional other approaches rely on spectral-hole burning [25] or emitting photons on an ultra-narrow clock transition into the mode of a high Q resonator [26]. This progress was paralleled by a better understanding of the relevant noise processes and their experimental characterization [27] and theoretical modelling [28].

Spectroscopy on highly charged ions together with the associated technology enter now the field of frequency metrology and deliver exciting results [29]. They have the prospect to lead to alternative frequency standards and to push the field of precision tests of theoretical approaches.

Clocks and oscillators in the optical and microwave domain are by far the most accurate measuring instruments and therefore are also utilized for investigating fundamental questions like possible variations of fundamental constants [30] or the effects of dark energy [31]. Nuclear transitions might extend these opportunities [32]. Femtosecond combs [33] have been further miniaturized, have been tested in microgravity and have found novel applications as astro-combs.

Finally, I would like to express my cordial thanks to the sponsors, the Governing Board, the International Steering Committee, the Local Organizing Committee, the session chairs and the reviewers who helped with the realization of the symposium and the preparation of these proceedings.

The Editor

Fritz Riehle

References

[1] See: www.ptb.de/8fsm2015

[2] *Le Journal de Physique Colloques - Third Symposium on Frequency Standards and Metrology J. Phys. Colloques* **42** No. C8 (Décembre 1981) http://jphyscol.journaldephysique.org/en/articles/jphyscol/abs/1981/08/contents/contents.html

[3] *Proc. 4th Symposium on Frequency Standards and Metrology, Ancona, 1988* ed A De Marchi (Springer Verlag Berlin ISBN 3-540-50818-X)

[4] *Proc. 5th Symposium on Frequency Standards and Metrology, Woods Hole, 1995 ed J C Bergquist (World Scientific Singapore ISBN 981-02-2527-X)*

[5] *Proc. 6th Symposium on Frequency Standards and Metrology, St Andrews, 2001* ed P Gill (World Scientific New Jersey ISBN 981-02-4911-X)

[6] *Proc. 7th Symposium on Frequency Standards and Metrology, Pacific Grove, 2008* ed L Maleki (World Scientific New Jersey ISBN – 10 981-283-821-X)

[7] Wineland D 2016 The evolution of the Frequency Standards and Metrology symposium and its physics *J. Phys. Conf. Ser.* **723**

[8] Szymaniec K , Lea S N , Gibble K, Park S E, Liu K, Głowacki P 2016 NPL Cs fountain frequency standards and the quest for the ultimate accuracy *J. Phys. Conf. Ser.* **723**
[9] Gharavipour M, Affolderbach C, Kang S, Bandi T, Gruet F, Pellaton M, Mileti C 2016 High performance vapour-cell frequency standards *J. Phys. Conf. Ser.* **723**
[10] Kitching J et al 2016 NIST on a Chip: Realizing SI units with microfabricated alkali vapour cells *J. Phys. Conf. Ser.* **723**
[11] Knappe S, Alem O, Sheng D, Kitching J 2016 Microfabricated Optically-Pumped Magnetometers for Biomagnetic Applications *J. Phys. Conf. Ser.* **723**
[12] Dehmelt H 1981 Coherent spectroscopy on single atomic system at rest in free space II *J. Phys. Colloques* **42** C8-299 - 305
[13] Zanon-Wilette T, Minissale M, Yudin V I, Taichenachev A V 2016 Composite pulses in Hyper-Ramsey spectroscopy for the next generation of atomic clocks *J. Phys. Conf. Ser.* **723**
[14] Gill P 2016 Is the time right for a redefinition of the second by optical atomic clocks? *J. Phys. Conf. Ser.* **723**
[15] Fujieda M, Gotoh T, Amagai J 2016 Advanced two-way satellite frequency transfer by carrier-phase and carrier-frequency measurements *J. Phys. Conf. Ser.* **723**
[16] Vogt S et al 2016 A transportable optical lattice clock *J. Phys. Conf. Ser.* **723**
[17] Nemitz N et al 2016 Frequency ratio of Yb and Sr clocks with 5×10^{-17} uncertainty at 150 seconds averaging time *Nature Photon.* **10** 258 - 261
[18] Margolis H S, Gill P 2016 Determination of optimized frequency and frequency ratio values from over-determined sets of clock comparison data *J. Phys. Conf. Ser.* **723**
[19] Lisdat C et al 2015 A clock network for geodesy and fundamental science *arXiv:1511.07735*
[20] Flury J 2016 Relativistivc Geodesy *J. Phys. Conf. Ser.* **723**
[21] Numata K, Kemery A, Camp J 2004 Thermal-Noise Limit in the Frequency Stabilization of Lasers with Rigid Cavities *Phys. Rev. Lett.* **93** 250602
[22] Häfner S et al 2015 8×10^{-17} fractional laser frequency instability with a long room-temperature cavity *Opt. Lett.* **40** 2112-2115
[23] Matei D G 2016 A second generation of low thermal noise cryogenic silicon resonators *J. Phys.Conf. Ser.* **723**
[24] Cole G D, Zhang W, Martin M J, Ye J, Aspelmeyer, M 2013 Tenfold reduction of Brownian noise in optical interferometry *Nature Photonics* **7** 644-650
[25] Cook S, Rosenband T, Leibrandt D R 2015 Laser frequency stabilization based on steady-state spectral-hole burning in Eu^{3+}:Y_2SiO_5 *Phys. Rev. Lett.* **114** 253902
[26] Meiser D, Ye J, Carlson D R, Holland M J 2009 Prospects for a Millihertz-Linewidth Laser *Phys. Rev. Lett.* **102** 163601
[27] Zhang W et al 2011Advanced noise reduction techniques for ultra-low phase noise optical-to-microwave division with femtosecond fiber combs *IEEE Trans. Ultrason. Ferroelectr. Freq. Control*, **58** 900-908
[28] Benkler E, Lisdat C, Sterr U 2015 On the relation between uncertainties of weighted frequency averages and the various types of Allan deviations *Metrologia* **52**, 565
[29] Crespo López-Urrutia J R Frequency metrology using highly charged ions 2016 *J. Phys. Conf. Ser.* **723**
[30] Huntemann N, Sanner C, Lipphardt B, Tamm C, Peik E 2016 Single-Ion Atomic Clock with 3×10^{-18} Systematic Uncertainty *Phys. Rev. Lett.* **116** 063001
[31] Derevianko A 2016 Atomic clocks and dark-matter signatures *J. Phys. Conf. Ser.* **723**
[32] Stellmer S, Schreitl M, Kazakov G, Yoshimura K, Schumm T (2016) Towards a measurement of the nuclear clock transition in ^{229}Th *J. Phys. Conf. Ser.* **723**
[33] Cole D C, Beha K M, Diddams S A, Papp S B 2016 Octave-spanning supercontinuum generation via microwave frequency multiplication *J. Phys. Conf. Ser.* **723**

Journal of Physics: Conference Series **723** (2016) 011002 doi:10.1088/1742-6596/723/1/011002

Governing Board

Claude Audoin, France

James C. Bergquist, USA

Patrick Gill, Great Britain

Andrea de Marchi, Italy

Lute Maleki, USA

Jacques Vanier, Canada

Symposium Chair

Fritz Riehle, Germany

International Steering Committee

Claude Audoin, France

Sébastien Bize, France

James Bergquist, USA

Andrea de Marchi, Italy

Kurt Gibble, USA

Patrick Gill, Great Britain

Feng-Lei Hong, Japan

Massimo Inguscio, Italy

Hidetoshi Katori, Japan

John Kitching, USA

Tianchu Li, China

Long-Shen Ma, China

Alan Madej, Canada

Lute Maleki, USA

Vitalij G. Pal'chikov, Russia

Gerard Petit, France

Pascal Rochat, Switzerland

Christophe Salomon, France

Stephan Schiller, Germany

Michael Tobar, Australia

Wim Ubachs, The Netherlands

Thomas Udem, Germany

Jacques Vanier, Canada

Bruce Warrington, Australia

Jun Ye, USA

Journal of Physics: Conference Series **723** (2016) 011002 doi:10.1088/1742-6596/723/1/011002

Local Organizing Committee

Ekkehard Peik, PTB, Germany

Andreas Bauch, PTB, Germany

Christian Lisdat, PTB, Germany

Piet Schmidt, PTB, Germany

Nadine Weber, PTB, Germany

Ingrid Herrmann, PTB, Germany

Christine Werner, PTB, Germany

Journal of Physics: Conference Series **723** (2016) 011003 doi:10.1088/1742-6596/723/1/011003

Sponsors:

Physikalisch-Technische Bundesanstalt (PTB), Braunschweig, Germany

---

Helmholtz-Fonds e.V., Germany

National Institute of Standards and Technology (NIST), USA

First TF, Paris, France

Observatoire de Paris, SYRTE, France

Centre National de la Recherche Scientifique, France

Laboratoire national de métrologie et d‘essais, France

Deutsche Forschungsgemeinschaft, Germany

Journal of Physics: Conference Series **723** (2016) 011004 doi:10.1088/1742-6596/723/1/011004

Symposium Photographs

Seminaris SeeHotel, Potsdam. Venue of the 8th Symposium on Frequency Standards and Metrology

In his keynote lecture, Dr Dave Wineland recalls earlier breakthroughs presented at previous symposia.

Journal of Physics: Conference Series **723** (2016) 011004 doi:10.1088/1742-6596/723/1/011004

The audience

Journal of Physics: Conference Series **723** (2016) 011004 doi:10.1088/1742-6596/723/1/011004

Journal of Physics: Conference Series **723** (2016) 011004 doi:10.1088/1742-6596/723/1/011004

Excursion I: The great refractor in the Einstein Tower

Visiting the Einstein Tower

Excursion II: Visiting palace Sanssouci in Potsdam in the rain

Excursion III: Visiting Einstein's summerhouse in Caputh

At the banquet

Handing over the prizes

Journal of Physics: Conference Series **723** (2016) 011004 doi:10.1088/1742-6596/723/1/011004

The first prize: the cuckoo clock is also a frequency standard

Journal of Physics: Conference Series **723** (2016) 011005 doi:10.1088/1742-6596/723/1/011005

Participants of the 8FSM, 2015

Name	Country
Moustafa Abdel Hafiz	France
Michel Abgrall	France
Claude Audoin	France
Anne Amy-Klein	France
Geoffrey Barwood	United Kingdom
Ahmad Bawamia	Germany
Scott Beattie	Canada
Matthias Beck	Germany
Kyle Beloy	United States
Erik Benkler	Germany
Anthony Bercy	France
James Bergquist	United States
Nicolò Beverini	Italy
Sébastien Bize	France
Dmitri Boiko	Switzerland
Rodolphe Boudot	France
Denis Brazhnikov	Russia
Pierre Brochard	Switzerland
Eric Burt	United States
James Camparo	United States
Jingbiao Chen	China
Garrett Cole	United States
Shon Cook	United States
Jose Crespo Lopez-Urrutia	Germany
Magnus Danielson	Sweden
Andrea De Marchi	Italy
Luigi De Sarlo	France
Argyrios Dellis	United States
Andrei Derevianko	United States
Laurent Devenoges	Switzerland
Scott Diddams	United States
Alexandre Didier	France
Elizabeth Donley	United States
Albane Douillet	France
Pierre Dube	Canada
Kjeld Eikema	Netherlands
Sergey Ermak	Russia
Jeremy Everard	United Kingdom
Isaac Fan	Germany
Fang Fang	China
Zhanjun Fang	China
Jakob Flury	Germany
Christian Freier	Germany
Miho Fujieda	Japan
Naceur Gaaloul	Germany
Vladislav Gerginov	Germany
Kurt Gibble	United States
Patrick Gill	United Kingdom
Vincent Giordano	France

Przemyslaw Glowacki	Germany
Rachel Godun	United Kingdom
Christian Grebing	Germany
Gesine Grosche	Germany
Hua Guan	China
Mikhail Gubin	Russia
Stephane Guerandel	France
Hong Guo	China
Hidekazu Hachisu	Japan
Wolfgang Hänsel	Germany
James Hanssen	United States
John Hartnett	Australia
Archita Hati	United States
Ian Hill	United Kingdom
Jonathan Hirschauer	United States
Ronald Holzwarth	Germany
Feng-Lei Hong	Japan
Matthew Hummon	United States
Nils Huntemann	Germany
Tetsuya Ido	Japan
Antoine Jallageas	Switzerland
Steven Jefferts	United States
Bin Jian	Canada
Haifeng Jiang	China
Yanyi Jiang	China
Mark Kasevich	United States
Hidetoshi Katori	Japan
Yann Kersalé	France
Attila Kinali	Germany
John Kitching	United States
Svenja Knappe	United States
Jeroen Koelemeij	Netherlands
Sebastian Koke	Germany
Evgeny Kovalchuk	Germany
Jochen Kronjaeger	United Kingdom
Alexander Kuhl	Germany
Clement Lacroute	France
Arnaud Landragin	France
Won-Kyu Lee	Republic of Korea
Thomas Legero	Germany
David Leibrandt	United States
Ian Leroux	Germany
Filippo Levi	Italy
Yige Lin	China
Thomas Lindvall	Finland
Burghard Lipphardt	Germany
Christian Lisdat	Germany
Peng Liu	China
Xiaochi Liu	United States

Qu Liu	China
Olivier Lopez	France
J. Mauricio López Romero	Mexico
Zehuang Lu	China
Andrew Ludlow	United States
Longsheng Ma	China
Lothar Maisenbacher	Germany
Lute Maleki	United States
Helen Margolis	United Kingdom
Jonas Matthias	Germany
Tanja Mehlstäubler	Germany
Sinda Mejri	France
Salvatore Micalizio	Italy
Gaetano Mileti	Switzerland
Jacques Millo	France
Jacques Morel	Switzerland
Atsuo Morinaga	Japan
Piotr Morzynski	Poland
Moritz Nagel	Germany
Janko Nauta	Germany
Nils Nemitz	Japan
Alexander Nevsky	Germany
Nathan Newbury	United States
Maksim Okhapkin	Germany
Atsushi Onae	Japan
Scott Papp	United States
Sang Eon Park	Republic of Korea
Sayan Patra	Netherlands
Marek Peca	Czech Republic
Ekkehard Peik	Germany
Steven Peil	United States
Achim Peters	Germany
Michael Petersen	Switzerland
Gerard Petit	France
John Prestage	United States
Franklyn Quinlan	United States
Ernst Rasel	Germany
Fritz Riehle	Germany
Felix Rohde	Germany
Peter Rosenbusch	France
Daniele Rovera	France
Enrico Rubiola	France
Christophe Salomon	France
Wolfgang Schaefer	Germany
Stephan Schiller	Germany
Stephane Schilt	Switzerland
Dennis Schlippert	Germany
Piet Schmidt	Germany
Lisa Schmöger	Germany

Harald Schnatz	Germany
Christian Schubert	Germany
Thilo Schuldt	Germany
Peter Schwindt	United States
Benedikt Seiferle	Germany
Yong Shen	China
Shinsuke Shichi	Japan
Yeshpal Singh	United Kingdom
Sergey Slyusarev	Russia
Lukasz Sobolewski	Poland
Christian Stenzel	Germany
Uwe Sterr	Germany
Sergey Strelkin	Russia
Thomas Südmeyer	Switzerland
Denis Sutyrin	Germany
Krzysztof Szymaniec	United Kingdom
Akifumi Takamizawa	Japan
Masao Takamoto	Japan
Christian Tamm	Germany
Robert Tjoelker	United States
Michael Tobar	Australia
Lutz Trahms	Germany
Jacques Vanier	Canada
Stefan Vogt	Germany
Lars von der Wense	Germany
Yanhui Wang	China
Stefan Weyers	Germany
David Wineland	United States
Szymon Wojtewicz	Poland
Peter Wolf	France
Xinye Xu	China
Zhen Xu	China
Jun Ye	United States
Dai-Hyuk Yu	Republic of Korea
Thomas Zanon-Willette	France
Martin Zelan	Sweden
Jie Zhang	China
Hao Zhang	United States
Klaus Zipfel	Germany
Hongxin Zou	China
Magdalena Zych	Australia

Peer review statement

All papers published in this volume of *Journal of Physics: Conference Series* have been peer reviewed through processes administered by the proceedings Editor. Reviews were conducted by expert referees to the professional and scientific standards expected of a proceedings journal published by IOP Publishing.

The evolution of the Frequency Standards and Metrology symposium and its physics

David Wineland

NIST, Time and Frequency Division,
325 Broadway
Boulder, CO, 80305-3328, USA

david.wineland@nist.gov

Abstract. In this paper I recall some of the earlier developments in the physics and measurement of atomic frequency standards and clocks, as revealed in presentations at past Frequency Standards and Metrology (FSM) symposia in this series. Coverage is not meant to be comprehensive; unfortunately, I can select only a relatively small number of representative examples of work that I felt stimulated many of the major areas of interest since the first meeting in 1971.

1. Introduction

Although I didn't attend the first Symposium on Frequency Standards and Metrology[1], I have been fortunate to attend all subsequent symposia, starting in 1976. I very much like this meeting, in part because it is held relatively infrequently and each new installment brings significant advances and renewed excitement in the field. This was the intention of the founders of the meeting, Claude Audoin (LHA), Helmut Hellwig (NBS), and Jacques Vanier (Laval University). It is interesting to highlight some of these advances; here, I will focus on the physics of high-performance frequency standards.

In recalling past conferences, I was struck by the fact that many of today's current high-interest topics have been with us for a long time, of course with dramatic advances. Also, some overriding themes have persisted such as efforts to: reduce first- and second-order Doppler shifts; increase resolution and Q through velocity selection, confinement, and cooling; improve the spectral purity of probe oscillators; reduce noise to the limits allowed by quantum mechanics; connect a broad range of frequencies in a coherent way; and search for new physics with precision frequency comparisons. I will attempt to highlight these topics with some illustrative examples taken from the earlier symposia; more recent results are much better covered in these proceedings. This approach is inherently flawed because it doesn't give proper credit to all the groups working on these subjects; moreover, it suffers from my inherently personal (and probably too-narrow) view. For this, I apologize.

[1] The first meeting was actually called a "Seminar."

2. 1971, University of Laval, Quebec; Chair, Jacques Vanier

Not surprisingly, there were a number of reports on Cs beam standards and H masers in this first meeting. For Cs beams, Dr. Gerhard Becker (PTB) set the benchmark, reporting $\sigma_y(\tau) \cong 3 \times 10^{-12} \tau^{-1/2}$ and an inaccuracy of 4×10^{-13}. This standard had a Ramsey cavity length L = 0.8 m and also featured hexapole state selection and detection magnets, in order to focus the beam and mitigate uncertainties from the distributed cavity phase shift. Interestingly, distributed cavity phase shifts have continued to be a concern even today. Hydrogen masers featured prominently with reports from NRC, Goddard, Harvard, Orsay, Laval, NPL, and SAO. Although the difficulty of detecting hydrogen atoms was well known, an interesting report was given by Harry Peters (Goddard), who described a hydrogen beam magnetic resonance apparatus, where detection downstream was provided by a Penning ion gauge.

One talk (J. DePrins, University of Brussels) described results with an ammonia beam maser. I found this interesting given the historical significance of ammonia being the first maser/laser type device, but it was the only talk on ammonia masers at FSM symposia. An interesting report was presented by Hans Schuessler (Texas A&M) on trapped ion spectroscopy. This was not long after the experiments he carried out on $^3He^+$ hyperfine spectroscopy (~ 8.67 GHz) in the group of Hans Dehmelt (University of Washington) [1]. This was a rather heroic experiment in which the ions were polarized through spin-exchange with a polarized Cs beam and detected through state-dependent charge transfer ($^3He^+ + Cs \rightarrow {}^3He + Cs^+$) in conjunction with measuring the change in $^3He^+$ ion number by sensing ion currents in the trap electrodes. Along with the H-maser, this experiment showed a key advantage of confinement - that the atoms' velocity and first-order Doppler shift can average to zero. In this experiment however, the ions were rather hot, which limited the accuracy to about 1 part in 10^9 from the second-order Doppler or time-dilation shift.

Today, given the recent dramatic advances with optical clocks, some of our younger colleagues might think that the subject of optical clocks is fairly new. However, already by 1971, this area was being actively pursued. One report was presented by Dick Barger describing his work with Jan Hall [2] on stabilization of a standing wave, HeNe laser at 3.39 μm by saturated absorption of CH_4 contained in an intra-cavity cell. Here also the first-order Doppler shift is essentially eliminated because the saturated absorption feature is due to both forward and backward traveling waves interacting with the same group of atoms whose velocity along the axis of the laser cavity is zero. This particular work was significant because the reproducibility of the frequency was at the level of 10^{-11}, two orders of magnitude better than the wavelength standard based on a 606 nm line in Kr, which was limited in accuracy because the lineshape was asymmetric. If frequency measurements could be extended to the visible, then the wavelength standard could be eliminated by defining the speed of light. Other talks on stable lasers included those by Charles Freed, Tony Siegman, Helmut Hellwig, and Shaoul Ezekiel.

An interesting talk was given by Judah Levine who reported a strain meter based on a 30 m reference cavity (at 3.39 μm) where the mirrors were mounted to bedrock in an abandoned mine near Boulder, Colorado. One goal of the project was to measure earth vibrations and tides but the device was also sensitive to strains induced by underground nuclear explosions. Data was shown for such an event from the underground test site in Nevada. The signature of these signals could be distinguished from naturally occurring earthquakes, an important factor in insuring the viability of the future nuclear test ban treaty. The possible application to detection of gravity waves was also noted, and although the sensitivity was well below that of today's gravitational wave detectors, the required technology was already being developed in these early experiments.

3. 1976 Copper Mountain, Colorado; chair: Helmut Hellwig

Cesium standard inaccuracy had steadily improved (to around 10^{-13}) and there were several reports on advances in hydrogen masers. One of the most interesting of these was a preliminary report by Bob Vessot on the results of a sub-orbital rocket flight ("Gravity Probe A") that carried a maser [3]. During the approximately 2 hour, 10,000 km high flight, the frequency of this maser was compared to one on the earth's surface. Because of the different time dependences of the relativistic frequency shifts caused by changes in the gravitational potential and time dilation, both effects could be confirmed with an uncertainty less than 10^{-4}. This was a pioneering experiment showing the viability of carrying precise atomic clocks into space, a topic of ever increasing interest up to this day. Moreover, from a basic physics standpoint, it is still the most accurate direct measurement of the gravitational potential red shift. As a side note, the transponder technique used to determine and correct for the first-order Doppler shift between masers is now commonly used to stabilize the phase between separate sites in optical clock and frequency standard comparisons [4].

Optical frequency standards continued to improve. An interesting effect in saturated-absorption spectroscopy was the observation [5] of the splitting of the feature due to photon recoil [6] where the position of the two lines is governed by the condition where the transverse velocity of either the upper or lower state in the transition is at zero velocity. For many years, Christian Bordé, who has attended these symposia, has been a key person in explaining how the effects of recoil must be accounted for in atomic clocks, at all ranges of frequency. Recoil effects are of course an integral part of the atom interferometer experiments whose dramatic advances are reported at the current Symposium. To increase resolution in saturated absorption experiments, some laboratories started pushing towards laser beams with larger waists and heavier absorbers such as SF_6 and OsO_4, for increased interaction times. In the early 70's, the dye laser was becoming an extremely useful tool in AMO physics; for optical frequency standards, it's wide tunability meant that atomic or molecular reference transitions need not be in near coincidence with an existing fixed frequency laser line and one could now focus on choosing a reference that had particularly low systematic offsets. Ted Hänsch described his early very impressive experiments on the 1S – 2S two-photon Doppler-free transition in hydrogen, using a pulsed dye laser at 486 nm (peak power ~ 50 kW) doubled to 243 nm in a lithium-formate crystal.

Frequency chains were featured by a number of laboratories. Here, through a sequence of lasers of increasing frequency referenced to each other with harmonic mixers, phase coherence could be maintained. At the highest frequencies, metal-on-metal diodes were used but these lost efficiency as one approached the visible range of the spectrum. Therefore, a difficult step was to go from frequencies corresponding to wavelengths of a few micrometers to the visible range. As possible solutions to this problem, Veniamin Chebotayev discussed 4-wave mixing experiments in Ne to sum three infrared laser beams to the visible ($\lambda \cong 0.65$ μm) and also the use of nonlinear crystals (e.g. $LiNbO_3$). In any case, with the Cs to 3.39 μm frequency chain, it was now possible to connect the ^{86}Kr wave length standard to Cesium via a comparison of the ^{86}Kr wavelength to the 3.39 μm wavelength in a shared reference cavity [7]. This led to a measurement of the speed of light to about 3 parts in 10^9 prior to the second Symposium [8] (see also [9]).

Even at this relatively early stage in the Symposium series, the benefits of low temperature reference cavities were demonstrated. Sam Stein presented results of a microwave parametric oscillator whose signal component was stabilized to a Nb reference cavity that had a Q of around 10^{10}, yielding an instability $\sigma_y(100\ \text{s}) \cong 6 \times 10^{-16}$. Of course cold, cavity-stabilized microwave and optical oscillators continue to be intensely studied today.

Cliff Will and Robert Pound gave very interesting lectures proposing various fundamental investigations. These included tests of the equivalence principle for clocks whose frequencies depended in different ways on the fundamental forces, a laboratory test of frame dragging due to a nearby rotating object, space-born detection of gravitational waves, and searches for a time rate of change in the ratios of the fundamental constants. These topics endure today, benefitting from ever-increasing sensitivity. Prof. Pound pointed out the high Q's available in Mössbauer spectroscopy and gave an example of the 92 keV transition in ^{67}Zn whose Q could be measured using one ^{67}Zn sample as the source and another ^{67}Zn sample as the absorber, a direct comparison of two essentially identical samples. Here, the relative frequencies of the two samples could be tuned via the first-order Doppler shift by moving one of the samples relative to the other. The linewidth measured in this way was 0.8 μm/s corresponding to a Q of around 3×10^{14}. As a personal aside, when I was a graduate student at Harvard working on masers in Norman Ramsey's group, Prof. Pound would tease us in a light-hearted way about our meager Q's of only around 10^9. I had always hoped we might compete using atoms, and finally in 2011 our lab at NIST was able to compare the $^1S_0 - {}^3P_0$ transitions of two $^{27}Al^+$ ions in a similar way to Mössbauer spectroscopy and demonstrate a Q of 3.4×10^{16} [10].

For atomic ions at this symposium, Guenter Werth reported on laser optical pumping in Ba^+ ions; however, more interesting to me was that between this and the previous symposium, he and Fouad Major had measured the 40.5 GHz hyperfine transition in $^{199}Hg^+$, where optical pumping and state detection could be accomplished with fluorescence from a $^{202}Hg^+$ resonance lamp [11]. This work was extended by Len Cutler and colleagues at HP and John Prestage, Lute Maleki, and colleagues at JPL, where it has undergone dramatic improvements, continuing today. Hans Dehmelt described his "shelved electron amplifier" detection idea [12] where state-dependent fluorescence on an allowed transition can be used to discriminate between two levels of a clock transition with essentially 100 % efficiency. In this case, noise in the measurement of transitions reduces to the fundamental quantum "projection" noise associated with the fluctuations in the measured state for a superposition of two clock states, the so-called "standard quantum limit" for uncorrelated atoms. Impressively, with the current precise control of technical noise in the clock experiments, the quantum noise can still dominate on even large numbers of atoms ($N \sim 10^6$) where the quantum noise-to-signal ratio scales as $N^{-1/2}$. Dehmelt also described early laser-cooling proposals [13, 14].

4. 1981, Aussois, France; chair: Claude Audoin

Cesium beam standards started appearing where state selection and detection could now be accomplished with compact lasers rather than large magnets. New cavity designs were explored with the goal of further suppressing distributed cavity phase shifts. As inaccuracies became smaller, one now needed to include Stark shifts from blackbody radiation (-1.7×10^{-14} at 300 K); an effect that has become increasingly important over the years for both Cs standards and the precise optical clocks of today. Walter Hardy (UBC) gave a very interesting talk proposing H-masers operated below 1 K where the temperature dependences of collision shifts due to Helium in the gas phase and liquid Helium on the glass bulb would cancel at a certain temperature. Hans Dehmelt described the initial laser cooling experiments on trapped ions by groups at Heidelberg and NIST. The cooling was important for suppressing time dilation, and the Heidelberg group showed the first pictures of individual atoms in a trap [15].

Optical frequency standards continued to advance with the use of significantly expanded laser beams and heavy molecules, to increase interaction times. These improvements exposed the difficulty of dealing with shifts due to wavefront curvature; however, some experiments were now using separated optical beams to observe Ramsey fringes, where loss of signal due to atomic/molecular beam divergence could be suppressed with multiple optical beam paths. Progress on frequency chains had been dramatic since the last Symposium with reports from IEN (E. Bava), NRC (K. M. Baird), NPL (D. J. E. Knight), Novosibirsk (V. Chebotayev) and NBS (K. Evenson). Despite the many stages required in the chains, this was a triumph of frequency metrology since it was now possible to relate optical frequencies to the frequency of Cesium in a phase coherent way.

5. 1988, Ancona, Italy; chair Andrea De Marchi

The Ancona meeting again saw many improvements across all areas – and was particularly memorable for some of us because Andrea De Marchi provided an excellent wine that came from the vinyards of his family. For Cesium, the use of lasers for state preparation and detection was becoming more common place. Bill Phillips reported on "sub-Doppler" laser cooling of Na atoms. This cooling was significantly better than that predicted by Doppler cooling, an effect Bill referred to as "a strong violation of Murphy's law." It wasn't that the simple theory of Doppler cooling was wrong but it applied only to two-level systems. As explained by Claude Cohen-Tannoudji, Steve Chu, and their colleagues, in sub-Doppler cooling, dynamic optical pumping of the ground state hyperfine levels was important, leading to the so-called "Sisyphus effect" cooling and much lower temperatures. The achievement of much lower temperatures than those provided by Doppler cooling was crucial for the enormous improvement in performance of current fountains, to prevent the atomic sample from expanding too much during free fall.

For hydrogen masers, the cryogenic masers proposed by Walter Hardy at the previous symposium had been subsequently built and successfully operated at UBC and SAO. These devices involved wonderful new physics but ultimately for frequency standards applications, the collision shifts proved to be too difficult to control.

In the area of stable lasers, Dieter Hils and Jan Hall reported locking separate HeNe lasers beams (633 nm) to adjacent fringes of the same cavity. By observing the beat frequency between these two beams, a relative linewidth of 30 mHz was observed. Although not providing absolute frequency stability, this experiment showed that state-of-the art electronics was sufficient to achieve extremely narrow laser beam linewidths, providing the optical cavities were stable. Therefore, the problem of providing precise oscillators was transferred to providing stable reference cavities, which of course occupies the efforts of many groups in our field to this day.

A number of advances were made in optical frequency standards. As one example, the Novosibirsk group (S. N. Bagayev, R. P. Chebotayev et al.) described a CH_4 saturated absorption cell for the 3.39 μm He-Ne laser that had a length of about 8 m, was cooled by LN_2, and was probed by a beam waist w_0 = 30 cm, giving a linewidth of 250 Hz. Similar experiments were carried out in Paris and Boulder. As another example, a PTB/NRC team (H. Helmke, J. Ishikawa, F. Riehle) reported a four-travelling-wave Ramsey/Bordé interferometer for the dye-laser excited 657 $^1S_0$ – $^3P_1$ transition in a Calcium beam, achieving a linewidth of around 20 kHz.

For trapped ions, groups at the University of Washington, (H. Dehmelt et al.), Heidelberg (P. Toschek et al.) and NIST (Ion Storage group) had been able to clearly observe quantum jumps between two states, as indicated by the presence or absence of fluorescence from a strong transition that involved one of the states. This demonstrated Dehmelt's electron-shelving amplifier scheme and showed that clock transitions could be detected, limited only by quantum projection noise. Using Dehmelt's detection scheme, Jim Bergquist showed spectra on the $^2S_{1/2}$ – $^2D_{5/2}$ clock transition from a single trapped Doppler-cooled $^{198}Hg^+$ ion which consisted of the Doppler-free "recoilless" central feature (optical version of the Mössbauer effect) and first-order Doppler-effect generated motional sidebands. John Bollinger discussed the evaluation of the first laser-cooled clock, based on a ~ 303 MHz transition in $^9Be^+$ ions probed with a 550 s Ramsey time leading to a linewidth under 1 mHz. (Subsequently Peter Fisk (CSIRO) used a 600 s Ramsey time on the 12.6 GHz transition in $^{171}Yb^+$ to achieve a Q of 1.5×10^{13} [16]). I now find it amusing that in his introduction to the trapped ion session, Hans Dehmelt, in discussing a path to 10^{-18} uncertainties had written: "All this is no wonder, because high resolution spectroscopy on anything else but on an individual isolated metastable atomic particle at rest in the laboratory will always be affected by shifts of various kinds." Of course, these days, the neutral atom lattice-clock experiments seem to do quite well overcoming this apparent difficulty.

6. 1995, Woods Hole, Massachusetts, Chair: Jim Bergquist
2001, St. Andrews, Scotland; Chair: Patrick Gill
2008, Asilomar, California; Chair: Lute Maleki

By 1995, the first atomic Cesium fountains by the groups of Steve Chu and Andre Clairon had been demonstrated and at the meeting, Andre Clairon reported a Cesium uncertainty of 3×10^{-15}. Ideas of surpassing the standard quantum limit in signal-to-noise ratio emerged (e.g., spin squeezing), and although the basic effects have subsequently been observed, they are yet to be incorporated in a high precision clock. M. Kourogi presented his work on "Optical Frequency Comb Generation," in which an intercavity phase modulator is modulated at the free spectral range of the cavity to dramatically increase the range of sidebands (to ~ 7.5 THz at ~ 785 nm). Although impressive, this was to be dramatically superseded by the advent of optical combs.

The last two meetings brought us close to where we are today. Cs (and Rb) fountains were operated at the projection noise limit reaching inaccuracies of ~ 10^{-15}. For optical frequency standards, Ted Hänsch, Jan Hall, and their colleagues had achieved optical combs that spanned an octave. This now meant that optical frequencies could be tied to microwave frequencies in a single step - the "frequency chain" had been simplified in dramatic fashion and within a year or so of the initial demonstrations, many labs were able to build and use these devices to make precise measurements of optical frequencies. We've seen a gradual but strong shift to optical clocks based on trapped atoms. The development of laser-cooled neutral atom clocks (Sr, Ca, Yb, Hg...) where the atoms are trapped in optical-dipole-force lattices, strongly suppresses all Doppler shifts. For ions, a number of candidates are being studied including In^+, Yb^+, Sr^+, Ca^+, Hg^+, and Al^+. Stable reference cavities are now operated at low temperatures and use crystalline materials such as silicon. Interestingly, mirror substrates and coatings cause significant noise which has led to the development of crystalline coatings. The state-of-the-art for optical clocks and frequency standards has recently been reviewed in [17].

Many other topics including searches for time variation of the fundamental constants, tests of Local Lorentz invariance and the equivalence principle are actively pursued. The performance of atom interferometers has increased dramatically in very large and small sizes. Advanced clocks are scheduled for space, and long-distance frequency transfer over fibers is now possible. Miniature clocks based on MEMS technology have found interesting applications. From the level of interest and excitement at the 2015 Symposium, we can look forward to an equally interesting next Symposium.

I thank Jim Bergquist, John Bollinger, Jan Hall, and Judah Levine for helpful comments on the manuscript. This is a contribution of the US National Institute of Standards and Technology; not subject to U. S. copyright.

References

[1] Schuessler H A, Fortson E N and Dehmelt H G 1969 *Phys. Rev.* A**187** 5
[2] Barger R L and Hall J L 1969 *Phys. Rev. Lett.* **22** 4
[3] Vessot R F C, Levine M W, Mattison E M, Blomberg E L, Hoffman T E, Nystrom G U, Farrel B F, Decher R, Eby P B, Baugher D R, Watts J W, Teuber D L and Wills F D 1980 *Phys. Rev. Lett.* **45**, 2081
[4] Bergquist J C, Itano W M and Wineland D J 1994 in *Frontiiers in Laser Spectroscopy, Proc. Int. School of Physics <<Enrico Fermi>>* CXX, ed. by T. W. Hänsch and M. Inguscio (North Holland, 1994) p. 359
[5] Hall J L, Bordé C J and Uehara K 1976 *Phys. Rev. Lett.* **37**, 1339
[6] Kol'chenko A P, Rautian S G and Sokolovskii R I 1968 *Sov. Phys. JETP* **28**, 986
[7] Barger R L and Hall J L 1973 *Appl. Phys. Lett.* **22** 196
[8] Evenson K M, Wells J S, Petersen F R, Danielson B L, Day G W, Barger R L and Hall J L 1972 *Phys. Rev. Lett.* **29** 1346
[9] Bay Z, Luther G G and White J A 1972 *Phys. Rev. Lett.* **29**, 189
[10] Chou C W, Hume D B, Thorpe M J, Wineland D J and Rosenband T 2011 *Phys. Rev. Lett.* **106** 160801
[11] Major F and Werth G 1973 *Phys. Rev. Lett.* **30**
[12] Dehmelt H G 1975 *Bulletin Am. Phys. Soc.* **20** 60
[13] Hänsch T W and Schawlow 1975 *Opt. Commun.* **13** 68
[14] Wineland D J and Dehmelt H G 1975 *Bulletin Am. Phys. Soc.* **20** 637
[15] Neuhauser W, Hohenstatt M, Toschek P E and Dehmelt H 1980 *Phys. Rev.* A **22** 1137
[16] Fisk P T H, Sellars M J, Lawn M A, Coles C, Mann A G and Blair D G 1995 *IEEE Trans. Instrum. Meas.* **44**, 113
[17] Ludlow A D, Boyd M M, Ye J, Peik E and Schmidt P O 2015 *Rev. Mod. Phys.* **87**, 637

Systematic Effects in Atomic Fountain Clocks

Kurt Gibble

Department of Physics, The Pennsylvania State University, University Park, USA

E-mail: kgibble@psu.edu

Abstract. We describe recent advances in the accuracies of atomic fountain clocks. New rigorous treatments of the previously large systematic uncertainties, distributed cavity phase, microwave lensing, and background gas collisions, enabled these advances. We also discuss background gas collisions of optical lattice and ion clocks and derive the smooth transition of the microwave lensing frequency shift to photon recoil shifts for large atomic wave packets.

1. Introduction

Cesium fountain clocks currently establish the frequency accuracy of International Atomic Time, TAI. The accuracy of fountains has improved by about a factor of two in the last five years. It is noteworthy that the recent accuracy improvements are largely due to new treatments that can rigorously evaluate previously significant systematic errors. Most prominent are frequency shifts due to distributed cavity phase [1-5], microwave lensing [4-8], and background gas collisions [9-14]. We review each of these and describe approaches to evaluate these frequency shifts. Regarding microwave lensing, here we explicitly show the connection of the microwave lensing frequency shift to photon recoil shifts, addressing some recent controversy about the evaluation of this frequency shift [8,15-19]. Our treatment of background gas collision shifts applies to optical lattice clocks, and may be useful for ion clocks and other precision measurements. We discuss these and close with a summary of several implementable design considerations for fountains that facilitate accuracy evaluations.

2. Frequency Shifts due to Distributed Cavity Phase

Distributed cavity phase (DCP) shifts are a first order Doppler shift due to small travelling waves in the microwave clock cavity and the motion of the cold atoms. Known for 40 years, measurements and calculations only recently agreed [1-5]. To accurately calculate the microwave fields, we first express the field as a superposition $\mathbf{H}(\mathbf{r})= \mathbf{H}_0(\mathbf{r})+(\Delta\omega/\Gamma+i)\mathbf{g}(\mathbf{r})$ of a pure standing wave $\mathbf{H}_0(\mathbf{r})$, for no cavity losses, and a field due to the feeds and wall losses $\mathbf{g}(\mathbf{r})$ [1,2]. Because the losses are small, it is important to separately solve finite element models of $\mathbf{H}_0(\mathbf{r})$ and $\mathbf{g}(\mathbf{r})$. Second, $\mathbf{g}(\mathbf{r})$ is decomposed into an azimuthal Fourier series $g_z(\mathbf{r})=\Sigma_m\, g_{z,m}(\rho,z)\cos(m\phi)$ [1,2]. Because the atoms are restricted to the center of the cavity and $g_{z,m}(\rho,z)$ is proportional to ρ^m for small ρ, only m=0, 1, and 2 DCP shifts are significant. Three densely-meshed 2D finite-element solutions for $g_{z,m}(\rho,z)$ thus provide the full 3D solution and require dramatically fewer resources than a direct 3D solution [1,2].

The symmetry of each azimuthal Fourier component corresponds to particular effects in the fountains that produce DCP shifts. The transverse m=0 phase variations are negligible, but the longitudinal phase gradients are large and, in combination with the transverse variation of the Rabi pulse area, lead to large shifts in most clocks near $(4,8,12)\pi/2$ pulses. These shifts cannot be reduced with more azimuthally distributed feeds – more longitudinally distributed feeds are required [2,20].

Journal of Physics: Conference Series **723** (2016) 012002 doi:10.1088/1742-6596/723/1/012002

Since the symmetry of the m=0 phase gradients suppresses DCP shifts for $\pi/2$ pulses, the small m=0 DCP uncertainty often comes from potential top-bottom resistance inhomogeneities in the cavities [4,5,7], which can be most sensitively probed near $(2,6)\pi/2$ pulses. The m=1 variations are phase gradients, which combine with fountain tilts to produce DCP shifts that are naturally large [2-5]. The m=1 gradients from the feeds usually have a large dependence on microwave amplitude and it is helpful to use this to vertically align the fountain, along a single axis with two independent feeds [3,4, 17] and along two horizontal axes with four independent feeds [2,20,21]. To exclude m=1 phase gradients from resistance inhomogeneities, the tilt sensitivity of the fountain is measured for $\pi/2$ pulses and nulled by adjusting the relative feed amplitudes [2,3]. Measurements at a higher amplitude ($5\pi/2$) do not generally exclude large m=1 shifts for $\pi/2$ pulses. Quadrapoler m=2 phase variations produce DCP shifts primarily via detection inhomogeneities and offsets of the initial cloud position and are generally small enough that they can be accurately calculated [3-5,7]. Aligning the cavity feeds at 45° relative to the detection laser beams usually gives negligible m=2 DCP errors.

Stringent verification of our complete model led to significantly smaller systematic errors [3-5] and gave confidence to construct improved cavity designs with much smaller longitudinal phase gradients [2,20,21]. The most important feature to add is feeding clock cavities with 4 or more independent cables so that the fountain can be precisely aligned to be vertical along two horizontal axes. Several of the latest fountains have used the new cavities, and are consequently realizing more precise vertical alignments and observing much smaller microwave-amplitude-dependent frequency shifts [21].

3. Microwave Lensing Frequency Shifts

The microwave photon recoil shift of $\delta\omega/\omega=1.5\times10^{-16}$ is now comparable to fountain clock inaccuracies and most accurate fountains contributing to TAI correct for this frequency bias. Because atoms in a fountain are localized to less than a microwave wavelength, the resonant microwave dipole forces do not yield resolved photon recoils, but instead act as weak focusing and defocusing lenses on the atom wave packets. The resulting shift depends on the clock geometry and is typically 6×10^{-17} to 9×10^{-17} for fountains [4-8,17], and larger for the microgravity clock PHARAO, 11.4×10^{-17} [22].

There is currently some controversy since the recent NIST treatment [16,18] predicts a much smaller shift, 1.6×10^{-17}, disagreeing with the prior results [4-7,17]. The NIST treatment and two recent fountain accuracy evaluations have considered that the microwave lensing shift goes to zero in the limit of zero microwave amplitude [12,15,16,19]. In [8] we explicitly derived that, while the microwave lensing of wave packets goes to zero in the limit of zero amplitude, the frequency shift is non-zero because the perturbation of the transition probability δP is proportional to the lensing and the frequency shift is δP divided by the Ramsey fringe amplitude, which also goes to zero. Here, we give a simplified derivation of the microwave lensing frequency shift that is specific to weak fields. This is a fundamental limit and demonstrates the smooth connection of microwave lensing shifts to well-known photon-recoil shifts as atomic wave packets become large. Since recoil shifts are non-zero for weak fields, this connection supports the microwave lensing shift being non-zero in this limit and serves as a valuable check of numerical factors and the sign of calculated microwave lensing shifts.

We consider a small atom cloud on the fountain axis and 1D standing waves $\cos(k_{\alpha(\beta)}x)$, with general wave vectors in the first(second) Ramsey interactions. Following [6,8], but using the bare atom basis $|g\rangle$ and $|e\rangle$, small Ramsey pulse areas ϕ_α and ϕ_β imply that we need to only treat opposing single photon-recoils $\pm k_{\alpha(\beta)}$, $\cos(k_{\alpha(\beta)}x)=\frac{1}{2}\exp(ik_{\alpha(\beta)}x)+\frac{1}{2}\exp(-ik_{\alpha(\beta)}x)$, which destructively interfere for small x to produce the lensing shift [6]. In momentum space, the detected excited-state wave function, just after the second Ramsey pulse ϕ_β, is $\psi_e(k,t_2=t_1+T)=-(i/2)\exp[i(kx-\hbar k^2t_2/2m-\chi/2)]\,[\phi_\alpha\exp(-i\omega_R T+i\chi)\cos(k_\alpha x-kv_R T)+\phi_\beta\cos(k_\beta x)]$, using the microwave photon-recoil velocity $v_R=\hbar k_\alpha/m$, the photon recoil shift $\omega_R=\hbar k_\alpha^2/2m$, and χ as the phase shift of the second Ramsey pulse [8]. Constructing a wave packet by integrating over k with a weight $\exp(-k^2\sigma_x^2/2)$ gives $\psi_e(k,t_2)=\tilde{A}\exp\left(-x^2/2\sigma_x\tilde{w}_2\right)[\phi_\alpha\exp(-i\omega_R T\,\tilde{w}_1/\tilde{w}_2+i\chi)\cos(k_\alpha x\,\tilde{w}_1/\tilde{w}_2)+\phi_\beta\cos(k_\beta x)]$, where the complex wave

Journal of Physics: Conference Series **723** (2016) 012002 doi:10.1088/1742-6596/723/1/012002

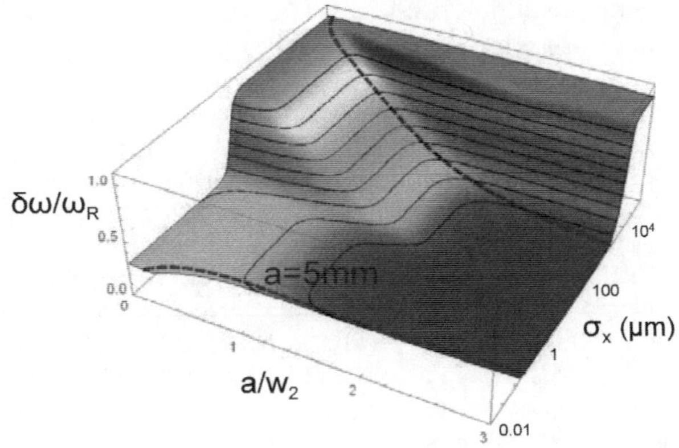

Fig. 1. The microwave lensing shift from (1) smoothly transitions to a photon recoil shift $\omega_R=\hbar k_\alpha^2/2m$ for large initial wave packet sizes σ_x and large apertures a. Here we consider cesium atoms and weak Ramsey pulses at t_1=0.21s and t_2=0.71s, with $k_{\alpha,\beta}=2\pi\times9.1926$ GHz/c. The dashed lines depict an a=5mm aperture versus σ_x. For a small initial wave packet size σ_x, the quantum spread of the wave packet w_2 is large so the atoms are strongly clipped by the clock cavity aperture at t_2, yielding a microwave lensing frequency shift $\omega_R t_1/t_2$ [6,8]. As σ_x grows, the spread w_2 decreases, yielding minimal clipping of atoms and no frequency shift [6]. If σ_x would grow beyond a very large 5mm, the aperture would again clip the atom wave packets and the frequency shift goes to the value of the recoil shift ω_R.

packet sizes are $\tilde{w}_{1,2}=i\hbar t_{1,2}/m\sigma_x+\sigma_x$ at the two Ramsey interactions. The excited state probability $|\psi_e(k,t_2)|^2$ yields a Ramsey fringe with a frequency shift:

$$\delta\omega=\frac{1}{T}\frac{\phi_\alpha\phi_\beta\int_{-a}^{a}\cos(k_\beta x)\sum_\pm \pm e^{-\frac{x^2\pm v_R xT+\frac{1}{2}v_R{}^2T^2}{w_2^2}}\sin\left(w_{12}^2\frac{k_\alpha x\pm\omega_R T}{w_2^2}\right)dx}{\phi_\alpha\phi_\beta\int_{-a}^{a}\cos(k_\beta x)\sum_\pm e^{-\frac{x^2\pm v_R xT+\frac{1}{2}v_R{}^2T^2}{w_2^2}}\cos\left(w_{12}^2\frac{k_\alpha x\pm\omega_R T}{w_2^2}\right)dx} \quad (1)$$

Here, $w_2=|\tilde{w}_2|$, $w_{12}^2=\hbar^2 t_1 t_2/m^2\sigma_x^2+\sigma_x^2$, and $2a$ is the width of the detection aperture. The numerator is the perturbation of the transition probability due to lensing and the denominator is the Ramsey fringe amplitude. Both go to zero as $\phi_\alpha\phi_\beta$, yielding a non-zero frequency shift in the limit of zero microwave amplitude [6,8,17], in contrast to [12,15,16]. Additionally, the NIST treatment gives zero frequency shift for all σ_x and a/w_2 in Fig. 1 since the atoms are centered on the cavity axis and [18] neglects the variation of the dipole force over the wave packets [19]. Fig. 1 shows that detecting only the center of a wave packet ($a/w_2\ll1$) yields a microwave lensing shift of $\omega_R w_{12}/w_2$, which becomes very nearly equal to the value of the photon recoil shift when wave packets spread minimally during the fountain time, $w_{12}\approx w_2$, for $\sigma_x>10\mu$m. Note that this is not actually the limit of recoil shifts as the shift is independent of k_β and the recoils are not resolved as $w_{1,2}$ are less than a microwave wavelength. In the limit of large wave packets, $k_\alpha=k_\beta$, $\sigma_x\to\infty$, and $k_{\alpha,\beta}a\gg2\pi$, (1) gives the recoil shift, as expected.

4. Frequency Shifts due to Background Gas Collisions

Precision measurements with cold atoms, including fountains, optical lattice, and quantum-logic ion clocks, have a different sensitivity to background gas collisions than room-temperature clocks because background gas collisions prevent cold atoms from being detected. Therefore, room-temperature background gas collision shifts [23] do not apply to cold-atom clocks. We have derived analytic

Journal of Physics: Conference Series **723** (2016) 012002 doi:10.1088/1742-6596/723/1/012002

expressions for background gas collision shifts of cold atoms and have shown that there is a highly useful proportionality between background gas shifts and atom loss rates [9]. The shifts are essentially independent of the background gas, except potentially for helium. Lifetime measurements can therefore set accurate uncertainties due to background gas frequency shifts [10-14].

The potential momentum transferred to laser-cooled atoms in fountains and lattice clocks from a background gas collision is large. In fountains, atoms that acquire velocities greater 3 cm/s are not detected and for lattice clocks, ≈10 cm/s. These velocities are sufficiently low that the scattering is quantum mechanical and a partial wave expansion is useful. The quantum scattering can be separated into two contributions – the usual scattered current j_{sc} and the interference between the scattered amplitude and the unscattered amplitude in the forward direction [9]. The interference current j_{int} describes the loss of atoms due to background gas scattering, and for clocks, importantly includes a phase shift of the clock coherences. For fountains and lattice clocks, the detected scattered current is negligible and only the interference current is significant. The interference current gives a frequency shift proportional to $\sin(2\delta_{g\ell})-\sin(2\delta_{e\ell})$, which is approximately $2(\delta_{g\ell}-\delta_{e\ell})\cos(2\delta_{g\ell})$ for small $\delta_{g\ell}-\delta_{e\ell}$, where $\delta_{g,e\ell}$ are the ℓth partial wave phase shift. For small angular momenta ℓ, corresponding to hard short-range collisions, the phase shifts become large and the frequency shift averages to zero. Only weak long-range collisions with $\delta_{g,e\ell}\lesssim 1$ contribute to the collision shift. The short-range collisions give large momentum transfers and have cross sections proportional to $\sin^2(2\delta_{g,e\ell})$, which averages to ½, not 0. Including both, we get a relationship between the frequency shift and the loss of cold atoms [9]:

$$\langle j_{e,\mathrm{int}}\rangle \approx -0.38n\frac{(\mu k_B T)^{3/10}}{\hbar^{2/5} m_p^{3/5}}\left[5\left(C_{6e}^{2/5}-C_{6g}^{2/5}\right)\sin(\chi)+13.8\left(C_{6e}^{2/5}+C_{6g}^{2/5}\right)\cos^2\left(\frac{\chi}{2}\right)\right]. \quad (2)$$

Here, $-C_{6g,e}/R^6$ is the van der Waals interaction of the clock states with the background gas atom of mass m_p, and μ is the reduced mass. In room temperature clocks, the entire scattered current is detected and, combined with j_{int}, gives a frequency shift proportional to $\sin(2\delta_{g\ell}-2\delta_{e\ell})$, which does not average to zero for $\delta_{g,e\ell}>1$ if $\delta_{g\ell}-\delta_{e\ell}$ is small [9]. For microwave clocks, both clock states have the same electronic structure and $(C_{6e}-C_{6g})/C_{6g}$ can be bounded by $\hbar\omega[E_1^{-1}+(E_1+E_p)^{-1}]$, where E_1 and E_p are the lowest resonant excitations of the clock and perturber [9]. For microwave clocks, this leads to a simple expression of $-\delta\omega/\omega\approx 2\times10^{-16}\,\Delta A$, where ΔA is the loss of Ramsey fringe amplitude during the Ramsey interrogation time [9].

While ion clocks trap atoms sufficiently strongly to retain atoms that have hard collisions with background gas atoms or molecules, quantum-logic detection [24] can exclude ions that change their trap state. This gives a tremendous immunity to background gas shifts, because the scattering phase shifts are given by strong long range $-C_4/R^4$ interactions, which are identical for both clocks states. The frequency shift is again given by C_{6e}-C_{6g}, which gives very small $\delta_{g\ell}-\delta_{e\ell}$ for $\delta_{g,e\ell}\lesssim 1$ [9].

5. Conclusions

The many collective experiences described above have lead to a number of ways to facilitate accuracy evaluations of fountains. Many are easy to implement, especially during the design stage. Explicitly designing a method to insure that the atom cloud travels vertically through the fountain can improve and reduce the time required for accuracy evaluations. The best current recommendation is to use at least 4 independent cables to have opposing cavity feeds along both horizontal tilt axes [2, 21]. Orienting four cavity feeds to be at 45° to the orthogonal detection laser beams and fluorescence collection minimizes m=2 DCP shifts, likely to a negligible level. Together, these allow measurements of m=1 DCP shifts and negligible m=2 shifts so that only the small m=0 shifts needs to be calculated, combined with limits on top-bottom resistance inhomogeneities, which are measured near (2,6) π/2 pulses. We note that finite-element calculations are much less intensive for cylindrically symmetric cavities, and further helped if they have no sharp corners and exclude atom trajectories that experience large microwave fields near cavity walls [1,2]. The improved and recently-implemented cavity designs [20,21] reduce the longitudinal phase gradients, and even allow for some adjustment. Such a cavity could enable a convincing measurement of the microwave lensing frequency shift at elevated microwave amplitudes, such as $5\pi/2$ pulses, especially when combined with two state preparation cavities that can vary the atomic density and prepare the atoms in either the F=3 or 4 m_F=0 state [6,8].

Several recommendations are related to the vacuum chamber. Elegant methods [25,26] have been used to evaluate the cold-collision frequency shift [27]. Using a small MOT, which could be loaded from a slow beam, has produced small uncertainties [12,21] and avoids slow drifts of the atomic distribution, which produces an unstable effective fountain-tilt and potential m=1 DCP shift [3,7]. A 111 launch [27] has some advantages; opposing laser beams can be aligned irrespective of the other beams and generating the beams is marginally simpler. Constructing the vacuum chamber with two ion pumps, with associated valves, helps to assess the background gas collision shift [10]. To reduce microwave leakage shifts, it is advantageous to control the electromagnetic mode structure in the fountain, from the clearing pulse, through the interrogation time, to the state detection. A series of cutoff waveguides, possibly including detuned microwave cavities [28], can be used. More precise measurements of blackbody coefficients and temperature measurements, as in recent lattice clocks [14,29], may be useful in the future. Finally, juggling atoms [30] significantly improves the short-term stability [31], but this technique requires more effort than the methods above.

In summary, fountain accuracies have advanced significantly over the last few years by rigorously evaluating the systematic errors due to distributed cavity phase, microwave lensing, and background gas collisions. The m=0 and 2 DCP shifts can be calculated and frequency measurements with tilted fountains at optimal amplitude set the m=1 uncertainties. We also have calculated microwave lensing frequency shifts and here have shown a direct connection to photon-recoil shifts for large wave packets. This connection gives valuable confidence to the corrections that are applied to the majority of the accurate fountain standards contributing to TAI. The treatment of background gas collisions for fountains is also applicable to lattice clocks [13,14], and potentially to quantum-logic ion clocks [24].

Looking forward, the anticipated future redefinition of the SI second on an optical or higher frequency transition is an important motivation to continue to operate and improve cesium standards. At present, a number of systematic are becoming more difficult, especially considering the averaging times required by many fountains to reach their reported accuracies, and therefore the rate of future improvements is not clear. One could reasonably argue that fountains will not improve dramatically beyond the current 10^{-16} level. However, if a redefinition is still years away [32], the current limits are not fundamental and history suggests that fountain accuracies will continue to improve significantly.

I am delighted to acknowledge many collaborators who have contributed to the progress described here, especially S. Bize, V. Gerginov, J. Guéna, P. Laurent, S. Lea, R. Li, N. Nemitz, S-E. Park, P. Peterman, C. Salomon, R. Schröder, K. Szymaniec, S. Weyers, and R. Wynands. Financial support from NASA, the NSF, First TF, la ville de Paris, NPL, and PTB is gratefully acknowledged.

References

[1] Li R, and Gibble K 2004 Phase variations in microwave cavities for atomic clocks *Metrologia* **41**, pp 376-86
[2] Li R, and Gibble K 2010 Evaluating and minimizing distributed cavity phase errors in atomic clocks *Metrologia* **47** pp 534–51
[3] Guéna J, Li R, Gibble K, Bize S and Clairon A 2011 Evaluation of Doppler shifts to improve the accuracy of primary fountain clocks *Phys. Rev. Lett.* **106** p 130801
[4] Li R, Gibble K and Szymaniec K 2011 Improved accuracy of the NPL-CsF2 primary frequency standard: evaluation of distributed cavity phase and microwave lensing frequency shifts *Metrologia* **48** pp 283-9
[5] Weyers S, Gerginov V, Nemitz N, Li R and Gibble K 2011 Distributed cavity phase frequency shifts of the caesium fountain PTB-CSF2 *Metrologia* **49** pp 82–7
[6] Gibble K 2006 Difference between a Photon's Momentum and an Atom's Recoil *Phys. Rev. Lett.* **97** p 073002
[7] Guéna J *et al.* 2012 Progress in atomic fountains at LNE-SYRTE *IEEE Trans. Ultrason.*

Ferroelectr. Freq. Control **59** pp 391–410

[8] Gibble K 2014 Ramsey spectroscopy, matter-wave interferometry, and the microwave-lensing frequency shift *Phys. Rev. A* **90** p 015601

[9] Gibble K 2013 Scattering of cold atom coherences by hot atoms: frequency shifts from background gas collisions *Phys. Rev. Lett.* **110** p 180802

[10] Szymaniec K, Lea S N and Liu K 2014 An Evaluation of the Frequency Shift Caused by Collisions With Background Gas in the Primary Frequency Standard NPL-CsF2 *IEEE Trans. Ultrason. Ferroelectr. Freq. Control* **61** pp 203-6

[11] Guéna J, Abgrall M, Clairon A and Bize S 2014 Contributing to TAI with a secondary representation of the SI second *Metrologia* **51** pp 108-20

[12] Levi F, Calonico D, Calosso C E, Godone A, Micalizio S, and Costanzo G A, 2014 Accuracy evaluation of ITCsF2: a nitrogen cooled caesium fountain *Metrologia* **51** pp 270–84

[13] Falke S *et al.* 2014 A strontium lattice clock with 3×10^{-17} inaccuracy and its frequency," *New J. Phys.* **16** p 073023

[14] Nicholson T L *et al.* 2015 Systematic evaluation of an atomic clock at 2×10^{-18} total uncertainty *Nature Comm.* **6** p 6896

[15] Ashby N, Barlow S, Heavner T P and Jefferts S R 2014 Frequency shifts in NIST Cs primary frequency standards due to transverse RF field gradients *arXiv:1404.4101v2*

[16] Heavner T P *et al.* 2014 First accuracy evaluation of NIST-F2 2014 *Metrologia* **51** pp 174–182

[17] Gibble K 2015 Comment on 'First accuracy evaluation of NIST-F2' *Metrologia* **52** pp 163-6

[18] Ashby N, Barlow S, Heavner T P and Jefferts S R 2015 Frequency shifts in NIST Cs primary frequency standards due to transverse RF field gradients *Phys. Rev. A* **91** p 033624

[19] Gibble K 2015 Comment on "Frequency shifts in NIST Cs primary frequency standards due to transverse rf field gradients" *arXiv:1505.00691*

[20] Gibble K, Lea S N and Szymaniec K 2012 A microwave cavity designed to minimize distributed cavity phase errors in a primary cesium frequency standard *2012 Conf. Prec. Electromag. Meas.* pp 700 – 701

[21] Szymaniec K, Lea S N, Gibble K, Park S E, Liu K and Glowacki P 2016 NPL Cs fountain frequency standards and the quest for the ultimate accuracy *this issue*

[22] Peterman P, Gibble K, Laurent P and Salomon C 2016 Microwave Lensing Frequency Shift of the PHARAO Laser-cooled Microgravity Atomic Clock *Metrologia* **53** pp 899-907

[23] See Beer C W and Bernheim R A 1976 Hyperfine pressure shift of ^{133}Cs atoms in noble and molecular buffer gases *Phys. Rev. A* **13** pp 1052-7

[24] Chou C W, Hume D B, Koelemeij J C J, Wineland D J and Rosenband T 2010 Frequency Comparison of Two High-Accuracy $^{+}$Al Optical Clocks *Phys. Rev. Lett.* **104** p 070802

[25] Szymaniec K *et al.* 2007 Cancellation of the Collisional Frequency Shift in Caesium Fountain Clocks *Phys. Rev. Lett.* **98** p 153002

[26] Pereira Dos Santos F, Marion H, Bize S, Sortais Y, Clairon A and Salomon C 2002 Controlling the Cold Collision Shift in High Precision Atomic Interferometry *Phys. Rev. Lett.* **89** p 233004

[27] Gibble K and Steven Chu S 1993 Laser-Cooled Cs Frequency Standard and a Measurement of the Frequency Shift due to Ultracold Collisions *Phys. Rev. Lett.* **70** pp 1771-4

[28] Fertig C and Gibble K 2000 Measurement and Cancellation of the Cold Collision Shift in an ^{87}Rb Fountain Clock *Phys. Rev. Lett.* **85** pp 1622-5

[29] Beloy K *et al.* 2014 An atomic clock with 1×10^{-18} room-temperature blackbody Stark uncertainty *Phys. Rev. Lett.* **113** p 260801

[30] Legere R and Gibble K 1998 Quantum Scattering in a Juggling Atomic Fountain *Phys. Rev. Lett.* **81** pp 5780-3

[31] Fertig C, Rees J I and Gibble K 2001A juggling Rb fountain clock and a direct measurement of population differences *Proc. 2001 IEEE Freq. Contr. Symp.*, 6-8 June, pp 18 -21

[32] Gill P 2016 Is the time right for a redefinition of the second by optical atomic clocks? *this issue*

NPL Cs fountain frequency standards and the quest for the ultimate accuracy

K Szymaniec[1], S N Lea[1], K Gibble[2], S E Park[3], K Liu[4] and P Głowacki[5]

[1]National Physical Laboratory, Teddington, UK
[2]The Pennsylvania State University, University Park, USA
[3]Korea Research Institute of Standards and Science, Daejeon, Korea
[4]National Institute of Metrology, Beijing, China
[5]Poznań University of Technology, Poznań, Poland

E-mail: ks1@npl.co.uk

Abstract. NPL operates a system of two primary caesium fountain clocks consisting of a fully characterised standard NPL-CsF2 together with a new standard NPL-CsF3, which has recently become operational. Both fountains feature a single-stage vapour-loaded magneto-optical trap as the source of cold atoms and an approximate cancellation of the potentially large cold collision frequency shift. As a result, the collision-shift type-B uncertainty is less than 10^{-16}. Subsequently, more subtle systematic effects, including the frequency shifts from distributed cavity phase, microwave lensing and collisions with background gas have also been evaluated at the level of 10^{-16} or below. Now, as several systematic effects contribute to the fountains' uncertainty budgets similarly, further significant improvement of their accuracies is expected to be even more difficult. The short-term stability of these standards is also a significant factor limiting the overall precision as many days or even weeks of averaging is required for the type-A statistical uncertainty to approach the declared type-B systematic uncertainty. Going forward, further improvements in the reliability and robustness of operation of fountain standards is one of our priorities.

1. Introduction

Operation of atomic fountain frequency standards has been one of the first and most successful examples of applications of the laser cooling technology. A prototype atomic fountain was first demonstrated in 1989 [1] and, by 1995, a metrological fountain made the first contribution to TAI [2]. A number of time and frequency laboratories have further developed and improved the design and operation of fountain standards. Inaccuracies of fountains primary frequency standards (PFS) have improved by more than an order of magnitude, approaching an uncertainty of $\delta\nu/\nu$ of 10^{-16}. Today, the calibration of TAI rests almost entirely on the evaluations performed by about a dozen PFS run by several National Measurement Institutes around the world. With the operation of fountains becoming more reliable, a number of labs have also been using them to discipline clocks, forming local timescales [3]. The availability of fountain standards is instrumental for the development of optical clocks, the anticipated next generation of primary frequency standards. Here, they provide the most accurate frequency to measure the frequency of an optical clock transition, but they also provide a practical, highly stable and accurate reference for numerous test measurements. Frequency measurements of spectrally narrow transitions in microwave fountain and optical standards have placed the tightest constraints on the time variation of fundamental constants [4].

Most Cs fountain PFS share the key aspects of the design of various subsystems [5]. At the same time, different and sometimes complementary approaches have been adopted to address particular issues like cold collisions, blackbody radiation, or microwave cavity design. This is a healthy situation increasing the confidence and the robustness of the realization of the SI definition. In this paper we give an overview of the Cs fountain development at NPL over the last decade. In the following section we describe the main specific features of the NPL approach to the physics package design and to the operation protocol. In section 3 we report on recent enhancements of the short-term stability and, in section 4, we discuss the accuracy of our PFS and comment on a long-term behaviour of one of our fountains. Finally, we give an outlook for the future operation and planned improvements.

2. NPL Cs fountains – main features

NPL currently operates two Cs fountain PFS, NPL-CsF2 and NPL-CsF3, which share major design features. A prototype system, NPL-CsF1, operated from 2004 until 2007 and was later decommissioned and parts of its physics package and optical bench were reused to construct NPL-CsF3. A key feature of the NPL Cs fountains is a single-stage vapour-loaded magneto-optical trap (MOT) as source of cold atoms. Because early studies of cold collisions in Cs had reported large density dependent frequency shifts [6], MOT's have often been avoided in PFS designs to reduce the atomic density during the interrogation time. Instead, atoms are often collected and launched with optical molasses (e.g. loaded from another cold atom source, such as a two-dimensional MOT). The desire for high short-term stability requires many atoms, which in turn requires accurate density ratios to determine and correct relatively large collisional shifts. Along this line, an elegant approach is to accurately select half of the atoms with an adiabatically chirped pulse [7]. Alternatively, we have shown that the collisional frequency shift can be cancelled by launching a small cloud from a MOT [8]. This requires the clock state populations to produce frequency shifts with opposite signs [9], which occurs for Cs atoms at nano-kelvin temperatures [10]. Although such temperatures are not achieved by the polarisation gradient cooling in fountains, the collision energies in an expanding small cloud become very low during the fountain trajectory as correlations between position and velocity develop. The collisional shift is precisely cancelled for the proper clock state population ratio after the first Ramsey interaction (see figure1).

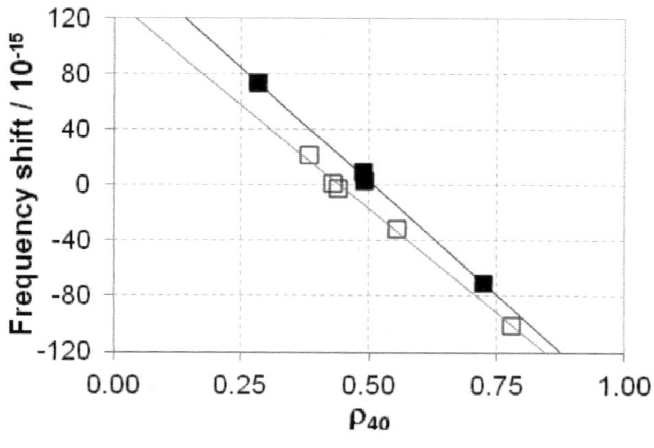

Figure 1. Collisional frequency shifts from measured frequency differences at high and low atomic density in NPL-CsF3, as a function of the |F=4, m_F=0⟩ population fraction (ρ_{40}) after the first Ramsey interaction. Full (open) squares correspond to a launched cloud size of 0.9 mm (1.0 mm).

To achieve a better short-term stability, one may attempt to increase the detection signal-to-noise ratio by loading more atoms in the MOT. This, however, may eventually lead to an increase of the initial cloud size and increase the effective collision energy. An alternative is to optically pump atoms distributed throughout all m_F states into $m_F = 0$. This technique has been demonstrated for thermal beam clocks, but has not been widely implemented for normal operation. In principle, one can expect a 9-fold increase of the $m_F = 0$ population from 9 m_F states. In practice, our enhancement is about a factor of 5, due to photon scattering during the optical pumping pulse and a subsequent heating and accelerated expansion of the atomic cloud, which is truncated by apertures in the fountain [11].

A major novelty introduced in NPL-CsF3 is a non-cylindrical microwave cavity. Its shape was designed to minimize the distributed cavity phase (DCP) frequency shift [12], especially longitudinal phase gradients (related to the $m = 0$ component of the azimuthal expansion of the cavity field) and their dependence on microwave amplitude [13]. A further reduction of the shift is achieved by feeding the cavity by four rectangular waveguides with coupling holes placed symmetrically about the mid-plane. This also significantly reduces the $m = 1$ contributions to the DCP shift. Care was taken to assure homogeneity of the detection (probing) beam across the active area, which, together with orienting the beam direction at 45° with respect to the cavity feeds, makes the $m = 2$ DCP shift negligible. The new design also allows for a larger opening in the cavity endcaps, approximately doubling the number of atoms passing through the cavity on the descent. Preliminary measurements of the $m = 0$ and $m = 1$ DCP shifts as a function of the field amplitude show good agreement with model predictions (figure 2).

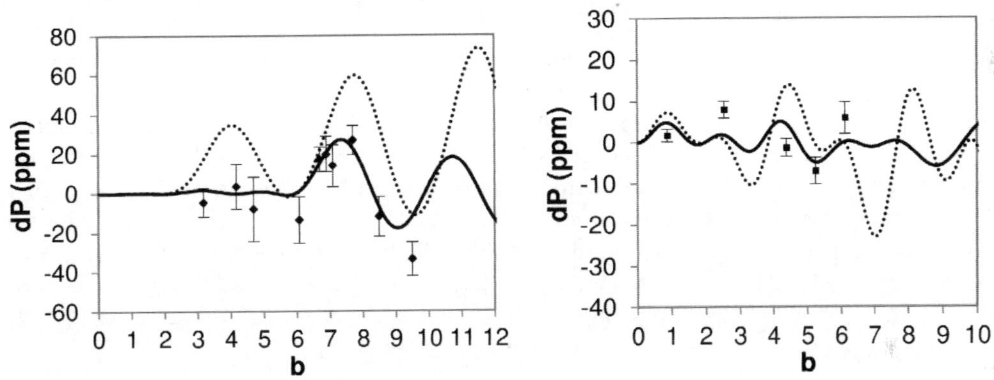

Figure 2. Calculated (solid line) and measured (full symbols) perturbations to the clock transition probability **dP** due to DCP for the new cavity installed in NPL-CsF3. Also shown are the much larger calculated shifts for the cylindrical cavity of NPL-CsF2 (dotted lines). The microwave field amplitude is **b**, with **b** approximately equal to 1 for $\pi/2$ pulse areas. The left plot is the $m = 0$ DCP component and the right is the $m = 1$, increased by feeding fully asymmetrically and tilting the fountain by 1 mrad.

In both NPL fountains, the temperature of the flight tube (and thus of the microwave cavity) is set and stabilised by a water jacket installed around the tube. This helps to accurately determine the blackbody radiation frequency shift as well as to tune the Ramsey cavity to resonance. The cavities are constructed so that their resonant frequencies are at temperatures within a few degrees of the lab temperature.

3. Short-term stability

By launching atoms from the MOT, and including the optical pumping, several millions of atoms in the clock state arrive in the detection region after the Ramsey interrogation. A signal-to-noise ratio of around 2500, limited by the quantum projection noise, was observed when exciting the atoms with $\pi/4$ pulses on resonance so the excitation is insensitive to the local oscillator (LO) phase noise. This sets a limit for the attainable short-term stability in the range of low 10^{-14} at 1 s. However, under normal operating conditions, when using a room temperature quartz-based LO, the stability is poorer, about 1.5×10^{-13} (1 s), owing to the Dick effect. Recently, an interrogation signal synthesized from an ultra-stable laser via a femtosecond optical frequency comb has been used, demonstrating four times lower instability of 3.7×10^{-14} (1 s) (figure 3). Based on the highest atom number observed in detection, an even lower instability of 2.5×10^{-14} (1 s) can be expected.

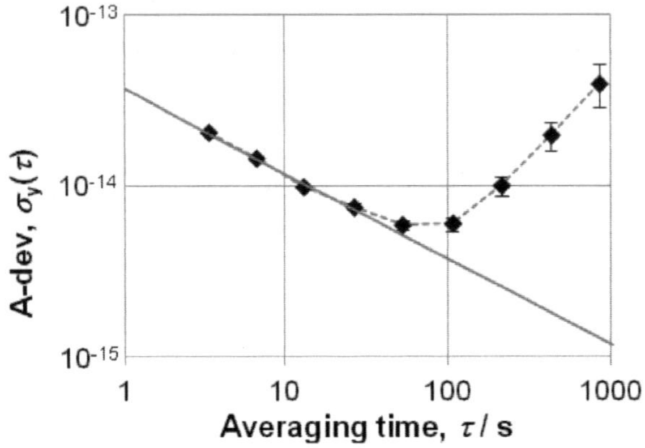

Figure 3. Short-term stability of NPL-CsF3 using an LO based on an ultra-stable laser. The LO was free running and drifted at a rate of approximately 10^{-16} s^{-1}. The solid line represents the demonstrated stability of 3.7×10^{-14} at 1 s.

The collisional frequency shift may vary significantly in the vicinity of the cancellation point (figure 1) and therefore a residual shift is measured and corrected during the normal operation of the fountain as frequency standard. The number of atoms is alternated by varying a microwave field amplitude in the state selection cavity, instead of changing the MOT loading, to preserve the cloud parameters (initial size and temperature). From the measurements for the high and low atom numbers (i.e. high and low density), we extract a zero-density frequency $F_{ext} = (kF_L - F_H)/(k-1)$; F_H (F_L) is the frequency measured for the high (low) density of the atomic cloud and k is the high-to-low density ratio. An uncertainty σ_{ext} of F_{ext} can be expressed as [14]:

$$\sigma_{ext}^2(\tau) = \left(\frac{k}{k-1}\right)^2 \sigma_L^2(\tau_L) + \left(\frac{1}{k-1}\right)^2 \sigma_H^2(\tau_H) + \left(\frac{F_L - F_H}{(k-1)^2}\right)^2 \sigma_k^2; \quad (1)$$

where $\sigma_L(\sigma_H)$ and $\tau_L(\tau_H)$ are the short-term stability and averaging time for operation at low (high) atom number; σ_k is an estimated fractional deviation of the density ratio from the measured atom number ratio. The last term is a type B uncertainty and, close to the cancellation point, can be made small compared to other type B contributions (see table 1 below). The first two terms are included in the type A (statistical) uncertainty, increasing the effective short-term instability. If the detection noise is dominated by QPN, σ_{ext} is minimized for $k = 4$ at $\sigma_{ext} = 3\sigma_H$. This yields 2.5×10^{-16} after 1 day of averaging for the lowest expected $\sigma_H = 2.5 \times 10^{-14}/\tau^{1/2}$.

4. Accuracy evaluations and long term operation

NPL-CsF2 was first operational in 2009 and its total type B uncertainty was initially $u_B = 4.1 \times 10^{-16}$ [15]. Over the following years the systematic effects with large uncertainties were re-evaluated, leading to a reduction of u_B by a factor of two, as in 2013 [16, 17]. The optical pumping stage was introduced during this time, but no modifications to the physic package were made.

The major reduction of u_B came from a reassessment of the distributed cavity phase frequency shift based on the model of [13] and a series of measurements verifying its predictions for the microwave cavity used in NPL-CsF2. Similarly, the mechanical effect of the microwave photons on the atomic wave functions, microwave lensing, was evaluated, extending the treatment in [18]. Finally, the uncertainty related to frequency shifts due to collisions with room temperature background gas was lowered, following [19], which showed that a limit on the frequency shift can be set by measuring the atom loss rates in the cloud due to background gas collisions during the Ramsey interaction time. Loss rates in NPL-CsF2 were measured by increasing the background gas pressure and extrapolating to the normal operating conditions [17].

The new PFS setup NPL-CsF3 became operational in 2014 and is currently undergoing an accuracy evaluation. Preliminary values for the major uncertainties are listed in Table 1 and the total uncertainty is expected to be noticeably below 2×10^{-16}. The most significant reduction is expected to result from implementing the novel microwave cavity and reducing the total DCP uncertainty.

Systematic effect / uncertainty (10^{-16})	NPL-CsF2 (2009)	NPL-CsF2*) (2013)	NPL-CsF3 (2015)
2nd order Zeeman	0.8	0.8	0.5
Blackbody radiation	1.1	1.0	*0.6*
Cold collisions	1.0	0.4	0.4
Background gas	1.0	0.3	*0.3*
Microwave leakage	1.0	0.6	*0.5*
Distributed cavity phase	3.0	1.1	*0.5*
Microwave lensing	1.5	0.3	0.3
Gravity	0.5	0.5	0.5
Other	0.4	0.3	0.3
Total u_B	**4.1**	**2.0**	***1.4***

Table 1. Successive improvements of accuracy of the NPL caesium fountain standards following re-evaluations of NPL-CsF2 and construction of NPL-CsF3. Underlined are uncertainties that were reduced the most. In italic are preliminary or expected values for NPL-CsF3, subject to verification. *) The re-evaluation of NPL-CsF2 involved new analyses and measurements of systematic effects, but no modifications to the physics package; a marginal reduction of the uncertainty due to the blackbody radiation follows from using coefficients published in [20] for the calculation of the frequency offset.

The declared type B uncertainty should be consistent with long-term frequency comparisons with other PFS. We have analysed the data contributed to BIPM by NPL-CsF2 over more than 5 years and covering over 1300 days of measurement. We have compared the standard rate realized by NPL-CsF2 and that of the time scale TT (calculated by BIPM based on relevant contributions by all PFS) for all the periods when the NPL-CsF2 data were available. The average difference is 1.1×10^{-16}, within the u_B (1σ) of NPL-CsF2.

5. Conclusions and outlook

We have developed and now operate a system of highly accurate and stable caesium fountain primary frequency standards. NPL-CsF2 has been contributing regularly to the TAI evaluations for several years and NPL-CsF3, which has been constructed more recently, is currently undergoing a full performance characterisation. Both devices feature a relatively simple physics package with a vapour loaded MOT as the cold atom source and operate with nearly zero collisional frequency shift. Boosting the detected signal by optical pumping and using an optical frequency-comb based microwave source, which derives its stability from an ultra-narrow laser, has reduced the short-term stability to few parts in 10^{14} at 1 s. Improved accuracy evaluations for NPL-CsF2 significantly reduced the type B uncertainty to 2×10^{-16}, among the lowest values reported. These reductions followed analyses based on new models of frequency shifts due to distributed cavity phase, microwave lensing, and background gas collisions, combined with supporting measurements, without modifying the physics package.

The novel microwave cavity realized in NPL-CsF3 and other minor modifications, compared to NPL-CsF2, promise further accuracy improvements. However, we appreciate that bringing the total u_B significantly below 10^{-16} may be difficult because several systematic effects contribute to the uncertainty budget at a similar level. Anticipated future efforts are likely to improve the reliability and robustness of operation. Although a routine use of the optically based local oscillator would shorten the averaging time required for u_A to match the level of u_B to 3 days from the currently needed 3 weeks for NPL-CsF2, the effective statistical resolution of Cs fountain clocks remains a practical limitation to their performance.

We plan to enhance the generation of the timescale UTC(NPL) by regularly disciplining an existing maser ensemble to the local PFS's. Running at least two independent PFS's provides a minimum operational redundancy. It would also be useful to compare, as often as possible, the NPL fountains to similar devices at other NMIs for verification and to avoid operational errors. This is anticipated, for example, using the growing network of optical fibre links between European time and frequency labs.

We acknowledge financial support from National Measurement System (UK) and National Science Foundation (US).

References

[1] Kasevich M, Riis E, Chu S and De Voe R 1989 *Phys. Rev. Lett.* **63** 612
[2] Clairon A, Laurent P, Santarelli G, Ghezali S, Lea S N and Bahoura M 1995 *IEEE Trans. Meas. Instrum.* **44** 128
[3] Bauch A, Weyers S, Piester D, Staliuniene E and Yang W 2012 *Metrologia* **49** 180
[4] Godun R M, Nisbet-Jones P B R, Jones J M, King S A, Johnson L A M, Margolis H S, Szymaniec K, Lea S N, Bongs K and Gill P 2014 *Phys. Rev. Lett.* **113** 210801
[5] Wynands R and Weyers S 2005 *Metrologia* **42** S64
[6] Gibble K and Chu S 1993 *Phys. Rev. Lett.* **70** 1771
[7] Pereira Dos Santos F, Marion H, Bize S, Sortais Y, Clairon A and Salomon C 2002 *Phys. Rev. Lett.* **89** 233004
[8] Szymaniec K, Chałupczak W, Tiesinga E, Williams C J, Weyers S and Wynands R 2007 *Phys. Rev. Lett.* **98** 153002
[9] Gibble K and Verhaar B J 1995 *Phys. Rev. A* **52** 3370
[10] Leo P J, Julienne P S, Mies F H and Williams C J 2001 *Phys. Rev. Lett.* **86** 3743
[11] Szymaniec K, Noh H R, Park S E and Takamizawa A 2013 *Appl. Phys. B* **111** 527
[12] Gibble K, Lea S N and Szymaniec K 2012 *Proc. Conference on Precision Electromagnetic Measurement (Washington DC, USA)* 700
[13] Li R and Gibble K 2010 *Metrologia* **47** 534
[14] Szymaniec K, Park S E 2011 *IEEE Trans. Meas. Instrum.* **60** 2475
[15] Szymaniec K, Park S E, Marra G and Chałupczak W 2010 *Metrologia* **47** 363
[16] Li R, Gibble K and Szymaniec K 2011 *Metrologia* **48** 283
[17] Szymaniec K, Lea S N and Liu K 2014 *IEEE Trans. UFFC* **61** 203
[18] Gibble K 2006 *Phys. Rev. Lett.* **97** 073002
[19] Gibble K 2013 *Phys. Rev. Lett.* **110** 180802
[20] Rosenbusch P, Zhang S and Clairon A 2007 *Proc. IEEE International Frquency Control Symposium and 21st European Frequency and Time Forum (Geneva, Switzerland)* 1060.

Journal of Physics: Conference Series **723** (2016) 012004 doi:10.1088/1742-6596/723/1/012004

The USNO rubidium fountains

Steven Peil, James Hanssen, Thomas B. Swanson, Jennifer Taylor, and Christopher R. Ekstrom

United States Naval Observatory, 3450 Massachusetts Ave. NW, Washington, DC 20392 USA

steven.peil@usno.navy.mil

Abstract. Four rubidium fountains at the U.S. Naval Observatory (USNO) have been in operation for 4.5 years. Each fountain has demonstrated periods of stability marked by Total or Theo deviation below 10^{-16}. Occasional frequency changes, on order of 1.5 times per year per fountain, introduce deviations from white-frequency noise behavior. Averaged together, the four fountains form an ensemble with a white-frequency noise level of 10^{-13} and excellent long-term stability as compared to the primary frequency standards contributing to TAI. Progress on using the clocks at USNO for improving limits on coupling of fundamental constants to gravity by measuring the universality of the gravitational redshift for different types of clocks is discussed.

1. Introduction

Primary frequency standards based on atomic fountains were introduced in the mid-1990s. These laser cooled systems marked a major technological advance over the thermal beam standards of the time, reaching about 50 times better accuracy. On the other hand, thermal-beam systems, along with hydrogen masers, are still the predominant technology for continuously running clocks used for timescale generation. When laser cooled frequency standards first started to show promise, the U.S. Naval Observatory (USNO) initiated a program to develop continuously running atomic fountain clocks to supplement its ensemble of commercial cesium beams and hydrogen masers for precise timing. This program has culminated in six operational rubidium fountain clocks, designated NRF2 through NRF7, at two locations – NRF2, NRF3, NRF4 and NRF5 are at USNO in Washington, DC, and NRF6 and NRF7 are at the Alternate Master Clock facility in Colorado. The ensemble of four fountains in Washington, DC has been declared fully operational and has been running for 4.5 years, and each of the four fountains has been contributing to EAL (Echelle Atomique Libre) without interruption for the past 3.5 years. The results, analysis and discussion presented here focus on these four clocks.

2. Rubidium Fountain Overview

The design of the USNO fountains has been discussed in detail elsewhere [1-3]. The systems use rubidium, which has a much smaller cold-collision shift than cesium, trapped directly from background vapor into a magneto-optical trap. External cavity diode lasers (ECDLs) provide the laser light, and compact, mechanically stable optical components are used to ensure long-term optical alignment. The fountains were built two at a time, using a revised design after the first two, NRF2 and NRF3, were built. Each fountain has a 5MHz output that is sent to a clock measurement system, which measures individual clocks against the USNO master clock (MC). The 5MHz signal is generated from a high resolution frequency synthesizer (AOG), which is steered roughly every 20s to the fountain frequency (with a gain that results in a loop time constant of 1 minute). A quartz crystal oscillator that is phase locked to a hydrogen maser provides the reference for both the AOG and for the 6.8GHz drive applied to the microwave cavity. The 6.8GHz drive is steered along with the AOG to keep it centered on the central Ramsey fringe. The steering configuration is illustrated in Fig. 1.

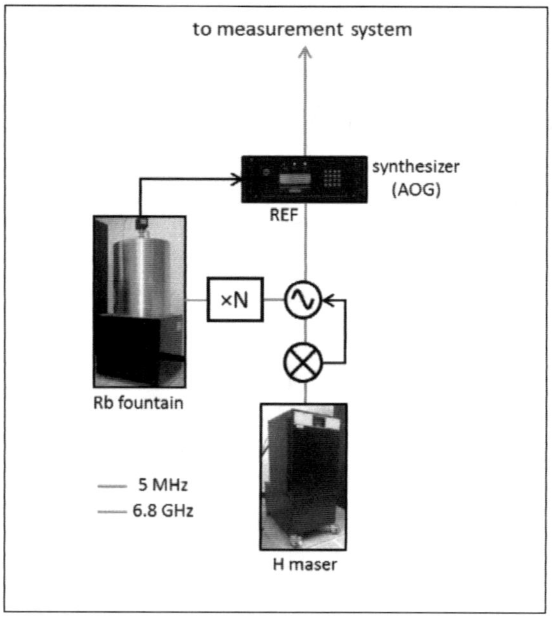

Fig. 1. (colour online). Illustration showing how a rubidium fountain is integrated into our clock measurement system. A 5MHz drive referenced to a hydrogen maser serves as the reference for a synthesizer (AOG) and for the microwave drive for the atoms. The fountain steers the AOG (and the microwave drive applied to the cavity) with a time constant of 1 minute, creating a continuous signal that is measured and compared to other clocks.

Each fountain has its own dedicated maser. Steering of the AOG is suspended if there is an interruption in normal operation, and the system relies on the frequency stability of the maser. Alarms are triggered that indicate attention is required, and steering automatically resumes when the problem is resolved. This graceful holdover is also implemented when scheduled maintenance is carried out, avoiding an interruption in the 5MHz output during remediation. A maintenance schedule has been established which calls for replacement of the ECDLs and the probe beam shutter every two years; the tapered amplifier chips age more slowly and can be replaced less frequently. User intervention can also be precipitated by processor errors and optical misalignment. Dedicated masers, stable optical components, and a well-regulated environment all contribute to an average uptime of greater than 99.5% for the four fountains.

3. Rubidium Fountain Performance

3.1. Individual fountains

The average short-term stability of the four fountains over 4.5 years of operation is 2×10^{-13} at 1s. For certain intervals of time, this white-frequency noise level dictates performance and measured fountain (in)stabilities can reach the mid 10^{-17}'s; in fact, each fountain has shown performance consistent with white-frequency noise for periods over 1 year.

In Fig. 2 we show the relative phase difference and the frequency stability of NRF4 and NRF5, which have been the most stable of the four clocks. Figure 2(a) shows performance over a particular 1.5-year interval (within the first 2 years of operation) during which the stability plot is consistent with white-frequency noise behavior out to the maximum averaging time of 6 months (for Theo). This performance is consistent with an average (in)stability per fountain below 5×10^{-17}. For this data set, the relative drift between the fountains is zero at the level of 6×10^{-19}/day. In Fig. 2(b) we show the entire phase record from the first two years of operation. The stability plot in this case shows a deviation from white-frequency noise, with a slope for long averaging times consistent with $\sqrt{\tau}$. The phase record indicates that after about 0.9 years there is a change in the relative frequency of the fountains. If we process the data and remove this frequency change, of size $<3\times10^{-16}$, the resulting frequency record integrates down, showing no signs of deviation from white-frequency noise. Thus, the appearance of non-stationary behavior is consistent with the occurrence of a single change in the relative frequency over 2 years.

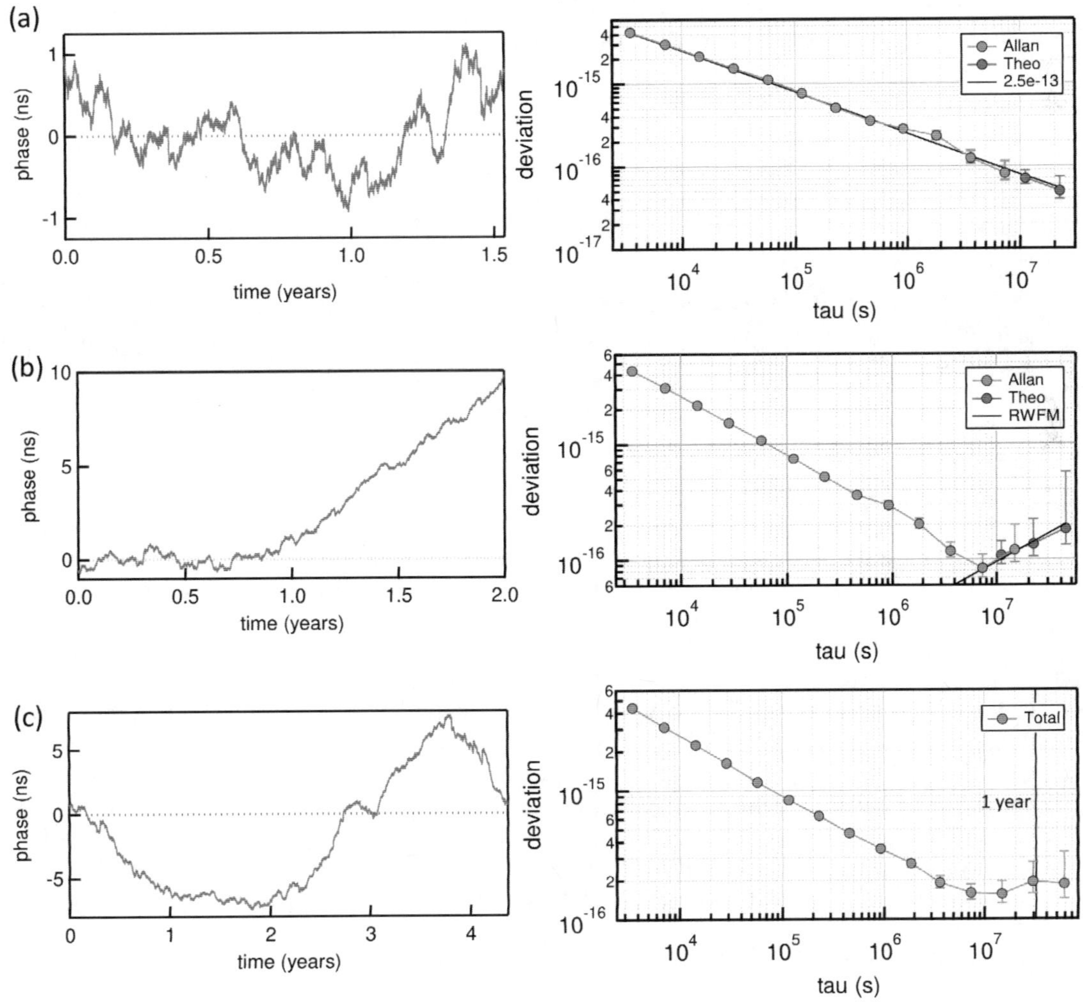

Figure 2. (colour online). Plots of relative phase (left) and stability (right) for NRF4 versus NRF5 for (a) a particular 1.5-year interval, (b) the first two years of operation, and (c) the entire 4.5 years of operation. The value of the frequency removed in each plot is different; the average frequency difference in (a) and (c) are zero and the change in frequency in (b) is emphasized. Solid lines on stability plots are references for (a) $1/\sqrt{\tau}$ and (b) $\sqrt{\tau}$ behavior.

The NRF4 versus NRF5 data for the entire 4.5 years the fountains have been operating are shown in Fig. 2(c). Over this time there have been several changes in the relative frequency, and the result is a stability plot more consistent with flicker frequency noise than with random walk. Even with these frequency changes, the phase record for NRF4 versus NRF5 shows that the fountains remained within 15 ns (±7.5 ns) of each other over 4.5 years.

A three-cornered-hat analysis shows that all of the fountains reaching stabilities below 5×10^{-16}, from 4×10^{-16} (NRF3) to 4×10^{-17} (NRF5), for averaging times of a year and longer; see Fig. 3(a). These data points are obtained by averaging the four different ways of calculating individual stabilities from four clocks (using three-cornered-hat analysis). For some of these calculations, negative variances were obtained for the best clocks (NRF4 and NRF5). This failure of the three-cornered hat technique is likely due to disparate stabilities rather than correlations between fountains. Representative error bars are included on the 1-year data points in Fig. 3(a).

3.2. Frequency changes

The source of the frequency changes in the fountains is presently unknown. Some seem to correspond with user intervention, but no decisive pattern has yet been determined. Over 4.5 years of operation, there have been on average 1.5 frequency changes per fountain per year, with an average magnitude of 6×10^{-16}, but including the sign of the shift, the average value is 6×10^{-18}.

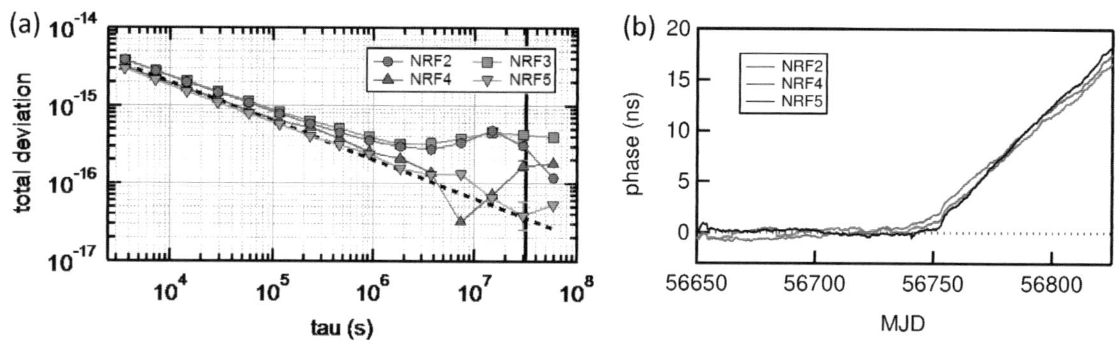

Figure 3. (a) (colour online). Three-cornered hat stability plots for the four fountains, using all of the 4.5 years of data. (b) Example of a frequency shift in NRF3, identified in plots of the relative phase of NRF3 compared to the other 3 fountains.

Obvious candidates for these frequency shifts are changes in the values of systematic biases. The largest drivers of frequency biases are the second-order Zeeman shift, the blackbody shift, and the gravitational redshift. Any changes in the gravitational redshift would be common to all of the fountains and therefore would not appear in frequency comparisons. Similarly, changes in room temperature would be common mode for pairs of fountains, which would yield a unique signature for the frequency changes that is not observed. Additionally, both temperature and humidity are continually monitored and show no changes that are correlated with fountain frequency changes.

We can also rule out the second-order Zeeman term as the source of our frequency shifts. We regularly measure the average magnetic field experienced by the atoms over their trajectory by measuring the field-sensitive transition frequency. We see no changes in magnetic field that correlate in size or time with the frequency changes observed in the fountains.

Another possible cause of frequency shifts that we have studied is the light shift. Laser light is fiber coupled to an optical table with beams splitters, acousto-optic modulators (AOMs), and shutters for the trapping and detection beams. Light is then fiber coupled from this table and delivered to the vacuum chamber. When the shutters block the laser light, scattered light that could conceivably couple to the vacuum chamber is still present. We have added an overall shutter at the output of the laser that prevents any light from reaching this optical table. The use of this additional shutter does not affect the fountain frequency at an observable level, and we therefore do not believe that the fountains are affected by light shifts.

Changes in the size of the cold collision shift over time are calculated to be far too small to be noticeable. Furthermore, changes in the density of the atoms involved in the fountain cycle should be reflected by changes in the number of atoms. We measure this number on each cycle and have not observed a correlation between a change in the number of atoms and a change in the frequency of a fountain.

Systematic biases that involve the microwave cavity and microwave drive are difficult to decouple. We investigated biases associated with the microwave cavity in our engineering prototype, NRF1, but did not carry out detailed measurements in the operational systems. Based on the size of cavity-related biases reported in primary frequency standards, including the distributed cavity phase shift, cavity pulling, and microwave leakage, it is unlikely that any of these are the source of the largest frequency shifts we have seen, though we cannot rule them out as contributing to the smaller shifts.

It is possible that subtle perturbations to the microwave drive are causing the frequency shifts we observe. During laboratory testing, we observed a coupling between RF signals going to the AOMs and the microwave drive going to the cavity in a particular fountain. Additional isolation reduced this coupling beyond our ability to observe; but completely eliminating it at the level we are concerned with is difficult to confirm. We remain suspicious that stray coupling to the microwave drive could be contributing to the shifts that we observe, and we hope to incorporate a microwave drive monitor in the future to investigate this possibility.

3.3. Fountain timescale

The primary motivation for building multiple fountains was to maximize the probability that at least one fountain was working at any given time. With the unexpected robustness in fountain operation that has been achieved, marked by an average uptime greater than 99.5%, the existence of four fountains allows us to mitigate the effects of the frequency shifts we observe. Intercomparisons of the relative phases of the four systems typically enables a frequency change to be unambiguously associated with a particular fountain, as illustrated in Fig. 3(b). A post-processed timescale can be created in which the effect of the frequency change is removed. This is a recharacterization of the rubidium fountains analogous to that applied to hydrogen masers when making a timescale. This rubidium fountain-only timescale should demonstrate better long-term stability than the individual clocks. Perhaps the ultimate reference for determining long-term frequency stability is the collection of primary frequency standards that calibrate TAI, and we have demonstrated in the past that our fountain timescale shows no drift with respect to the primary standards at a level of 1.2×10^{-18} per day [4]. The average short-term stability of the fountain timescale is 10^{-13} at 1 second, improving over the timescale derived from USNO's 70 commercial cesiums by a factor of 10, which translates to a factor of 100 improvement in the averaging time required to reach a particular level of stability.

4. Tests of Relativity

The behavior of the frequencies of atomic clocks in time and space has been used to constrain violations of the Einstein Equivalence Principle (EEP), which lays the foundation for General Relativity. One test of EEP includes looking for a difference in the gravitational redshift of clocks of different species as they experience a changing gravitational potential, typically the varying solar potential due to the Earth's elliptical orbit [5]. From the dependence of clock frequencies on fundamental constants, the redshift tests can be used to constrain the dependence of fundamental constants on gravity [6]. While optical clocks provide significantly higher precision, microwave clocks involve transitions that depend on more fundamental constants and are therefore essential for constraining those dependencies.

Measurements made with the rubidium fountains (and other clocks at USNO) after the first 1.5 years of continuous operation resulted in the most precise test of the universality of the gravitational redshift and enabled constraints on the coupling of fundamental constants to gravity to be improved by on average a factor of four [7]. With 4.5 years of data, we expect to be able to improve upon these measurements. The highest precision comparison we can carry out is a comparison of rubidium and hydrogen, using our four fountains and many masers. Looking for an annual oscillation in 4.5 years of data enables most long-term maser drift to be taken into account without affecting the result. An important technique that was used in our original analysis was to determine baselines for EEP-violating signals among different pairs of masers, which enabled a bias for the EEP-violating signal to be assigned to each maser. This eliminated 'false positives' that otherwise would have appeared in inter-species comparisons. A new EEP analysis is underway and will hopefully be finalized in the near future.

The clocks in these measurements experience the same gravitational potential; using clocks at different locations on the Earth puts them at different solar potentials and enables a distinct test of EEP. While the size of the driving term – the variation in gravitational potential due to the Earth's rotation in the Sun's potential well – is less than that due to the Earth's elliptical orbit, the putative signal would arise as a daily oscillation. Experiments planned for space considered carrying out this type of test using common view satellite comparisons of clocks separated by intercontinental distances [8]. Terrestrial measurements using optical clocks connected via fiber networks are a promising avenue for this type of test. We have considered using clocks at USNO in Washington, DC and in Colorado Springs, CO, which are compared regularly via two-way satellite time and frequency transfer. These locations give a difference in gravitational potential of more than 10^{-13}, which should lead to a limit on EEP violation for this type of test of about 3 orders of magnitude higher than the measurements with co-located clocks discussed above. The noise associated with satellite frequency transfer will likely limit the precision above this level, however.

5. Acknowledgements

We thank Paul Koppang and Jim Skinner for assistance with obtaining various types of clock data. Demetrios Matsakis has been involved with the study of EEP violation using clocks at different locations on the Earth.

References

[1] Peil S, Crane S, Swanson T B and Ekstrom C R, 2005 *Proc. 2005 IEEE Int. Frequency Control Symp.* (*Vancouver*) p 304
[2] Peil S, Crane S, Swanson T and Ekstrom C R, 2006 *Proc. 20th European Frequency and Time Forum* (*Braunschweig, Germany*, 2006) p 194
[3] Ekstrom C, Peil S, Swanson T and Crane S, 2009 *Frequency Standards and Metrology, Proc. 7th Symp.* (*Pacific Grove, CA*) (Singapore: World Scientific) p 308
[4] Peil S, Hanssen J L, Swanson T B, Taylor J and Ekstrom C R, *Metrologia* **51** (2014) p 263
[5] Bauch A and Weyers S, Phys Rev D **65** (2002) 081101(R)
[6] Blatt S, *et al.*, Phys Rev Lett **100** (2008) 140801
[7] Peil S, Crane S, Hanssen J, Swanson T and Ekstrom C R, Phys Rev A **87** (2013) 010102(R)
[8] Altschul B, *et al.*, Advances in Space Research **55** (2015) pp 501-524

Measurement of the Microwave Lensing shift in NIST-F1 and NIST-F2

S R Jefferts, T P Heavner, S E Barlow and N Ashby

NIST Time and Frequency Division, National Institute for Standards and Technology, Boulder, CO 80305, USA

Steven.Jefferts@nist.gov

Abstract. With several Primary Frequency Standards (PFS) across the world demonstrating systematic fractional frequency uncertainties on order of 1 x 10^{-16}, it is crucial to accurately measure or model even small frequency shifts that could affect the ultimate PFS uncertainty, and thus ultimately impact the rate of Coordinated Universal Time (UTC) which relies on precision PFS measurements. Recently there has been controversy about the physical causes and size of PFS frequency shifts due to microwave lensing effects. We present here the first measurements of microwave lensing frequency shifts in the PFS NIST-F1 and NIST-F2. The measured frequency shifts agree well with the recent theory of Ashby et al [1].

1. Introduction

Precise determination of Coordinated Universal Time (UTC) or closely related quantities is crucial for a broad range of modern technologies such as global navigation satellite systems, telecommunications, and many other applications. The rate of UTC, and thus its precision and accuracy, is ultimately determined by about 15 Primary Frequency Standards (PFS) across the world. With several of these PFS now demonstrating systematic fractional frequency uncertainties close to 1 x 10^{-16}, the ultimate accuracy of UTC and related quantities increasingly depends on the measurement and/or modelling of very small PFS frequency shifts. In recent years, the group at Penn State [2] has proposed a microwave lensing model for PFS that predicts a frequency shift in NIST-F2 of approximately $\mathbf{0.9 \times 10^{-16}}$, comparable to the type B fractional frequency uncertainty of $\mathbf{1.1 \times 10^{-16}}$. A different model by Ashby et al [1] predicts a much smaller microwave lensing frequency shift, less than $\mathbf{0.3 \times 10^{-16}}$, for both NIST-F1 and NIST-F2. The predictions of these models are sufficiently different as to potentially change the rate of UTC if applied to PFS, so it is crucial to determine as accurately as possible which model (if any) best reflects physical reality. Finally, as we note in [3], this shift is probably not amenable to calculation of a precision correction as it requires knowledge of too many unknown interaction potentials.

We describe measurements of microwave lensing frequency shifts in the PFS NIST-F1 and NIST-F2. The results of the measurement are in agreement with the prediction by Ashby et al (See figure 1). The measured lensing frequency shift is substantially smaller than the shift predicted by the model of [2]. In any case, the method used here to measure and correct the frequency bias: measuring frequency shifts at elevated odd-integer multiples of the optimum microwave levels and extrapolating to zero in order to estimate the shift at optimum microwave amplitude is a correct and robust method

Journal of Physics: Conference Series **723** (2016) 012005 doi:10.1088/1742-6596/723/1/012005

of dealing with this systematic frequency shift. Given the current state of the art of the theory of the lensing shift, the results of the measurements presented in this paper suggest that PFS frequencies should certainly not be corrected via calculation based on theory, rather any correction to the PFS frequency should rely on measured data.

2. Model and Results

Two, somewhat different, theories of the microwave lensing shift [1,2] predict frequency shifts in fountain style microwave clocks as a result of focusing and deflection of the atomic wave packets. This deflection and focusing, first predicted by Cook [4,5], results from small forces exerted on the atomic wavefunctions by the gradient of the microwave field during Ramsey excitation. Both of the two theories of microwave lensing predict a frequency shift where the frequency shift at elevated microwave amplitude shows a particular pattern. If data is taken only at a Rabi tipping angle $\phi = b\tau = n\pi/2$, with n=1,3,5,7,9 the shift scales as 1,-3,5,-7,9...etc. The frequency shift at n=3 is expected to be -3 times that at n=1. Similarly the frequency shift at n=5 is 5 times that at n=1, and the shift at n=7 is -7 times that at n=1 etc. Thus, for data taken at odd integer multiples of $\phi = \pi/2$, the pattern of frequency shifts is identical to that from microwave leakage below the Ramsey cavity [6].

Under particular circumstances microwave spurs can also mimic the predicted behaviour of the lensing frequency bias at higher microwave amplitudes [7]. As a result of these other microwave effects we have long measured our PFS frequency as a function of microwave amplitude looking for this pattern. We do this in spite of the fact that other factors preclude either microwave leakage frequency biases or spur induced frequency biases in either NIST-F1 or NIST-F2. Indeed, in both NIST-F1 and NIST-F2, microwave leakage is minimized by a large collection of techniques all of which are continuously employed so as to be sure the effect is mitigated. Schematically these fountains are quite similar and we refer to both fountains in what follows unless explicitly indicated otherwise. The minimization/elimination techniques for microwave leakage are explained in some detail in [6-9]. We review some of the major points here. In order that there be a microwave leakage frequency bias, there must, of course, be resonant microwaves present at the location of the atoms. We seek to prevent leakage induced frequency shifts using techniques that fall into two broad categories: first, don't generate any resonant microwaves and/or second, shield the atoms so that a microwave field cannot exist within the area the atoms occupy. As shown in [5] microwave leakage either before or after the two-pulse Ramsey interaction causes the pattern of frequency bias as predicted by [1,2] for measurements at odd integer multiples of optimum excitation strength. Because we state-select within the magnetic shields and finish that state-selection with an optical pulse while the atoms are in the below-cutoff waveguide between the two microwave cavities (one for state-selection and one for Ramsey interrogation, see ref [11], figure 2) we can eliminate microwave leakage before the Ramsey interaction as a source of frequency bias. Further, no field can exist within the below-cutoff waveguide except that from the Ramsey cavity. The microwave synthesizer which generates the microwaves for state-selection is attenuated and strongly detuned when the atoms are outside the state-selection area and therefore cannot induce a frequency bias. Microwave-leakage induced frequency-biases resulting from leakage above the Ramsey interaction cavity is precluded in our fountains by the design of the drift region and has a different behaviour vs. microwave amplitude in any case [5-8]. This leaves microwave leakage after the second Ramsey interaction as the only possible source of leakage induced frequency bias. This too is effectively impossible in our PFS; the microwave synthesizer that generates the Ramsey interaction microwaves is the only operational microwave source between the time the atoms enter the Ramsey cavity and the detection sequence and it is strongly detuned while the atoms are still in the below-cutoff waveguide just below the Ramsey cavity after the second Ramsey interaction (the same place the optical blast occurred on the way up). By the time the atoms exit the Ramsey-cavity below-cutoff waveguide there are therefore no resonant microwaves being generated in the experiment.

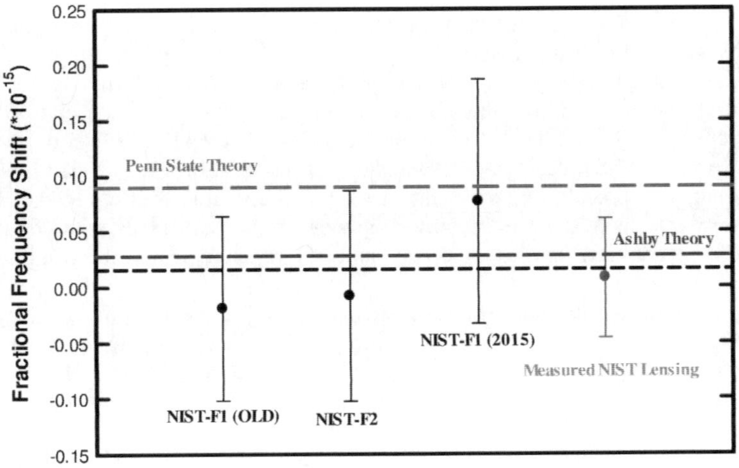

Figure 1 – Measured microwave lensing shifts (points with error bars) for NIST-F1 and NIST-F2 vs. predictions (no error bars) of various theories. The blue and black dashed lines are the results of predictions using the theory in [1] for a wavepacket size of the deBroglie wavelength (Blue) or minimum uncertainty wavepackets (Black) for the case of NIST-F1 and NIST-F2. The red dashed line is the prediction from [12]. We see good agreement between all of our data and with the theory in [1] and statistically-significant disagreement with the theory in [2,12,14]. Note that the result labelled "Measured NIST Lensing" is the weighted average of the previous 3 points labelled: NIST-F1(OLD), NIST-F2, and NIST-F1(2015). (colour online)

We have measured the spur levels and patterns in the frequency synthesizers used in both fountains [10]. None of the 4 synthesizers used in these experiments generates spurs that can cause these frequency shifts at levels of $\delta\nu/\nu > 10^{-17}$.

As a result of these considerations we can use previously measured data for NIST-F2 to estimate the lensing shift. Analysing these data for the pattern of frequency shifts predicted in [1,2] results in a measured factional frequency bias of $\frac{\delta\nu}{\nu_o} = (-0.008 \pm 0.095) \times 10^{-15}$ at n=1. Previously we have analysed these data using nonlinear weighted regression, recently we have developed a linearization technique that allows weighted linear regression to be used, and hence the change from the results reported in [10]. We take this to be a microwave lensing induced frequency bias. Ashby et al. [1], predict $\frac{\delta\nu}{\nu_o} = 0.029 \times 10^{-15}$ for this frequency bias in NIST-F2 while ref [12] predicts $\frac{\delta\nu}{\nu_o} = 0.09 \times 10^{-15}$ for NIST-F2 under similar conditions. Because these data are collected as a routine part of normal fountain operation for both NIST-F1 and NIST-F2, we have similar data for NIST-F1 as well. Treating those data in the same fashion as the NIST-F2 data, we can measure the lensing shift in NIST-F1. This results in a measured shift for NIST-F1 of $\frac{\delta\nu}{\nu_o} = (-0.019 \pm 0.083) \times 10^{-15}$ for older data (up until 2015). NIST-F1 was recently moved into a new location which resulted in a re-measurement of this bias. The re-measured point is also shown in figure 1 along with the earlier NIST-F1 and NIST-F2 measurements, as well as the average measured microwave lensing bias. Also shown in figure 1 are the predictions made by both the Ashby et al and Penn State theories. Because NIST-F1 and NIST-F2 have such similar microwave structures and Ramsey times, we can combine the measured lensing shifts for the two fountains and average the Ashby results (the Ashby theory gives $\frac{\delta\nu}{\nu_o} = 0.029 \times 10^{-15}$ for NIST-F2 and $\frac{\delta\nu}{\nu_o} = 0.027 \times 10^{-15}$ for NIST-F1) giving the result shown in figure 1. Further, we have scaled the prediction for the bias in NIST-F2 found in [12] by the ratio of Ashby's F1/F2 predictions to generate our best guess as to the F1 prediction using the theory of [2,12-13]. We are forced into this unusual step, because published formulae for frequency corrections using the theory in [2,12, 14] seem to be in serious error as discussed in Appendix A of this document and in [3, 11, 13]. In [12] it is claimed that Eq. (1) (of [12]) is used to calculate the lensing frequency shift in NIST-F2. This is unlikely because Eq. (1) of [12] seems to be dimensionally incorrect, see appendix A. We fail to understand how it is possible to calculate a shift using the published expressions for the shift contained in work of [2, 12]. As shown in figure 1, while there is good agreement between the Ashby theory and our measurements, our results are not in agreement with those claimed by [2,12,14]

All existing theories of the microwave lensing bias share a common defect; they predict a frequency shift due to lensing which results from a purely mechanical "clipping" of a portion of the atomic wavepacket for a particular atom. Essentially all atoms have their wavefunctions clipped and both superposition components of the wavefunctions are clipped almost equally (if they were, in fact, equal there would be no frequency shift from clipping). Such purely mechanical clipping is, of course, only a mathematical artifice, physically any such clipping is presumably the product of an interaction between the atom and the aperture doing the clipping and the resulting interaction Hamiltonian must be evaluated. As we pointed out in [3], phase shifts from wall interactions six or more orders of magnitude greater than those required for this frequency shift have been measured. Because the wavefunction of one of the two microwave superposition states is clipped a small amount more than the other, any differential phase shift of the superposition state as a result of the clipping could easily be imposed on the superposition by the clipping. The scale and sign of the effect could therefore be vastly different from that predicted in [1,2]. The only method of dealing with this shift in a precision measurement is, in fact, to measure it. The lensing shift cannot be calculated with sufficient robustness to be confidently used in a PFS even if the theories were in agreement. In the present case where the theories disagree, this is especially true.

3. Conclusions We present the first measurements of the microwave lensing frequency shift. Three independent measurements of the microwave lensing frequency bias in two different Primary Frequency Standards; NIST-F1 and NIST-F2, are discussed. These results are reinterpretations of previously measured data along with some newly measured data in the case of NIST-F1. All of the measurements are in good agreement with the predictions made using the theory of Ashby et al. [1]. The method used to extract the lensing bias from these measurements; using the predicted behaviour of the frequency shift vs. microwave amplitude at odd integers of optimum excitation along with measurements made at those odd integer multiples of the optimum excitation is a robust and correct method for measuring the actual frequency bias from this effect. Our result, measurement of the lensing bias, is the only published experimental data on this effect at this time and it disagrees with the theory of Gibble in [2, 12, 14]. It would seem therefore that fountain corrections should not be made on the basis of calculations made using the theories of microwave lensing presented in [2,12,14] until and unless experimental evidence allows for such corrections. We would speculate that, generally speaking, frequency corrections estimated on the basis of measurements are more likely to reflect the actual reality of that particular correction in any given fountain than a (necessarily oversimplified) theoretical calculation.

Acknowledgements: We gratefully acknowledge T. Parker and T. O'Brian for many helpful comments and discussions regarding this shift. We acknowledge K. Gibble for motivating us to return to this data for reconsideration.

Journal of Physics: Conference Series **723** (2016) 012005 doi:10.1088/1742-6596/723/1/012005

Appendix A We have suggested that there are multiple errors in the theory promulgated in [2, 12] as well as earlier work in [14]. Some of these errors are discussed in [3], here we discuss a few that we feel undermine the ability of the theory to be used to calculate "corrections" to PFS or SFS. This is not meant to be an exhaustive list, rather it is a limited sampling of the reasons we find the exposition of the theory lacking in sufficient rigor to be used in the calculation of corrections to PFS. Note that references to equation numbers are to the equation numbers contained in the reference under discussion.

Starting with the oldest (2006) of the references first [14], we note the following errors:

1 - First, the microwave magnetic field is defined as $B(\vec{r}) = \cos(k_{1x}x)\cos(k_{1z}z)$. The units on the two sides of the equation do not match, the left hand side (LHS) has units of magnetic field while the right hand side (RHS) has no units. Equations with different units on the RHS and LHS are, by definition, nonsense.

2 – Equations (3) and (4) are changes in probability that result from the resonant interaction of the atoms with the microwave field. However, the amplitude of the microwave field never appears in Eq. (3) or (Eq. (4), and, as a result the δP and ΔP_R as defined by the equations are *independent* of microwave field amplitude with the result that the resulting frequency shift is also independent of amplitude. This is unphysical.

3 – The entire mathematical development of theory is based on the idea that the individual atom wavepacket size is much smaller than the density distribution of atoms, this is incorrect. Consider a 1mm sized atomic cloud at launch with a 1μK temperature. Further consider that the atoms are described by minimum uncertainty wavepackets. At the end of a 0.6s ballistic flight (typical of a fountain PFS) the atomic density (cloud) size is about 0.7mm as a result of thermal spreading, while the individual wavepacket size is more than 1cm as a result of quantum mechanical spreading. Clearly, if the size of the atom-cloud density distribution matters (and the theory of [2,12,14] says it does), then so must the size of the wavepacket under the conditions that obtain in a fountain PFS. The Penn State group's theory incorrectly ignores the individual atoms in favour of the cloud density distribution. The individual atom's wavepacket sizes are never used in the calculation of the frequency shift.

Continuing with later versions of the theory (2014) we see in Ref [2], the following errors:

1 – In Eq. (4) we see the supposed frequency shift is identically zero at all times.

2 – In Eq. (6) and Eq. (7) the units on the LHS do not match those in the RHS of the equations.

Finally in the latest version of the theory (Ref [12], 2015) put forward by the Penn State group we see the following:

1 – In Eq. (2), reproduced here $\phi_1(\vec{r}) = \int_{-\infty}^{\infty} H_z(\vec{r}, z)dz$. We therefore apparently have an angle (presumably in radians) which has units of magnetic field times length on the RHS of the equation. Thus, according to [12], angle(radians)=tesla-meters or, assuming magnetic field units are in Hz (ie. Rabi frequency), then angle(radians)=velocity or some such.

2 – In Eq (1) of [12], given as $\delta\nu/_{\nu} =$

$$\frac{\phi_1}{sin(\phi_1)} \frac{a(t_{2L}-t_1)}{k(t_2-t_1)}$$
$$\times \frac{\int_0^{2\pi} d\phi_{2LO} \int_{r_{1L}<a} \frac{r_{2LO}(t_1-t_{1L}) + r_{1L}(t_{2L}-t_1)\cos(\phi_{2LO})}{r_1} J_1(kr_1)\psi_o{}^2(\vec{r}_{2LO},\vec{r}_{1L})|_{r_{2LO}=a} d\overrightarrow{r_{1L}} d\phi_{2LO}}{\iint_{r_{2LO}<a,\ r_{1L}<a} \psi_o{}^2(\vec{r}_{2LO},\vec{r}_{1L}) d\overrightarrow{r_{1L}} d\overrightarrow{r_{2LO}}}$$

We identify a number of issues. The first term, $\frac{\phi_1}{sin(\phi_1)}$, has no units, the second term $\left(\frac{a(t_{2L}-t_1)}{k(t_2-t_1)}\right)$, has units of l^2, The term $\frac{r_{2LO}(t_1-t_{1L})+r_{1L}(t_{2L}-t_1)\cos(\phi_{2LO})}{r_1}$, has units of time. Note this apparently must be true as $\vec{r_1}$ is defined as $\vec{r_1} = \frac{\overrightarrow{r_{2LO}}(t_1-t_{1L})+\overrightarrow{r_{1L}}(t_{2L}-t_1)}{(t_{2L}-t_{1L})}$, so that $r_{2LO}(t_1-t_{1L})$ must logically be a length times a time or $\vec{r_1}$ has units of $l \times s^{-1}$, either interpretation will leave the units in the above equation being incorrect. Returning to the equation above, the meaning of expressions like $\iint_{r_{2LO}<a,\ r_{1L}<a} d\overrightarrow{r_{1L}} d\overrightarrow{r_{2LO}}$ are undefined anywhere in [12] (as in fact are most of the symbols in Eq (1)). The integrals could reasonably be supposed to carry units of $l, l^2, or\ l^3$, depending on the dimensionality of the (undefined) vector integration, $\int d\vec{r}$. Combining all of these terms leaves the fractional frequency shift in the above equation having overall units of $l \times t, t, or\ t/l$ for the integrals having dimensionality of $l, l^2, or\ l^3$ respectively. Under any set of assumptions about the undefined notation, the fractional frequency shift should not have any units. Further errors in the same equation involve ψ_o^2. The definition of ψ_o^2 in Ref [12] is:

$$\psi_o^2(\vec{r}_{2LO},\vec{r}_{1L}) = e^{-\frac{r_{2LO}^2 w_{1L}^2 + r_{1L}^2 w_{2L}^2 - 2r_{2LO}r_{1L}\left(w_\Theta^2 + u^2 t_{2L}t_{1L}\cos(\phi_{2LO})\right)}{w_0^2 u^2 (t_{2L}-t_{1L})^2}}.$$

The ψ in question here is not an atomic wave function and does not describe any atom, it is, in fact, related only to the atomic cloud density distribution, and then only for an atomic cloud centred on the cavity. Further, atomic wavefunctions are generally normalized, this is not. We would contend that the use of ψ for something which is not an atomic wavefunction within the present context is a wholly unnecessary obfuscation of the theory in question. The theory in [1] deals with atoms transiting the cavity at any radial position and with transverse velocity, that is, atoms as they actually exist in fountain PFS.

Space considerations mean that we have attempted to illustrate only a subset of the errors in the theory of the lensing shift propounded by the Penn State group. These errors span the full range of time in which the Penn State group has published papers on the theory. The errors persist from the earliest paper in 2006 [14] to the most recent in 2015 [12]. We feel that in light of these errors it is impossible to use the formulae in these references to actually calculate a correction to a PFS.

References

[1] Ashby N, Barlow S, Heavner T and Jefferts S 2015 Frequency shifts in NIST cs primary frequency standards due to transverse rf field gradients, *Phys. Rev.* A **91** 033624.
[2] Gibble K 2014 Ramsey spectroscopy, matter-wave interferometry, and the microwave lensing frequency shift *Phys. Rev* A **90** 015601.
[3] Jefferts S, Heavner T, Barlow S and Ashby N 2015 Comment on 'Ramsey spectroscopy, matter-wave interferometry, and the microwave lensing frequency shift' *Phys. Rev* A **91** 067601.
[4] Cook R J 1978 Theory of Atomic Motion in a Resonant Electromagnetic Wave *Phys. Rev. Lett.*

41 1788.

[5] Cook R J 1987 Optical Stern-Gerlach Effect *Phys. Rev.* A **35** 3844.

[6] Shirley J H, Levi F, Heavner T P, Calonico D, Dai-HyukYu and Jefferts S R 2006 Microwave leakage-induced frequency shifts in the primary frequency standards NIST-F1and IEN-CS1 *IEEE Trans. Utrason.* **53** 2376-2385.

[7] Levi F, Shirley J H, Heavner T P, Dai-Hyuk Yu and Jefferts S R 2006 Power dependence of the frequency bias caused by spurious components in the microwave spectrum in atomic fountains *IEEE Trans. Ultrason.* **53** 1584-1589.

[8] Jefferts et al. 2002 Accuracy evaluation of NIST-F1 *Metrologia* **39** 321-336.

[9] Heavner et al. 2005 NIST-F1: recent improvements and accuracy evaluations *Metrologia* **42** 411-422.

[10] Heavner, T P, Jefferts, S R, Donley, E A, Parker, T E, Levi, F 2005 A New Microwave Synthesis Chain for the Primary Frequency Standard NIST-F1 *Proc. 2005 Joint Mtg. IEEE Freq. Cont. Symp. PTTI* 308-311.

[11] Heavner T P, et al. 2014 First accuracy evaluation of NIST-F2 *Metrologia* **51** 174-182.

[12] Gibble K 2015 Comment on 'First accuracy evaluation of NIST-F2', *Metrologia* **52** 163-166.

[13] Jefferts S R et al. 2016 Reply to "Comment on 'First accuracy evaluation of NIST-F2'" *to be submitted to Metrologia.*

[14] Gibble K 2006 *Phys Rev Lett* **97** 073002.

Journal of Physics: Conference Series **723** (2016) 012006 doi:10.1088/1742-6596/723/1/012006

High performance vapour-cell frequency standards

M Gharavipour, C Affolderbach, S Kang, T Bandi, F Gruet, M Pellaton and G Mileti

Laboratoire Temps-Fréquence, Institut de Physique,
Université de Neuchâtel, Switzerland

Email: gaetano.mileti@unine.ch

Abstract. We report our investigations on a compact high-performance rubidium (Rb) vapour-cell clock based on microwave-optical double-resonance (DR). These studies are done in both DR continuous-wave (CW) and Ramsey schemes using the same Physics Package (PP), with the same Rb vapour cell and a magnetron-type cavity with only 45 cm^3 external volume. In the CW-DR scheme, we demonstrate a DR signal with a contrast of 26% and a linewidth of 334 Hz; in Ramsey-DR mode Ramsey signals with higher contrast up to 35% and a linewidth of 160 Hz have been demonstrated. Short-term stabilities of $1.4\times10^{-13}\ \tau^{-1/2}$ and $2.4\times10^{-13}\ \tau^{-1/2}$ are measured for CW-DR and Ramsey-DR schemes, respectively. In the Ramsey-DR operation, thanks to the separation of light and microwave interactions in time, the light-shift effect has been suppressed which allows improving the long-term clock stability as compared to CW-DR operation. Implementations in miniature atomic clocks are considered.

1. Introduction

Since the first invention of lamp-pumped rubidium (Rb) clocks in the 1960's [1,2], the demand for a stable, reliable, and compact Rb frequency standard with low power consumption has been growing, for various applications fields such as industry, telecommunication, navigation, space applications (e.g. GPS [3], GALILEO [4] and COMPASS [5]). Recently, thanks to the semiconductor laser technology, laser-pumped Rb clocks have been considered in several studies, for their better optical pumping efficiency results in short-term frequency stabilities at least one order of magnitude better than their lamp-pumped counterparts [6-9]. This opened up also several new schemes for vapour-cell atomic clocks, such as the continuous-wave (CW) Rb clock [9], coherent population trapping (CPT) Cs clock [10] and pulsed optical pumping (POP) Rb clock [11]. In this paper we report our studies on high-performance Rb frequency standards based on the laser-microwave double-resonance (DR) scheme, using both CW and Ramsey interrogation. Both CW-DR and Ramsey-DR schemes require two resonant electromagnetic fields of light and microwave. In the case of CW-DR, the laser optically pumps the vapour of the Rb atoms by creating a polarization of the atomic ground-state; simultaneously the microwave field, generated from a quartz oscillator and is near-resonant with the ^{87}Rb hyperfine clock transition (energy levels $5^2S_{1/2}\,|F_g=1, m_F=0\rangle \leftrightarrow |F_g=2, m_F=0\rangle$), is applied. The opacity of the Rb vapour to the transmitted light signal as a function of microwave frequency is a measure of the atomic ground state polarization known as CW-DR signal [9]. In CW-DR, the steps of optical pumping, microwave interrogation and optical detection are applied simultaneously (figure 1(b)). In the case of Ramsey-DR, these three sequences are separated in time, see figure 1(c): first, during optical pumping a strong laser pulse creates a significant atom population imbalance between

Journal of Physics: Conference Series **723** (2016) 012006 doi:10.1088/1742-6596/723/1/012006

the two ground-state levels of ^{87}Rb. Then, in absence of the light, the microwave radiation probes the Rb clock transition using the Ramsey scheme, i.e. by two coherent $\pi/2$ microwave (Rabi) pulses of same pulse duration T_1, amplitude and frequency, but separated by the Ramsey time T_{Ramsey}. Finally, light of same frequency as during optical pumping but with much weaker intensity serves for optical detection and produces a narrow clock-resonance Ramsey signal [12, 13]. The setup design in the case of the CW-DR scheme is less complex with a lower volume and a lower power consumption than in the case of Ramsey-DR. Indeed, this latter requires an acousto-optical-modulator (AOM) that is used as a fast optical switch. However, in the case of the Ramsey-DR scheme, the light shifts (LS) effect (see section 3.3) can be suppressed, thanks to the separation of the light-atom and microwave-atom interactions. Furthermore, the durations and intensities of the optical pumping and detection pulses can be optimized, separately.

2. CW-DR and Ramsey-DR Rb clock setup

The details of our experimental clock setups for both the CW-DR and Ramsey-DR cases were previously presented in [9] and [14, 15] respectively. The same setup (see figure 1(a)) is used for the CW-DR and Ramsey-DR schemes and contains three main components: 1) the frequency-stabilized compact laser head (LH) [16], 2) the Physics Package (PP, containing the Rb cell and magnetron cavity) with an overall volume < 0.8 dm^3, and 3) the local oscillator (LO).

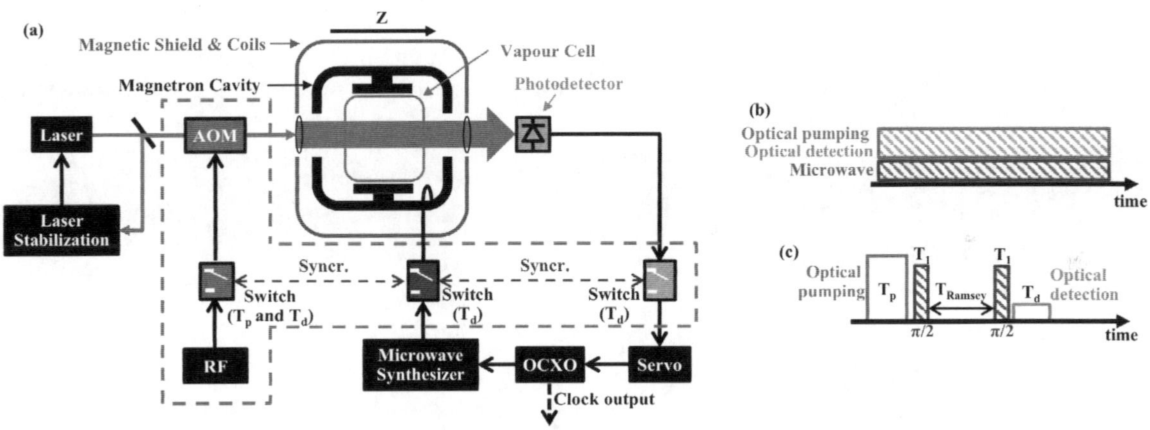

Figure 1. (a) Schematic setup for CW-DR and Ramsey-DR clocks. Components in the dash line are used only in the case of Ramsey-DR scheme. (b) CW-DR sequences. (c) Ramsey-DR sequences. (colour online)

The laser source is a distributed-feedback (DFB) laser diode emitting at 780 nm (Rb D2 transition). Its frequency is stabilized to the sub-Doppler $|5S_{1/2}, F = 1\rangle \rightarrow |5P_{3/2}, F = 1, 2\rangle$ cross-over resonance, using an evacuated ^{87}Rb reference cell that is integrated in the laser-head (LH) assembly [16]. In the case of the Ramsey-DR scheme, an AOM operated in double-pass (not required for CW-DR scheme) is implemented in the LH [17] and serves for: 1) laser output switching to control the durations of the pumping and detection pulses, 2) frequency shifting by -160 MHz, and 3) the control of the laser output power. The PP contains a home-made 25 mm diameter and length glass cell filled with an enriched ^{87}Rb vapour and 26 mbar total buffer-gas pressure (mixture of Argon and Nitrogen, P_{Ar}/P_{N2}=1.6). The cell has a 25 mm long stem acting as a reservoir for the Rb atoms. The Rb vapour cell is placed in a compact magnetron-type cavity with a total external volume of 45 cm^3. The cavity resonates at the ^{87}Rb clock's transition frequency of ≈6.835 GHz, in a TE_{011}-like mode with a relatively low loaded quality factor Q≈200 [18]. This magnetron-type resonator is a good candidate to be also used in the case of Ramsey-DR scheme as it shows a highly uniform microwave magnetic field distribution across the vapour cell. This results in a high contrast Ramsey signal since most of the Rb atoms experienced well approximated π/2 -pulses [15]. A C-field coil around the resonator generates a static magnetic field (typical amplitude ≈40 mG) oriented parallel to the laser propagation vector (z direction). This static field is required to lift the degeneracy of ^{87}Rb hyperfine ground states into their respective Zeeman levels. A two-layer magnetic shield surrounds the whole system to prevent the external magnetic field fluctuations. The photodetector collects the detection signal after the cell. The LO consists of an oven controlled crystal oscillator (OCXO), the microwave synthesizer to generate the ≈6.835 GHz radiation for ^{87}Rb clock transition interrogation and the servo loop [19].

3. Experimental Results

3.1. DR and Ramsey signals

As explained in the previous section, in the case of the CW-DR scheme both resonant optical and microwave fields act simultaneously (figure 1(b)). The light power transmitted through the cell (as function of the microwave frequency) constitutes the clock signal used to adjust the frequency of the LO. For this scheme the optimized DR signal is shown by the solid red line in figure 2. Here, the laser intensity is 0.46 μW/mm^2 and the microwave input power is -34 dBm. The signal can be well fitted by a Lorentzian line shape with a full width at half maximum (FWHM) $\Delta\nu_{1/2}$=334 Hz, which is limited by optical and microwave power broadening, and a contrast of C=26% [9].

The pulse sequences for the case of the Ramsey-DR scheme are shown in figure 1(c), using the following parameters: 1) optical pumping pulse time T_p=0.4 ms with a laser power of about 15 mW, 2) two π/2 microwave pulses with the duration of T_1=0.4 ms and input power (to the cavity) of –20 dBm, 3) Ramsey time T_{Ramsey}=3 ms, and 4) optical detection pulse time of T_d=0.7 ms with a laser power of about 0.1 mW. The Ramsey pattern obtained in this situation is given by the dashed blue line in figure 2. Its central fringe has a contrast of C=35% and a linewidth of $\Delta\nu_{1/2}$=160 Hz, consistent with the theoretical prediction of $\Delta\nu_{1/2}=1/(2\cdot T_{Ramsey})$ [15].

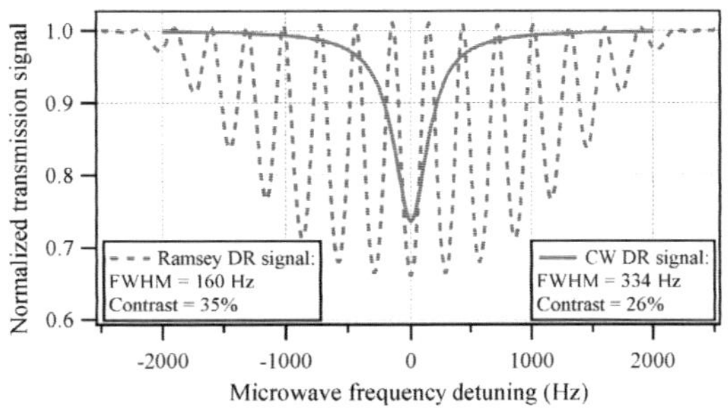

Figure 2. Double resonance signal in the case of the CW-DR scheme (solid red line) compared to Ramsey fringes in the case of Ramsey-DR (dashed blue line) [9, 15]. (colour online)

3.2. Short-term clock stability

The signal-to-noise ratio (S/N) limits the short-term stability (at τ=1 to 100 s) of a passive Rb clock and can be estimated by [9]:

$$\sigma_{S/N}(\tau) = \frac{N_{psd}}{\sqrt{2}\cdot D\cdot \nu_{Rb}}\tau^{-1/2}, \qquad (1)$$

where N_{psd} is the power-spectral density of the detection noise when microwave and pump laser are switched on (in closed-loop condition) and D is the discriminator slope of the error signal. Here, N_{psd} includes shot-noise, laser's amplitude-modulation (AM) noise and FM-to-AM noise conversion in the vapour cell [20]. Moreover, in the case of the Ramsey-DR scheme, contributions of AOM's additional noises are also added [15]. For the CW-DR scheme the clock's shot-noise limit is estimated as $4.9\times10^{-14}\ \tau^{-1/2}$, by replacing N_{psd} in (1) with the shot-noise level. In the Ramsey-DR scheme with optical detection the shot-noise limit can be estimated by [11]:

$$\sigma_{sn}(\tau)=\frac{1}{\pi Q_a R_{sn}}\sqrt{\frac{T_c}{\tau}}, \quad (2)$$

where $Q_a \approx 4.3\times10^7$ is the quality factor of the clock transition and $T_c = 4.95$ ms is the cycle time of the pulse sequence. The signal-to-noise ratio R_{sn} for our clock is at the level of 30000 [14] and the final expected shot-noise stability limit is $\sigma_{sn}(\tau) \approx 1.7\times10^{-14}\,\tau^{-1/2}$.

The total short-term clock stability of $\sigma_y(\tau)$ includes, in addition to the S/N limit, the contributions from the LO's phase noise through the Dick effect [21] and from laser intensity and frequency fluctuations via the light-shifts (LS), and can be expressed as [7]:

$$\sigma_y(\tau)=\sqrt{\sigma^2_{S/N}(\tau)+\sigma^2_{Dick}(\tau)+\sigma^2_{LS}(\tau)}. \quad (3)$$

Table 1 shows the measured short-term instability budget for both CW-DR and Ramsey-DR schemes. The dominant source for this stability is the optical detection noise which mainly originates from the technical laser noise and FM-to-AM noise conversion in the cell [15, 18].

Table 1. Short-term instability budget comparison for both CW-DR and Ramsey-DR Rb clocks.

Instability	**Shot-noise limit ($\tau^{-1/2}$)**	Signal-to-noise $\sigma_{S/N}$ ($\tau^{-1/2}$)	Dick effect ($\tau^{-1/2}$)	Light Shift ($\tau^{-1/2}$)	**Total ($\tau^{-1/2}$) (theory)**	**Total ($\tau^{-1/2}$) (measured)**
CW-DR [9]	**4.9×10^{-14}**	1.2×10^{-13}	7.5×10^{-14}	4×10^{-14}	**1.4×10^{-13}**	**1.4×10^{-13}**
Ramsey-DR [17]	**1.7×10^{-14}**	2.0×10^{-13}	$\leq1\times10^{-13}$	3×10^{-15}	**2.2×10^{-13}**	**2.4×10^{-13}**

3.3. Light shifts (LS)

Light shifts (or AC Stark shifts) are the shifts of the atomic energy levels arising when a near-resonant optical field interacts with an atom [22, 23]. With the clock frequency ν_{clock}, we define the intensity LS coefficient $\alpha=\delta\nu_{clock}/\delta I_L$ (at fixed laser frequency ν_L) and the frequency LS coefficient $\beta=\delta\nu_{clock}/\delta\nu_L$ (at fixed laser intensity I_L). Measured LS coefficients in the CW-DR scheme are $\alpha=1.0\times10^{-12}$ /% and $\beta=1.2\times10^{-11}$ /MHz [9], more than one order of magnitude higher than $\alpha=-2.1\times10^{-14}$/% and $\beta=4.6\times10^{-13}$ /MHz in the Ramsey-DR case [17]. In principle, LS can be fully suppressed in a Ramsey scheme. The residual LSs found in Ramsey-DR mode might be due to the position shift effect (non-uniform spatial distribution of microwave field and laser absorption rate) [12], residual coherence at the beginning of the first Ramsey pulse [24] and residual non-zero light power during the Ramsey phase.

3.4. Microwave power shift (PS)

In both cases of optical-microwave DR interrogation schemes, the spatial inhomogeneity of the optical and microwave fields as well as the residual gradients of the static magnetic field can cause a position dependent inhomogeneous shift on the clock hyperfine frequency. This depends on the microwave power and is called microwave power shift [25]. The microwave power shift coefficients for CW-DR and Ramsey-DR are measured 2.2×10^{-12} /µW and 1.8×10^{-12} /µW, respectively [9, 17].

3.5. Systematic shifts and long-term clock stability

We evaluate the limitations on the medium- to long term frequency stability of our Rb clocks arising from systematic shifts. The contributions from the main physical perturbations such as intensity and frequency LS, microwave power shift, vapour cell and stem temperature coefficients (TC) are measured. Table 2 summarizes and compares the long-term instability budgets for our CW and Ramsey DR clocks, at $\tau=10^4$ s. Measurements of the long-term frequency stability show that in the case of our current vapour cell with a long stem (25 mm), in both CW and Ramsey schemes the total clock instabilities are limited to $\approx6\times10^{-14}$, mainly because of the stem temperature coefficient (TC), although in Ramsey-DR scheme the instability contributions due to LS are reduced by more than one order of magnitude compared to the CW-DR scheme (from 3.8×10^{-14} to 2.3×10^{-15}). Therefore a new vapour cell with ten times smaller stem volume was designed, resulting in its potential instability contribution reduced to 5.5×10^{-15}. With this cell, we expect our Ramsey-DR clock to reach 7.2×10^{-15} at $\tau=10^4$ s.

Table 2. Comparison of clock stability limitations at $\tau = 10^4$ s, calculated from the main systematic shift coefficients and parameters instabilities.

Effect or source	CW-DR		Ramsey-DR	
	Long stem [9]	Short stem	Long stem [17]	Short stem
Intensity LS	9.9×10^{-15}		1.1×10^{-15}	
Frequency LS	3.7×10^{-14}		2×10^{-15}	
Cell TC	$\leq 3.5\times10^{-15}$		$\leq 3.5\times10^{-15}$	
Microwave PS	8.9×10^{-16}		1.9×10^{-15}	
Stem TC	5.6×10^{-14}	5.5×10^{-15}	5.6×10^{-14}	5.5×10^{-15}
Total $\sigma_{tot}=[\Sigma_i \sigma_i^2]^{1/2}$	**6.8×10^{-14}**	**3.9×10^{-14}**	**5.6×10^{-14}**	**7.2×10^{-15}**
Total (measured)	**6×10^{-14}**	----	**6×10^{-14}**	----

4. Applications in miniature DR clocks

The suppression of light-shift effects in the Ramsey-DR scheme could also be beneficial for improving the long-term stability of miniature or chip-scale atomic clocks [26]. We therefore investigate on the approach of Ramsey-mode operation of a miniature DR clock as e.g. described in [27], based on a 50 mm^3 internal volume microfabricated Rb cell and a micro-loop-gap resonator, by numerical simulations using the model described in [15]. First results show that Ramsey signals with a narrow central fringe (760 Hz FWHM) and > 30% contrast may be expected. Such a miniature Ramsey DR clock could be a good candidate for achieving short-term stabilities of $\leq 1\times10^{-11}\ \tau^{-1/2}$ and $< 1\times10^{-12}$ at one day, for e.g. applications such as in smart grids.

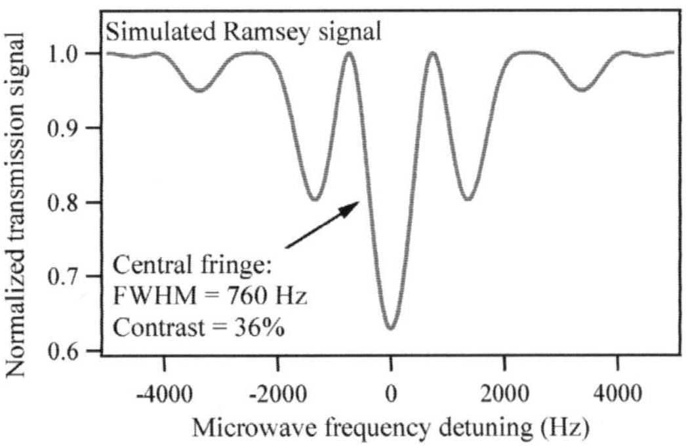

Figure 3: Ramsey DR signal simulated for a micro-fabricated Rb vapour cell such as used in [27]. T_{Ramsey}=0.2ms and T_1=0.4ms. (colour online)

5. Conclusion

We have studied compact Rb atomic clocks based on both CW-DR and Ramsey-DR schemes, using the same clock setup with a clock resonator Physics Package of only 0.8 litres overall volume. CW-DR and Ramsey signals with contrasts of 26% and 35%, and FWHMs of 334 Hz and 160 Hz, respectively, are shown. The measured short-term clock stabilities are $\approx 2\times10^{-13}\ \tau^{-1/2}$ (at τ=1 to 100 s) in both schemes, with optical detection noise being the main limitation. On medium- to long-term scales ($\tau\approx10^4$ s) the main limiting effects in the CW-DR scheme are the LS and stem TC, which can be reduced by about one order of magnitude by passing to the Ramsey-DR scheme and implementing a new cell design, respectively. A final clock stability of $< 1\times10^{-14}$ at 10^4 s up to 1 day is expected.

Acknowledgments
This work was supported by the Swiss National Science Foundation (SNSF grant no. 140712) and the European Metrology Research Programme (EMRP project IND55-Mclocks). The EMRP is jointly funded by the EMRP participating countries within EURAMET and the European Union. We also acknowledge previous support from the European Space Agency (ESA) and the Swiss Space Office (Swiss Confederation). We thank A. K. Skrivervik, C. Stefanucci (both EPFL-LEMA) for their support on the microwave cavity, and C. Calosso (INRIM, Italy) for the microwave LO.

References

[1] Camparo J 2007 *Phys. Today*, **60** 33
[2] Kastler A 1950 *J. Phys. Radium*, **11** 255
[3] Dupuis R T, Lynch T J and Vaccaro J R 2008 *Proc. IEEE Int. Frequency Control Symposium* edited by Jadusliwer B. *(Honolulu, Hawaii, USA 19-21 May 2008)* pp. 655-660
[4] Waller P, Gonzalez S, Binda S, Sesia I, Hidalgo I, Tobias G and Tavella P 2010 *IEEE Trans. Ultrason. Ferroelct. Freq. Control* **57** 738
[5] Chunhao H, Zhiwu C, Yuting L, Li L, Shenghong X, Lingfeng Z and Xianglei W 2013, *Int. J. Navig. Observ.* **2013** 371450
[6] Camparo J and Frueholz 1986 *J. Appl. Phys.* **59** 3313
[7] Mileti G, Deng J, Walls F, Jennings D and Drullinger R 1998 *IEEE J. Quantum Electron.* **34** 233
[8] Vanier J 2005 *Appl. Phys.* B **81** 421
[9] Bandi T, Affolderbach C, Stefanucci C, Merli F, Skrivervik A K and Mileti G 2014 *Ultrasonics, Ferroelectrics, and Frequency Control, IEEE Transactions on* **61** 1769-1778
[10] Danet J M, Kozlova O, Yun P, Guérandel S and De Clercq E 2014 *EPJ Web of Conf.* **77** 00017
[11] Micalizio S, Calosso C E, Godone A and Levi F 2012 , Metrologia **49** 425-436
[12] Micalizio S, Godone A, Levi F and Calosso C 2009 *Phys. Rev.* A. **79** 013403
[13] Godone A, Micalizio S, Levi F and Calosso C 2006 *Phys. Rev.* A **74** 043401
[14] Kang S, Affolderbach C, Gruet F, Gharavipour M, Calosso C E and Mileti G 2014 *Proc. 26th European Frequency and Time Forum (EFTF) (Neuchatel, Switzerland, 23-26 June 2014)* pp. 544-547
[15] Kang S, Gharavipour M, Affolderbach C, Gruet F and Mileti G 2015 *J. Appl. Phys.* **117** 104510
[16] Gruet F, Pellaton M, Affolderbach C, Bandi T, Matthey R and Mileti G, 2012 *Proc. Int. Conf. on Space Optics (ICSO) (Ajaccio, Corsica, 9-12 October 2012)* no. 48
[17] Kang S, Gharavipour M, Gruet F, Affolderbach C and Mileti G. 2015 *Proc. Joint Conf. IEEE Int. Frequency Control Symposium & European Frequency and Time Forum (EFTF) (Denver, Colorado, USA 12-16 April 2015)*, pp. 800-803
[18] Stefanucci C, Bandi T, Merli F, Pellaton M, Affolderbach C, Mileti G and Skrivervik A K 2012 *Rev. Sci. Instrum.* **83** 104706
[19] Calosso C E, Micalizio S, Godone A, Bertacco E K and Levi F 2007 *IEEE Trans. Ultrason., Ferroelectr., Freq. Control.* **54** 1731
[20] Camparo J C 1998 *J. Opt. Soc. Am. B* **15** 1177–1186
[21] Deng J Q, Mileti G, Drullinger R E, Jennings D A, and Walls F L 1999, Phys. Rev. A **59** 773
[22] Cohen-Tannoudji C and Barrat J 1961 P *J. Phys. Radium* **22** 329
[23] Mathur B S, Tang H and Happer W 1968 *Phys. Rev.* **171** 11
[24] Godone A, Micalizio S and Levi F 2004 *Phys. Rev.* A **70** 023409
[25] Vanier J and Audoin C 1989 *The Quantum Physics of Atomic Frequency Standards* vol. 1, (Bristol, UK: Adam Hilger)
[26] Knappe S 2008 MEMS atomic clocks, *Comprehensive Microsystems* (vol. 3) ed. Y B Gianchadani et al. (Amsterdam: Elsevier) pp.571-612
[27] Violetti M, Pellaton M, Merli F, Zürcher J–F, Affolderbach C, Mileti G and Skrivervik A K 2014 *IEEE Journal of Sensors* **14** 3193

Journal of Physics: Conference Series **723** (2016) 012007 doi:10.1088/1742-6596/723/1/012007

Reaching $5.0\times10^{-13}\tau^{-1/2}$ short term frequency stability of the integrating sphere cold atom clock

P Liu, Y L Meng, J Y Wan, X M Wang, Y N Wang, L Xiao, H D Cheng* and L Liu†

Key Laboratory of Quantum Optics and Center of Cold Atom Physics,
Shanghai Institute of Optics and Fine Mechanics, Chinese Academy of Sciences,
Shanghai 201800, China
E-mail: †liang.liu@siom.ac.cn, *chenghd@siom.ac.cn

Abstract. We present an improvement of short term frequency stability of the integrating sphere cold atom clock after increasing the intensities of clock signals and optimizing the feedback loop of the clock. A short term frequency stability of $5.0\times10^{-13}\tau^{-1/2}$ has been achieved and the limiting factors have been analyzed.

1. Introduction

The integrating sphere cold atom clock (ISCAC) has potential applications in space because of its compact volume and great performance in microgravity environment [1]. For an ISCAC, atoms are cooled by diffuse laser light [2,3], which makes it simple and robust because any careful alignment of laser beam and light polarization adjustment are not necessary. Here we use a cylindrical microwave cavity whose inner surface was sanded and electroplated with a layer of silver to meet the requirement of diffuse reflectance. The cooling lights are directly injected into the microwave cavity and the diffuse light is formed by the reflection of the inner sanded surface. The microwave cavity serves two functions for the cold atom clock, that is, cooling atoms and microwave interrogation with cold atoms. The frequency stability of $7.3\times10^{-13}\tau^{-1/2}$ has been achieved with this setup and the limiting factors have been analyzed in our previous work [4]. It is found that the main noise sources are from the AM and PM noises of the probe laser, the feedback electronics noise, the atomic shot noise, and the local oscillator phase noise. In order to improve the frequency stability further, we reduce the effect of AM and PM noises by improving the intensities of clock signals. Furthermore, the feedback electronics noise is reduced by changing the feedback loop of the ISCAC.

Previously, the error signals for frequency correction were directly sent to the local oscillator (LO) by an analog circuit. The electronic noises of the feedback loop degraded the frequency stability of the ISCAC. In this work, the error signals are sent to the frequency auxiliary output generator which is phase locked to the LO.

In this paper, the first part is dedicated to generally describe the clock system. Then, we present the improvement recently made on the probe laser and the servo loop for the LO. The third section is devoted to show the improved performance of the ISCAC.

2. The set up and the improvement on the clock

The core of the ISCAC setup is the cylindrical microwave cavity which is also used for cooling atoms. Figure 1 shows the schematic of the ISCAC's physical package. The microwave cavity has a loaded quality factor of about 11000 and is adjusted to resonate at the transition frequency of 6.834 GHz between the two rubidium atom's ground states on the TE_{011} electromagnetic mode. The cavity is placed in a vacuum chamber which is covered by an Al barrel. The solenoid coils intertwining around the barrel is used to provide the bias magnetic field. All these devices are covered by five-layer of magnetic shields to attenuate the external magnetic fields.

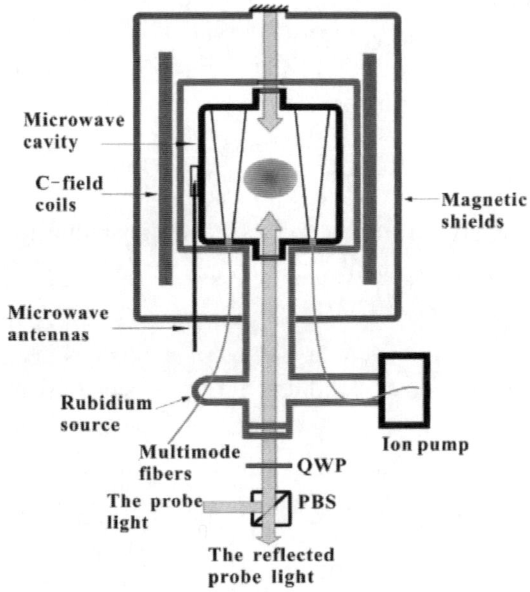

Figure 1. Schematic of the physical package (colour online). QWP-quarter wave plate, PBS-polarizing beam splitter.

The cooling, pumping, and repumping lights are injected into the microwave cavity with four multimode fibers set in the bottom of the cavity [5]. The probe light propagates along the axis of the cylindrical microwave cavity and is retroreflected by a mirror horizontal set on the top of the physical package. The standing wave probe light is used to avoid heating up the cold atoms.

There are four phases in operating the ISCAC. The ^{87}Rb atoms are first cooled by diffuse light (T_c=95 ms), and most of the cold atoms are accumulated in the central area of the microwave cavity [6]. A short pulse (T_p=0.5 ms) of pumping light is then injected into the cavity to populate the atoms from F=2 to F=1 state. Then the cold atoms are interrogated by two microwave pulses (τ =3 ms) separated by a free evolution process (T_f =20 ms). At last, a probe light (T_d =11.5 ms) which is locked to the transition of $5^2S_{1/2}\,|F=2\rangle \rightarrow 5^2P_{3/2}\,|F'=3\rangle$ is used to detect the population of the state $|F=2, m_F=0\rangle$. The time sequence and servo loop of the clock system are shown in figure 2.

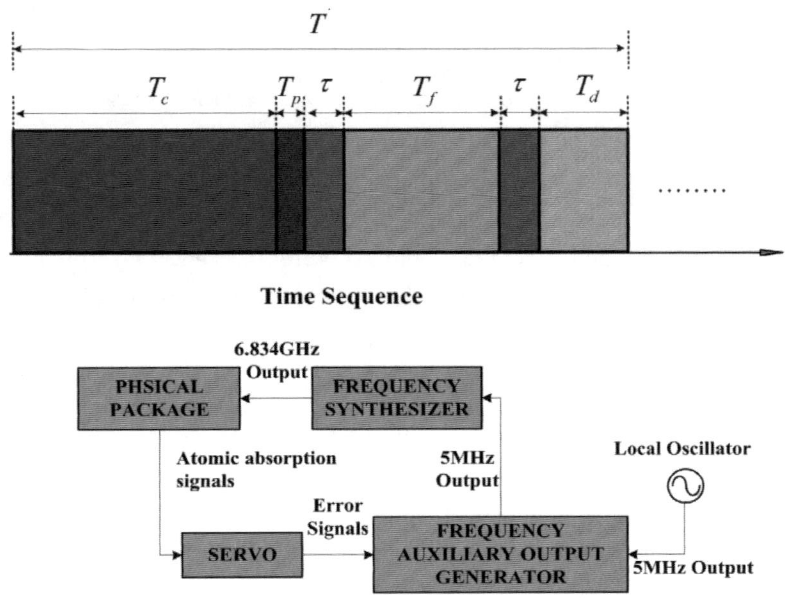

Figure 2. The time sequence of operating the integrating sphere cold atom clock and the architecture of the frequency lock loop (colour online). Where T=133 ms, T_c=95 ms, T_p=0.5 ms, τ =3 ms, T_f=20 ms and T_d=11.5 ms.

The cycle time of the ISCAC is 133 ms, rather than 92 ms used before [4]. This is because large noise contributions near 10 Hz, which can degrade the clock performance (AM noise), appeared in the spectral density of the Relative Intensity Noise (RIN) of the probe laser (figure 3). However, the atoms' free evolution time Tf keeps the same value of 20 ms. This is a trade-off between narrow line-width and large signal.

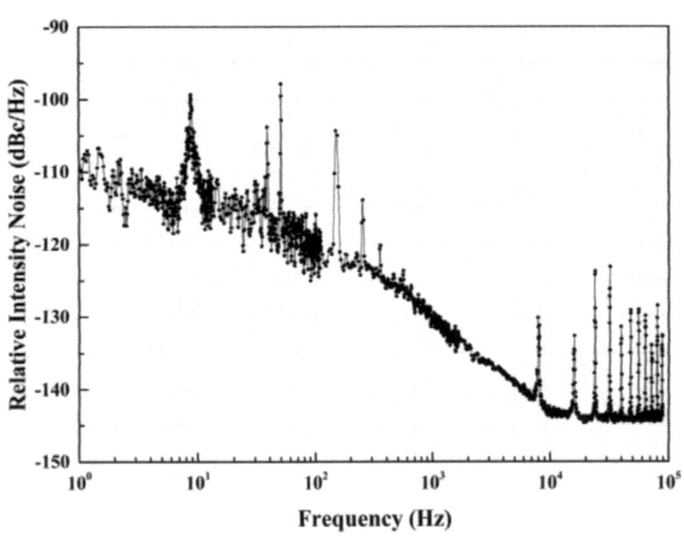

Figure 3. The Relative Intensity Noise of the background probe laser after passing through the physical package.

During the Ramsey interrogation process, the two microwave pulses are provided by a low phase noise frequency synthesizer (figure 2). Using the frequency multiplier chain in the frequency synthesizer, the 5 MHz signal coming from the output of the frequency auxiliary generator which is locked to the LO by the inner phase-locked loop is synthesized to 6.834 GHz. Through the clock signal detection at each side of the full width at half maximum (FWHM) of the Ramsey fringe, we can obtain the error signals to steer the output of the frequency auxiliary generator by serial communication. Thus, the noise from the analog feed-back loop which has an impact on the short term frequency stability of about $1.8\times10^{-13}\tau^{-1/2}$ can be eliminated.

3. Results and discussions

Figure 4 shows the Allan deviations of the ISCAC. The black curve ($7.3\times10^{-13}\tau^{-1/2}$) is the result we have measured [4] before these modifications while the red curve ($5.0\times10^{-13}\tau^{-1/2}$) is the frequency stability of the improved clock. The improvement on the short term stability is mainly due to the increasing of the clock signals and the switch of feeding the LO with a digital control loop.

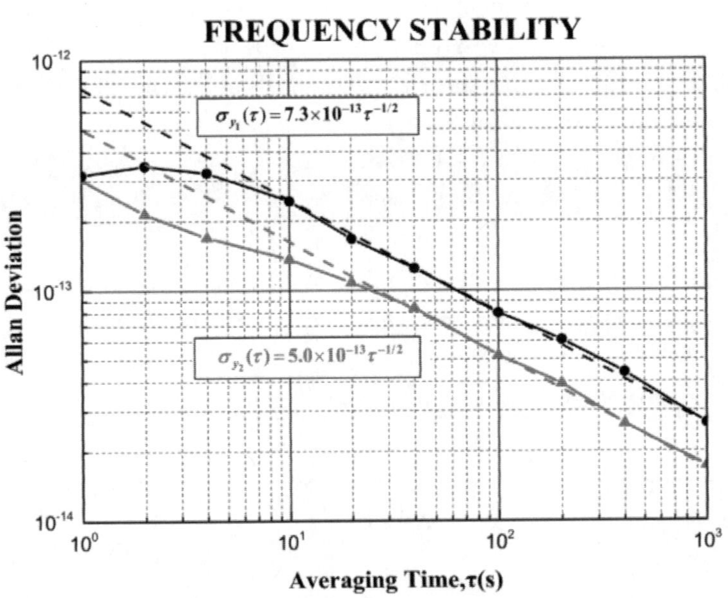

Figure 4. The short term frequency stability of the integrating sphere cold atom clock (colour online). The red curve which shows better frequency stability is the result after the modifications on the clock.

Limited by the laser power, the previous optical system can only provide total power of about 165 mW as the cooling light which didn't reach the saturation intensity of the system. With the new laser power amplifier module, we can increase the cooling light power to 240 mW. The time of laser cooling stage is also increased from 55 ms to 95 ms. Thus, the intensity of the clock signals changed to about two times of the previous results. Figure 5 shows the intensities of the central Ramsey fringes under these two experimental conditions.

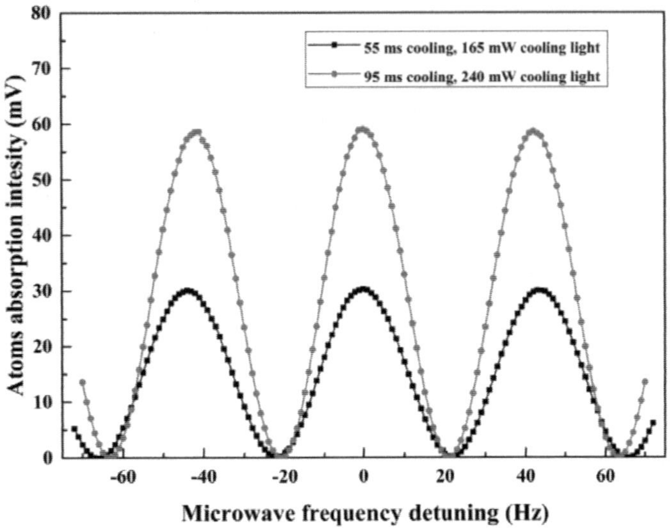

Figure 5. The central Ramsey fringes of the clock under different experimental conditions where the curve with red dots is the recent result (colour online).

The shot noise of the cold atoms absorption is proportional to $1/\sqrt{N}$ where N is the number of the cold atoms experience the "0-0" transition between the two ground states. Thus, the limitation of the short term frequency stability of atomic shot noise will decrease at the mean time. The shot noise contribution to the frequency stability of the clock is given as [7]

$$\sigma^{sn}{}_{y}(\tau) = \frac{1}{\pi Q R_{sn}} \sqrt{\frac{T}{\tau}} \tag{1}$$

where Q is the quality factor of the atomic resonance, T is the cycle time and R_{sn} is the signal to noise ratio. Table 1 is the list of the noise budget and their instabilities which have impact on the short term frequency stability of the ISCAC.

Table 1. Noise sources of the ISCAC and their instabilities.

Noise sources	Instability $\sigma_y(\tau)$
AM and FM noises of probe laser	$2.0\times10^{-13}\tau^{-1/2}$
Atomic shot noise	$3.2\times10^{-13}\tau^{-1/2}$
Local oscillator phase noise	$3.3\times10^{-13}\tau^{-1/2}$
Total $\sqrt{\sum_i \sigma_{y_i}^2(\tau)}$	$5.0\times10^{-13}\tau^{-1/2}$

4. Conclusions

In this paper, we reported the recent progress in the ISCAC. A short term frequency stability of $5.0\times10^{-13}\tau^{-1/2}$ has been achieved after we increased the power of the cooling light and used a digital control servo loop to eliminate the noise from the previous analog feedback loop. The intensities of clock signals have increased almost two times which directly lead to a better performance on the short term frequency stability of the cold atom clock.

Acknowledgments
This work was supported by National High Technology Research and Development Program of China (No. 2012AA120702) and Youth Innovation Promotion Association CAS.

References

[1] Esnault F X, Holleville D, Rossetto N, Guerandel S and Dimarcq N 2010 *Phys. Rev. A* **82**, 033436
[2] Guillot E, Pottie P E and Dimarcq N 2001 *Opt. Lett.* **26**, 1639
[3] Cheng H D, Zhang W Z, Ma H Y, Liu L and Wang Y Z 2009 *Phys. Rev. A* **79**, 023407
[4] Liu P, Meng Y L, Wan J Y, Wang X M, Wang Y N, Xiao L, Cheng H D and Liu L 2015 *Phys. Rev. A* **92**.062101
[5] Meng Y L, Zheng B C, Liu P, Wan J Y, Xiao L, Wang X M, Gao Y C, Cheng H D and Liu L 2014 *Acta Optica Sinica* **34**, 0902001
[6] Meng Y L, Cheng H D, Zheng B C, Wang X C, Xiao L and Liu L 2013 *Chin. Phys. Lett.* **30**, 063701
[7] Micalizio S, Godone A, Levi F and Calosso C 2009 *Phys. Rev. A* **79**, 013403

Status of the atomic fountain clock at the National Research Council of Canada

S Beattie [1], J Alcock[1], B Jian[1,2], M Gertsvolf[1] and J Bernard[1]

[1] National Research Council of Canada, Ottawa, Ontario K1A 0R6, Canada
[2] Department of Physics and Astronomy, York University, Toronto, Ontario M3J 1P3, Canada

E-mail: Scott.Beattie@nrc-cnrc.gc.ca

Abstract. Despite the rapid advances in optical frequency standards, caesium fountain clocks retain a critical role as the most accurate primary frequency standards available. At the National Research Council Canada, we are working to develop a second generation caesium fountain clock. Work is currently underway to improve several systems of FCs1, such as the laser system and microwave local oscillator, which will be incorporated into its refurbished version, FCs2. In addition, we have added an optical pumping stage which has increased the detected atom number by over a factor of six. In collaboration with the National Physical Laboratory (NPL), we are planning on replacing the physics package of FCs1. We will report on several recent improvements to FCs1, along with our progress in the development of FCs2.

1. Introduction

Current state of the art caesium fountain clocks can reach uncertainty levels of a few parts in 10^{16}[1-3]. Though this uncertainty is one to two orders of magnitude higher than that of the top performing optical clocks, fountain clocks are extremely robust primary standards and represent the most accurate realization of the International System (SI) second. At the National Research Council of Canada (NRC), we are currently working to rebuild our current caesium fountain FCs1 with several design changes. With these design changes and several technical improvements, we hope to achieve state of the art accuracy in its refurbished form, FCs2.

Here, we report on several technical improvements made to NRC FCs1 that have allowed us to increase the signal to noise and stability of our fountain. These improvements include changes to the laser setup, as well as the implementation of an optical pumping stage. We also describe our proposed design changes for FCs2, and report on our progress.

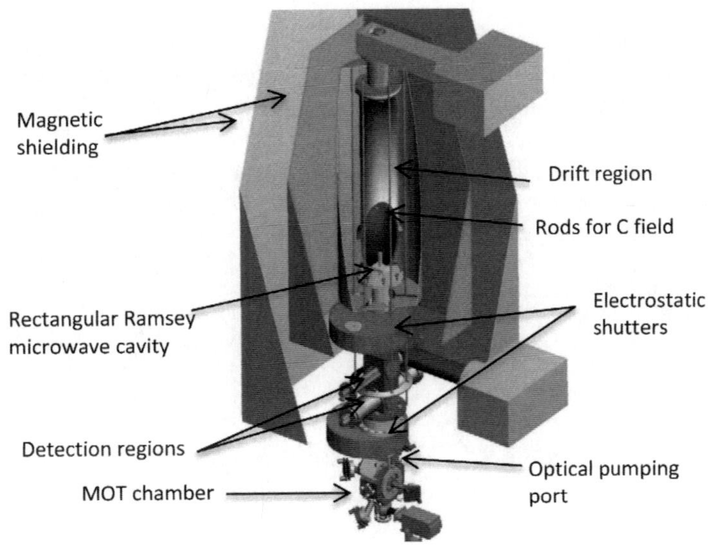

Figure 1. Physics package for NRC FCs1.

2. Experimental methods

2.1. Physics package

Figure 1 shows a schematic of the physics package for NRC-FCs1. It uses a (1,1,0) magneto-optical trap (MOT) beam geometry with beam diameters of ~ 30 mm. The C field is generated using 4 parallel, current-carrying rods which create a magnetic field transverse to the atoms' trajectory. The physics package also includes electrostatic shutters designed to allow the juggling of atoms to reduce the Dick effect. These shutters, however, have been problematic and have never been used in the operation of FCs1.

In FCs2, we will use a (0,0,1) MOT beam geometry. We will also use the more conventional technique of providing a C field along the vertical axis of the atom's trajectory, via a solenoid wound around the drift tube. The electrostatic shutters will be also eliminated in FCs2, simplifying the system and allowing a more compact design.

2.2. Laser setup

Until recently, the laser system has relied on master and repumper external cavity diode lasers, as well as four slave diode lasers to amplify the cooling and detection light. The slave diode lasers presented the weak point in the stability of the laser system, as they would regularly need to be adjusted. We have recently replaced the slave diode lasers with 2 tapered amplifier systems (Sacher TEC 400). These systems have proven to be extremely robust and have dramatically improved the stability of the system. They require only occasional adjustment to their seed beam alignment. In addition, they require less space on the optical table, and allow the total beam path lengths to be reduced, further increasing the stability of the system.

2.3. Optical Pumping

In a typical experimental launch sequence, the atoms are cooled and launched from a moving optical molasses in the F=4 state (roughly equally populated m_F states) before they pass through a state selection microwave cavity which applies a π pulse of microwaves on the $|F=4,m_F=0\rangle \rightarrow |F'=3,m_F'=0\rangle$ transition. Atoms left in the F=4 ground state are eliminated by applying a "pusher" beam of light to cycle atoms on the $|F=4\rangle \rightarrow |F'=5\rangle$ transition. This procedure gives a clean sample of atoms only in the $|F=3,m_F=0\rangle$ clock state, but it reduces the atom number by roughly 8/9=89%.

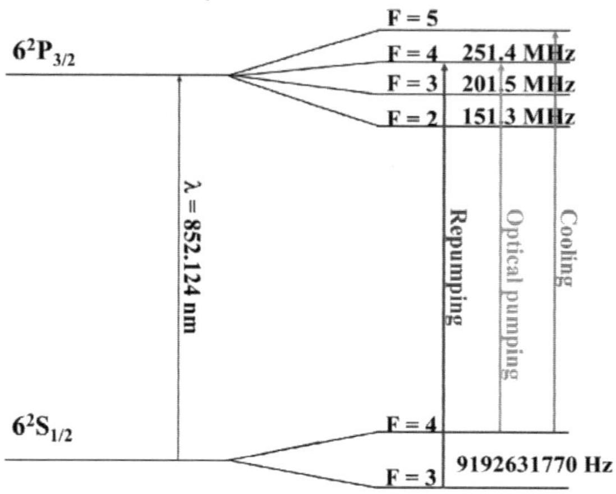

Figure 2. Partial energy level diagram of ^{133}Cs.

We have recently implemented an optical pumping stage which allows us to cycle atoms into the $|F=4,m_F=0\rangle$ state. In this case, after the atoms are launched but before they reach the state selection cavity, we apply optical pumping light which is tuned to the F=4→F=4' transition, as shown in Fig. 2. If the light is π polarized, transitions with $\Delta m=0$ are forbidden. We also apply repump light, tuned to the F=3→F'=4 transition. Whenever an atom is repumped into the $|F=4,m_F=0\rangle$ state, it will become dark to the optical pumping light and therefore the atoms will be preferentially pumped into this state.

We have found that this technique has resulted in an increase in detected atom number of over a factor of 6. One limitation to the technique is that the extra scattered photons will heat the atomic cloud, increasing losses at the apertures of microwave cavities.

3. Results

With our recent improvements to the stability and detected atom number, we have achieved a signal to noise ratio (SNR) of over 1000. This puts the fountain into a regime where it is limited by the short term stability of the local oscillator. The local oscillator is a NIST spaceclock microwave synthesizer locked to a 5 MHz maser reference signal. To test if the fountain was limited by the short term stability of our microwaves, we measured the standard deviation of our detected population fraction, $F=N_{F=4}/(N_{F=3}+N_{F=4})$. The log of the detected noise is plotted against the log of the total atom number (controlled by varying the loading time), as shown in Fig. 3.

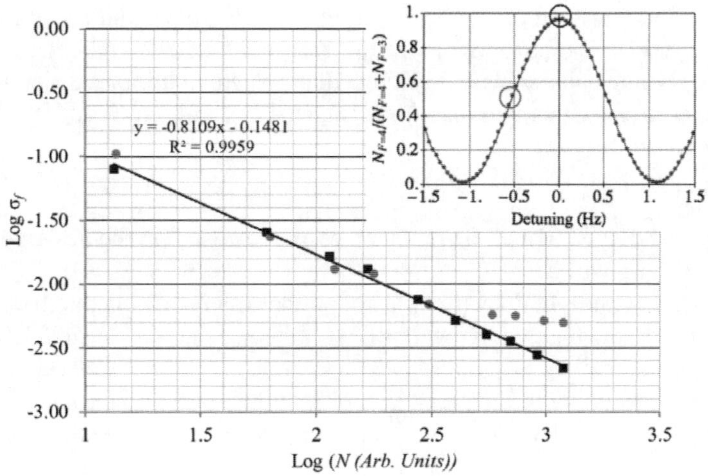

Figure 3. Log of the standard deviation of the measured population fraction vs the log of total detected atom number (in arb. units). The blue squares show data taken on resonance, while the red circles show data taken with the microwaves detuned to achieve equal populations in F=3 and F=4, as shown in the inset.

Here we took data at the peak of the central Ramsey fringe (blue), where noise in the local oscillator should have no effect due to the zero slope of the Ramsey fringe, as well as on the side of the Ramsey fringe where any noise in the frequency of the microwaves would impact the measured fraction. As shown in Fig. 3, we find that while the blue data points show a continuing decrease with increasing atom number, the red data points level off. This is evidence that the short term stability is currently limited by the local oscillator.

4. Progress towards FCs2

In collaboration with the UK National Physical Laboratory (NPL), we are currently developing a new physics package for FCs2. This fountain will reuse many elements from FCs1 such as the laser system and microwave sources. Therefore we are making several improvements to those systems before a new physics package is ready to be installed.

4.1. Microwave local oscillator

As shown above, the short term stability is limited by the local oscillator. This will certainly still be the case when we integrate our newly designed physics package, which has a significantly higher collection efficiency for the detection optics. To improve the stability of the local oscillator, we have purchased an ultra-high performance crystal oscillator from Microsemi (1000C model) to serve as the 5 MHz reference for our microwave synthesizer. The phase noise is -130 dBc at 1 Hz and the instability is $\sigma_y < 2\times10^{-13}$ from 0.1s to 100s. This performance could show a significant improvement over the maser which is currently used as a 5 MHz reference. In future, we hope to implement an optical local oscillator, derived from an ultra-stable laser [4] locked via a frequency comb.

4.2. Laser system

As described briefly above, our laser system is based on two external cavity diode lasers to provide the radiation for repumping and cooling/optical pumping/detection. We have replaced the slave diodes with tapered amplifier systems which have dramatically reduced the maintenance and downtime for the system.

Currently, we have visible noise on our detection signals due to an oscillation on the frequency of our cooling laser (Vortex 7017). We hope to replace this laser with a newer model, not susceptible to the mechanical resonance which causes this ~ 2 kHz oscillation. In addition, we plan to implement a fast feedback to the laser current, which will result in a higher-bandwidth lock and better compensate for this type of modulation on the laser.

4.3. Physics Package

The physics package of NRC FCs1 will be replaced with one based on the design of NPL CsF3. The most significant differences will be;

- (0,0,1) geometry with light delivered via six optical fibers, attached to the vacuum chamber by cage systems. This will ease alignment of MOT beams and the verticality of the atom cloud launch.
- Solid angle of detection will increase to ~ 30%, an increase of over a factor of two.
- The C-field will be parallel to the atoms' motion, in contrast to the transverse C-field of NRC-FCs1.
- The cavity will be made with cylindrical geometry, with four input ports to facilitate the characterisation of the distributed cavity phase shift [5].

5. Conclusions

We have presented improvements of the signal to noise and stability of FCs1, with a SNR of over 1000 and stability limited by the our microwave local oscillator. We have successfully added an optical pumping stage which has increased our detected atom number by over a factor of six. With these improvements, along with the current work to further upgrade our local oscillator, laser system, and physics package, we hope to achieve state of the art performance with FCs2.

References

[1] Heavner T P, Donley E A, Levi F, Costanzo G, Parker T E, Shirley J H, Ashby N, Barlow S E and Jefferts S R, 2014 *Metrologia* **51** 174
[2] Weyers S, Gerginov V, Nemitz N, Li N and Gibble K 2012 *Metrologia* **49** no. 1 pp. 82-87
[3] Szymaniec K, Park S E, Marra G and Chałupczak W 2010 *Metrologia* **47** 363
[4] Dubé P, Madej A A, Bernard J E, Marmet L and Shiner A D 2009, *Appl. Phys. B* **95** pp. 43-54
[5] Gibble K, Lea S and Szymaniec K 2012 Conference on Precision Electromagnetic Measurements (CPEM).

Operation of NIM5 fountain with 1.5×10^{-15} uncertainty and design of new NIM6 in NIM

F Fang*, N Liu, K Liu, W Chen, R Suo and Tianchu Li

Time and Frequency Division, National Institute of Metrology, Beijing, China

*E-mail: fangf@nim.ac.cn

Abstract. The cesium fountain primary frequency standard NIM5 started to operate since 2008 and started to report to BIPM since 2014. The major constrains of NIM5 is a relatively large background signal at the detection and microwave leakages due to the Ramsey cavity. A new fountain clock NIM6 is under construction. Besides some improvements on the vacuum system, a new Ramsey cavity and a microwave synthesizer are made to reduce the Type B uncertainty. Another feature of NIM6 is collecting atoms from a MOT loading optical molasses to get more atoms with a more uniform density distribution. With a new frequency synthesizer based on cryogenic sapphire oscillator (CSO), NIM6 is aiming to reach the quantum projection noise, thus leading to a reduced Type A uncertainty compared with NIM5.

1. Introduction

The cesium fountain primary frequency standard NIM5 started to operate since 2008 and 9 evaluations were reported to BIPM in 2014 and 2015 with a typical fractional frequency instability of 3×10^{-13} $(\tau/s)^{-1/2}$ and a Type B uncertainty of 1.4×10^{-15}, which was dominated by the microwave-related frequency shifts. Since its first operation, NIM5 underwent a series of improvements, including laser locking, adding a microwave interferometric switch and monitoring the transit phase in real time, reducing the microwave leakage effect by selecting atom signals in certain range, and so on. The major constrains of NIM5 now is a relatively large background signal at the detection and microwave leakages. A new cesium fountain clock NIM6 is under construction. Besides some improvements on the vacuum system, Ramsey cavity and microwave synthesizer to reduce the Type B uncertainty. NIM6 will collect atoms from a MOT loading optical molasses and increase number of atoms with optical pumping to get a better signal to noise ratio. The atom density will be more uniform compared with a 2D MOT loading optical molasses, and the diameter of the cloud can be adjusted by the intensity and detuning of lights during the post cooling to keep the collisional shift low. With a new cryogenic sapphire oscillator (CSO) based microwave frequency synthesizer, NIM6 is aiming to reach the quantum projection noise, thus lead to a reduced Type A uncertainty compared with NIM5.

2. The current status of NIM5

The NIM5 consists of a physical package, a laser-optics system and a microwave-electronics control system on two racks. The detailed description of NIM5 design is in the reference [1]. In a routine operation, atoms are collected in the optical molasses for 600 ms, then accelerated for 1 ms in a moving molasses, and further cooled for another 1.5 ms by adiabatically reducing the cooling beam intensities and red offsetting the frequencies by 60 MHz. The atoms are launched 810 mm above the centre of OM and cooled to a temperature of about 2 μK. The $|F=4, m_F=0\rangle$ atoms are transferred to the $|F=3, m_F=0\rangle$ state by a microwave pulse in the state selection cavity. Atoms remaining in the $|F=4\rangle$

state are removed by the detection beam on the way up. A photo of NIM5 and a typical Ramsey fringe pattern is shown in the figure 1.

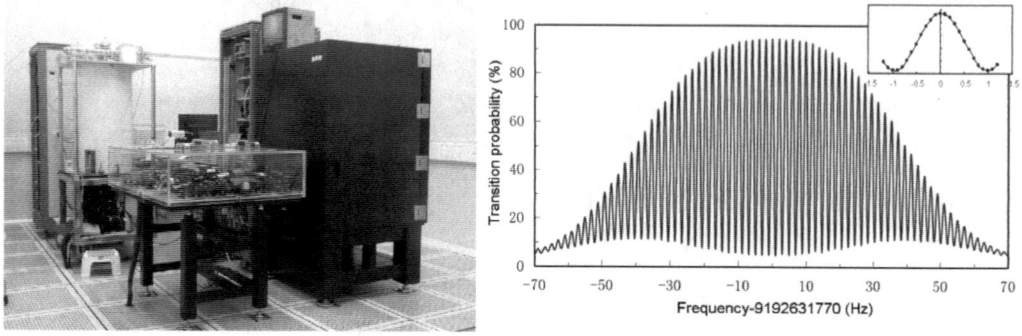

(a) (b)

Figure 1. (a) A photo of NIM5, (b) A typical Ramsey fringes with a frequency scanning step of 0.1 Hz and no averaging. The inset shows the central Ramsey fringe, and the FWHM width is around 1 Hz.

2.1. *Statistical (type A) uncertainty of NIM5*

In a routine evaluation period, NIM5 is running at high and low densities alternatively and typical relative frequency instabilities against a hydrogen maser are expressed in the figure 2.

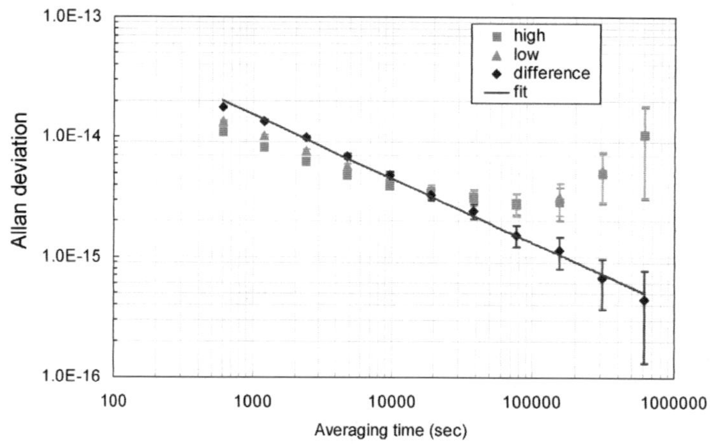

Figure 2. Standard Allan deviation $\sigma_y(\tau)$ of NIM5 measured against the H271 H-maser over a period of 15 days. The triangles and squares denote the stabilities at low and high atom density respectively. The diamonds represent the stabilities of the frequency difference between the low and high densities.

Collisional free frequency f_0 at zero density is derived by extrapolating the experimental data using:

$$f_0 = \frac{k\bar{f}_{\rm L} - \bar{f}_{\rm H}}{k-1} \tag{1}$$

Here k is the ratio between high and low densities (n_H/n_L), and f_H and f_L are the measured frequencies at high and low densities respectively. The uncertainty in eliminating the collisional frequency shift is derived as [2]:

$$\sigma_0^2 = \left(\frac{k}{k-1}\right)^2 \sigma_{\rm L}{}^2(\tau_{\rm L}) + \left(\frac{1}{k-1}\right)^2 \sigma_{\rm H}{}^2(\tau_{\rm H}) + \left(\frac{\bar{f}_{\rm L} - \bar{f}_{\rm H}}{(k-1)^2}\right)^2 \sigma_k{}^2 \tag{2}$$

Here, the first two terms are related to the statistical frequency uncertainties at the low and high densities respectively, and the last term is the cold-collision-induced type B uncertainty, which includes the nonlinearity between the measured atom numbers and the average density as stated in the references [3, 4].

In the figure 2, it shows that for averaging time above 10^4 seconds, the instability is limited by the H-maser, with the Allen deviation drifting up for both cases. The contributions from the H-maser are common-mode, and rejected in the differential measurement provided that the switching rate between different densities is faster than the H-maser drifting. The frequency difference between high and low densities falls according to the square root of the averaging time. Assuming it keeps square root of averaging time law, 15 days of averaging will give instabilities of 0.26×10^{-15} and 0.31×10^{-15} for low and high densities. The combined type A uncertainty of NIM5 is calculated to be 0.67×10^{-15}.

2.2. *A typical Type B uncertainty of NIM5*

The major contributions to the frequency shifts are from the second-order Zeeman shift, microwave related frequency shift, the collisional shift (mentioned in the above section), AC Stark shift due to black-body radiation and the gravitational redshift [5-8]. A summary of a typical systematic frequency shift evaluations for NIM5 is listed in the Table 1 and the combined relative Type B uncertainty for NIM5 is evaluated to be about 1.4×10^{-15}.

Table 1. A typical uncertainty budget of NIM5

Physical Effect	Bias (10^{-15})	Uncertainty(10^{-15})
2nd order Zeeman	73.4	0.2
Collisional shift	-1.1*	0.2
Microwave interferometric switch	0.0	1.2
Microwave leakage	0.0	<0.1
DCP	0.0	0.5
Microwave spectral impurities	0.0	<0.1
Blackbody radiation	-16.2	0.1
Gravitational red shift	11.8	0.1
Majorana transition	0.0	<0.1
Light shift	0.0	<0.1
Rabi and Ramsey pulling	0.0	<0.1
Cavity pulling	0.0	<0.1
Collision with background gases	0.0	<0.1
Total	67.9*	1.4

* The collisional shift is calculated at low density.

The dominate frequency uncertainty of NIM5 comes from the microwave-power-related frequency shift. We found NIM5 has a relatively large microwave leakage, and an interferometric RF switched is applied to reduce this effect. The detailed evaluations of the NIM5 are stated in detail in [1].

3. The design of NIM6

A cutaway figure of the NIM6 physical package is shown in the figure 3. The entire physics package is enclosed in a layer of soft iron and the flight tube is surrounded with another three layers of μ metal shielding with a measured shield factor of about10^4. Atoms are collected in the lower MOT chamber and then launched to the upper optical molasses chamber with a small angle (10°) to reduce the background Cs atoms flying into the detection chamber directly. The lower MOT chamber is pumped by a 20 l/s ion pump and the upper OM chamber is pumped by a 40 l/s ion pump. A getter pump is on the top of the flight tube to keep the ultra-low pressure in the atom interrogation region. A new TE_{011} Ramsey cavity has been made with 4 rectangular waveguides to feed the microwaves in. A measured loaded Q factor is about 15000. Also, the new cavity is well sealed with indium to reduce the microwave leakage.

3.1. Isothermal liner heat pipe

The new fountain will be operated in a lab with temperature fluctuations less than 0.3 K, and no active temperature control system will be added outside the flight tube. Instead, an isothermal liner

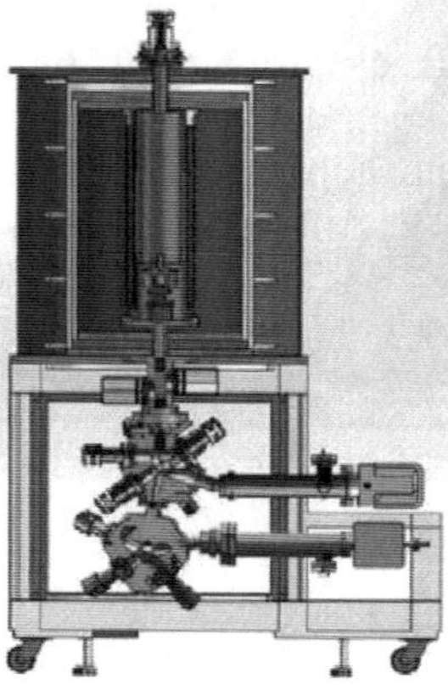

Figure 3. Schematic of the NIM6 cesium fountain clock.

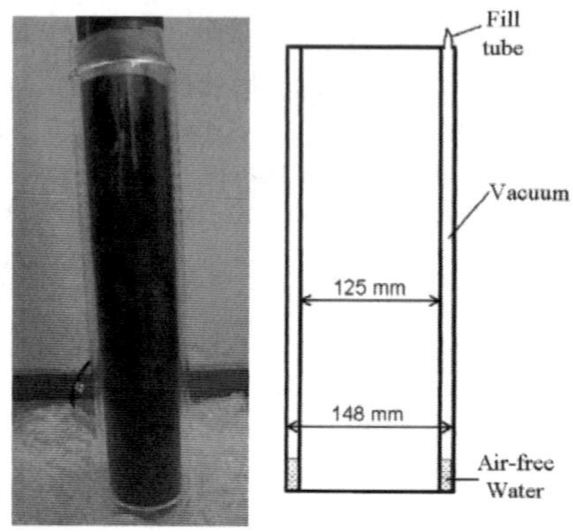

Figure 4. Picture of a mock up isothermal liner made of glass.

will surround the flight tube. It is a special type of heat pipe, made of a thin glass layer of vacuum tube filled with pure water. A mock up isothermal liner heat pipe is shown in the figure 4. High purity air-free water is filled in glass vacuum tube. When there is a temperature gradient, liquid water around the hot part absorbs heat, transforms into vapour and flows toward the condensate section (cold part). This phase transitions inside the heat pipe redissipate the heat, block corresponding changes in the inner chamber temperature, regulate thermal transport and dampen temperature oscillations. The performance of this mock-up isothermal liner is tested [9].It reduces the temperature fluctuations and improves the temperature gradients. With precision PT standard thermometers which have a temperature uncertainty of less than 10 mK, hopefully, the total temperature uncertainty will be less than 50 mK, keeping blackbody-radiation-induced frequency shift less than one part in 10^{16}.

3.2. MOT loading optical molasses

In NIM6, Cs atoms will be collected in the lower MOT chamber and launched to the centre of the OM chamber. The separation between these two centres is 280 mm, the flying time is about50 ms with a launching velocity of 5.5 m/s. The final temperature of the cloud after launching can be adjusted by the intensity and detuning of the post cooling beams to make sure the cloud expanding enough when reaching the OM centre to keep collisional shift low. The velocity and temperature of atoms will be re-adjusted in the optical molasses region. Then, atoms are launched vertically to the flight tube. The advantage of this design is not only being able to collect more atoms compared to a direct optical molasses loading like NIM5, the background Cs gas in the detection chamber is also reduced due to a differential pumping. Furthermore, the atom density distribution is more uniform than loading OM from a 2D-MOT.Another feature here is that the cooling beams for MOT and OM are applied at different times. The lights for the MOT cooling and OM cooling can be provided by only one tapered amplifier with a total output power of about 800 mW.

3.3. Home-made microwave inspection system

The major uncertainty of NIM5 comes from an RF interferometric switch in the microwave synthesizer chain. A triggered-phase transient analyzer (TPTA) [10] was developed in our laboratory to measure the phase difference of the two Ramsey pulses. A schematic of TPTA is shown in figure 5a. The TPTA has the ability to measure the phase variations of two pulsed microwave switch a fixed interval in one cycle. Therefore, by measuring the phase discontinuity between two switching on states in one fountain cycle, the frequency uncertainty induced by interferometric switch or other components which affect the instability of the interrogation microwave phase can be checked. The figure 5b shows the measurement noise floor of this inspection system and the results with two SDI microwave synthesizers.

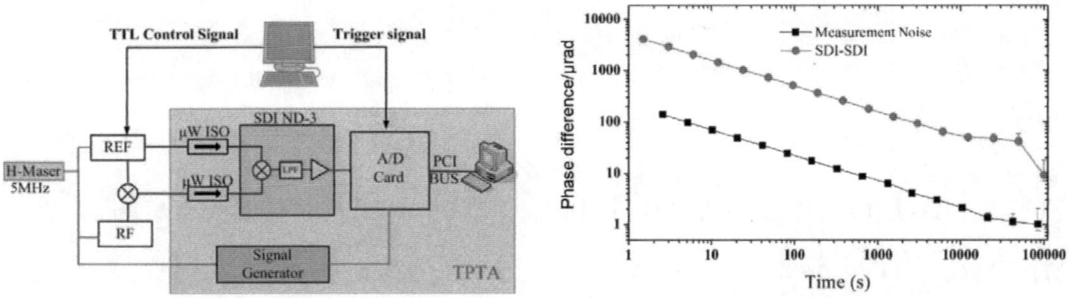

(a) (b)

Figure 5 (a) The schematic of Triggered-Phase Transient Analyzer (TPTA).It consists of the microwave demodulator (ND-3 and microwave isolators), analog-to-digital sampling board, and a high performance computer. (b) Noise floor of the measurement system.

4. Summary and future plans

With NIM5 is operational. A new cesium fountain clock NIM6 is under construction in the National Institute of Metrology China, and aiming to operate in 2016. Besides some improvements on the design of the Ramsey cavity to reduce the distributed cavity phase shift and microwave leakage, NIM6 is also aiming to collect more atoms from a MOT loading optical molasses (OM) and optical pumping [11,12], leading to a better signal to noise ratio at the detection. A new cryogenic sapphire oscillator (CSO) based frequency synthesizer [13] and ultra-stable microwave generated from ultra-stable laser are also under developing to reduce the microwave phase noise in order to reach the quantum projection noise, thus leading to a lower Type A uncertainty of the new fountain.

Acknowledgments

This work is supported by the Ministry of Science and Technology of China (2013YQ090943) and Chinese NSF (11174260).

References

[1] Fang F, Li M, Lin P et al 2015 *Metrologia* **52** 454
[2] Szymaniec K and Park S P 2011 *IEEE Trans. Instrum. Meas.* **60** 2475
[3] Szymaniec K, Chalupczak W, Whibberley P B et al 2010 *Metrologia* **47** 363
[4] Santarelli G, Laurent P, Lemonde P, and Clairon A 1999 *Phys. Rev. Lett.* **83** 4619
[5] Gerginov V, Nemitz N, Weyers S et al 2010 *Metrologia* **47** 65
[6] Vian C, Rosenbusch P, Marion H et al 2005 *IEEE Trans. Instrum. Meas.* **54** 833
[7] Heavner T, Donley E and Levi F et al 2014 *Metrologia* **51** 174
[8] Levi F, Calonico D, Calosso C et al 2014 *Metrologia* **51** 270
[9] Fang F, Liu K, Yan X et al 2013 *Joint UFFC, EFTF and PFM Symposium* 232
[10] Zhang Y, Liu K, Fang F et al 2014 *IFCS Symposium* 376
[11] Szymaniec K, Noh HR, Park S E et al 2013 *Appl. Phys. B* **111** 527
[12] Takamizawa A, Yanagimachi S, Tanabe T et al 2014 *IEEE Trans. UFFC* **61** 1463
[13] Hartnett J G, Nand N R and Lu C 2012 *Appl. Phys. Lett.* **100** 183501

Journal of Physics: Conference Series **723** (2016) 012010 doi:10.1088/1742-6596/723/1/012010

Status and prospect of the Swiss continuous Cs fountain FoCS-2

A Jallageas[1], L Devenoges[2], M Petersen[1], J Morel[2], L-G Bernier[2], P Thomann[1] and T Südmeyer[1]

[1] Laboratoire Temps-Fréquence, Université de Neuchâtel, Neuchâtel, Switzerland

[2] Photonics, time and frequency laboratory, Federal Institute of Metrology (METAS), Bern, Switzerland

E-mail: antoine.jallageas@unine.ch

Abstract. The continuous cesium fountain clock FoCS-2 at METAS presents many unique characteristics and challenges in comparison with standard pulsed fountain clocks. For several years FoCS-2 was limited by an unexplained frequency sensitivity on the velocity of the atoms, in the range of $140 \cdot 10^{-15}$. Recent experiments allowed us to identify the origin of this problem as undesirable microwave surface currents circulating on the shield of the coaxial cables that feed the microwave cavity. A strong reduction of this effect was obtained by adding microwave absorbing coatings on the coaxial cables and absorbers inside of the vacuum chamber. This breakthrough opens the door to a true metrological validation of the fountain. A series of simulation tools have already been developed and proved their efficiency in the evaluation of some of the uncertainties of the continuous fountain. With these recent improvements, we are confident in the future demonstration of an uncertainty budget at the 10^{-15} level and below.

1. Introduction

FOCS-2 is the only Cs fountain clock worldwide taking advantage of a continuous atomic beam. This characteristic brings a series of remarkable advantages compared to the classical pulsed fountains. For instance, a frequency stability below the standard Dick limit has been demonstrated using a commercial synthesizer [1] and the low density of the cold atomic beam maintains the collisional shift at a minimum [2]. Furthermore, the fundamentally different principle of operation contributes to the metrological diversity. Several upgrades to the original design have been implemented in the last few years, such as the design and the implementation of a new microwave Ramsey cavity with reduced phase gradients [3]. These improvements led us to focus on the evaluation of electromagnetic perturbations in the system and its effects on the clock frequency. Figure 1 shows a schematic representation of FoCS-2.

2. Shift due to microwave currents

In recent years, many in-depth analyses were carried out on pulsed fountains to evaluate the impact of microwave frequency shifts due to microwave leaks in the atomic resonator [5–7]. In the case of FoCS-2 several investigations led us to postulate that the velocity dependent $140 \cdot 10^{-15}$ fractional frequency shift is caused by a peculiar type of microwave leakage. The unwanted electromagnetic field is generated by spurious currents circulating on the surface of the coaxial cables which lie very close to the apogee of the trajectory. When the atomic velocity is varied,

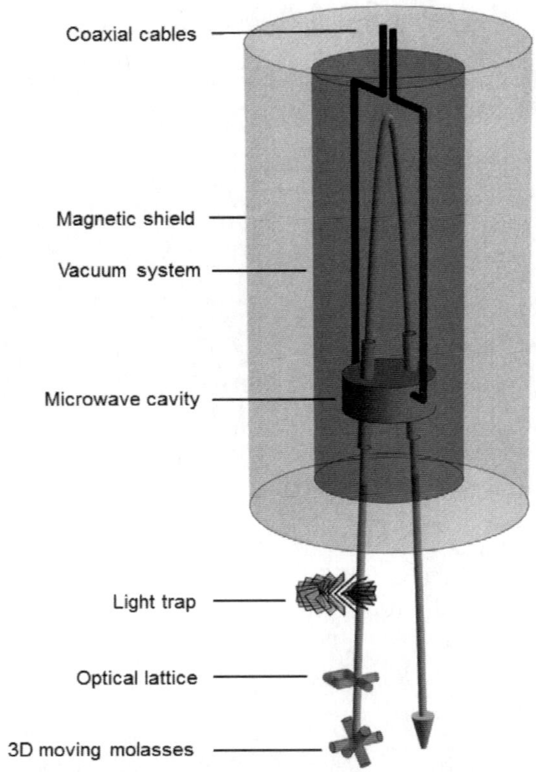

Figure 1: Scheme of the continuous atomic fountain clock FoCS-2. A 2D-MOT (not shown) loads the 3D moving molasses. There, the atoms are cooled to 75 μK and launched at a speed of 4 m/s. Then, the atomic beam is collimated by transverse Sisyphus cooling before the atoms are pumped into $|F=3, m=0\rangle$ with the state preparation stage. Through the microwave cavity, the atoms experience a first $\pi/2$-pulse and evolve freely for 0.52 s before they undergo a second $\pi/2$-pulse on the way down. Finally, the transition probability between $|F=3, m=0\rangle$ and $|F=4, m=0\rangle$ is measured by fluorescence detection of the atoms in $F=4$. A light trap suppresses the stray light from the atomic cooling and state preparation stages from reaching the free evolution zone situated above [4].

the distance between the apogee and the coaxial cables is varied hence the velocity dependent frequency shift. In order to avoid this effect a coating of microwave-absorbing graphite was added to the surface of the coaxial cables and on the inner wall of the vacuum chamber. Additionally, several pieces of graphite were placed near the cavity (as shown in figure 2).

Figure 3 shows the substantial decrease in frequency dependency on the velocity from $140 \cdot 10^{-15}$ to $3 \cdot 10^{-15}$ after the implementation of the graphite absorbers. The reduction by a factor 50 of this effect supports the theory of microwave currents and opens up the possibility of reaching the expected performance of the clock. That being so, different sustainable solutions to fully eliminate this shift are under preparation. One consists in a graphite cylindrical shield placed above the microwave cavity to protect the free evolving atoms by absorbing the spurious microwave radiation.

2.1. Optical transport of microwave radiation

Another solution towards minimizing microwave leak issues consists in transporting the two microwave signals to the microwave cavity optically, using optical fibers, and by performing the optical-electrical conversion directly inside the microwave cavity couplers [8]. Figure 4 shows a schematic of the experimental setup for one of the two microwave feeds. The optical power generated by an extended cavity laser diode (ECDL) emitting at 1550 nm is modulated at 9.19263177 GHz using a Mach-Zehnder Modulator (MZ), driven by a low noise frequency synthesizer. The modulated light is transported to the microwave cavity using a single-mode fiber. The optical-electrical conversion is performed using unbiased PIN photo-diodes (PD)

Figure 2: Photo of the cavity with microwave cables coated with graphite. Pieces of graphite can also be seen around the cavity for additional absorption of the microwave radiation.

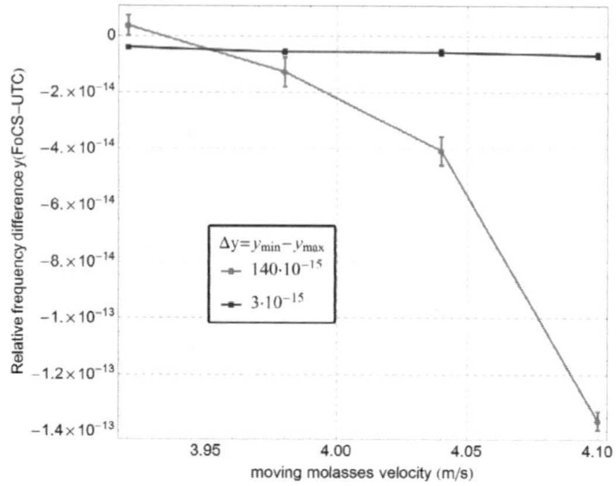

Figure 3: Relative frequency offset of FoCS-2 as a function of the average atomic launch velocity varying from 3.90 m/s to 4.10 m/s. Red dots show the measured frequency with the unwanted background microwave radiation. The blue squares show the significantly decreased change in frequency which was obtained using graphite coating cables.

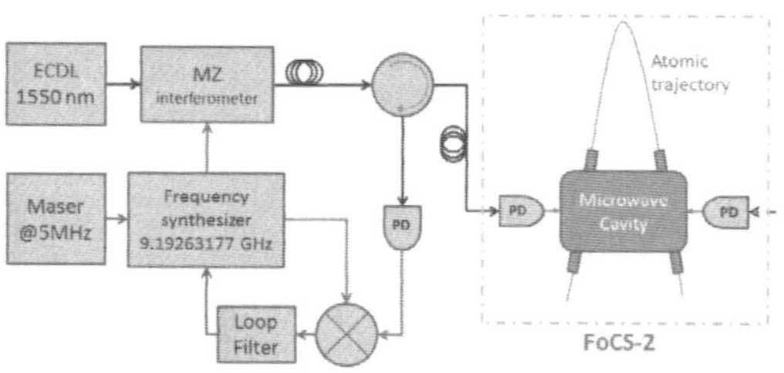

Figure 4: Schematic of the optical transport of the microwave radiation from the synthesizer to the cavity and the opto-electrical conversion made on a fast photo-diode.

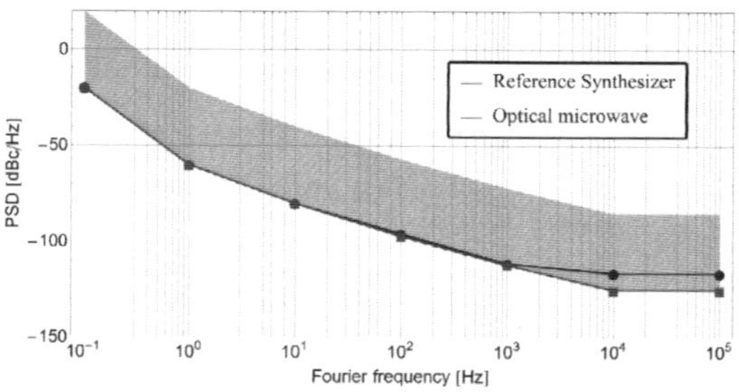

Figure 5: Phase noise measurement of the LO at the output of the synthesizer (Squares) and after optical transport over a 2 m long optical fiber (Dots).

directly imbedded in the cavity microwave feeders. A feedback loop, acting on the measured back-reflected light from the PD, allows the compensating of possible thermally induced phase shifts in the fiber.

Therefore, the optical microwave transfer system must preserve the noise properties and spectral purity of the high performance local oscillator (LO) at 9.19263177 GHz. We have measured the phase stability and the excess phase noise on a simplified open loop version of the test setup. A phase drift produces frequency offsets similar to the effect of an end-to-end cavity phase shift. A five day long measurement showed no phase drifts above 10 μrad/s ($1.5 \cdot 10^{-16}$) across any one hour time period. A continuous fountain is much less sensitive to the Dick effect as compared to a pulsed clock [9], which significantly lowers the constraint on the phase noise performances of the LO. We show in figure 5 the measured phase noise of the LO with and without optical transfer through a 2 meter long optical fiber. The shaded area highlights the

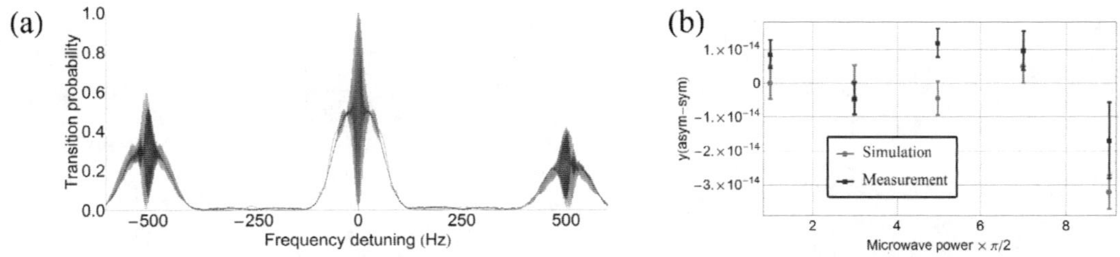

Figure 6: (a) Comparison of the measured (blue curve) and the simulated (red curve) Ramsey fringes for the $|F = 4, m = 0, \pm 1\rangle$ Zeeman sublevels. The asymmetries of the fringes measured on $|F = 4, m = \pm 1\rangle$ are due to small magnetic inhomogeneities present near the atomic trajectories' apogee. The side lobes on the numerical data are artifacts due to the limited number of trajectories used in the simulation. (b) Microwave power-dependent frequency shifts for different pulse areas in the experimental and numerical cases. The vertical scale shows the frequency difference between asymmetric and symmetric feeding of the microwave cavity.

noise level where the short term stability of the continuous fountain is still limited by the atomic shot-noise. These preliminary results suggest that this technique is a very promising solution for solving the issue of microwave leakages encountered in many fountains and will be further investigated.

3. Finite element modeling and Monte Carlo simulations

3.1. Numerical evaluation of the microwave cavity related frequency shifts

The microwave cavity of FoCS-2 was designed to have a low mode density and a reduced time-averaged power flow across the interaction zones [3]. Its dimensions were optimized such that the TE_{105} electromagnetic mode would be resonant at $\nu_0 = 9.19263177$ GHz. Four cutoff waveguides confine the field to the cavity. A specific diaphragm design at their upper and lower ends eliminates atomic trajectories close to the edges of the cavity [10]. We used a full 3D finite elements model (FEM) to compute simultaneously the amplitude H and the phase ϕ of the electromagnetic field throughout the interaction zones. As a first approach to evaluate the FEM and for possible later use in evaluating the gradient phase shift [11], we have solved Maxwell's equations numerically for symmetric and asymmetric cavity feeding to reproduce the power measurements that were performed experimentally. Using these results and a theoretical model of 2×3 Zeeman sublevels atoms following trajectories through the FoCS-2 fountain, we developed a Monte Carlo simulation allowing us to quantify the phase shifts in the unique FoCS-2 cavity. We present in figure 6 the first results computed to validate the numerical models. Figure 6.a shows the computed Ramsey fringes for the $|F = 4, m = 0, \pm 1\rangle$ Zeeman sublevels together with the experimental data. The free parameters of the Monte-Carlo simulations were adjusted to reproduce the real atomic trajectories. Figure 6.b shows a rough comparison of the total microwave power-dependent frequency shifts for experimental and numerical cases. We present the frequency difference between the asymmetric and symmetric microwave feeding to eliminate the contribution of other experimental effects. This good agreement between simulations and measurements allows us to use these numerical tools to support the experimental evaluation of FoCS-2.

3.2. Magnetic field homogeneity

The homogeneity of the magnetic field inside the fountain is important for guaranteeing the efficiency of the state preparation and to avoid Majorana transitions. Zero-crossing of the field can induce atomic state coherences and frequency shifts [12]. Direct measurements of the

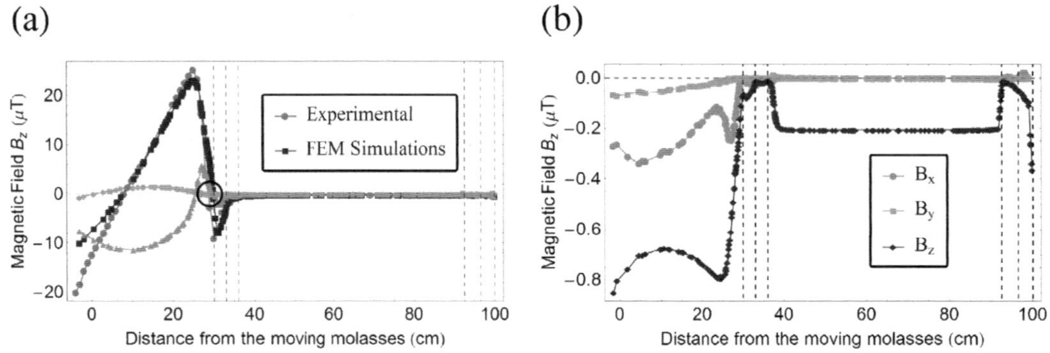

Figure 7: (a) Comparison of the vertical magnetic field component B_z from FEM simulations (blue points) with direct measurements (red points) at different heights in the vacuum system. The green and yellow points show the simulated results of the X and Y components of the magnetic field. The black circle highlights a zero-crossing point of the three components B_x, B_y and B_z just before the free evolution region shielded with three layers of mu-metal (vertical dashed lines). (b) Simulation of the three components of the magnetic field after optimization of the compensations coils currents. With this new configuration we suppressed the zero-crossing point to avoid parasitic Majorana transitions.

magnetic field between the optical molasses and the C-field region are not possible under normal clock operation. We developed a finite element model of the magnetic field throughout the whole fountain to predict its behavior when adjusting the external electric coils [13]. We implemented the C-field and the X-Y-Z compensation coils, ionic pump magnets, Earth magnetic field as well as the four layers of mu-metal shielding in the 3D model. All external field values were carefully measured in the laboratory and reproduced in the simulation. Figure 7.a shows the comparison between the model and the measured B_z field at different heights in the vacuum system. In-situ measurements were done with a one-axis flux-gate probe inserted inside the vacuum chamber while it being opened for maintenance. The good agreement between the data gives us a good confidence in the prediction of the magnetic field with the simulation although in this particular configuration it highlights a zero-crossing point. Figure 7.b shows the new magnetic field configuration without zero-crossing point after numerical tuning of the compensation coils.

4. Current status

Recent improvements on the lasers and on the optical setup have led to a better white frequency noise level. The FoCS-2 Allan deviation is now $8 \cdot 10^{-14}$ at one second (see figure 8).

After the significant decrease in the effect of the microwave surface-currents on the microwave cables and the improvement of the stability, it has been possible to improve the investigation on the uncertainty budget. A preliminary version of the uncertainty budget is shown in table 1. Some effects such as the distributed cavity phase shift and the collisional shift are not included as these have not been evaluated yet. The uncertainty on the light shift measurement due to detection light is shown as the limiting factor. This is due to relatively short measurements. Longer measurement series are expected to easily bring down the uncertainty. Eventually, a second light trap can be inserted between the microwave cavity and the detection region. The second-order Zeeman shift is estimated by two methods time-resolved Zeeman spectroscopy and Fourier transform of Ramsey fringes measured on Zeeman sensitive components. The uncertainty is limited by the atomic vertical velocity distribution (basically the vertical temperature). It is expected to be further improved by detailed modeling of the effect of the atomic velocity distribution on the two methods. The end-to-end shift was evaluated using a reversal mechanism that switches the two interaction zones [14]. The uncertainty can be further improved through longer measurements.

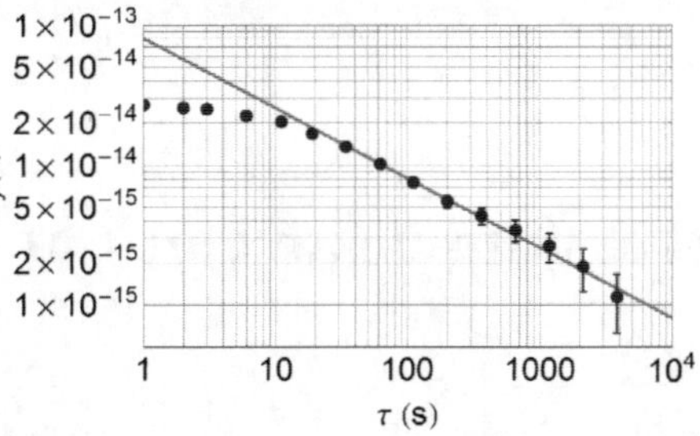

Figure 8: The relative frequency stability (Allan deviation) of the FoCS-2 frequency compared to the local H-maser. The red line shows the best fit at $8 \cdot 10^{-14}\,\tau^{-1/2}$.

Physical effect	Shift	Uncertainty
Relativistic effects	59.8	< 0.1
Second-order Zeeman	25.1	0.1
Blackbody radiation	−16.9	< 0.1
Light shift preparation	−0.2	< 0.1
Light shift detection	−0.2	1.8
Rabi pulling	0.0	< 0.1
Second order Doppler	−0.1	< 0.1
End-to-End phase shift	4.8	0.6
Total	72.3	1.9

Table 1: Present status of FoCS-2 uncertainty budget ($\times 10^{-15}$).

5. Conclusion

The understanding of microwave surface current perturbations and its significant reduction as well as the development of new numerical tools are significant steps toward the complete metrological evaluation of FoCS-2 at the 10^{-15} level and below. We report here a preliminary uncertainty budget of $1.9 \cdot 10^{-15}$ (excluding certain important effects) and a short term stability of $8 \cdot 10^{-14}$ at 1 s, which confirms the relevance of the continuous atomic clock. Our first priority for the time being is to finalize a robust solution for removing the microwave perturbations and to complete the uncertainty budget advancing towards a future contribution to TAI.

References

[1] Devenoges L, Stefanov A, Joyet A, Thomann P and Domenico G D 2012 *IEEE Trans. Ultrason. Ferroelectr. Freq. Control* **59** 211–216 ISSN 08853010

[2] Dudle G, Mileti G, Joyet A, Fretel E, Berthoud P and Thomann P 2000 *IEEE Trans. Ultrason. Ferro electr. Freq. Control.* **47** 438–442

[3] Devenoges L, Bernier L G, Morel J, Di Domenico G, Jallageas A, Petersen M and Thomann P 2013 *2013 Jt. Eur. Freq. Time Forum Int. Freq. Control Symp. EFTF/IFC 2013* **1** 1

[4] Füzesi F, Jornod A, Thomann P, Plimmer M D, Dudle G, Moser R, Sache L and Bleuler H 2007 *Rev. Sci. Instrum.* **78** 103–109 ISSN 0034-6748

[5] Shirley J H, Levi F, Heavner T P, Calonico D, Yu D H and Jefferts S R 2006 *IEEE Transactions on Ultrasonics, Ferroelectrics, and Frequency Control* **53** 2376–2384 ISSN 08853010

[6] Weyers S and Schröder R 2006 *IEEE Transactions on Ultrasonics, Ferroelectrics, and Frequency Control* 173–180

[7] Guena J, Abgrall M, Rovera D, Laurent P, Chupin B, Lours M, Santarelli G, Rosenbusch P, Tobar M E, Li R, Gibble K, Clairon A and Bize S 2012 *IEEE Trans. Ultrason. Ferroelectr. Freq. Control* **59** 391–410 ISSN 08853010 (*Preprint* 1204.3621)

[8] Jallageas A, Petersen M, Devenoges L, Bernier L G and Morel J 2015 *Jt. Conf. IEEE IFCS EFTF, April 12-16 (2015).*

[9] Joyet A, Mileti G, Dudle G and Thomann P 2001 *IEEE Transactions on Instrumentation and Measurement* **50** 150–156 ISSN 00189456

[10] Li R and Gibble K 2010 *Metrologia* **47** 534–551 ISSN 0026-1394 (*Preprint* 1008.1505)

[11] Guéna J, Li R, Gibble K, Bize S and Clairon A 2011 *Phys. Rev. Lett.* **106** 130801 ISSN 0031-9007 (*Preprint* 1103.1633)

[12] Wynands R, Schröder R and Weyers S 2007 *IEEE Trans. Instrum. Meas.* **56** 660–663 ISSN 00189456

[13] Jallageas A, Devenoges L, Petersen M, Südmeyer T and Morel J 2014 *28th Eur. Freq. Time Forum, June 23-26*

[14] Di Domenico G, Devenoges L, Joyet A, Stefanov A and Thomann P 2011 *Freq. Control Eur. Freq. Time Forum (FCS), 2011 Jt. Conf. IEEE Int.* 1–5 ISSN 1075-6787

Journal of Physics: Conference Series **723** (2016) 012011 doi:10.1088/1742-6596/723/1/012011

Recent progress in optically-pumped cesium beam clock at Peking University

C Liu, S Zhou, J Wan, S Wang and Y Wang

School of Electronics Engineering and Computer Science,
Peking University, Beijing, China

E-mail: wangyanhui@pku.edu.cn

Abstract. A compact, long-life, and low-drift cesium beam clock is investigated at Peking University, where the atoms are magnetic-state selected and optically detected. Stability close to that of the best commercial cesium clocks has been achieved from 10 to 10^5 s. As previously shown, the short-term stability is determined by atomic shot noise or laser frequency noise. The stabilizations of microwave power and C-field improve the long-term stability, with the help of a digital servo system based on field-programmable gate array.

1. Introduction

Magnetic state selection is a simple and stable method in a cesium beam clock, even though, compared with optical selection, it decreases the number of atoms that approaches the detection region. Therefore, cesium clocks with magnetic state selection have been studied and applied for more than fifty years [1]. On the other hand, optically pumped cesium clocks have shown better short-term frequency stability [2, 3], indicating that laser is an alternative method of state preparation and detection. To avoid the life limitation due to the electron multiplier, we try to use optical detection instead of the traditional detection method in our cesium beam clock [4], which is also aiming at a compact device for long-term time-keeping. Operating in the atomic shot noise region, with frequency stability of $9\times10^{-12}\cdot\tau^{-1/2}$ that reaches 3×10^{-14} at 10^5 s, the clock is potentially suitable for terrestrial and space-borne application. Noise analysis shows that the short-term stability is limited at higher oven temperature, by noise induced by laser frequency deviation, which makes demands of the laser source. The laser frequency noise is coupled by the relatively narrow line shape of absorption of the atoms. Digital circuits based on field-programmable gate array (FPGA) help controlling the servo loops.

2. Experimental set-up

As shown in figure 1, in a sealed vacuum tube, two thermal Cs beams, collimated by collimators, emit from an oven which is temperature-controlled. They cross the inhomogeneous magnetic field generated by a set of magnets, partly deflected, depending on the state of atoms, towards the microwave interrogation region. The two beams separate in the direction parallel to the C-field, hence not distinguishable in the figure. An E-bend Ramsey cavity, whose field-free length is 17 cm, makes microwave interrogation possible. Coils surrounding the cavity generate a homogeneous magnetic field, the C-field, which is needed to separate the $m_F=0$ clock transition from then magnetic-field sensitive transitions. Two apertures on the cavity wall, each with size of 1.8×6 mm^2, determine the cross section of the beams. The beams passing the cavity reach the optical-detection region, which is surrounded by a pair of spherical mirrors. Laser beam (not shown), pointing parallel to the C-field, irradiates the atoms, whose fluorescence is reflected by the mirrors and detected by a photo diode. The microwave interrogation region and optical detection region are independently shielded.

Digital servo system based on FPGA, as shown in figure 2, which is an improved version of that described in [5], is implemented. It has three servo loops, local-oscillator loop, microwave-power loop, and C-field loop, working sequentially. Microwave frequency is modulated by a 78Hz square wave, with the help of a commercial microwave synthesizer.

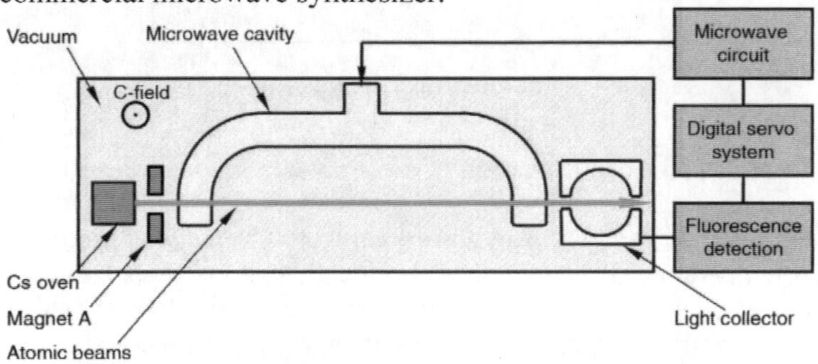

Figure 1. Experimental set-up. (colour online)

Figure 2. The printed circuit board of the servo system. (colour online)

The 852-nm laser beam is generated by either an external cavity diode laser (ECDL) or a distributed feedback (DFB) diode laser, whose nominal line widths are 100 kHz and 2 MHz respectively. The frequency of the laser is locked to the F=4-F'=5 transition, which is a cyclic transition, of Cs D2 line with an independent saturated absorption spectroscopy. The modulation frequency of the laser, for the purpose of its frequency locking, is 5 kHz. Incident light power to the tube is 3 mW for ECDL and 10 mW for DFB.

3. Results

Allan deviation calculated from a 15-day continuous measurement without C-field stabilization, during which the oven temperature is 95 ^{O}C and light source is ECDL, is illustrated in figure 3, where the fitted Allan deviation is $9\times 10^{-12}\cdot\tau^{-1/2}$ from 10 to 10^5 s. The short-term stability is at the same level as the best commercial magnetic-state-selection clocks, e.g. 5071A, high performance. For longer integration time than 10^5 s, the stability is degraded. A previous measurement without microwave power locking loop demonstrated a long-term stability limit of 2×10^{-13}, which motivated us to incorporate the microwave power stabilization in the current stage of development.

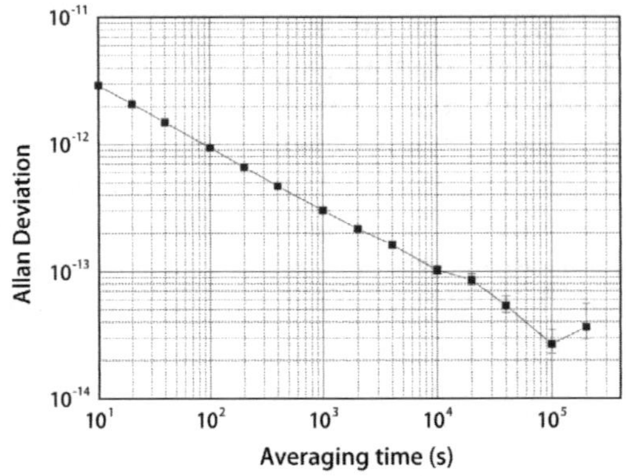

Figure 3. The Allan Deviation of a 15-day continuous measurement.

The 10-s stability is plotted in figure 4, with respect to Ramsey signal amplitude that is varied by controlling the oven temperature. It is seen that short-term stability with ECDL, which is locked to F=4-F'=5 cyclic transition, saturates when oven temperature is above 105 ^{O}C. This indicates that noise due to laser frequency deviation becomes dominant at high oven temperature, which is explained in detail in [4]. When the light source is replaced by the DFB laser locked to the same transition, the short-term stability is saturated to a larger value, at a temperature slightly lower than that of ECDL, because the DFB's laser frequency noise becomes dominant more easily.

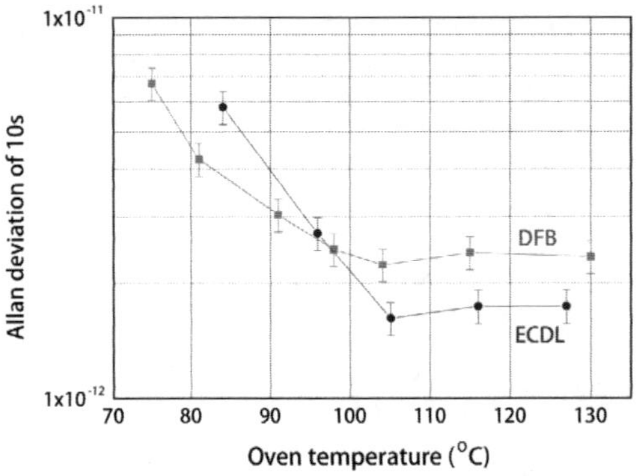

Figure 4. The short-term stability depending on the oven temperature and laser choice, red for DFB and blue for ECDL. (colour online)

Long-term stability is limited by several possible causes, including light shift, C-field shift and shift related to the microwave cavity. The frequency difference with respect of the fluorescence power depending on the incident power, with ECDL, is plotted in figure 5. The output frequency is significantly influenced, with total variation of 3×10^{-11} from 0 to 20 mW, by change of incident power. It is seen that when the fluorescence intensity is small, the frequency shifts are close under different oven temperatures, indicating that the light shift caused by fluorescence is the main shift. On the other hand, when the fluorescence intensity is large, the shifts have differences, because the scattered light becomes intense. When oven temperature is lower, extra incident light power is needed, because of saturation, to assure the same fluorescence, leading to a larger frequency shift caused by scattered light. Near the operation conditions, the measured slope of frequency shift with respect to incident light power is 1.3×10^{-12} / mW. In order to achieve the clock's long-term stability in 10^{-15} range, the laser's power perhaps needs to be stabilized. The C-field stabilization is considered to be necessary for the frequency varies strongly with C-field due to second-order Zeeman frequency shift. The cavity is tuned to the clock frequency before the measurement to minimize the cavity-pulling frequency shift. The cavity loaded quality factor is approximately 1000 and the estimated frequency shift is 3×10^{-13} for cavity mistuning of 1 MHz. However, the performance of the cavity is still unknown during a long-term operation.

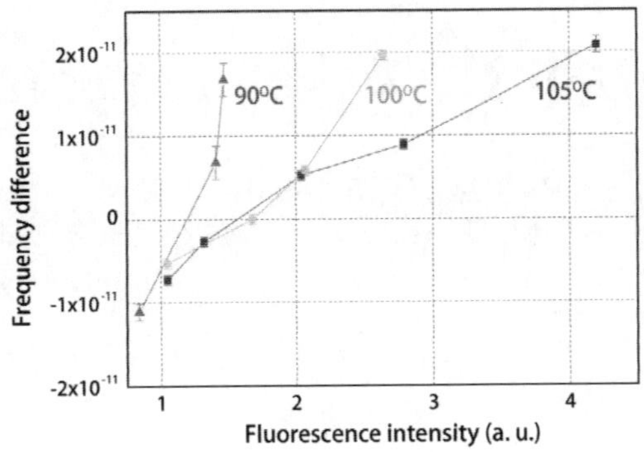

Figure 5. Light shift at different oven temperatures, red, green and blue for 90, 100 and 105 °C. (colour online)

4. Conclusion

The application of optical detection in place of the traditional detection method in cesium clocks allows a long life of the clock in principle. However, the optical detection leads to some effects, e.g. light shift, that affect the performance of the clock. Moreover, the long-term operation of laser diode has to be taken care of. Besides, the digital circuits need further modification to complete the laser frequency stabilization and automatic control.

Acknowledgments
This project is supported by the National Natural Science Foundation of China (Grant No. 11174015).

References

[1] Cutler L S 2005 *Metrologia* **42**(3) S90-9
[2] Lecomte S, Haldimann M, Ruffieux R, Berthoud P and Thomann P 2007 *Proceedings of the 2007 IEEE International Frequency Control Symposium-jointly with the 21st European Frequency and Time Forum* (Geneva) pp 1127-31
[3] Cao Y, Zhao X, Yang L, Chen H and Zhang S 2014 *Frequency Control Symp. (FCS), 2014 IEEE Int.* (Taipei) pp 401-4
[4] Liu C, Zhou S and Wang Y 2015 *2015 Joint Conference of the IEEE International Frequency Control Symposium & the European Frequency and Time Forum (FCS)* (Denver) pp 483-6
[5] Zhou S , Liu C, Wang Y and Zhang D 2015 *2015 Joint Conference of the IEEE International Frequency Control Symposium & the European Frequency and Time Forum (FCS)* (Denver) pp 471-3

Double-modulation CPT cesium compact clock

Peter Yun[1], Sinda Mejri[1], Francois Tricot[1], Moustafa Abdel Hafiz[2], Rodolphe Boudot[2], Emeric de Clercq[1], Stéphane Guérandel[1]

[1] LNE-SYRTE, Observatoire de Paris, PSL Research University, CNRS, Sorbonne Universités, UPMC Univ. Paris 06, 61 avenue de l'Observatoire, 75014 Paris, France

[2] FEMTO-ST, CNRS, UFC, 26 chemin de l'épitaphe 25030 Besançon cedex, France

E-mail: enxue.yun@obspm.fr

Abstract. Double-modulation coherent population trapping (CPT) is based on a synchronous modulation of Raman phase and laser polarization, which allows the atomic population to accumulate in a common dark state. The high contrast signal obtained on the clock transition with a relative compact and robust laser system is interesting as basis of a high performance microwave clock. Here we study the parameters of a double-modulation CPT Cs clock working in cw mode. The optimal polarization modulation frequency and cell temperature for maximum contrast of clock transition are investigated. The parameters of the detection are also studied. With the optimal parameters, we observe a CPT signal with contrast of 10% and linewidth of 492 Hz, which is well suited for implementing a cw atomic clock.

1. Introduction

Vapor cell frequency standards [1, 2] are of prime interest in all applications where frequency stability and size are the major criterion, such as satellite on-board clocks. This field has known outstanding progress since a few years with laboratory clocks exhibiting frequency stabilities at 1 s in the $1 - 3 \times 10^{-13}$ range, i.e. a gain of almost two order of magnitudes on present Rb clocks. These new clocks are based on the double-resonance technique [3], pulsed optical pumping and microwave interrogation [4, 5], or coherent population trapping (CPT) [6, 7]. CPT clocks need a circularly polarized laser beam to create a state superposition, the dark state, in the ground state of alkali-metal atoms [8, 9]. High performance CPT clocks need more sophisticated schemes in order to increase the contrast by avoiding the optical pumping and the trapping of the atoms into the end Zeeman sublevels. The lin$\perp$lin method [10, 11] needs two phase-locked lasers to provide two linearly and orthogonally polarized laser beams. The push-pull optical pumping (PPOP) [12, 13] reduces this complexity by using a single laser plus a Michelson type set-up, which introduces a phase shift between two counter-rotating circularly polarized beams. Without this phase shift the dark state built by a polarization is destroyed by the other one. In our double-modulation scheme [14], both the Raman phase and laser polarization are synchronously modulated to realize a time-version of push-pull optical pumping. Contrary to the PPOP configuration which separates and recombines the laser beams, our scheme provides a simple and robust laser bench by utilizing a polarization switch.

In our scheme, the polarization of a bichromatic laser beam is modulated between two opposite circular polarizations [15]. An appropriate modulation of the phase between the two components of the bichromatic laser beam, the so-called Raman phase, is applied synchronously. The two CPT dark states, produced successively by the alternate polarizations, add constructively. Owing to this additional Raman phase modulation, the common dark state is decoupled from the laser at all times. For these two reasons, i.e., the atomic population accumulated into the clock states and the constructive build-up of two CPT dark states, a high contrast signal of the clock transition is observed. Another advantage of this scheme is the absence of $\Delta m_F = 2$ spurious CPT transitions [16, 17], here m_F is the magnetic quantum number. Moreover, our scheme provides compact and robust system for CPT atomic clock, in which only one-laser system is utilized, the polarization switch should be less vibration sensitive than a Michelson type set-up. Here we study the detail of double-modulation CPT clock work in cw mode. Experimental parameters, e.g., cell temperature, modulation frequency, detected window duration and delay are investigated. The short term frequency stability based on the measured relative intensity noise of our laser beam is estimated.

2. Experimental set-up

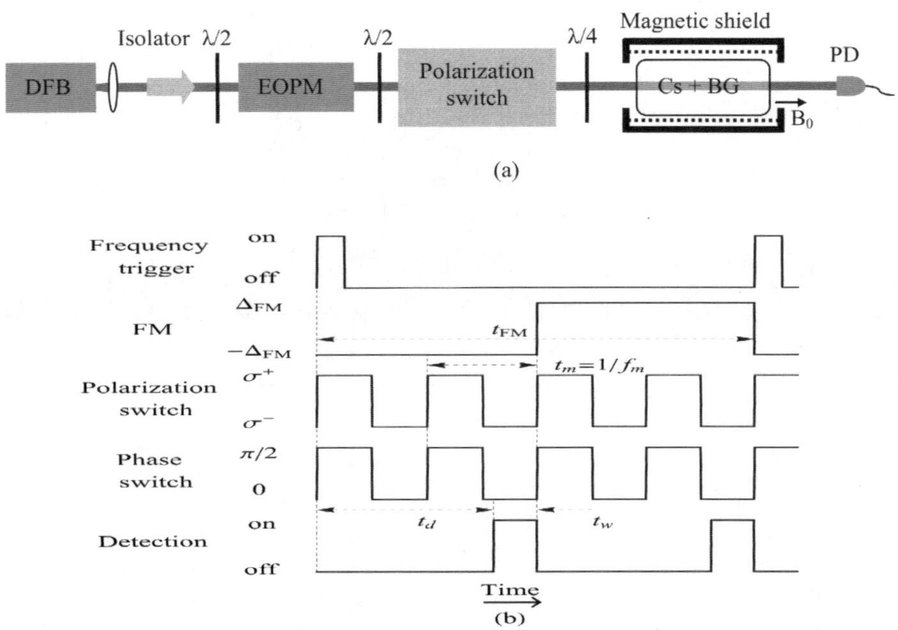

Figure 1: (a)(Color online) Double-modulation CPT setup. (b) Pulse sequence. The abbreviation, DFB: distributed feedback diode laser, EOPM: electro-optic phase modulator, BG: buffer gas, PD: photodiode. FM: frequency modulation.

As shown in figure 1(a), a distributed feedback diode (DFB) emits a monochromatic laser beam tuned on the D_1 line of Cesium. After passing through a fiber electro-optic phase modulator (EOPM), driven at 4.6 GHz with 26 dBm microwave power, a bichromatic laser is formed by the ± 1 first sidebands of the modulated spectrum. The $-1(+1)$st sideband is tuned to the transitions $|6^2S_{1/2}, F = 4(3)\rangle \leftrightarrow |6^2P_{1/2}, F' = 4\rangle$. A $0 - \pi/2$ phase modulation is added to the 4.6 GHz microwave, thus the optical Raman phase is modulated between 0 and π. Joint with a polarization modulator, synchronously modulated with the Raman phase, we get a double-modulated Raman laser. The polarization modulator is formed by an amplitude electro-optic modulator (EOM, Thorlabs: EO-AM-NR-C2), which shows 18 dB extinction ratio of polarization and 2.5 μs rising and falling time. A quarter-wave plate converts the perpendicular linear polarizations in opposite circular polarizations. The double-modulated Raman laser beam is then expanded to a diameter of 10 mm before the glass cell. The cylindrical Cs vapor cell, 20 mm diameter and 50 mm long, is filled with 21 Torr of mixed buffer gas (argon and nitrogen). Unless otherwise specified, the cell temperature is stabilized at 37.5 °C. A solenoid coil surrounded by two magnetic shields provides a uniform magnetic field of 4.82 µT in order to remove the Zeeman degeneracy.

The pulse sequence displayed in figure 1(b) is designed for cw double-modulation CPT, in which the $0-\pi/2$ microwave phase and the $\sigma^- - \sigma^+$ polarization are synchronously modulated at frequency f_m to prepare constructive CPT states. The signal is recorded by a photodiode with a time delay t_d after the beginning of the sequence and during a time t_w.

To obtain a discriminator signal, the usually employed frequency modulation (FM) technique is adopted, i.e., in the two successive half periods of FM, the microwave frequency is up and down shifted Δ_{FM} relative to the un-modulated one f_0 respectively, the correspondent photo-detector (PD) signals are subtracted, an error signal is then obtained after a low pass filter when we scan the microwave frequency f_0.

3. Results

3.1. CPT spectrum and contrast

A typical Zeeman CPT spectrum is demonstrated in figure 2(a). The seven peaks observed are the seven allowed transitions between the hyperfine ground states of same magnetic quantum number m_F, $m_F = -3$ to $+3$. The central and highest peak is the clock transition $(0-0)$. This clearly shows that the atomic population is effectively collected in the clock states. The distortion of side peaks is explained by magnetic field inhomogeneity.

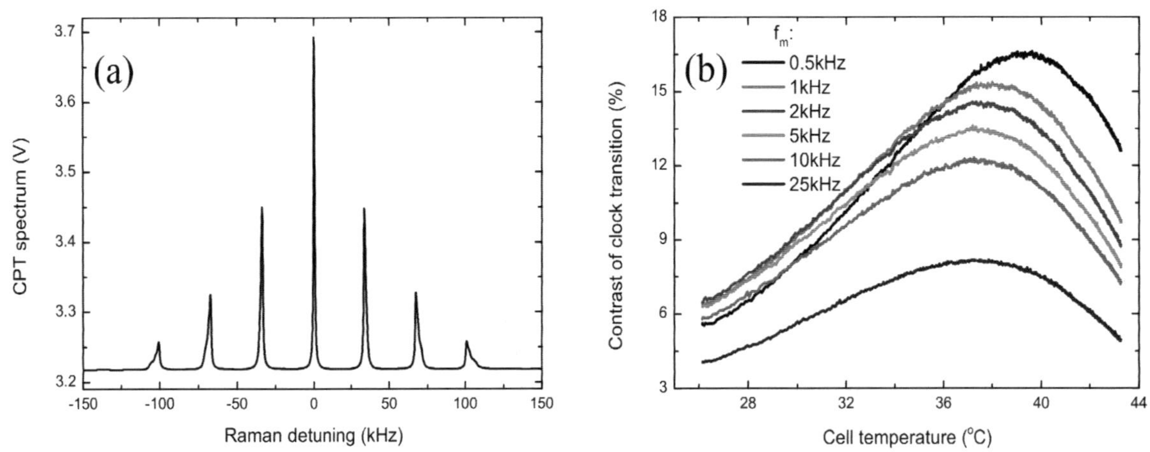

Figure 2: (a)Zeeman CPT spectrum obtained with a cw double-modulation with $f_m = 2$ kHz. (b)(Color online) CPT contrast vs cell temperature at various f_m. In both cases, $t_d = 3.5$ ms, $t_w = 0.5$ ms. Laser power: 1 mW.

The contrast of the clock transition, defined as CPT amplitude divided by the background level, reaches its maximum at the cell temperature of 37.5 °C in the case of $f_m = 2$ kHz as shown in figure 2(b). The optimal temperature shifts to a lower value as f_m increased.

We study the effect of modulation frequency f_m on the clock contrast at different detection window times as presented in figure 3(a). The highest CPT contrast is obtained around 1 kHz with a broad range of modulation frequency. This releases the requirements on the polarization switch, thus it is possible to further reduce the size of our laser system by utilizing a liquid crystal polarization switch. When f_m is in the range from 500 Hz to 20 kHz, the contrast increases for short detection windows t_w. It begins to saturate at longer detection delay time t_d as illustrated in figure 3(b).

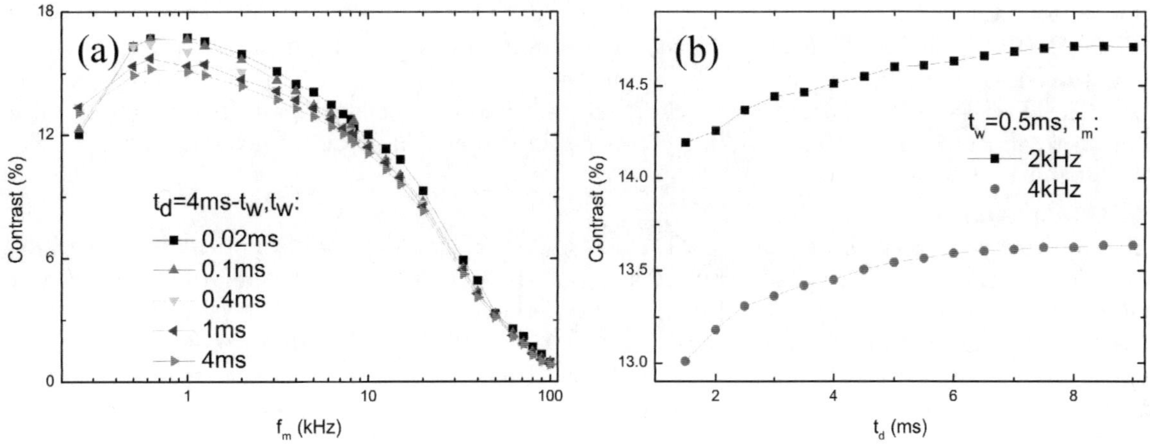

Figure 3: (Color online) (a) Contrast of clock transition as function of polarization modulation frequency f_m for several detection window durations t_w, $t_d = 4\,\text{ms}{-}t_w$. (b) Contrast vs t_d at two f_m values (2 and 4 kHz). In both cases, Laser power: 1 mW.

3.2. Clock signal and laser noise

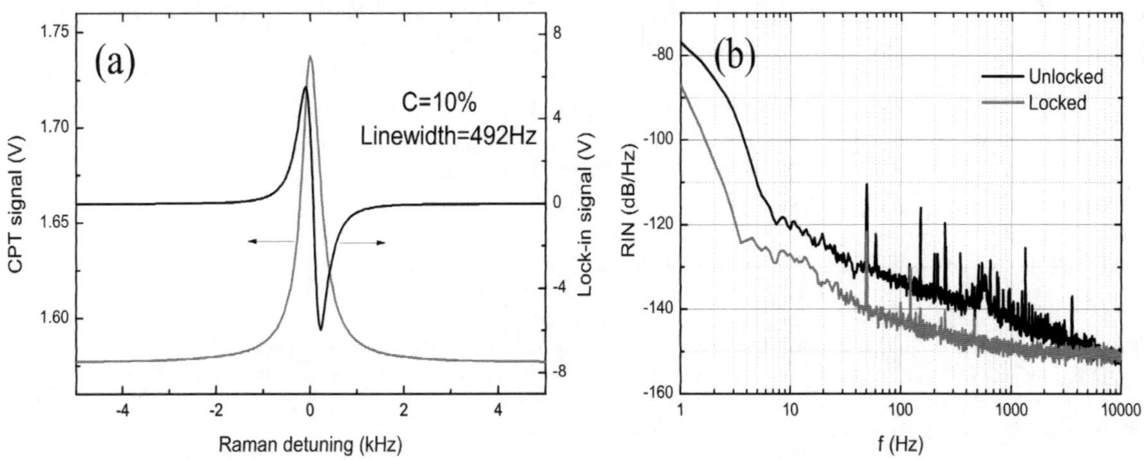

Figure 4: (Color online) (a)CPT signal and lock-in output signal of clock transition with $f_m = 2\,\text{kHz}$, $t_d = 3.5\,\text{ms}$, $t_w = 0.5\,\text{ms}$, $FM = 125\,\text{Hz}$, $\Delta_{FM} = 320\,\text{Hz}$. Laser power: 0.5 mW. (b) Relative intensity noise of our laser beam with and without power locking.

The signal of the clock transition, recorded with a laser power of 500 µW, is demonstrated in figure 4(a). The linewidth is 492 Hz and the contrast 10%. The error signal (figure 4(a)) is obtained after demodulation of the clock signal, square-wave frequency modulated as described in figure 1(b). The chosen modulation frequency, 125 Hz, is a compromise between a fast modulation reducing the noise level, and the time needed to accumulate the atomic population in the clock states.

As the laser relative intensity noise (RIN) usually is the main limitation to the short term frequency stability of CPT clock [6, 7, 11], a laser power locking set-up is employed. It utilizes an acousto-optic modulator, not drawn in figure 1(a), positioned between the EOPM and the polarization modulator. The RIN of the laser beam measured after the EOPM with and without power locking are displayed in figure 4(b). In the locked laser power case, the measured RIN is $-142\,\mathrm{dB/Hz}$ at 125 Hz. With the CPT signal obtain in figure 4(a) and if this RIN level could be maintained after travelling through the atomic cell, the RIN limited frequency stability would be at level of 4×10^{-14} at 1 second. However, the light-atom interaction in the vapor cell is known to add tens of dB to the intensity noise. This point is now under investigation.

4. Conclusion

To implement a high performance CPT clock based on double-modulation method, we have investigated the optimal cell temperature and polarization and phase modulation frequency in cw mode. The optimal modulation frequency around several kHz implies we can utilize a liquid crystal polarization switch to further reduce the size of our clock. The duration and delay of the detection window are also addressed. The observed high contrast (10%) and narrow linewidth (492) Hz clock signal should lead to a frequency stability of 4×10^{-13} at 1 second if the laser RIN can be kept below $-122\,\mathrm{dB/Hz}$ after crossing the vapor cell.

Acknowledgments

We would like to thank David Holleville, Thomas Zanon-Willette, Natascia Castagna, Jean-Marie Danet, Olga Kozlova, Stephane Trémine and Luca Lorini for helpful discussions. We are grateful to José Pinto Fernandes, Michel Lours and Electronic service at SYRTE for technical assistance and the realization of various electronic devices, Pierre Bonnay and Annie Gérard for manufacturing Cs cells. We thank Candy Xiong for reading of the manuscript. P. Y. is supported by the Facilities for Innovation, Research,Services, Training in Time & Frequency (FIRSTTF). This work is supported in part by ANR and DGA(ISIMAC project ANR-11-ASTR-0004). This work has been funded by the EMRP program (IND55 Mclocks). The EMRP is jointly funded by the EMRP participating countries within EURAMET and the European Union.

References

[1] Camparo J 2007 *Physics Today* **60** 33
[2] Godone A, Levi F, Calosso C E, Micalizio S 2015 *Riv. Nuovo Cimento* **38** 133-171
[3] Bandi T, Affolderbach C, Stefanucci C, Merli, Skrivervik A K And Mileti G 2014 *IEEE Trans. Ultrason. Ferroelectr. Freq. Cont.* **61** 1769
[4] Micalizio S, Calosso C E, Godone A and Levi F 2012 *Metrologia* **49** 425-436
[5] Kang S, Gharavipour M, Affolderbach C, Gruet F and Mileti G 2015 *J. Appl. Phys.* **117** 104510
[6] Danet J M, Kozlova O, Yun P, Guérandel S and de Clercq E 2014 *EPJ Web of Conferences* **77** 00017
[7] Abdel Hafiz M and Boudot R 2015 *J. Appl. Phys.* **118** 124903
[8] Alzetta G, Gozzini A, Moi L and Orriols G 1976 *Nuovo Cimento B* **61** 5
[9] Vanier J 2005 *Appl. Phys. B* **81** 421
[10] Zanon T, Guérandel S, de Clercq E, Holleville D, Dimarcq N and Clairon A 2005 *Phys. Rev. Lett.* **94** 193002
[11] Danet J M, Lours M, Guérandel S and de Clercq E 2014 *IEEE Trans. Ultrason. Ferroelectr. Freq. Cont.* **61** 567
[12] Jau Y Y, Miron E, Post A B, Kuzma N N and Happer W 2004 *Phys. Rev. Lett.* **93** 160802
[13] Liu X C, Mérolla J M, Guérandel S, Gorecki C, de Clercq E and Boudot R 2013 *Phys. Rev. A* **87** 013416
[14] Yun P, Danet J M, Holleville D, de Clercq E and Guérandel S 2014 *Appl. Phys. Lett.* **105** 231106
[15] Huang M and Camparo J C 2012 *Phys. Rev. A* **85** 012509
[16] Yun P, Mejri S, Tricot F, Holleville D, de Clercq E and Guérandel S 2015 *Proc. Joint Conf. IEEE Int. Frequency Control Symp. - European Time and Frequency Forum IFCS-EFTF* (April 12-16 Denver, USA) p 797
[17] Kozlova O, Danet J M, Guérandel S, and de Clercq E 2014 *IEEE Trans. Instrum. Meas.* **63** 1863

A CPT-based Cs vapor cell atomic clock with a short-term fractional frequency stability of $3 \times 10^{-13}\ \tau^{-1/2}$

Moustafa Abdel Hafiz[1], Xiaochi Liu[1⋆], Stéphane Guérandel[2], Emeric De Clercq[2], Rodolphe Boudot[1]

[1]FEMTO-ST, CNRS, UFC, 26 chemin de l'épitaphe 25030 Besançon cedex, France.
[2]LNE-SYRTE, Observatoire de Paris, PSL Research University, CNRS, Sorbonne Universités, UPMC Univ. Paris 06, 61 avenue de l'Observatoire, 75014 Paris, France.
[⋆]now in National Institute of Standards and Technology (NIST), Boulder, CO 80305, USA.

E-mail: `rodolphe.boudot@femto-st.fr`

Abstract. This article reports on the development and short-term fractional frequency stability of a continuous-regime (CW) Cs vapor cell atomic clock based on coherent population trapping (CPT). The push-pull optical pumping technique is used to increase the number of atoms that participate to the clock transition, yielding a typical CPT resonance contrast of 25 % for a CPT linewidth of about 450 Hz. The clock short-term fractional frequency stability is measured to be $3 \times 10^{-13}\ \tau^{-1/2}$ up to 100 seconds averaging time, in correct agreement with the signal-to-noise ratio limit. The mid-term frequency stability results are currently mainly limited by laser power effects. The detection of high-contrast narrow Raman-Ramsey fringes is demonstrated with this setup by making the atoms interact with a light pulse sequence.

1. Introduction

The development of high-performance, compact and low-power consuming atomic clocks is of great interest in numerous scientific and industrial fields including new-generation satellite-based navigation systems, network synchronization, timekeeping applications or defense systems. Vapor cell atomic clocks, basically based on the combination of a laser system and thermal alkali atoms confined in a glass vapor cell, are in that sense an attractive technology to propose simultaneously simple architecture and high-performance. Over the last years, thanks to the use of better performing laser sources and dedicated techniques, laboratory-prototype vapor cell atomic clocks, based on optical-microwave double resonance technique [1, 2] or coherent population trapping (CPT) [3], have demonstrated short-term fractional frequency stability in the $1-4\times10^{-13}\ \tau^{-1/2}$ range [4, 5, 6, 7, 8]. These performances are about two orders of magnitude better than current commercially-available Rb clocks and tend to be competitive with those of passive hydrogen masers.

In the frame of the MClocks project [9], we recently started the development of a compact Cs CPT atomic clock based on the push-pull optical pumping (PPOP) technique, pioneerly proposed by Jau et al. [10]. In recent literature, we reported the spectroscopy of high-contrast CPT resonances in Cs vapor cells [11] and demonstrated the possibility to detect high-contrast Raman-Ramsey fringes [12]. This article aims to report short-term frequency-stability results of

Journal of Physics: Conference Series **723** (2016) 012013 doi:10.1088/1742-6596/723/1/012013

our CPT-PPOP based Cs vapor cell atomic clock. A promising short-term fractional frequency stability of $3 \times 10^{-13}\ \tau^{-1/2}$ up to $\tau = 100$ s integration time in CW regime, in correct agreement with the noise budget, is demonstrated.

2. Experimental setup

Figure 1 presents the Cs CPT clock experimental set-up.

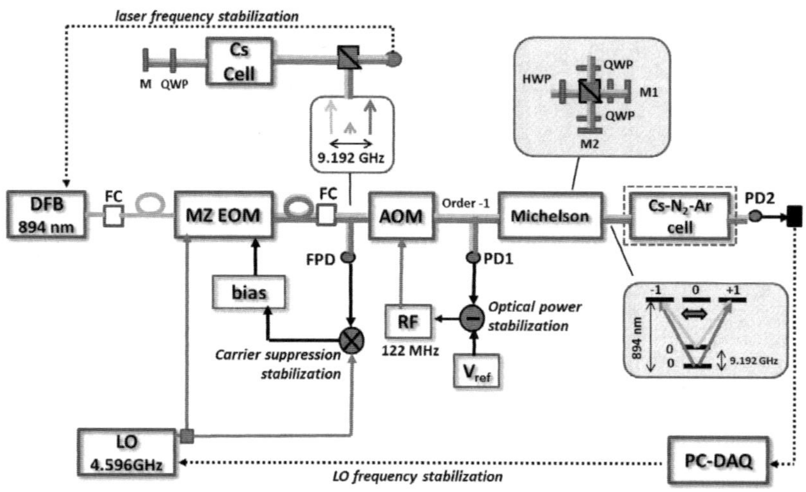

Figure 1. Schematic of the Cs vapor cell atomic clock based on push-pull optical pumping. DFB: Distributed feedback diode laser, FC: fiber collimator, MZ EOM: Mach-Zehnder electro-optic modulator, AOM: acousto-optic modulator, LO: microwave local oscillator, FPD: fast photodiode, PD: photodiode, RF: radiofrequency source, V_{ref}: reference voltage, Michelson: Michelson-type delay-line and polarization orthogonalizer system, PC-DAQ: personal computer - data acquisition card, M: mirror, QWP: quarter-wave plate, HWP: half-wave plate. An inset shows the CPT diagram involved in the push-pull optical pumping technique. A second inset shows the Michelson delay-line system principle.

The laser source is a commercially-available GaAs semi-conductor 1 MHz-linewidth distributed-feedback (DFB) diode laser tuned at 894 nm on the Cs D_1 line [13]. A Faraday optical isolator (not shown on Fig. 1) is placed at the output of the DFB laser in order to avoid optical feedback. A MZ EOM (Photline NIR-MX800-LN-10) driven at 4.596 GHz with a low noise microwave frequency synthesizer is used to generate phase-coherent optical sidebands frequency-separated by 9.192 GHz. The local oscillator consists of a high-spectral purity 100 MHz quartz-crystal oscillator phase-locked to a high-stability 10 MHz quartz source and then frequency-multiplied to 4.596 GHz. At the output of the EOM, optical carrier rejection is actively stabilized thanks to an original microwave synchronous detector presented in [11]. The laser beam that carries both optical lines frequency-split by 9.192 GHz, is sent into an annex reference Cs cell to stabilize the laser frequency using a saturated absorption spectroscopy setup. The laser is frequency-stabilized by modulating the DFB laser current at 62 kHz and demodulating it with a lock-in amplifier LA1. The laser frequency servo bandwidth is about 1.5 kHz. At the output of the EOM, an AOM-based laser power stabilization system with a servo bandwidth of about 20 kHz is implemented. The order -1 at the output of the AOM is sent to the CPT cell to shift the laser frequency by -122 MHz and to compensate for the buffer-gas induced optical frequency shift. At the output of the AOM, a Michelson delay-line and polarization orthogonalizer system is used to produce the PPOP scheme [10]. At the output of the Michelson system, the diameter beam is expanded to 2 cm. The CPT cell consists of a 2-cm diameter and 5-cm long cylindrical Cs vapor cell filled with a N_2-Ar buffer gas mixture of total pressure 15.3 Torr and partial pressure ratio $r = P_{Ar}/P_{N_2} = 0.4$. In clock operation, the cell temperature T_{cell} is stabilized at about 35°C where the CPT signal height is maximized. The presence of buffer gas in the CPT cell causes a collisional frequency shift of the Cs clock transition of about 9.3 kHz. A static magnetic field of 89 mG parallel to the laser beam propagation direction is applied to split the hyperfine ground-state Zeeman transitions. The ensemble is surrounded by a double-layer mu-metal magnetic shield. The CPT resonance is monitored by detecting the laser power transmitted through the cell using the low-noise photodiode PD2. The resulting signal is analyzed by a computer that drives the local oscillator frequency. For frequency stability measurements, the clock output signal is compared to the signal of a state-of-the-art reference active hydrogen maser that exhibits a fractional frequency stability of 8×10^{-14} and 3×10^{-15} at $\tau = 1$ s and 100 s respectively [14].

3. Frequency stability

Figure 2(a) reports the typical clock signal for an incident laser power P_L of 475 μW in the 5-cm long cell.

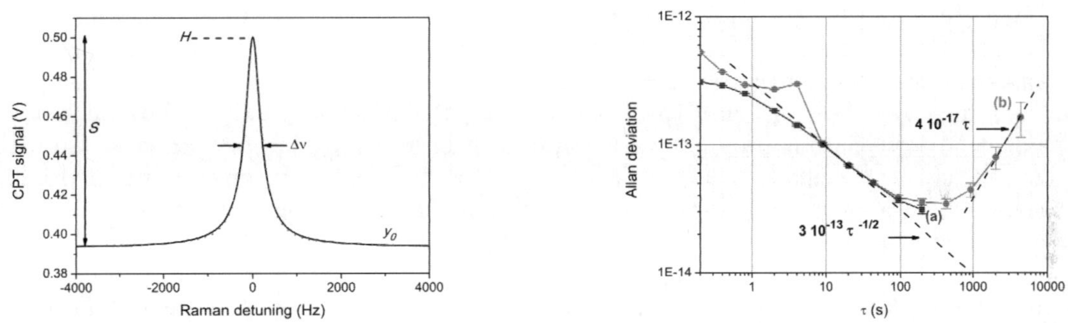

(a) (b)

Figure 2. (a): Clock CPT resonance signal. Experimental data of the direct CPT resonance are fitted by a lorentzian fit function. The total laser power incident on the cell is 475 μW. The CPT linewidth is 453 Hz. The resonance contrast is 26.4 %. (b): Allan deviation of the clock frequency. The laser power incident on the cell is 475 μW. The static magnetic field is 89 mG. (a): EOM bias servo loop off, (a): EOM bias servo loop activated. The first dashed line is a curve with a 3×10^{-13} $\tau^{-1/2}$ slope. The second dashed line is a linear curve with a slope of about 4×10^{-17} τ slope.

The clock resonance is well-fitted by a lorentzian function. The CPT linewidth $\Delta\nu$ is 453 Hz. The clock signal S, defined on Fig. 2(a) as $S = H - y_0$, is 0.104 V. The resonance contrast C, defined as the ratio between the CPT signal S and the dc background y_0, is 26.4 %. The discriminator slope $D = S/\Delta\nu$ is measured to be 2.3×10^{-4} V/Hz. Figure 2(b) shows the measured short-term fractional frequency stability of the CPT clock. In normal operation (curve b), the latter is measured to be 3×10^{-13} $\tau^{-1/2}$ up to about 100 s averaging time, in correct agreement with SNR measurements. The bump between 1 and 10 s is attributed to the EOM bias voltage servo loop gain. We suspect that this issue could prevent to observe the expected calculated short-term stability of 2×10^{-13} (see table 1). In the case where this servo is off (Fig. 2(b), curve a), the bump between 1 and 10 s clearly disappears. For integration times higher than 10 s, the observed difference between the two stability curves are within the measurement error bars. After 100 s (see curve b), the clock frequency drifts to reach 2×10^{-13} at 5000 s.

Table 1. Main contributions to the clock short term frequency stability at $\tau = 1$ s.

Noise Source	σ (1 s)
Shot noise	8×10^{-15}
Detector noise	2×10^{-14}
LO phase noise	1.1×10^{-13}
Total laser noise (AM and FM-AM)	1.7×10^{-13}
Laser current driver noise	1.2×10^{-14}
Total $\sigma_y(1s) = \sqrt{\sum \sigma_y^2}$	2×10^{-13}

For further investigation, we measured main clock frequency shifts to explain mid-term frequency stability limitations of the CPT-PPOP clock. These studies are detailed in [15]. The latter is presently strongly limited by the laser intensity fluctuations through the light shift effect. Other relevant mechanisms could be the laser frequency variations, the Zeeman shift and length fluctuations of the Michelson delay-line system.

4. Prospects for future improvements

The intermodulation effect is found to be an important contribution to the clock short-term noise budget. A low phase noise microwave synthesizer based on a high-performance 100 MHz oven-controlled quartz oscillator (OCXO) will be implemented nextly on the experiment [16, 17]. This source should allow to reduce the intermodulation effect [18] contribution at the level of about 3×10^{-14} at $\tau = 1$ s, a value closer to the clock shot noise limit.

Laser power fluctuations are a strong limitation to the clock mid-term frequency stability. A proposal to reduce laser intensity effects is to use a Ramsey-type temporal pulsed interrogation [12]. It has been demonstrated that pulsed interaction reduces strongly the intensity-light shift sensitivity [19, 4], prevents the CPT line broadening and allows the detection of narrow and high-contrast Ramsey fringes if combined with optimized CPT pumping schemes [20, 12, 21]. Figure 3 shows an example of a CPT-Ramsey fringe detected with the present setup by making the atoms interact with a light pulsed sequence [20]. The central Ramsey fringe linewidth is here 166 Hz. In pulsed configuration, the clock frequency sensitivity to laser power fluctuations was found to be reduced by a factor of about 10 compared to the CW regime. Further studies are in progress to evaluate the full potential of this technique.

Another way to improve the mid and long term frequency stability of the PPOP Cs CPT clock would be the use of a vapor cell with optimized N_2 - Ar buffer gas mixture for reduced temperature sensitivity of the clock frequency. From [22, 23], a pressure ratio $r = P_{Ar}/P_{N_2} = 0.56$ should allow to shift the inversion temperature close to 35°C.

5. Conclusions

We reported a CPT-based Cs vapor cell atomic clock, using the push-pull optical pumping technique, operating in the continuous regime, with a short-term fractional frequency stability at the level of $3 \times 10^{-13}\ \tau^{-1/2}$ up to $\tau = 100$ s. The short-term and mid-term performances of the clock are presently mainly limited by laser power effects. Further studies and techniques, including Ramsey-type pulsed interrogation or the use of an optimized buffer-gas cell, are under investigation to improve further the clock performances.

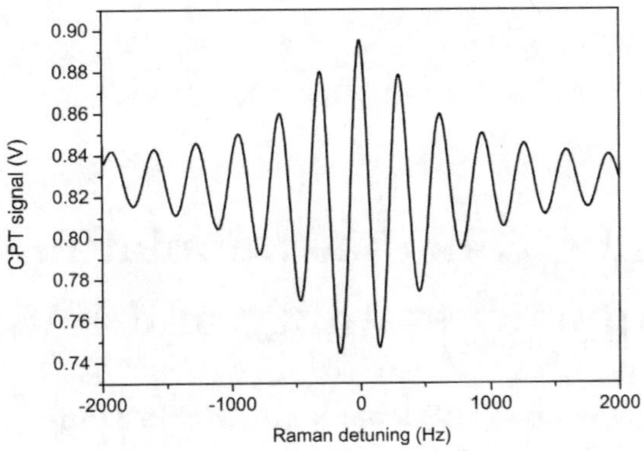

Figure 3. Example of a CPT-Ramsey fringe. The free evolution time is 3 ms. The pumping time is 2 ms. The total laser power incident on the cell is 1.5 mW.

Acknowledgments
This work has been funded by the EMRP program (IND55 Mclocks). The EMRP is jointly funded by the EMRP participating countries within EURAMET and the European Union. This work was partly supported by LNE, LabeX FIRST-TF and ANR-DGA ISIMAC project (ANR-11-ASTR-0004).

References

[1] Vanier J and Mandache C 2007 *Appl. Phys. B* Lasers Opt. **85** DOI: 10.1007/s00340-007-2643-5
[2] Camparo J 2007 *Phys. Today* **60** 11 33
[3] Vanier J 2005 *Appl. Phys. B Lasers Opt.* **81** DOI: 10.1007/s00340-005-1905-3
[4] Micalizio S, Calosso C E, Godone A and Levi F 2012 *Metrologia* **49** 425
[5] Stefanucci C, Bandi T, Merli F, Pellaton M, Affolderbach C, Mileti G and Skrivervik A K 2012 *Rev. Sci. Instrum.* **83** 104706
[6] Kang S, Gharavipour M, Affolderbach C, Gruet F and Mileti G 2012 *Journ. Appl. Phys.* **117** 104510
[7] Boudot R, Guérandel S, De Clercq E, Dimarcq N and Clairon A 2009 *IEEE Trans. Instrum. Meas.* **58** 4 1217
[8] Danet J M, Lours M, Guérandel S and De Clercq E 2014 *IEEE Trans. Ultrason. Ferroelec. Freq. Contr.* **61** 4 567
[9] http://www.inrim.it/Mclocks
[10] Jau Y Y, Miron E, Post A B, Kuzma N N and Happer W 2004 *Phys. Rev. Lett.* **93** 160802
[11] Liu X, Mérolla J M, Guérandel S, Gorecki C, De Clercq E and Boudot R 2013 *Phys. Rev. A* **87** 013416
[12] Liu X, Guérandel S, Mérolla J M, De Clercq E and Boudot R 2013 *Opt. Express* **21** 10 12451
[13] Liu X and Boudot R 2012 *IEEE Trans. Instr. Meas.* **61** 10 2852
[14] http://www.t4science.com/documents/iMaser-Clock-Spec.pdf (iMaser-ST 3000 specifications)
[15] Abdel Hafiz M and Boudot R 2015 *Journ. Appl. Phys.* **118** 124903.
[16] François B, Calosso C E, Danet J M and Boudot R 2014 *Rev. Sci. Instr.* **85** 094709
[17] François B, Calosso C E, Abdel Hafiz M, Micalizio S and Boudot R 2015 *Rev. Sci. Instr.* **86** 094707
[18] Audoin C, Candelier V and Dimarcq N 1991 *IEEE Trans. Instrum. Meas.* **40** 2 121
[19] Castagna N, Boudot R, Guérandel S, De Clercq E, Dimarcq N and Clairon A 2009 *IEEE Trans. Ultrason. Ferroelec. Freq. Contr.* **56** 2 246
[20] Zanon T, Guérandel S, De Clercq E, Holleville D, Dimarcq N and Clairon A 2005 *Phys. Rev. Lett.* **94** 193002
[21] Yun P, Danet J M, Holleville D, De Clercq E and Guérandel S 2014 *Appl. Phys. Lett.* **105** 231106
[22] Kozlova O, Guérandel S and De Clercq E 2011 *Phys. Rev. A* **83** 062714
[23] Kozlova O, Danet J M, Guérandel S and De Clercq 2014 *IEEE Trans. Instr. Meas.* **637** 1863

Journal of Physics: Conference Series **723** (2016) 012014 doi:10.1088/1742-6596/723/1/012014

Atomic clock using coherent population trapping in a cesium cell: frequency stability and limitations

Sinda Mejri, Francois Tricot, Jean-Marie Danet[1], Peter Yun, Emeric De Clercq and Stephane Guerandel

LNE-SYRTE, Observatoire de Paris, PSL Research University, CNRS, Sorbonne Universités, UPMC Univ. Paris 06, 61 avenue de l'Observatoire, 75014 Paris, France.

[1] Present address Syrlinks, 35170 Bruz, France

E-mail: sinda.mejri@obspm.fr

Abstract. Toward the next generation of compact devices, atomic clocks based on coherent population trapping (CPT) offer a very interesting alternative. We present a review of our studies on the short and mid term stability of a compact high performance atomic clock based on CPT in view of portable applications.

1. Introduction

Compact and high stability clocks are needed in many on-board applications like for example global navigation satellite system (GNSS). The present stability of vapor cell atomic clocks used in Galileo program are 5×10^{-12} at 1 s [1]. A number of innovative schemes for compact clock based on the double-resonance technique [5], pulsed optical pumping and microwave interrogation [6, 7], or coherent population trapping [2, 3], are showing performances at the level of hydrogen maser. The project presented here describes a compact atomic clock laboratory prototype which uses an original atoms interrogation technique called coherent population trapping (CPT)[8]. Contrary to the double resonance clock used today for GPS, the microwave pulses are optically carried. This allows the suppression of the microwave cavity and leads to a high miniaturization potential. In this paper, we present the implementation and the metrological characterization of a CPT clock prototype developed at LNE-SYRTE. Our prototype intends to explore what would be the ultimate stability of a CPT based device. To this aim, we combine a high signal amplitude and narrow resonance width by using a so-called double-Λ scheme and Ramsey interrogation technique [9, 10]. The reported metrological characterization includes some technical solutions for the effects that limit the frequency stability of our clock, such as the Dick effect and the laser amplitude noise.

2. Experimental setup

The experimental setup of the clock is shown in Figure 1. In order to pump the atoms in the clock dark state (linear superposition of $|F = 3, m_F = 0>$ and $|F = 4, m_F = 0>$) we use two phase-locked external cavity diode lasers tuned to the Cs D1 line at 894 nm. The master laser (M) is frequency locked to the ($F = 4 - F' = 4$) transition via saturated absorption in an auxiliary evacuated Cs cell. The scheme contains a frequency modulated acousto-optic modulator (AOM), in order to avoid modulating the master laser frequency. It also allows shifting the laser frequency compared to the Cs reference for compensating the buffer gas shift in the clock cell. The slave laser (S) is phase locked to the master laser with a frequency offset tunable near 9.192 GHz, i.e. around the ($F = 3 - F' = 4$) transition, by comparison with a low-noise synthesized microwave signal. The frequency synthesizer is built as follows. A 100 MHz Rakon quartz oscillator (LO) is phase locked to the 100 MHz signal generated by a H-maser of the laboratory. The LO signal is frequency multiplied to produce a 9400 MHz signal. The last one is mixed with a 9.392 GHz signal issued from dielectric resonator oscillator (DRO). The beat-note signal of 7.368 MHz is low pass filtered, amplified, and compared to a 7.368 MHz signal delivered by a direct digital synthesizer. The resulting error signal is then used to lock the DRO through a second-order loop whose bandwidth is about 180 kHz.

The two superposed laser beams are switched on and off by an AOM before travelling across the Cs cell. The transmitted power is recorded by a low noise Si photodiode (PD). The resulting digitalized signal is processed by the computer (PC) which drives the 7.4 MHz synthesizer as well as the AOM switch. A digital loop is used in order to lock the synthesizer frequency. To do so, the interrogation frequency (IF) hold at resonance, and the phase of the synthesizer is square-wave modulated of $\pm\pi/2$ during the period in the dark. The error signal is computed by the PC which can lock the LO frequency or equivalently the synthesizer frequency. Its frequency is proportional to the frequency difference between the maser and the Cs clock and the Allan variance is computed from its record.

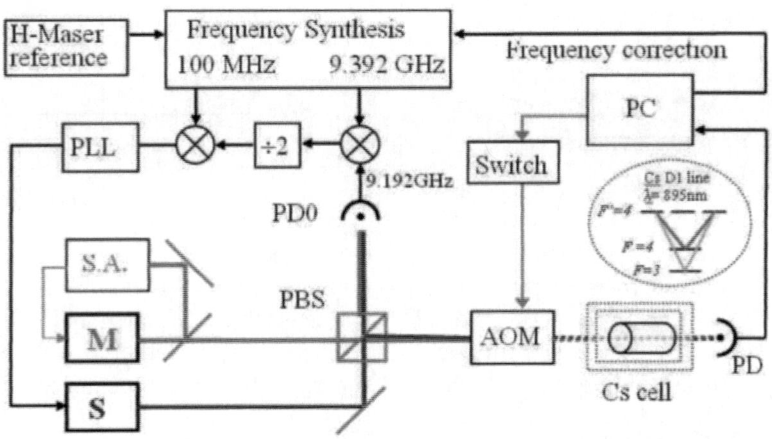

Figure 1: Scheme of the experimental setup; The master laser (M) is frequency locked on Cs resonance by a saturated absorption scheme (SA). The slave laser (S) is phase-locked on the master laser. PBS polarization beam splitter, PLL phase-lock loop. The inset shows the involved Cs energy levels.

The Cs cell, 5 cm long and 2 cm diameter, is filled with a 20 torr $N_2 - Ar$ buffer gas mixture ($P(Ar)/P(N_2) = 0.4$) [11]. The cell is temperature stabilized around 29°C which is the lowest temperature sensitivity point. It is located inside a solenoid and surrounded by two magnetic shields. A coaxial nonmagnetic wire is used with a homemade modified diode laser current source for heating the cell. Since the temperature fluctuations can be transferred to the clock transition frequency through the atomic collisions, the cell is temperature controlled within 100 μK which limits the temperature effect at the level of few 10^{-15}.

3. Short term limitations and optimization

As mentioned in the introduction, narrow linewidth can be obtained by combining coherent population trapping and Ramsey interrogation technique. The CPT clock operation continuously repeats the following phases (see Figure 2 (a) bottom): a first pulse pumps the atoms in the dark state, which is a steady-state. The detection window is at the beginning of the next pulse. The laser pulse duration is typically 2 ms for our setup separated by a Ramsey time T_R of 4 ms. τ_d typically around 10 μs is the delay time after which the clock transition is observed through the transmission signal. The observed signal is averaged during a time τ_m typically 25 μs. The experimental Ramsey fringes are shown in Figure 2(a).

The Figure 2 (b) shows the CPT clock stability when the local oscillator (LO) is locked to the central fringe of the Ramsey fringes. In order to have a better understanding of the effects leading to this performance, in the following part we will describe the models, measurements and technical choices used to get a total contribution of those noise sources to the short term stability around 3×10^{-13} at 1 s.

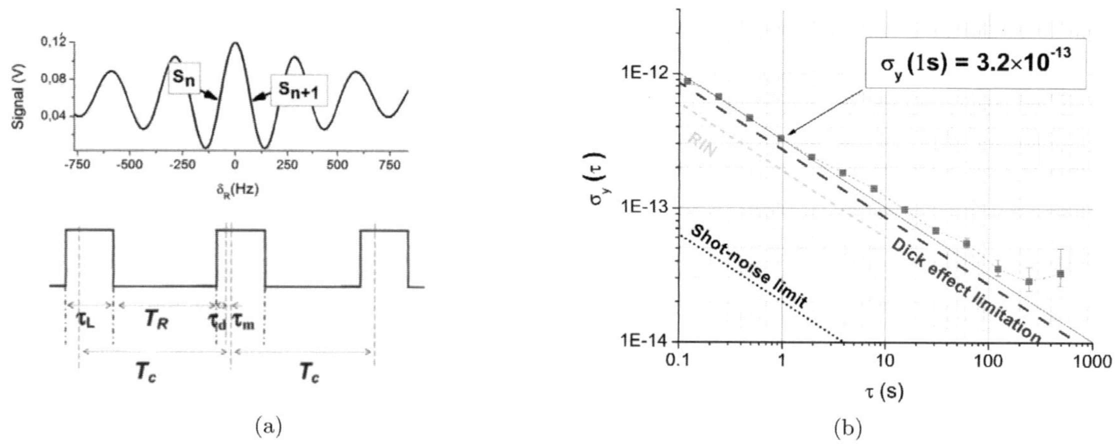

Figure 2: (a) Top: Ramsey fringes. S_n and S_{n+1} are the signal values measured on each side of the central fringe used to calculate the error signal. $\Delta\nu = 125$ Hz, SNR $= 42000$ in 1 Hz, $T_c = 6$ ms. Bottom: Time sequence of a CPT Ramsey clock (not to scale). τ_L is the length of each laser pulse, T_R is the free evolution time. The signal is detected after a time τ_d at the begining of each pulse, and averaged during a time τ_m. (b) CPT clock frequency stability: In blue the calculated Dick effect, in green the estimated RIN contribution. The red curve represent the measured stability. The shot noise limit at 1 s is at the level of 2×10^{-14}.

3.1. Short term limitations

Assuming the frequency stability to be limited by two dominant noise sources, the relative laser intensity noise (RIN) and LO noise, the Allan variance can be written:

$$\sigma^2(\tau) = \sigma^2_{Dick}(\tau) + \sigma^2_{SNR}(\tau) \tag{1}$$

with $\sigma^2_{Dick}(\tau)$ the contribution of the local oscillator phase noise and $\sigma^2_{SNR}(\tau)$ the contribution of the signal-to-noise ratio. The Dick effect is due to the down-conversion of the local oscillator intrinsic frequency noise at Fourier frequencies higher than the interrogation frequency [12]. The noise contribution to the Dick effect of each component of the microwave chain has been evaluated [13]. It turned out that the optical phase lock loop (PLL) noise gives the main contribution, evaluated at the 2.1×10^{-13} level. The total Dick limited stability is estimated at the level of $2.7 \times 10^{-13}\tau^{-1/2}$. In order to reduce this effect, current work is in progress to redesign our PLL and increase the bandwidth of our lock.

As already mentioned, the signal-to-noise ratio contribution due to the laser intensity noise is also an important effect. Using an intensity stabilization system, its contribution can be lowered until 1.7×10^{-13} at 1 s. A second way to reduce the RIN contribution is, based on the idea that if the intensity noise dominates the noise budget, the fluctuations on the detected signal might be proportional to the intensity fluctuations before the cell. This has been successfully checked on our setup. Such system is already implemented on other compact clock setups [17]. The stability presented here has been measured without the normalization scheme, giving a good hope for improvement.

3.2. Optimization

In this section we present the optimization of the two main limiting effects on the short term stability. The optimization parameters are the detection time τ_m and the free evolution time T_R. We have shown previously [13] that both Dick effect and laser intensity noise decrease by increasing τ_m. Therefore it seems that the clock frequency performances would be improved for longer detection times. However, our detection is destructive because the atomic phase originating the detected signal is progressively destroyed all along the laser detection pulse. The more τ_m increases, the more the fringe amplitude reduces and the instability of the clock increases. A similar study have been done on T_R. Indeed when T_R increases, the signal decreases because of the atomic coherence relaxation.

The figure 3 (a) presents the evolution of the SNR contribution to the frequency stability as a function of T_R and τ_m. The best compromise is for $T_R = 3$ ms and $\tau_m = 150$ μs. The gain from the current time sequence parameters ($T_R = 4$ ms and $\tau_m = 25$ μs) is about 30% on the current SNR. The Dick effect decreases as $T_R^{-1/2}$ and as $\tau_m^{-1/4}$ [13]. The increase of the evolution time T_R allows to improve the duty cycle. In this case the servo-loop of the LO adds less noise and the Dick effect decreases. Increasing τ_m allows to apply a temporal filter that reduces the influence of Fourier high frequency (Figure 3 (b)).

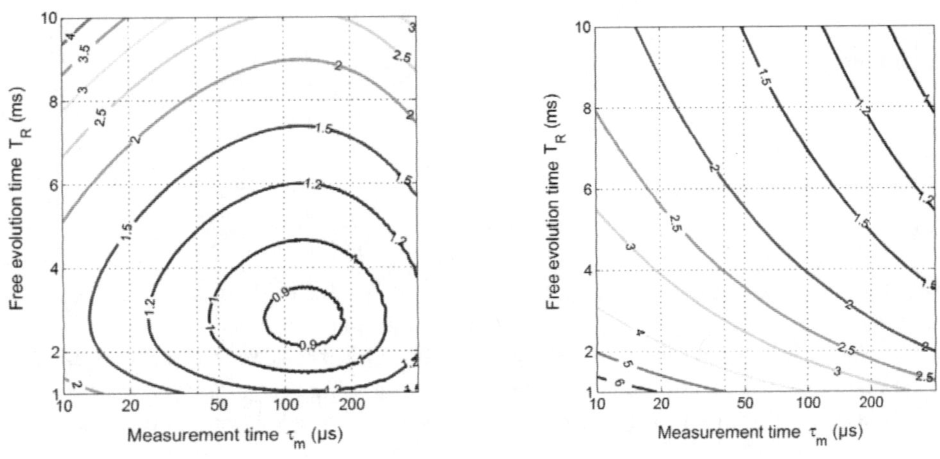

Figure 3: Influence of the free evolution time T_R and detection duration τ_m. (a) Contribution of the SNR to the frequency stability $\sigma(1s) \times 10^{13}$. The gain from current time sequence parameters ($T_R = 4$ ms, $\tau_m = 25$ μs) to ($T_R = 3$ ms, $\tau_m = 150$ μs) is around 30% on the current SNR. (b) Contribution of the Dick effect to the frequency stability $\sigma(1s) \times 10^{13}$.

4. Laser intensity effects

The effect of the laser light is known as one of the major effects limiting the mid term stability of a CPT based clock [15, 16, 14]. Presently our setup shows a mid term stability at the level of $2 - 3 \times 10^{-14}$ for an averaging time of about 4000 s [14]. The main sources of degradation at mid term are the sensitivity of the clock transition to the temperature, magnetic field and laser intensity. The temperature sensitivity of the clock transition can be greatly reduced by a proper choice of the composition of the mixture of gases [11]. For our clock, the shift is limited to the 10^{-14} level for 1 mK temperature variation [14].

Due to the double-Λ excitation scheme, the resonance ($m = -1 - m = +1$) between neighboring Zeeman sublevels of the clock transition leads to a shift of the locked frequency. The power sensitivity of the clock transition in the double-Λ scheme has been measured as a function of the magnetic field. Our clock operates at the magnetic field which maximizes the amplitude of the clock signal, i.e. $B = 22$ μT. The bottom part of Figure 4 (a) shows the frequency shift due to each laser intensity as a function of the magnetic field. A zero sensitivity point is shown around $B = 8$ μT. Unfortunately, operating the clock at this magnetic field value reduces the amplitude signal by 10% (Figure 4 (a) Top). The effect of such a reduction on the signal amplitude on the short term stability is under evaluation.

Figure 4 (b) presents the combined contribution of the power laser and magnetic field effects on the total instability of the clock. By operating the clock at $B = 8$ μT (green curve), the limitation of the power shift can be reduced to the level of 5×10^{-15} at 2000 s.

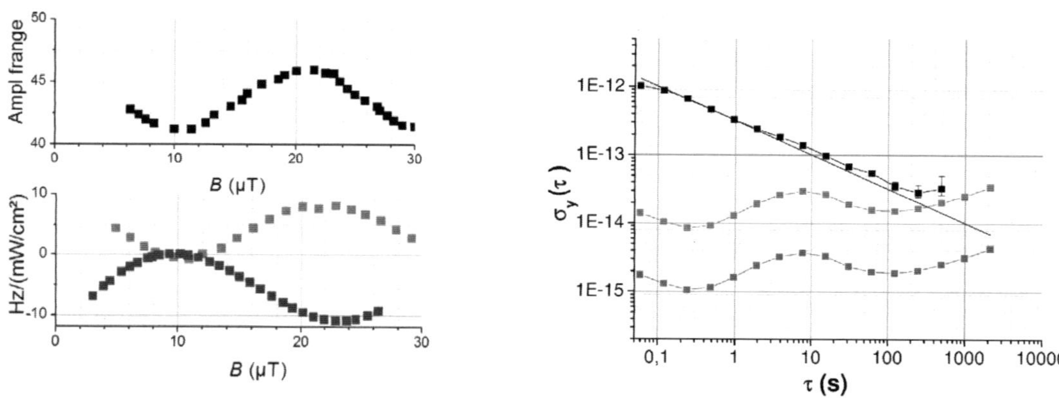

(a) (b)

Figure 4: (a) Top: CPT signal amplitude as a function of the magnetic field. Bottom: Laser intensity shift as a function of the magnetic field (Red master laser, blue slave laser). (b) Contribution of the power shift effect on the instability of the clock. Red: $B = 22$ μT, green: $B = 8$ μT. Black measured frequency stability. By operating the clock with $B = 8$ μT, the limitation of the power shift can be reduced to the level of 5×10^{-15} at 2000 s.

5. Conclusions

In summary we have presented a pulsed CPT vapor cell clock prototype. The studies of two main frequency limitations on the short term stability are reported: laser intensity noise and LO noise. The short term stability is at the present time around 3.2×10^{-13} level at 1 s. This stability can be improved by optimizing the time sequence of the clock. The laser intensity noise can be further reduced by combining the intensity stabilization system with the normalization scheme. The mid term stability is still limited after 200 s integration around $2-3\times10^{-14}$. The neighboring transitions impact the resonance frequency. This effect varies with the magnetic field value. A low sensitivity point has been shown. Using this value to operate the clock could lead to a drastic reduction of the power sensitivity of the clock. Experimental studies are carried in order to confirm this behavior.

Acknowledgments
We would like to thank Olga Kozlova and Rodolphe Boudot for helpful discussions. We are grateful to Michel Lours, Jose Pinto Fernandes and technical service at SYRTE for their assistance and advice, Pierre Bonnay and Annie Gerard for manufacturing Cs cell. Sinda Mejri is supported by CNES (Centre National d'Etudes Spatiales). Francois Tricot is supported by DGA and THALES.

References
[1] Waller P, Gonzalez F, Binda S, Sesia I, Hidalgo I, Tobias G and Tavella P 2010 *Trans. Ultrason. Ferroelect. Freq. Contr.* **57** 738
[2] Danet J M, Koslova O, Yun P, Guerandel S and de Clercq E 2014 *EPJ Web of Conferences* **77** 00017
[3] Abdel Hafiz M and Boudot R 2015 *J. Appl. Phys.* **118** 124903
[4] Esnault F-X, Holleville D, Rossetto N, Guerandel S and Dimarcq N 2010 *Phys. Rev. A* **82** 033436
[5] Bandi T, Affolderbach C, Stefanucci C, Merli F, Skrivervik A K Mileti G 2014 *IEEE Trans. Ultrason. Ferroelectr. Freq. Cont.* **61** 1769
[6] Micalizio S, Calosso C E, Godone A and Levi F 2012 *Metrologia* **49** 425
[7] Kang S, Gharavipour M, Affolderbach C, Guet F and Mileti G 2015 *J. Appl. Phys* **117** 00017
[8] Vanier J 2005 *Applied Physics B* **81** 421
[9] Zanon T, Guerandel S, de Clercq E, Holleville D, Dimarcq N and Clairon A 2005 *Phys. Rev. Lett* **94** 193002
[10] Zanon T, Tremine S, Guerandel S, de Clercq E, Holleville D, Dimarcq N and Clairon A 2005 *IEEE Trans. Instrum. Meas.* **54** 776
[11] Kozlova O, Guerandel S, de Clercq E 2011 *Phys. Rev. A* **83** 062714
[12] Dick G.J 1987 *Proc. 19 th Precise Time and Time Interval (PTTI) Applications and Planning Meeting* Redondo Beach CA, USA 133
[13] Danet J.M, Lours M, Guerandel S and de Clerq E 2014 *IEEE Ultrason. Ferroelectr. Freq. Control* **61**(4).
[14] Kozlova O, Danet J M, Guerandel S and de Clercq E 2014 *IEEE Trans. Instrum. Meas.* **63** 1863
[15] Miletic D, Affolderbach C, Hasegawa M, Boudot R, Gorecki C, Mileti G 2012 *Appl. Phys. B*, **109** 89
[16] Knappe S, Wynands R, Kitching J, Robinson H G, and Hollberg L 2001 *J. Opt. Soc. Am. B* **18** 1545
[17] Esnault F.X 2009 *PhD Thesis, Universite Pierre et Marie Curie*, Paris.

Pulsed Optically Pumped Rb clock

S Micalizio[1], F Levi[1], A Godone[1], C E Calosso[1], B François[1], R Boudot[2], C Affolderbach[3], S Kang[3], M Gharavipour[3], F Gruet[3], G Mileti[3]

[1]Physical Metrology Division, INRIM, Strada delle Cacce 91, 10135 Torino, Italy
[2]Time&Frequency Dpt, FEMTO-ST, CNRS, Université de Franche Comté, Besançon, France
[3]Laboratoire Temps-Fréquence, Université de Neuchâtel, Neuchâtel, Switzerland

Abstract. INRIM demonstrated a Rb vapour cell clock based on pulsed optical pumping (POP) with unprecedented frequency stability performances, both in the short and in the medium-long term period. In the frame of a EMRP project, we are developing a new clock based on the same POP principle but adopting solutions aimed at reducing the noise sources affecting the INRIM clock. At the same time, concerning possible technological applications, particular care are devoted in the project to reduce the size and the weight of the clock, still keeping the excellent stability of the INRIM clock. The paper resumes the main results of this activity.

Corresponding author: s.micalizio@inrim.it

1. Introduction

Development of high stability, low consumption, compact clocks is an important challenge for today's technology, both for space and for ground applications. Several different approaches have been pursued in the last years with interesting results on the short term stability [1]. The pulsed optically pumped (POP) Rb Clock with the possibility to separate in time optical pumping and microwave interrogation, is one of the most promising techniques. In fact, in the pulsed scheme, other than reducing by several orders of magnitude the light shift, it is possible to optimize both the microwave amplitude and the detection laser power to achieve the best signal-to-noise ratio. While several different approaches have demonstrated good short term stabilities in 10^{-13} range at 1s, to our knowledge the only approach that could tame at the same time also the medium-long term stability is the POP.

The first prototype of the POP was developed at INRIM under an ESA contract aimed at identifying possible candidates to replace the passive H-maser currently used for on board Galileo GNSS satellites. A short term stability $\sigma_y(\tau) = 1.7\times10^{-13}\ \tau^{-1/2}$ and a long term drift of 7×10^{-15}/day were obtained with this prototype [2]. Aiming to a further improvement of the clock performances, a new system is being developed taking in particular care the design of the microwave synthesis chain, the thermal stability and uniformity of the Rb cell. Another issue is the overall size of the clock; in this regard, a very compact laser head source has been developed and characterized [3] and a magnetron cavity [4] will be used in the physics package.

In this paper, we will discuss the major causes of noise that affect the short and long term stability of the laboratory prototype of the POP clock and how these effects will be mitigated in the new prototype we are developing.

Journal of Physics: Conference Series **723** (2016) 012015 doi:10.1088/1742-6596/723/1/012015

2. The POP Rb clock: laboratory prototype

The prototype developed at INRIM is based on a pulsed approach in which the clock transition is detected by observing Ramsey fringes on a laser absorption signal. The experimental set-up is represented in figure 1 and is composed of three parts: physics package, optics and electronics. A detailed description of the prototype can be found in [2].

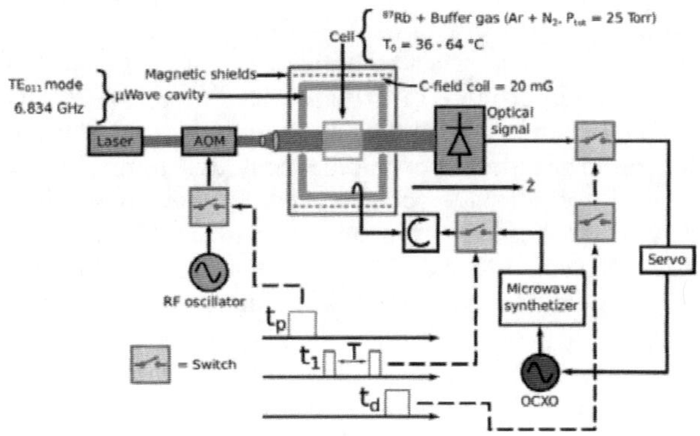

Figure 1. Schematic of the POP clock experimental setup; the timing sequence is also reported; t_p pumping time, t_1 microwave pulse duration, t_d detection time, T free evolution time.

The POP operation is composed of three phases. Initially, the atoms are pumped by a strong laser pulse which generates a large imbalance between the atomic populations of the two ground state hyperfine levels defining the clock transition. Then, the atomic sample experiences two separated microwave pulses according to the Ramsey interrogation scheme. Finally, the atoms that have made the clock transition are detected with a weak laser pulse.

With this approach, Ramsey fringes and frequency stability as those reported in figure 2 have been obtained. A H-maser has been used as a reference in the stability measurement.

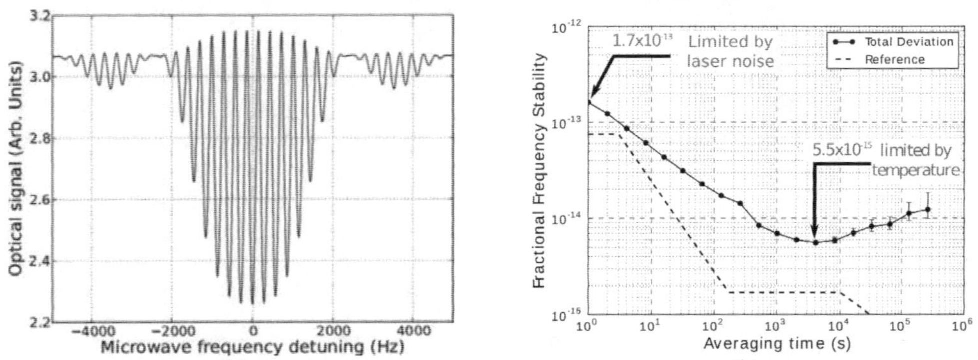

Figure 2. (a) Ramsey fringes detected in POP clock with optical detection; (b) frequency stability expressed in terms of Allan deviation. The main limiting factors are also reported.

The short-term stability is mainly limited by the transfer of laser noise to the atoms through the PM-AM conversion. The medium long term stability is instead limited by temperature fluctuations.

In next sections we report progress in the development of the new POP clock prototype.

3. Compact cell-cavity system

At the heart of the clock, a Rb vapour cell holds the atomic sample, and is in turn surrounded by a microwave cavity resonator. The microwave resonator serves for applying a well-controlled microwave field and at the same time acts as a thermal enclosure. Here we describe the cell-cavity system that is based on previous prototypes realized at UniNe-LTF [4] and is currently being assembled. Previous Rb cell realizations for the POP clock had evidenced enhanced temperature sensitivity (ETS) of the clock transition, due to the cells' stem serving as reservoir of atomic Rb [5]. Since the related temperature coefficient $\partial f_{clock}/\partial T_{stem}$ scales proportional to the stem volume, the ETS can be reduced by shrinking the cell stem. The cell used here (see figure 3a) has a diameter and length of 25 mm each, and is filled with ^{87}Rb and an Ar-N_2 buffer-gas mixture (pressure ratio P_{Ar}/P_{N2}=1.6). The cell' stem volume is reduced by more than two orders of magnitude compared to the cells used in [5], resulting in a similar reduction of the ETS. This allows to improve the clock stability and/or to relax the temperature stability requirements for the cell stem. In order to achieve a very compact clock realization, the microwave resonator used to excite the clock transition should add as little additional volume as possible to the cell itself. This is achieved here by a cavity design based on the loop-gap resonator approach, where in our case the resonant structure is formed by six metallic electrodes placed inside a cylindrical conducting shield [3]. This allows to reduce the internal volume of the cylindrical shield to V_{inner} = 28 cm^3, i.e. by more than a factor of three compared to a standard TE_{011} cavity used in previous studies. The outer dimensions of the resonator are 35 mm length by 40 mm diameter, giving an external volume of V_{outer} = 45 cm^3, for a cell volume of V_{cell} = 12 cm^3. The overall design view of this cavity is shown in figure 3b), two cut-off tubes are realize to allow maximum optical clearance whiles avoiding microwave leackage, finally an optical detector unit is directly attached to the cavity. The cavity resonates at the Rb clock transition frequency of ≈ 6.835 GHz, and has a low quality factor; Q≈200. This low Q factor does however not impeach obtaining high-quality Ramsey fringes and thus good clock short-term stability [6], but reduces clock instability contributions relevant on longer time scales arising from cavity pulling or aging of the cavity's resonance frequency.

a)

b)
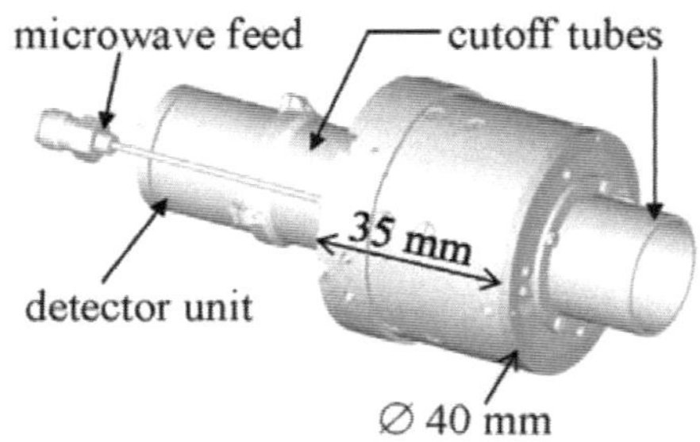

Figure 3. a) Photograph of the Rb vapour cell. The ruler shown indicates centimetres. b) 3D design view of the compact microwave cavity integrating the cell shown in sub-panel a). Both the cell and microwave cavity were designed and realized at UniNe-LTF.

4. Compact laser system

To achieve a compact optical laser system, a stabilized laser head was developed and realized at UniNe-LTF. This laser system includes all optical components required for the clock operation, and notably integrates an acousto-optical modulator (AOM) for fast switching of the optical output, and sub-Doppler spectroscopy optical circuit to lock the laser onto the Rb D_2 transitions observed in a low vapor cell. This laser system has a volume of only 1 litre, and is extensively described in [3].

5. Low-phase noise synthesis chain

In passive atomic clocks (and particularly in the pulsed ones), the interrogation is realized periodically and the control signal of the local oscillator (LO) is updated at equally spaced laps of time. The lack of information caused by the sampling process can lead to a limitation of the short term frequency stability. This effect is known as the intermodulation effect [7] and is coming from the down-conversion of the LO intrinsic frequency noise at Fourier frequencies higher than the interrogation frequency in the resonator bandwidth. This intermodulation effect is reported for pulsed system as the so-called Dick effect [8] and is a major limitation for high performance vapor cell atomic clocks. The contribution of Dick effect to the clock stability can be evaluated through the equation:

$$\sigma_y^{Dick}(\tau) = \left\{\sum_{k=1}^{\infty} sinc^2\left(k\pi \frac{T}{T_C}\right) S_y^{LO}(k\, f_C)\right\}^{1/2} \tau^{-1/2} \quad (1)$$

where $S_y^{LO}(k\, f_C)$ is the LO power spectral density of the microwave fractional frequency fluctuations and $f_C = 1/T_C$, T_C being the cycle time. Typically, in the POP operation $T_C \approx 4$ ms, so that the phase noise of the LO should be lower than –106 dBrad^2/Hz at $2\, f_C = 2/\, T_C = 460$ Hz to reject the short term frequency limitation of the Dick effect below the 10^{-13} level.
The local oscillator has been designed to be easily adapted also to Cs in a CPT clock. In this case, the synthesis chain deliver a low-phase noise signal at 4.596 GHz (half the Cs clock frequency) suited to drive an electro-optic modulator [9]. In the following parts, only the architecture and the results obtained for Rb synthesis chain are described.
The architecture of the synthesis chain for probing Rb-based vapor cell atomic clocks is reported on figure 4.

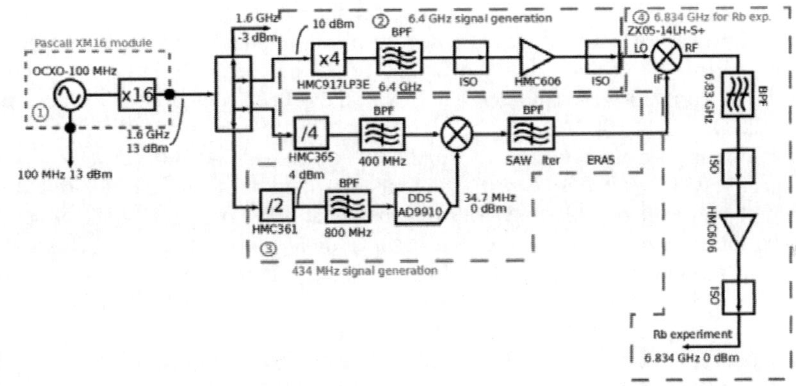

Figure 4. Architecture of the synthesis chain. Four main blocks are identified. 1: XM16 Pascall. 2: Frequency multiplication. 3: Frequency fine tuning. 4: Output signal of the synthesis chain.

The heart of the synthesis chain is the Pascall XM16 module that integrates a 100 MHz oven controlled quartz crystal oscillator (OCXOF-E-100) and its multiplication chain to 1.6 GHz free from excess noise, the latter is sent to a 4-ways power splitter. The 1.6 GHz signal is multiplied to 6.4 GHz by using active frequency multiplier (HMC917LP3E). The resulting signal is then bandpass filtered and amplified (HMC606) drives the LO port of the microwave mixer. The 1.6 GHz signal is also divided to 400 MHz (HMC365) and mixed with the 34 MHz output of the direct digital synthesizer (DDS). The 434 MHz is bandpass filtered by using a surface acoustic wave (SAW) filter. In the last arm, the 1.6 GHz is divided to 800 MHz (HMC361) to provide the clock to the DDS (AD9910). The 434 MHz signal drives the IF port of the microwave mixer. The resulting signal is bandpass filtered, amplified (HMC606), isolated and ready to be used. The DDS provides to the system the fine frequency tuning needed to probe the hyperfine atomic transition, particular attention is paid to the choice of the DDS so that its phase noise does not degrade the performances of the whole multiplication chain. Figure 5 reports the phase noise performances of the synthesis chain.

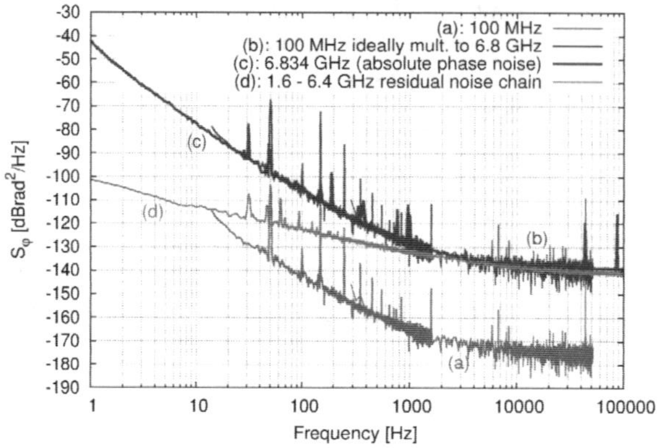

Figure 5. Phase noise performances of the synthesis chain and comparison with reference signals; (b) Dick effect contribution and limitation to the short term frequency.

Curve (a) reports the phase noise measurement of the 100 MHz signal. Particular attention must be paid while measuring the 100 MHz output of ultra low noise OCXOs (such as the Pascall) with a commercial cross-correlation measurement benches, because they do not have a low enough phase noise reference [10] to ensure the right measurement resulting in phase noise measurements exhibiting numerous bump and roll-off (a discrepancy of about 10 dB has been observed). A dedicated bench based on the cross-correlation techniques [11] has then been developed to ensure the right measurement of the OCXO at 100 MHz and at 1.6 GHz (see [12] for more information). The 100 MHz absolute phase noise is measured to be -143 and -174 dBrad^2/Hz at 100 Hz and 10 kHz Fourier frequencies respectively. Curve (c) is corresponds to the absolute phase noise of the 6.834 GHz output and is measured at the level of -105 and -138 dBrad^2/Hz at 100 Hz and 10 kHz offset frequencies respectively. The latter signal performances are then compared to curve (b) corresponding to the theoretical multiplication of the 100 MHz signal to 6.8 GHz, showing that no phase noise degradation occurs during the multiplication process. Curve (e) shows the residual phase noise of the synthesis chain and hence, corresponds to the ultimate phase noise performances achievable using this multiplication chain. Figure 6 reports the Dick effect evaluated according to equation (1) related to the multiplication of the 100 MHz OCXO to 6.34 GHz. On figure 6, the Dick effect is plotted as a function of the noise integration bandwidth. A floor is reached by σ_y once the integrated noise stops to contribute to the total Dick effect. The computation have been achieved by using fitted data. By using the XM16 based synthesis chain, the contribution of the Dick effect to the short term frequency stability has been evaluated at the level of 2×10^{-14}, a value close to the shot noise limit [2].

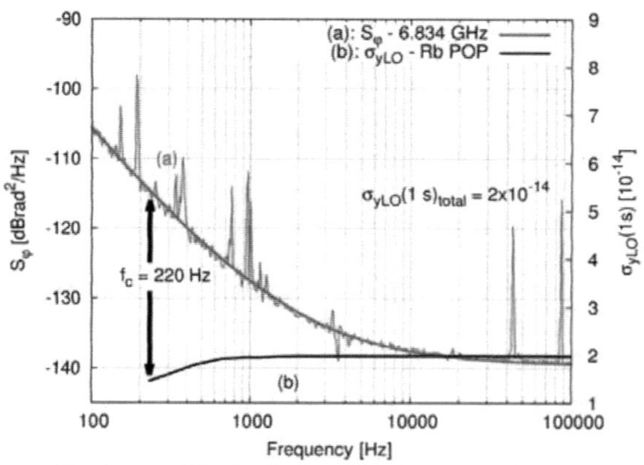

Figure 6. Dick effect contribution and limitation to the short term frequency.

6. Conclusions

The POP clock exhibits very interesting capabilities to be used in those applications where low power consumption, small size, reliability jointly to high frequency stability performances are required. A laboratory prototype has been already implemented and successfully tested at INRIM. The new standard we are going to realize aims at overcoming the limitations of the current prototype and, on the other hand, represents an intermediate step towards the implementation of a commercial device. A particular attention has then been paid to reduce the size, the weight and the power consumption, without losing in stability performances. A commercial prototype of the POP clock would be suitable in many different technological fields, but the features mentioned above make the POP clock particularly suited for space applications, such as GNSS.

Acknowledgments

This work has been funded by the EMRP project IND55 MClocks. EMRP is jointly funded by the EMRP participating countries within EURAMET and the European Union.

Work at LTF was also supported by the Swiss National Science Foundation (SNSF grant no. 140712), and previously by the European Space Agency (ESA) and the Swiss Space Office (Swiss Confederation). The LTF team thanks A.K. Skrivervik, C. Stefanucci, and A.E. Ivanov (EPFL-LEMA), M. Pellaton and T. Bandi for their contributions and collaboration on the cells and cavity.

[1] Godone A, Levi F, Calosso C E and Micalizio S 2015 *Rivista del Nuovo Cimento* **38** 133
[2] Micalizio S, Calosso C E, Godone A and Levi F 2012 *Metrologia* **49** 425
[3] Kang S, Gharavipour M, Gruet F, Affolderbach C and Mileti G 2015 *Proceedings of the joint International Frequency Control Symposium (IFCS) and European Frequency and Time Forum (EFTF),* Denver CO, USA, 800 – 803
[4] Stefanucci C, Bandi T, Merli F, Pellaton M, Affolderbach C, Mileti G and Skrivervik A K 2012 *Rev. Sci. Instrum.* **83** 104706
[5] Calosso C E, Godone A, Levi F and Micalizio S 2012 *IEEE Trans. Ultrason. Ferroel. Frequency Control* **59**, 2646 – 2654
[6] Kang S, Gharavipour M, Affolderbach C, Gruet F and Mileti G 2015 *J. Appl. Phys.* **117** 104510
[7] Kramer G 1974 *Digest of the Conf. on Prec. Electro. Meas.* 157-158
[8] Dick G J 1987 *Proc. Precise Time and Time interval* 133-147
[9] Abdel Hafiz M and Boudot R 2015 *Proc. of the IEEE EFTF-IFCS Conf. Joint Meeting* 71-7
[10] Nelson C W, Hati A and Howe D 2014 *Rev. Sci. Instr.* **85** 024705
[11] Rubiola E and Giordano V 2000 *Rev. Sci. Instr.* **71** 3085
[12] François B, Calosso C E, Abdel Hafiz M, Micalizio S and Boudot R 2015 *Rev. Sci. Instr.* **86** 94707

Self-generating magnetometer with laser pumping employment in "end resonance" wall coated vapor cell atomic clocks

A A Baranov, S V Ermak, R V Smolin and V V Semenov

Department of Quantum Electronics, Peter the Great St. Petersburg Polytechnic University, St. Petersburg, Russia

Vladimir_semenov@mail.ru

Abstract. This paper presents the results of two double resonance signals correlation investigation. These signals were observed synchronously in optically oriented Rb^{87} vapors with laser pumping in a dual scheme: low frequency M_x-magnetometer and microwave frequency discriminator. Analytical studies of the scalar and vector light shift components contribution to the frequency instability of the end resonance microwave transitions are presented. An experimental demonstration of the light shift components mutual compensation in optically pumped Rb^{87} atoms was provided. The results were processed in terms of Allan variance, which demonstrated an effect of decreasing frequency variation at averaging times more than 100 s for a joint scheme of the end resonance microwave transition and self-generating (M_x) magnetometer.

1. Introduction

The light shift of the energy sublevels caused by incident electromagnetic optical field is one of the major sources of instability of precision quantum devices such as alkali vapor cell frequency standards and quantum magnetometers with optical pumping.

The light shift theory developed in fundamental work [1] shows that frequency shift observed in experiments with alkali atoms vapors contains three components:

- The scalar shift, which is identical for all magnetic sublevels of a hyperfine structure F or F*.
- The vector light shift, caused by an effective magnetic field due to circular polarized light (an inverse Faraday Effect).
- Tensor light shift caused by ground state atoms alignment. It depends on a magnetic quantum number m_F.

Their weights are determined by pumping source spectrum and ground-state magneto-dipole transition type.

Scalar and tensor components are observed on a microwave transition between sublevels with different quantum number F. Vector and tensor components appear in quantum magnetometers working on Zeeman transitions between nearest sublevels in magnetic structure of a ground state in presence of a circular polarized light. The value of the tensor component as opposed to the scalar component depends on an angle θ between the C-field and the pumping light directions. When the tensor component is small enough to neglect, depending on the sign of the circular polarization of the pumping light (the direction of the effective magnetic field) the scalar and vector components of the light shift will either be folded or subtracted, which leads to the possibility of their mutual compensation, which in turn allows to reduce the influence of the pumping source on a quantum device frequency stability. It should be emphasized that such compensation is possible only for magnetically dependent transitions, whereas for microwave 0-0 transition used in atomic clocks vector component of the light shift is absent. In this paper we analyze the contribution of the vector and scalar components to the frequency shift of microwave magnetic dependent transitions, and provide an experimental demonstration of the compensation of these components with respect to the optical orientation of the Rb^{87} atoms with laser source tuned to the D_2 line of the doublet (figure 1a). D_2 line was chosen because of the opportunity to neglect the tensor component of the light shift, which in the case of the D_1-line pumping, plays an essential role [2].

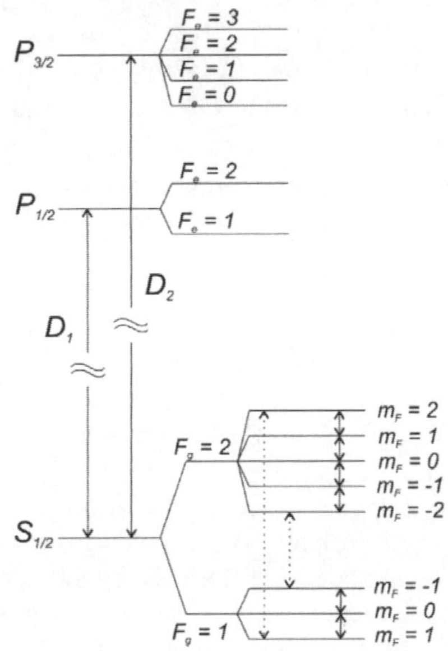

Figure 1a. Rb^{87} atom simplified energy levels system.

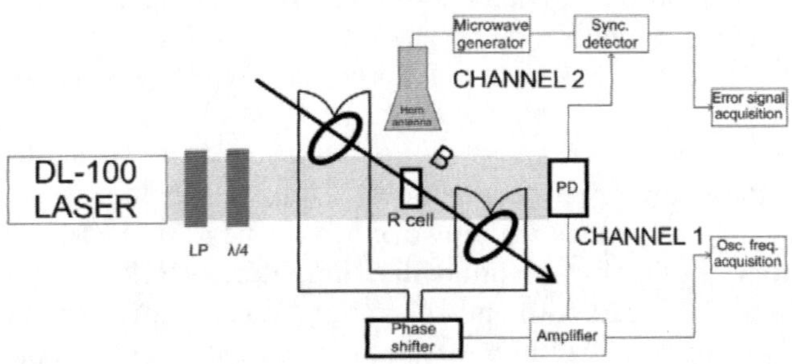

Figure 1b. Simplified experimental setup of the magnetometer joint scheme.

2. Theory

Analytical expressions for the components of the light shift for 0-0 microwave transition practiced in atomic frequency standards were obtained in [1]. In this paper, the analytical aspect was focused on the so-called "end resonance" transitions between hyperfine sublevels with extreme values of the magnetic quantum number [3]. Using the formalism [1], we can obtain the following expression for the tensor components of the light shift:

$$S_T = -\frac{2\pi}{hc}\left(1-3\cos^2\theta\right)\sum_{F_e}\sqrt{5}\left\{\frac{3m_F^2 - F_g\left(F_g+1\right)}{\left[\left(2F_g+3\right)\left(F_g+1\right)\left(2F_g+1\right)F_g\left(2F_g-1\right)\right]^{1/2}}\right\}\mathrm{Re}\,A^2\left(F_e,F_g\right)\left(-1\right)^{I+1/2-F_g}, \quad (1)$$

where

$$A^2(F_g,F_g)=\frac{3\lambda^2 r_0 f_{ge}}{4\pi^2}\left[\frac{Mc^2}{2RT}\right](2F_g+1)\sum_{F_g}(2F_e+1)W(11F_g,F_g;2F_e)W^2\left(J_eF_e\frac{1}{2}F_g;I1\right)Z(F_e,F_g)(-1)^{F_e-F_g-1} \quad (2)$$

$$Z(F_eF_g)\equiv Z\left[x(F_eF_g)+iy\right] \quad (3)$$

– form-factor of the Doppler broadened resonance line expressed in terms of plasma-dispersion function of the arguments x and y

$$x(F_eF_g)=\frac{1}{\nu}\left(\frac{Mc^2}{2RT}\right)^{1/2}\left\{\left[\nu-\nu_{cg}\right]-\left[\nu(F_eF_g)-\nu_{cg}\right]\right\} \quad (4)$$

$$y=\left(\frac{Mc^2}{2RT}\right)^{1/2}\left(\frac{1/2\tau+\gamma_c}{2\pi\nu}\right) \quad (5)$$

$W(11F_g,F_g;2F_e)$ - Racah coefficient, M – atomic mass, c– light velocity, R – gas constant, T – temperature, τ - excited state life time, γ_c – is determined by alkali atoms and buffer gas atoms colliding frequency, F_e and F_g – mechanical moments in the ground and excited states, f_{eg} – oscillator force, r_0 – classical electron radius, λ and ν pumping source wavelength and frequency, ν_{cg} – weighted frequency of electro-dipole transition.

S_T is a normalized to the pumping light intensity ground state frequency shift, which corresponds to the tensor component of the Light shift operator [1]. Equation (1) was derived from dipole matrix elements, which characterize the interaction between non-resonant optical field and Rb^{87} atoms. These elements were written as a product of Clebsch-Gordan coefficients and reduced matrix elements (Wigner-Eckart theorem) and further simplified with Racah-coefficients implementation. Final expression of the spectral response function (1) depends on an angle θ between magnetic field direction and light propagation, magnetic-dipole transition choice, absorption line profile and transition probability. Physically it is related to a transition resonance frequency shift caused by different transition probabilities for atoms with different magnetic quantum number projection.

Figure 2a and 2b demonstrate examples of Rb^{87} atoms spectral responses calculation under the influence of circularly polarized monochromatic radiation of the laser source tuned to the D_1 and D_2 line. The graphs show that the sign of the circular polarization of the pump light can significantly reduce the overall spectral response of an alkali atom due to the compensation of the scalar component of the light shift by the vector component.

3. Experiment

The light shift compensation effect was experimentally checked under the condition of the laser pumping of Rb^{87} by the long-wavelength component of the D_2 line in a dual scheme: self-oscillating M_x-magnetometer and end resonance microwave frequency standard. The experiments were performed on a laboratory setup of a quantum discriminator, which contains a well-known set of components [4] and a 1 cm^3 Rb^{87} cell with anti-relaxation coated walls. An experimental setup shown on Figure 1b combines two types of quantum magnetometers with a single optical trunk: the first is an M_x-magnetometer (self-oscillating magnetometer) and the second is a microwave "end resonance" magnetometer. The working frequency of the spin oscillator is determined by the weighted mean low frequency (Larmor frequency) of the magneto-dipole transitions between Zeeman sublevels in the Rb^{87} ground state structure. This frequency can be translated to magnetic field units using a well-known gyromagnetic equation $w=\gamma H$. Microwave frequency of the second magnetometer depends on the pumping light polarization choice and equals approx. 6.834 GHz. Error signal in this channel is used to determine the microwave generator and atomic transition frequency difference. As magnetic dependent "end resonance" transition was chosen, the error signal value is a measure of magnetic field variations. Resonance cell was placed in the center of the coils producing a working magnetic field (100 nT). To weaken the external magnetic interference and magnetic field inhomogeneity the whole system was placed in a two-layer magnetic shield with a shield factor 100. Microwave field, which induces transitions in the hyperfine structure of rubidium atoms was produced by frequency multiplication of the reference crystal oscillator followed by amplification of the signal in the microwave amplifier. A horn antenna was utilized to create a required microwave field distribution in the resonance cell.

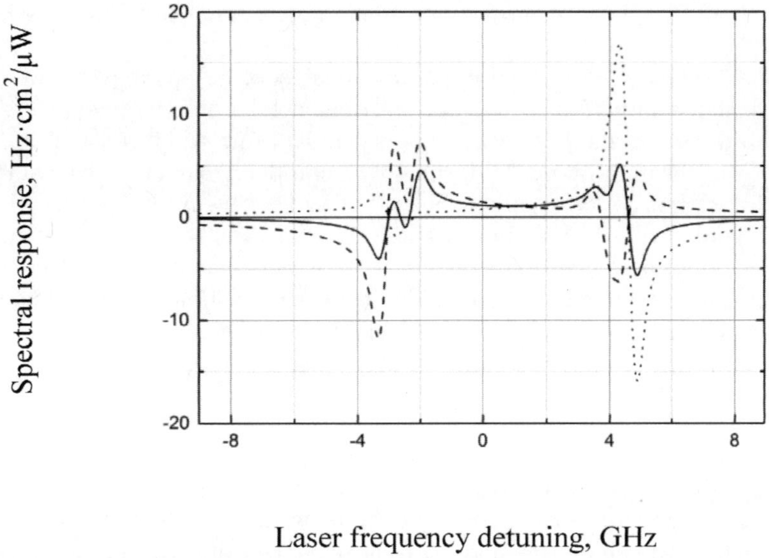

Figure 2a. Rb^{87} D_1-line spectral response: solid line – scalar component; dashed line – σ^- polarization (light shift components residual); dotted line - – σ^+polarization (light shift components superposition).

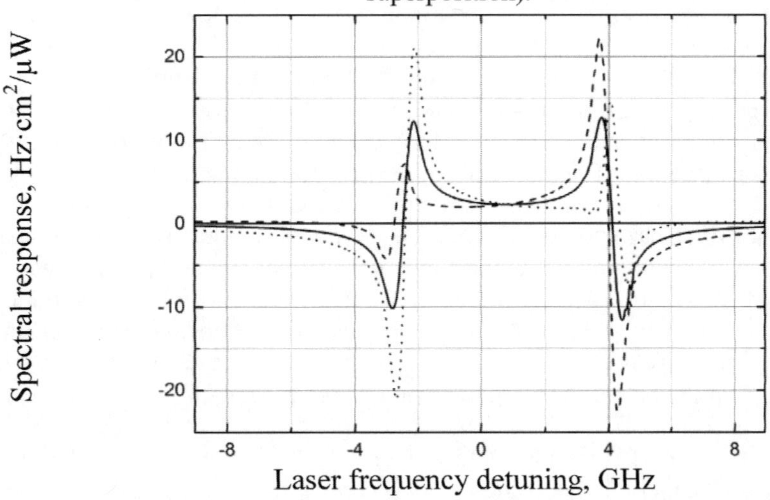

Figure 2b. Rb^{87} D_2-line spectral response: solid line – scalar component; dashed line – σ^- polarization (light shift components residual); dotted line - – σ^+polarization (light shift components superposition).

Along with the microwave signal a low frequency spin generator signal at Larmor precession frequency was induced at photodetector. The experiment showed significant interference of the LF and microwave signals, determined as changes in the dynamics of their intensity and spectral composition.

The difference between the synchronously recorded signals in a dual scheme normalized to uniform measurement units (nT) was processed to determine the Allan variance as a function of averaging time (an example of such a record is presented in figure 3).

The correlation coefficient between the detected signals in the dual scheme depends on the pumping rate and RF fields intensity in the vapor cell area. Depending on the intensity of the applied fields and the power of the spin generator self-oscillation correlation coefficient varied in a range of 0.3 - 0.98.

The optimal (the best long-term stability of the dual scheme) parameters of the fields were determined experimentally. In particular, for the wall coated cells optimal parameters in terms of the line broadening with respect to the microwave resonance dark line width (about 200 Hz) are: 20% - pumping light, 20% - of the microwave field, and 10% - spin RF generator. The last recommendation corresponds to half maximum of self-oscillations level in a spin system saturation mode.

Interference of microwave and LF signals appeared in a spin generator frequency shift due to microwave transition frequency variation which physics is not related to variations of the magnetic field. Such a shift is caused by coherence circulation between the sublevels of the hyperfine structure of alkali atoms "dressed" by LF field of the spin generator and depends on the sign of the circular polarization of the pumping light, the intensity of the applied field and magnetic dipole transition in the microwave absorption spectrum.

4. Conclusion

It was found that the absolute value of this shift for end resonance transition in the dual scheme is more than an order of magnitude smaller than in the analog tuned to 0-0 transition. In this case, for different pumping light polarization ($\sigma+$ and σ^-) frequency shift of the spin generator was observed with the opposite sign while increasing microwave field detuning from the resonance value.

The workability of the dual scheme was realized only when the laser is tuned to the long-wave component of the electric dipole transition, when the optical field interacts with the maximum number of atoms on the sublevels of the hyperfine structure. Tuning to the short-wavelength laser component of the D_2 line did not give a positive effect due to the breach of the self-excitation condition because of the low concentration of atoms caused by the effect of optical pumping.

Allan deviation plots show a relation between the Light shift of the resonance frequency, which depends on the circular polarization sign, and flicker floor value. SNR was controlled during experiment and its values were equal for left and right circular polarization which means that frequency drift was caused by AC Stark shift, otherwise, for example, in a case of residual magnetic fluctuations presence, flicker floor value should remain constant with polarization change.

As a consequence of scalar and vector components mutual compensation in a dual scheme, Allan variance decreases for hundreds of seconds averaging times in comparison with traditional 0-0 transition atomic clock (Figure 4). In this case, the flicker floor of Allan variance is realized in presence of $\sigma+$ pumping light at much higher averaging times than that observed in a conventional scheme. From the analysis of the effect of flicker processes on technical line width of the spin generator made in [5] follows that for large averaging times t technical line width (and therefore Allan variance) increases according to the law $(t^{n-1})^{1/2}$. Implementing this formula for the dual scheme case an estimated value of the coefficient n is: 2.15 for curve 1, 2.5 for curve 2 and 2.75 for curve 3.

The results allow to conclude that a greater long-term stability can be achieved in a dual scheme compared to traditional 0-0 transition scheme. The effect of the light shift components compensation by changing the sign of laser polarization provides an ability for a more detailed analysis of the pumping source role in the technical noise of a quantum device. In turn that expands the range of optimum designs of device's construction.

Dual scheme operating on a D_1 line can be found interesting to analyse because of sufficient weight of the tensor component of the light shift, which depends on the orientation of the magnetic field with respect to the optical axis [6]. A decrease in the orientation error of atomic devices is expected due to high degree of correlation of the light shift at microwave and Zeeman transitions, implemented in the dual scheme.

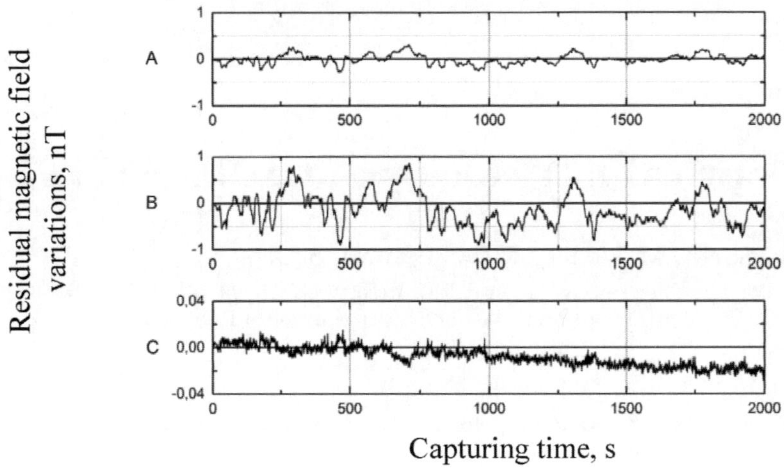

Figure 3. Dual scheme signals capturing: A – low frequency; B – microwave transition; C – residual noise of a dual scheme.

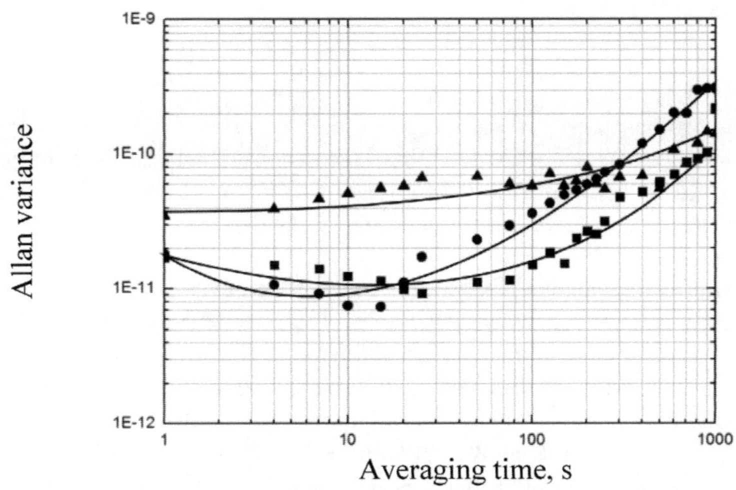

Figure 4. Allan variance for different pumping schemes: rounds (1) – 0-0 transition; triangles(3) – dual scheme (σ+ , optimal light); squares (2) – dual scheme (σ+ , non-optimal light, high quality factor)

References

[1] Happer W, Mathur B S 1967 *Phys. Rev.* **1**, v163.
[2] Semenov V V 1999 *Izv. Vuzov* Physics, **2**.
[3] Jau Y-Y, Post A B, Kuzma N N, Braun A M, et al. 2004 *Phys. Rev. Letters* **11**, 92.
[4] Rihle F 2004 *Frequency Standards: Basics and applications* (Weinheim: WILEY-VCH Verlag GmbH & Co. KGaA).
[5] Semenov V V, Zimnitsii P V, Smolin R V and Ermak S V 2014 *Technical Physics Letters* **3** 40 271-273.
[6] Baranov A A, Ermak S V and Semenov V V 2010 *Nauchno-technicheskie vedomosti SPbSTU* **3** 104.

A mercury optical lattice clock at LNE-SYRTE

L De Sarlo, M Favier, R Tyumenev, S Bize

LNE-SYRTE, Observatoire de Paris, PSL Research University, CNRS, Sorbonne Universités, UPMC Univ. Paris 06, 61 avenue de l'Observatoire, 75014 Paris, France

E-mail: `sebastien.bize@obspm.fr`

Abstract. We describe the development of an optical lattice clock based on mercury and the results obtained since the 7th SFSM. We briefly present a new solution for the cooling laser system and an improved lattice trap that allows us to interrogate a few thousand atoms in parallel. This translates into a fractional short term stability of 1.2×10^{-15} at the clock frequency of 1.129 PHz.

1. Introduction

Optical lattice clocks are based on the interrogation on a sample of cold atoms trapped in a laser standing wave at a suitable wavelength [1]. Thanks to the intrinsically low quantum projection noise associated to the interrogation of multiple particles, these devices are establishing themselves as a stability record breaking frequency standards. Leveraging on this unprecedented frequency stability, optical lattice clocks with an accuracy in the range of 10^{-18} are demonstrated in this conference and no limitation for their ultimate accuracy has been identified yet.

Among the atomic species employed in optical lattice clocks, mercury is particularly interesting because of a few remarkable properties. Mercury (Z=80) has an electronic structure dominated by two electrons in the 7s shell. This gives rise to a alkaline-earth-like structure, similar to that of Sr and Yb, the two other atomic species currently employed in optical lattice clocks. A level scheme of mercury is shown in the inset of figure 1.

Indeed, the clock transition in mercury has the same $^1S_0 \Rightarrow {}^3P_0$ character as in Sr and Yb. As in those atoms, the transition is strictly forbidden at zero magnetic field for bosonic isotopes (196,198,200,202) and only weakly allowed for the two fermionic isotopes (199,201) due to hyperfine mixing. The heavy atomic weight makes the $^1S_0 \Rightarrow {}^3P_1$ transition weakly allowed due to relativistic effects with a linewidth of around 1.2 MHz. This constitutes a first favorable property of mercury with respect to Sr and Yb in that such a linewidth allows for a simple laser cooling experimental scheme where atoms are directly loaded from the background pressure down to a temperature of about 30μK using a single MOT requiring only one laser source (see 3 below).

In addition to a level structure suitable for operating an optical lattice clock, mercury has an important advantage as far as the shift induced by thermal radiation is concerned. According to theoretical calculations of atomic static polarizabilities, the relative shift induced by thermal radiation at 300 K on the mercury clock transition is estimated to be -1.6×10^{-16} [2], 33 (16) times smaller than the corresponding shift for Sr (Yb). Note that the same scaling applies for differential DC Stark shift which can be introduced for instance by patch charges on the vacuum windows. A third advantage specific to the isotope ^{199}Hg is that, as ^{171}Yb but unlike ^{87}Sr, it has a total spin of 1/2. This means that, by symmetry, the tensor component of the lattice light shift vanishes.

On a broader scope, mercury atom is of high interest for high precision atomic physics with application to fundamental physics tests including the search for variations of fundamental constants [3, 4, 5].

However, in order to exploit the full potential of the mercury atom, one challenge has to be overcome: all the required laser sources are in the deep UV part of the spectrum. Reliable and powerful laser sources at the wavelengths of 253.7 and 362.6 nm for respectively cooling and trapping, is therefore the main enabling technology for a mercury optical lattice clock.

In this paper we briefly describe the status of the mercury optical lattice clock at LNE-SYRTE first by recalling the results obtained in our group since the last Symposium on Frequency Standards and Metrology and then presenting the latest improvements on the experimental setup and on the performances of the clock.

2. Feasibility of a Hg optical lattice clock

At the 7th Symposium on Frequency Standards and Metrology, we reported some initial steps toward the operation of an optical lattice clock using mercury [6]. We reported the operation of a tandem of 2D and 3D magneto-optical trap, the implementation of a high performance clock laser system [7]. With these, we were able to perform the first laser spectroscopy of the clock transition on free-falling atoms released from the MOT. This yield absolute frequency measurements of the clock transition in ^{199}Hg and ^{201}Hg with accuracies of a few kHz [8]. Since then, we implemented a dipole lattice trap near the predicted magic wavelength and demonstrated optical trapping of mercury. We performed the first lattice-bound spectroscopy of the ^{199}Hg $^1S_0 - {}^3P_0$ optical clock transition and the first experimental determination of the ^{199}Hg magic wavelength : 362.5697 ± 0.0011 nm [9, 10]. Finally, we performed an absolute frequency measurement of the ^{199}Hg clock transition frequency using lattice trapped atoms obtaining a fractional uncertainty of 5.7×10^{-15} improving our previous measurement by 3 orders of magnitude [11, 12]. The obtained value of 1128 575 290 808 155.1 Hz ±6.4(syst.) ± 0.3(stat) Hz is confirmed by the more accurate measurement reported three years later by the Katori's group [13]. This work clearly demonstrated the feasibility of a Hg lattice clock despite the considerable challenge due to deep UV laser sources.

3. An upgraded mercury lattice clock

After these first results we spent a significant amount of time improving the experimental setup in two critical subsystems: a new lattice trap allowing for a deeper lattice that will be described in Section 4 and a new cooling laser source that allows for a much more reliable operation of the clock that is described in the following. A detailed description of the other subsystems (e.g. vacuum system, clock laser) can be found in [7, 14]. A general overview of the mercury optical lattice clock setup is depicted in figure 1. Our new cooling laser source is based on a commercial fiber amplifier that amplifies the light of a custom-made extended cavity diode laser up to a power of around 12 W. This light is then frequency doubled in single-pass by focusing the output of the amplifier in a ppLN crystal and then further doubled by intracavity type I phase matching in a crystal of BBO enclosed in a bow-tie enhancement cavity.

The operating point of this second doubling stage requires a careful compromise between conversion efficiency and reliability. For an input power of 1.5 W at the wavelength of 508 nm the best efficiency we have achieved allows us in principle to obtain an output power of ~ 200 mW at 254 nm ($\eta \simeq 0.09\,\text{W}^{-1}$). As a matter of fact at this level of output power the operating lifetime of the BBO crystals is only a few days with maintainance operations required as often as every few hours. Experimentally we found out that decreasing the peak intracavity intensity at 508 nm to obtain only 70 mW of output power at 254 nm for an input power of 1.5 W ($\eta \simeq 0.03\,\text{W}^{-1}$),

Journal of Physics: Conference Series **723** (2016) 012017 doi:10.1088/1742-6596/723/1/012017

increases the cooling laser uptime to a few weeks with crystal lifetimes typically exceeding several months. This level of power is enough to operate the MOT at the lowest observed temperatures (30 μK) with a loading rate of about $\sim 1.3 \times 10^6$ atoms/s.

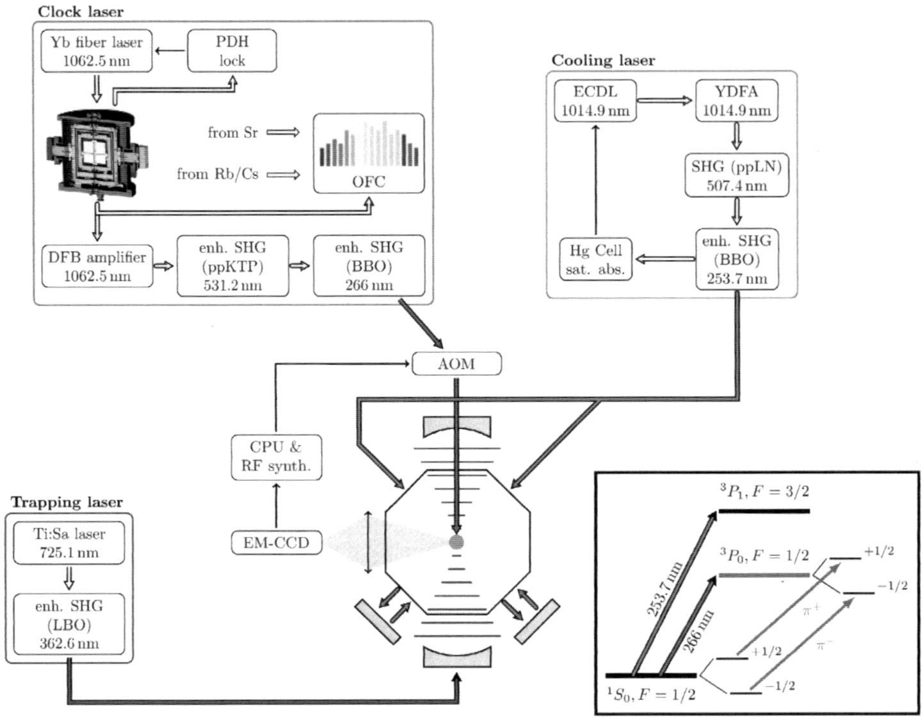

Figure 1: (Colour online) Schematics of the LNE-SYRTE's mercury optical lattice clock. One pair of MOT beams and the MOT coils are not shown. The inset shows a level scheme of mercury including the detail of the two Zeeman components of the clock transition.

4. A new lattice trap

The main limitation of the previous setup was the modest lattice trap depth of not more than 22 E_{rec}, where the recoil energy E_{rec} is the kinetic energy of a mercury atom with the momentum of a lattice photon. This limited trap depth gives poor collection efficiency of atoms from the MOT into the optical lattice (typically $> 0.5\%$) and limited confinement for the lattice-bound excitation of the clock transition [9]. The loading process is governed by two parameters: the trapping volume, and depth. Under the constraint of almost fixed laser power imposed by the damage threshold of the lattice optics a compromise shall be found between these two parameters in the form of an optimal value of the lattice beam waist [15]. Unfortunately this optimization is not easy on our experimental setup since the lattice standing wave is generated by a Fabry-Perot cavity whose mirror are also acting as vacuum windows (see figure 2a) and therefore changing the mirrors requires breaking the vacuum.

The initial size of the cavity mode was fixed based on estimation of the MOT size and temperature acquired in the earliest stage of the experiment. Based on the actual parameters of the MOT we decided to implement a new trap with a waist roughly two times smaller. This would maintain a good overlap between the MOT and the lattice while at the same time increasing the trap depth. Changing from a 250 mm radius of curvature to 150 mm, we decreased the waist from 120 μm to 69 μm expecting an improvement in the lattice depth by a factor of 3.

Journal of Physics: Conference Series **723** (2016) 012017 doi:10.1088/1742-6596/723/1/012017

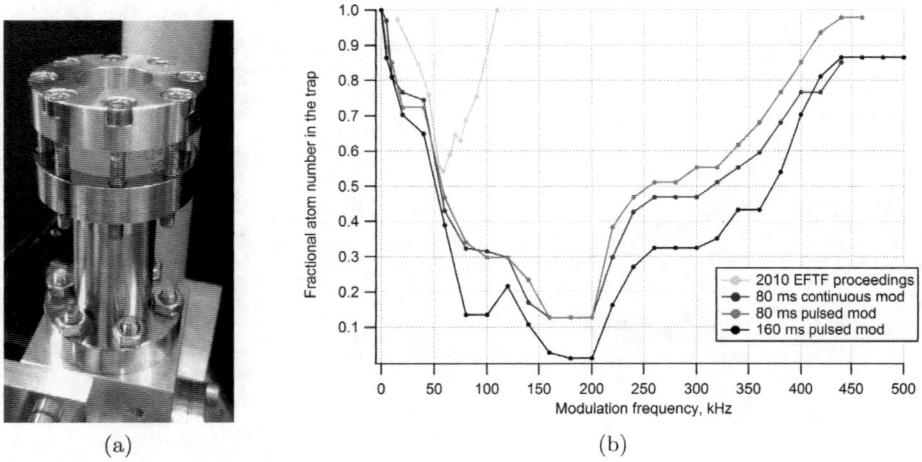

Figure 2: (Colour online) *(a)* Picture of the new lattice mirrors mounted on the vacuum system. *(b)* Parametric excitation in the lattice.

Another improvement to the lattice subsystem was to servo the wavelength of the lattice to a commercial wavemeter with an accuracy of a few MHz. The wavemeter is periodically calibrated using the 698 nm clock light of SYRTE's Sr optical lattice clock.

In order to characterise atoms in the new trap we performed a parametric excitation experiment [16, 17, 18]. After loading atoms in the lattice, we modulate for a given time the power of the lattice at a frequency ν_m and observe the number of atoms remaining in the trap. When the parametric resonance condition $\nu_m = 2\nu_{\text{trap}}$ is satisfied, the modulation excites the atomic sample inducing a loss of atoms. The results of the procedure are shown in figure 2b together with the results of the same procedure on the old lattice cavity normalized in such a way that the unperturbed signal is equal for the two curves. From the plots shown the increase in trapping oscillation frequency (and hence in lattice depth) is apparent: based on the peak at 180 kHz we infer a lattice depth of 56 recoils, consistent with the improvement expected from the waist reduction. We also observe a structure in the loss spectroscopy compatible with peaks at $\nu_{\text{trap}} \simeq 90$ kHz, $3\nu_{\text{trap}}$ and $4\nu_{\text{trap}}$ that we tentatively attribute to the trapping anharmonicity and to the population distribution in the radial vibrational states. We emphasize that this topic definitely requires further investigation.

5. Improved spectroscopy and short term stability

The improvements detailed in the previous section allowed us to perform high resolution spectroscopy as shown in figure 3a. These data are obtained using the following typical experimental cycle. Starting from a mercury vapor at room temperature, we cool and trap $\sim 10^6$ ^{199}Hg atoms at a temperature of $\sim 30\,\mu$K in a 3D-MOT in the presence of the lattice light at a typical depth of $U \sim 56E_{\text{rec}} \simeq h \times 0.43\,\text{MHz} \simeq k_B \times 20\,\mu$K. Following a compressed MOT phase the cooling light is turned off leaving the atoms in the lattice. The overall capture efficiency is around 2%, improved by a factor 10 with respect to the previous configuration, but still limited by the fraction of MOT atoms having an energy below the lattice depth. The typical duration of this capture phase is about 0.7 s.

Before applying a standard rectangular Rabi pulse, we perform initial state selection with a simple procedure that consist in two light pulses. A first short (15 ms) π-pulse of clock light tuned to the Zeeman component that is to be selected, transfers roughly half of all atoms to the

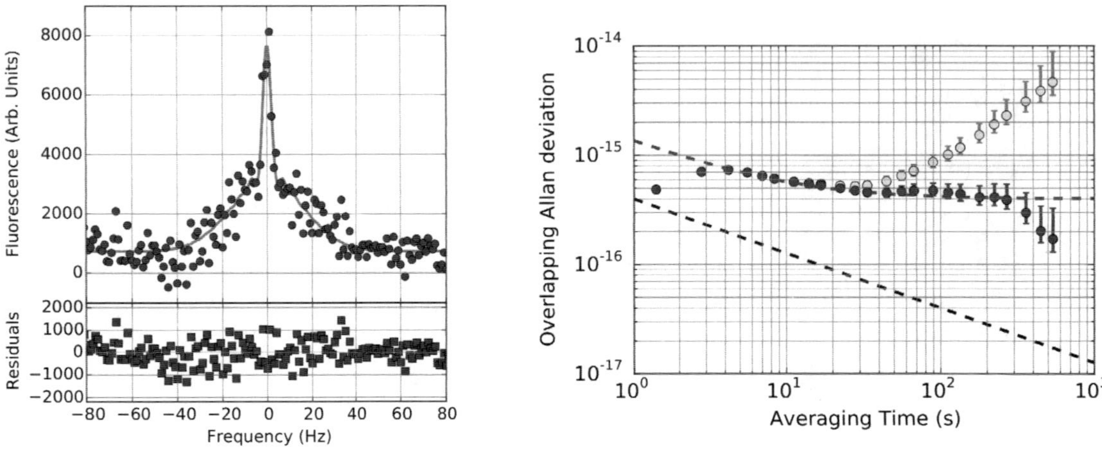

Figure 3: (Colour online) *(a)* Typical spectroscopy signal for one of the π components of the ^{199}Hg clock transition. Here Rabi time is 0.2 s and the fit function is the sum of a Rabi profile (fitted width 4.2 Hz) on a gaussian background. *(b)* Overlapping Allan deviation of the fractional frequency difference between the mercury clock laser and the mercury clock transition at 1.128 PHz in a closed integrator loop (gray) and dedrifted data (blue). Noise is modeled as the sum of the white noise of the clock and the flicker of the cavity (blue dashed line). The black dashed line corresponds to the Dick limit for short term stability.

excited clock state. A second pulse of light resonant with the cooling transition then removes atoms that are left in the ground state.

Following this procedure we perform spectroscopy on the clock transition by applying a Rabi pulse that is tuned to transfer at resonance all the atoms from the excited clock state to the ground state. We then detect the transferred atoms by collecting fluorescence of $^1S_0$ atoms from a molasses made by the four MOT beams depicted in figure 1, slightly ($\sim$ 2MHz $\sim 1.5\Gamma$) detuned from the cooling resonance.

The data shown in the figure 3a are taken for a Rabi time of 0.2 s. The fit to a Rabi model is consistent with the expected linewidth of 4.2 Hz and a gaussian background attributed to the flicker noise of the clock laser [19]. Note that, since the clock frequency is 1.13 PHz, this corresponds to an atomic quality factor of $Q \simeq 3.7 \times 10^{14}$ roughly the same as in the Sr case for a linewidth of 1.7 Hz. The peak signal to noise ratio is about 15.

When we operate in clock mode, we use the collected fluorescence to correct the frequency of the clock laser at each cycle. The equivalent detection noise is about 40 atoms but the spectroscopic signal to noise ratio is limited by shot to shot fluctuations in the total atom number typically one order of magnitude bigger. We note that we plan to negate the effect of this kind of fluctuations on the frequency discrimination by introducing a normalized detection.

We plot the overlapping Allan deviation of the relative frequency difference between the mercury clock laser and the mercury clock transition in a closed integrator loop in figure 3b for the typical Rabi time of 0.1 s. As we can see the data are well reproduced by a noise model comprising three terms : white frequency noise determined by the atomic resonator for averaging times between 2 and 8 seconds, flicker noise in frequency corresponding to the noise floor of our clock laser between 10 and 30 seconds, and finally a residual uncompensated frequency drift for longer averaging times.

By adjusting these noise form to the data we are able to infer the short term stability of our atomic frequency discriminator which is $1.2(1) \times 10^{-15}$ at 1s (almost a factor 5 improvement over our previous value [20]) and the noise floor of our clock laser which is $3.5(5) \times 10^{-16}$, in good agreement with the value inferred from the comparison of two similar laser [7]. As shown in figure 3b, the estimated stability limit from the Dick effect, as obtained from the parameters of reference [7], is significantly lower. Stabilities in the mid 10^{-16} range will be within reach when the detection noise will be lowered.

6. Latest developments and outlook

With the improvements reported in previous sections, we are characterizing systematic frequency shifts of the ^{199}Hg optical lattice clocks with individual uncertainties in the 10^{-17} range. Once completed, this work will allow us to perform optical to optical frequency ratio measurements against SYRTE's Sr lattice clocks [21] and to compared the result with the recent value of [13]. It will also yield refined measurements against Cs (and Rb) fountain clocks [22].

One of the advantages of Hg is the possibility to implement and use a high flux cold atom source based on a 2D-MOT loaded from thermal vapor. This possibility was so far underexploited due to the difficulty to obtain reliable laser sources at the cooling wavelength of 254 nm. The advent of fiber-based laser technology for this source which was described in section 3 has removed this obstacle. In the future, we hope to exploit this possibility to obtain stabilities at the 1×10^{-16} level at one second and to explore the limits of Hg optical lattice clocks which we expect to be below 1×10^{-18}.

Acknowledgments

We acknowledge the large number of contributions of SYRTE technical services. This work was supported by EMRP/JRP ITOC and by ERC Consolidator Grant AdOC. We acknowledge funding support from Centre National d'Études Spatiales (CNES), Conseil Régional Île-de-France (DIM Nano'K).

References

[1] Katori H, Takamoto M, Pal'Chikov V G and Ovsiannikov V D 2003 *Physical Review Letters* **91** 173005
[2] Hachisu H, Miyagishi K, Porsev S G, Derevianko A, Ovsiannikov V D, Pal'chikov V G, Takamoto M and Katori H 2008 *Phys. Rev. Lett.* **100**(5) 053001
[3] Angstmann E J, Dzuba V A and Flambaum V V 2004 *Phys. Rev. A* **70** 014102
[4] Tobar M E, Stanwix P L, McFerran J J, Guéna J, Abgrall M, Bize S, Clairon A, Laurent P, Rosenbusch P, Rovera D and Santarelli G 2013 *Phys. Rev. D* **87**(12) 122004
[5] Huntemann N, Lipphardt B, Tamm C, Gerginov V, Weyers S and Peik E 2014 *Phys. Rev. Lett.* **113**(21) 210802
[6] Petersen M, Millo J, Magalhaes D V, Mandache C, Dawkins S T, Chicireanu R, Lecoq Y, Guena J, Chapelet F, Rosenbusch P, Laurent P, Abgrall M, Rovera G D, Santarelli G, Clairon A, Bize S and Tobar M E 2009 *PROCEEDINGS OF THE 7TH SYMPOSIUM FREQUENCY STANDARDS AND METROLOGY* ed Maleki, L (WORLD SCIENTIFIC PUBL) pp 82–90
[7] Dawkins S T, Chicireanu R, Petersen M, Millo J, Magalhães D V, Mandache C, Le Coq Y and Bize S 2010 *Applied Physics B: Lasers and Optics* **99** 41–46
[8] Petersen M, Chicireanu R, Dawkins S T, Magalhães D V, Mandache C, Le Coq Y, Clairon A and Bize S 2008 *Physical Review Letters* **101** 183004 (*Preprint* 0807.3408)
[9] Yi L, Mejri S, McFerran J J, Le Coq Y and Bize S 2011 *Phys. Rev. Lett.* **106** 073005
[10] Mejri S, McFerran J J, Yi L, Le Coq Y and Bize S 2011 *Phys. Rev. A* **84** 032507
[11] McFerran J J, Yi L, Mejri S, Di Manno S, Zhang W, Guéna J, Le Coq Y and Bize S 2012 *Phys. Rev. Lett.* **108**(18) 183004
[12] McFerran J J, Yi L, Mejri S, Di Manno S, Zhang W, Guéna J, Le Coq Y and Bize S 2015 *Phys. Rev. Lett.* **115**(21) 219901
[13] Yamanaka K, Ohmae N, Ushijima I, Takamoto M and Katori H 2015 *Physical Review Letters* **114** 230801
[14] McFerran J J, Yi L, Mejri S, Zhang W, Di Manno S, Abgrall M, Guéna J, Le Coq Y and Bize S 2014 *Phys. Rev. A* **89**(4) 043432
[15] Kuppens S J M, Corwin K L, Miller K W, Chupp T E and Wieman C E 2000 *Phys. Rev. A* **62**(1) 013406

[16] Gehm M E, O'Hara K M, Savard T A and Thomas J E 1998 *Phys. Rev. A* **58**(5) 3914–3921
[17] Gehm M E, O'Hara K M, Savard T A and Thomas J E 2000 *Phys. Rev. A* **61**(2) 029902
[18] Wu J, Newell R, Hausmann M, Vieira D J and Zhao X 2006 *Journal of Applied Physics* **100** 054903
[19] Di Domenico G, Schilt S and Thomann P 2010 *Appl. Opt.* **49** 4801–4807
[20] McFerran J J, Magalhaes D V, Mandache C, Millo J, Zhang W, Le Coq Y, Santarelli G and Bize S 2012 *OPTICS LETTERS* **37** 3477–3479
[21] Le Targat R, Lorini L, Le Coq Y, Zawada M, Guéna J, Abgrall M, Gurov M, Rosenbusch P, Rovera D G, Nagórny B, Gartman R, Westergaard P G, Tobar M E, Lours M, Santarelli G, Clairon A, Bize S, Laurent P, Lemonde P and Lodewyck J 2013 *Nature Communications* **4** 2109 (*Preprint* 1301.6046)
[22] Guena J, Abgrall M, Rovera D, Laurent P, Chupin B, Lours M, Santarelli G, Rosenbusch P, Tobar M, Li R, Gibble K, Clairon A and Bize S 2012 *Ultrasonics, Ferroelectrics, and Frequency Control, IEEE Transactions on* **59** 391–409 ISSN 0885-3010

Journal of Physics: Conference Series **723** (2016) 012018 doi:10.1088/1742-6596/723/1/012018

Sr^+ single-ion clock

P Dubé[1], A A Madej[1,2] and B Jian[1,2]

[1] National Research Council Canada, Ottawa, Ontario, Canada K1A 0R6

[2] Department of Physics and Astronomy, York University, Toronto, Ontario, Canada M3J 1P3

E-mail: pierre.dube@nrc-cnrc.gc.ca

Abstract. The evaluated uncertainty of the $^{88}Sr^+$ ion optical clock has decreased by several orders of magnitude during the last 15 years, currently reaching a level of 1.2×10^{-17}. In this paper, we review the methods developed to control very effectively the largest frequency shifts that once were the main sources of uncertainty for the $^{88}Sr^+$ single-ion clock. These shifts are the micromotion shifts, the electric quadrupole shift and the blackbody radiation shift. With further improvements to the evaluation of the systematic shifts, especially the blackbody radiation shift, it is expected that the total uncertainty of the single-ion clock transition frequency will reach the low 10^{-18} level in the near future.

1. Introduction

The progress in the control and evaluation of the frequency shifts in optical clocks has been very rapid during the last decade. The reduction of uncertainties has often times been more than incremental, resulting from an understanding of the underlying physics that has led to breakthroughs in the control of the shifts. A number of optical atomic clocks based on neutral atoms in optical lattices or single ions in rf traps now have total evaluated uncertainties reaching the 10^{-17} to the 10^{-18} level [1, and references therein]. As time passes, more optical clocks are being reported with improved control of the systematic shifts at these levels of uncertainty.

In this paper we present an overview of the methods used to control the main systematic shifts of the $^{88}Sr^+$ ion that have allowed a reduction by four orders of magnitude of the total evaluated frequency uncertainty in the last 15 years [2–4]. The key methods are the minimization of micromotion, the electric quadrupole shift cancellation by Zeeman averaging of the magnetic sublevel energies [3, 5], and a high-accuracy measurement of the differential scalar polarizability $\Delta\alpha_0$ of the clock transition [4]. The latter allows a further reduction of the total micromotion shifts by more than two orders of magnitude and is an essential parameter in the evaluation of the blackbody radiation shift. An overview of the uncertainty budget is given to illustrate the relative contributions of the main shifts and the potential for the $^{88}Sr^+$ ion to realize the unperturbed clock transition frequency with a total uncertainty in the low 10^{-18} level.

2. $^{88}Sr^+$ ion clock

The optical frequency reference is realized by first trapping and cooling a single ion of $^{88}Sr^+$ in an rf quadrupole trap. As shown in figure 1, cooling is performed on the $5s\,^2S_{1/2} - 5p\,^2P_{1/2}$ transition at 422 nm. A repumper laser at 1092 nm prevents decay from the $^2P_{1/2}$ state to the metastable $^2D_{3/2}$ state during the cooling pulse.

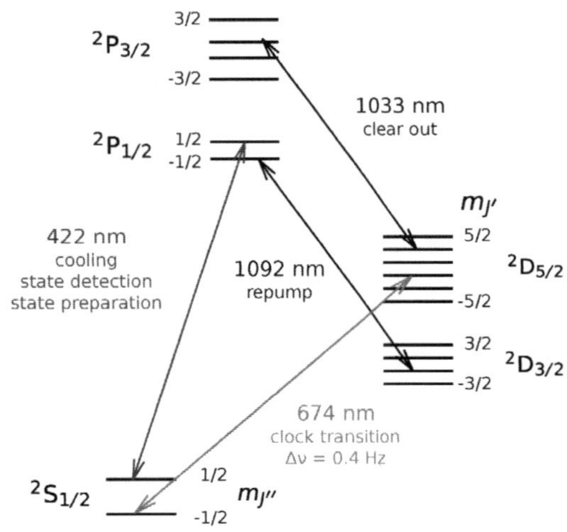

Figure 1. (color online) Partial energy level diagram of the $^{88}Sr^+$ ion.

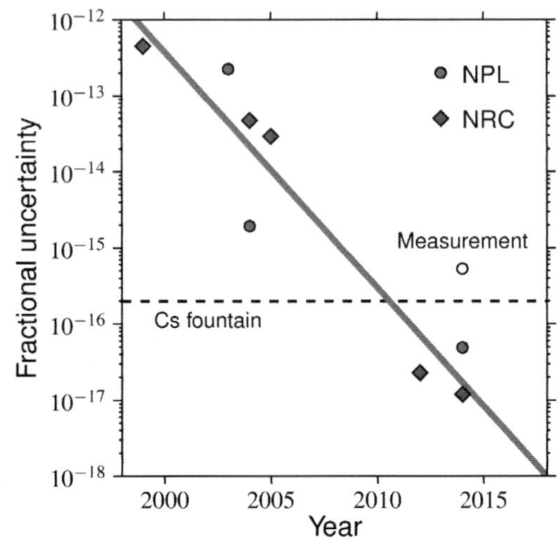

Figure 2. (color online) Evolution of the $^{88}Sr^+$ fractional frequency uncertainty.

The ion is typically cooled to temperatures of $\approx$ 2 mK in our setup [3, 6]. The reference frequency is realized by the 445 THz $5s\,^2S_{1/2} - 4d\,^2D_{5/2}$ electric-quadrupole transition. It has a natural linewidth of 0.4 Hz determined by the 0.4 s lifetime of the $^2D_{5/2}$ state [7]. A small magnetic field of $\approx 1\ \mu$T is applied to the ion to split the S–D spectrum into ten resolved Zeeman components and define the quantization axis for optical pumping. There are no first-order magnetic insensitive resonances in the spectrum. Any of the five pairs of symmetric Zeeman components can be probed to obtain the virtual line center of the clock transition [8]. As described below, the experiments are performed by measuring and averaging the line centers of six Zeeman components to remove other frequency shifts.

The narrowest linewidth observed in our system is 4.4(3) Hz for 200 ms pulse lengths [3]. This was measured by recording scans that took $\approx$ 100 s to complete. For these long measurement times, the linewidth is partially broadened by laser and linear Zeeman shift frequency noise. When locking the laser to the clock transition, the update time for a single Zeeman component line center is comparatively much shorter, typically 8 s, which reduces the frequency noise contribution to the linewidth. In addition, the pulse duration is decreased to 100 ms to allow for a faster servo response. The resulting Fourier-transform-limited linewidth is $\approx$ 8 Hz.

A 1033 nm laser is used to transfer the ion from the metastable $^2D_{5/2}$ state to the short-lived $^2P_{3/2}$ state after the ion state has been determined from the 422 nm fluorescence levels. This action quickly returns the ion to the ground state in preparation for the final optical pumping step with circularly polarized 422 nm light. State-preparation ensures that the ion interacts with the probe laser light at each cycle. The stability obtained is $3 \times 10^{-15}/\sqrt{\tau}$ for 100 ms pulses [9].

Amplified spontaneous emission (ASE) light sources at 1033 nm and 1092 nm were developed recently as alternatives to their laser counterparts. In addition to providing a very compact package with no need for frequency stabilization and external shuttering, the 1092 nm broadband source has yielded lower ion kinetic temperatures than the laser source under the same experimental conditions [6].

3. Progress in evaluating the $^{88}Sr^+$ clock transition uncertainty

Figure 2 is a plot of the fractional frequency uncertainty of the $^{88}Sr^+$ ion as a function of year of publication [2–5, 10–14]. The first data point (year 1999) in figure 2 is the uncertainty of the first and only Cs-based frequency synthesis chain measurement of $^{88}Sr^+$ [10]. The other data points report the evaluated uncertainty or the reproducibility of the clock transition frequency. As illustrated in figure 2, the uncertainty of the $^{88}Sr^+$ ion since 2012 has clearly surpassed that of Cs-based microwave standards. The most recent absolute frequency measurement of $^{88}Sr^+$ made with a cesium fountain clock (open circle in figure 2) is also shown for comparison [14].

4. Micromotion shifts

The micromotion shifts are caused by the interaction of the ion with the rf electric fields of the trap. The rf fields drive the ion motion at the trap frequency Ω, causing a second-order Doppler shift. They also cause scalar and tensor Stark shifts.

Micromotion shifts can be the most important sources of uncertainty in ion clocks when they are not properly minimized. For example, our first $^{88}Sr^+$ ion trap was not designed to control micromotion in three dimensions. As a consequence, its micromotion shifts are on the order of 30 Hz or $\approx 10^{-13}$ in fractional frequency units [4]. These large shifts were used for a high-accuracy measurement of the polarizability of the S–D transition as discussed in section 6. Micromotion caused by ion displacement is often called excess micromotion to distinguish it from the intrinsic micromotion caused by the thermal motion of the ion in the trapping potential [15, 16].

Our most recent trap of the endcap design [17] allows for proper minimization of excess micromotion. It has optical access ports for probing the ion motion along three mutually orthogonal directions and trim electrodes to adjust the ion position to the trap center. The micromotion-induced second-order Doppler and Stark shifts are reduced to the 10^{-17} level with minimization [2, 3]. Additional reduction is obtained by operating the trap at a special rf drive frequency as discussed in later sections.

5. Electric quadrupole shift

Unlike micromotion, the electric quadrupole shift (EQS) cannot be minimized with trim electrodes as it is an interaction between the $^2D_{5/2}$ state and an electric field gradient. The EQS is on the order of a few Hz in the endcap trap. It is the largest shift when micromotion is minimized.

The frequency shift Δf_Q of a magnetic sublevel $m_{J'}$ caused by the interaction between the quadrupole moment and an electric field gradient is given by [5, 18, 19]

$$\Delta f_Q = \frac{1}{4}\nu_Q \left(3\cos^2\theta - 1\right)\left[m_{J'}^2 - \frac{J'(J'+1)}{3}\right], \tag{1}$$

where ν_Q is a frequency proportional to the electric field gradient and the electric quadrupole moment [20]. θ is the angle between the electric field gradient principal axis and the quantization axis defined by the applied magnetic field [19]. $J' = 5/2$ refers to the upper $^2D_{5/2}$ state of the transition. The $^2S_{1/2}$ state has no quadrupole moment. Δf_Q describes the line center shift of symmetric Zeeman pairs since the two components of any pair have the same $m_{J'}^2$ value.

Two methods are commonly employed to cancel the EQS. One uses the property that the $(3\cos^2\theta - 1)$ term averages to zero when measurements are made for three mutually orthogonal magnetic field directions [19]. A limitation of this method is the requirement for a precise alignment of the magnetic field directions. On the other hand, it allows the use of a magnetic insensitive resonance to perform all the measurements.

The method we use is based on the observation that the term in square brackets in (1) vanishes when summing or averaging over the magnetic sublevels [3, 5, 21]:

$$\sum_{m_{J'}=-J'}^{J'} \left[m_{J'}^2 - \frac{J'(J'+1)}{3}\right] = 0. \tag{2}$$

This result is exact because the interaction Hamiltonian is traceless. Experimentally, (2) is realized by averaging the frequencies of six Zeeman components that connect to all of the $^2D_{5/2}$ state sublevels. The six components are grouped into three symmetric Zeeman pairs with $|m_{J'}| = 1/2$, $3/2$ and $5/2$. Each Zeeman pair performs the function of canceling the linear Zeeman shift. Averaging the line centers of the three pairs of Zeeman components cancels the electric quadrupole shift according to (2).

The level of cancellation is extremely high, similar to that obtained for the linear Zeeman shift. We have attributed an uncertainty of 3×10^{-19} in canceling this shift, based on the possible bias that could occur due to changes in the magnetic field direction or in the strength of the electric field gradient during measurements. A revised uncertainty for the current setup would be lower since the evaluation was made before the ion stability was improved with state-preparation [3, 9]. In our experiments, the lock to the ion cycles through the six Zeeman components for a continuous evaluation and cancellation of the electric quadrupole shift. The tensor Stark shifts are also canceled by this method [3, 5].

6. Blackbody radiation shift

The interaction of the clock transition with the blackbody radiation (BBR) field, $\langle E^2 \rangle_T$, causes a Stark shift given by [22, 23]:

$$\Delta\nu_{\text{BBR}} = -\frac{1}{2h}\langle E^2 \rangle_T \, \Delta\alpha_0 \, (1+\eta) \,, \tag{3}$$

where $\Delta\nu_{\text{BBR}}$ is expressed in Hz. h is Planck's constant and $\Delta\alpha_0 = \alpha_0(^2D_{5/2}) - \alpha_0(^2S_{1/2})$ is the differential static scalar polarizability of the clock transition. η is a dynamic correction that accounts for the response of the atomic energy levels to the BBR spectrum.

The uncertainty evaluation of $\Delta\nu_{\text{BBR}}$ depends on that of $\langle E^2 \rangle_T$, $\Delta\alpha_0$ and η. The values of $\Delta\alpha_0$ and η can be obtained from theoretical calculations or experimental measurements. The best theoretical calculations evaluate $\Delta\alpha_0$ with an uncertainty of 2×10^{-17} and η with an uncertainty of 8.4×10^{-20} [4, 24]. The current uncertainty estimate for $\langle E^2 \rangle_T$ in the endcap trap system is 1.1×10^{-17} [3].

Reduction of the $^{88}Sr^+$ uncertainty to the 10^{-18} level cannot be achieved without a more accurate determination of $\Delta\alpha_0$ and $\langle E^2 \rangle_T$. For the $^{88}Sr^+$ ion and other ions that have a positive scalar Stark shift, there is a convenient method for determining with high-accuracy the value of $\Delta\alpha_0$ [4]. The method is based on using large, applied micromotion shifts to compare the positive scalar Stark shift to the negative second-order Doppler shift. The trap drive frequency is then used to adjust the magnitude of the second-order Doppler shift for cancellation of the scalar Stark shift. This special trap drive angular frequency Ω_0 is directly related to $\Delta\alpha_0$ by the following relation [4]:

$$\Omega_0^2 \simeq -\frac{h\,\nu_0}{\Delta\alpha_0}\left(\frac{e}{mc}\right)^2 , \tag{4}$$

where ν_0 is the clock transition frequency in Hz, e the elementary charge, m the mass of $^{88}Sr^+$ and c the speed of light.

Ω_0 was determined experimentally by comparing two trapped ions. One was in a test trap of Paul trap design with micromotion shifts at the 10^{-13} level. The other was in a reference endcap trap where the total shifts were at the 10^{-17} level. These trap systems were described in section 4. Note that the micromotion induced tensor Stark shift in the test trap was canceled by the Zeeman averaging method of section 5.

Figure 3 shows the experimental data of the trap comparison. The two ion clock transition frequencies agree when the test trap drive frequency is $\Omega_0/2\pi = 14.404(10)$ MHz. The reference trap was operated with a drive frequency of 14.39 MHz during these experiments. This

Journal of Physics: Conference Series **723** (2016) 012018 doi:10.1088/1742-6596/723/1/012018

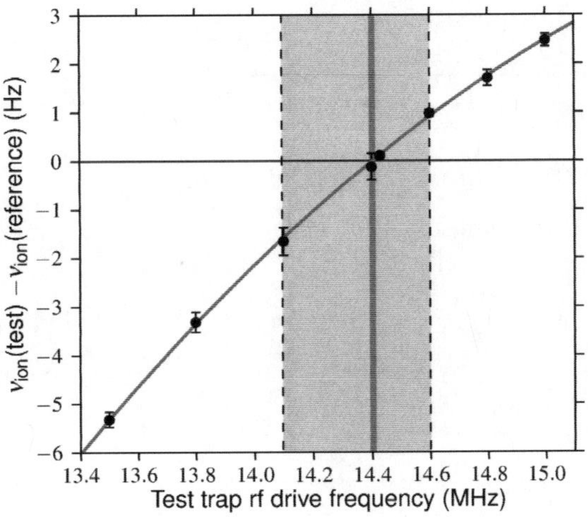

Figure 3. (color online) Comparison of the clock transition frequencies of two $^{88}Sr^+$ ions in different traps as a function of the rf drive frequency of the test trap [4]. The test trap has high micromotion shifts that vary with drive frequency. The narrow vertical band shows the value and uncertainty of the experimental determination of the drive frequency $\Omega_0/2\pi$ where the scalar Stark and second-order Doppler shifts cancel each other. The gray area is the predicted zone for the cancellation based on state-of-the-art theoretical calculations [24].

frequency was determined using (4) and a theoretical value of $\Delta\alpha_0$ [24]. From the experimental determination of Ω_0, we obtain $\Delta\alpha_0 = -4.7938(71) \times 10^{-40}\ \mathrm{J\,m^2/V^2}$ [4]. The reader is referred to previous work [4, 25] for a more accurate version of (4) that takes into account the second harmonic of ion motion. The BBR shift coefficient of the $^{88}Sr^+$ ion is calculated using (3) at 300 K, which gives $\Delta\nu_{\mathrm{BBR}}(300\ \mathrm{K}) = 0.247\,99(37)$ Hz. The fractional uncertainty of the frequency shift due to the uncertainty contributions of $\Delta\alpha_0$ has been reduced from 2×10^{-17} to 8.3×10^{-19} with this measurement. The evaluation of $\langle E^2 \rangle_T$ is now the most important source of uncertainty for our $^{88}Sr^+$ ion clock.

7. Uncertainty budget

Table 1 gives the current uncertainty budget for the $^{88}Sr^+$ ion in the reference endcap trap. It is a simplified and updated version of the budget given in [3]. The accurate measurement of $\Delta\alpha_0$ gives an improved evaluation of the BBR shift and a precise determination of the trap frequency for suppression of the micromotion shifts. By operating the endcap trap with a drive frequency of $\Omega_0 = 14.408(10)$ MHz, the net micromotion shifts are decreased from 10^{-17} to 10^{-19}. The test and endcap traps have slightly different values of Ω_0 because the ion motion direction and the Mathieu parameters are different in the two systems [25]. The total uncertainty is 1.2×10^{-17}, limited almost entirely by the BBR field evaluation. It is clear that a better evaluation of this field has the potential to reduce the total uncertainty of the $^{88}Sr^+$ ion to the low 10^{-18} level.

8. Conclusions

Several key methods were developed in recent years to deal very effectively with the most important sources of uncertainty in realizing the unperturbed frequency of the $5s\,^2S_{1/2} - 4d\,^2D_{5/2}$ clock transition of $^{88}Sr^+$. In addition to minimization of micromotion which is common to all ion optical clocks, Zeeman averaging is used to cancel the electric quadrupole shift and the tensor Stark shifts. We operate the ion trap with a special drive frequency Ω_0 that suppresses the net micromotion shifts by an additional factor of at least 200. The determination of $\Delta\alpha_0$ obtained from Ω_0 also provided an accurate value of the BBR shift coefficient, necessary for an accurate evaluation of the BBR shift. Further reduction of the total systematic shifts will require a better determination of the BBR field at the ion. The uncertainty budget indicates that a total uncertainty of $\approx 3 \times 10^{-18}$ should be attainable in the near future.

Table 1. Simplified uncertainty budget of the $^{88}Sr^+$ ion S–D frequency.

Source	mHz	Fractional
BBR field evaluation	4.9	1.1×10^{-17}
BBR coefficient ($\Delta\alpha_0$)	0.37	8.3×10^{-19}
Excess micromotion	0.05	1×10^{-19}
1092 nm ac Stark shift	1	2×10^{-18}
Second-order Doppler (thermal)	0.5	1×10^{-18}
Electric quadrupole shift	0.14	3×10^{-19}
Collisional shift	1	2×10^{-18}
Total uncertainty	5.1	1.2×10^{-17}

Acknowledgments
The authors gratefully acknowledge the important contributions of B. Hoger, W. Pakulski and R. Pelletier to the design and fabrication of several electronic components used in the single-ion clock system.

References

[1] Ludlow A D, Boyd M M, Ye J, Peik E and Schmidt P O 2015 *Rev. Mod. Phys.* **87** 637–701
[2] Madej A A, Dubé P, Zhou Z, Bernard J E and Gertsvolf M 2012 *Phys. Rev. Lett.* **109** 203002
[3] Dubé P, Madej A A, Zhou Z and Bernard J E 2013 *Phys. Rev.* A **87** 023806
[4] Dubé P, Madej A A, Tibbo M and Bernard J E 2014 *Phys. Rev. Lett.* **112** 173002
[5] Dubé P, Madej A A, Bernard J E, Marmet L, Boulanger J S and Cundy S 2005 *Phys. Rev. Lett.* **95** 033001
[6] Fordell T, Lindvall T, Dubé P, Madej A A, Wallin A E and Merimaa M 2015 *Opt. Lett.* **40** 1822–1825
[7] Letchumanan V, Wilson M A, Gill P and Sinclair A G 2005 *Phys. Rev.* A **72** 012509
[8] Bernard J E, Marmet L and Madej A A 1998 *Opt. Commun.* **150** 170–174
[9] Dubé P, Madej A A, Shiner A and Jian B 2015 *Phys. Rev.* A **92** 042119
[10] Bernard J E, Madej A A, Marmet L, Whitford B G, Siemsen K J and Cundy S 1999 *Phys. Rev. Lett.* **82** 3228–3231
[11] Margolis H S, Huang G, Barwood G P, Lea S N, Klein H A, Rowley W R C, Gill P and Windeler R S 2003 *Phys. Rev.* A **67** 032501
[12] Madej A A, Bernard J E, Dubé P, Marmet L and Windeler R S 2004 *Phys. Rev.* A **70** 012507
[13] Margolis H S, Barwood G P, Huang G, Klein H A, Lea S N, Szymaniec K and Gill P 2004 *Science* **306** 1355–1358
[14] Barwood G P, Huang G, Klein H A, Johnson L A M, King S A, Margolis H S, Szymaniec K and Gill P 2014 *Phys. Rev.* A **89** 050501
[15] Berkeland D J, Miller J D, Bergquist J C, Itano W M and Wineland D J 1998 *J. Appl. Phys.* **83** 5025–5033
[16] Keller J, Partner H L, Burgermeister T and Mehlstäubler T E 2015 *J. Appl. Phys.* **118** 104501
[17] Schrama C A, Peik E, Smith W W and Walther H 1993 *Opt. Commun.* **101** 32–36
[18] Abragam A 1978 *The Principles of Nuclear Magnetism* International series of monographs on physics (New York: Oxford University Press)
[19] Itano W M 2000 *J. Res. Natl. Inst. Stand. Technol.* **105** 829–837
[20] Barwood G P, Margolis H S, Huang G, Gill P and Klein H A 2004 *Phys. Rev. Lett.* **93** 133001
[21] Dubé P, Madej A A, Bernard J E and Shiner A D 2006 *Proceedings of the IEEE International Frequency Control Symposium and Exposition* (New York: IEEE) pp 409–414
[22] Itano W M, Lewis L L and Wineland D J 1982 *Phys. Rev.* A **25** 1233–1235
[23] Porsev S G and Derevianko A 2006 *Phys. Rev.* A **74** 020502
[24] Jiang D, Arora B, Safronova M S and Clark C W 2009 *J. Phys. B: At. Mol. Opt. Phys.* **42** 154020
[25] Dubé P, Madej A A, Tibbo M and Bernard J E 2014 *Proceedings of the 28th European Frequency and Time Forum* (Neuchâtel, Switzerland: IEEE) pp 443–446

Journal of Physics: Conference Series **723** (2016) 012019 doi:10.1088/1742-6596/723/1/012019

A low maintenance Sr optical lattice clock

I R Hill[1], R Hobson[1,2], W Bowden[1,2], E M Bridge[1,2], S Donnellan[1], E A Curtis[1] and P Gill[1]

[1]National Physical Laboratory, Hampton Road, Teddington, TW11 0LW, UK
[2]Clarendon Laboratory, University of Oxford, Parks Road, Oxford, OX1 3PU, UK

E-mail: ian.hill@npl.co.uk

Abstract. We describe the Sr optical lattice clock apparatus at NPL with particular emphasis on techniques used to increase reliability and minimise the human requirement in its operation. Central to this is a clock-referenced transfer cavity scheme for the stabilisation of cooling and trapping lasers. We highlight several measures to increase the reliability of the clock with a view towards the realisation of an optical time-scale. The clock contributed 502 hours of data over a 25 day period (84 % uptime) in a recent measurement campaign with several uninterrupted periods of more than 48 hours. An instability of 2×10^{-17} was reached after 10^5 s of averaging in an interleaved self-comparison of the clock.

1. Introduction

Ultra-precise optical clocks based on lattice-trapped cold atoms [1, 2, 3, 4, 5] offer a new vision of the SI second with accuracy and precision two orders of magnitude beyond the current realisation [6]. Such precision has profound implications for an Earth-based optical clock network which becomes sensitive to fluctuations in gravitational potential at the 1 cm level. Optical lattice clocks have potential for new applications in numerous precision measurement experiments to determine local gravity [7], perform relativistic geodesy and international leveling, form the basis of extremely narrow linewidth lasers [8], provide a route towards quantum computing [9], and search for possible time-variation of fundamental constants [10].

The reliable, low maintenance operation of these systems is crucial to their adoption in new technologies and in establishing a global optical time-scale. In particular, extended periods of operation place stringent requirements on the frequency stabilisation of multiple laser systems. We report here a simple scheme to transfer the intrinsic stability of the Sr clock to the cooling and trapping laser systems via a single tunable optical cavity; removing the need for regular frequency measurements of the optical lattice trap by femtosecond optical frequency combs, and ensuring a low frequency deviation of the cooling light without additional stabilisation to an atomic vapor. Methods to ensure the precision and robustness of the stabilisation are described, including additional servos for increased laser lock longevity, backed up by automatic relocking algorithms. We also give an overview of the experimental setup at NPL and its first operation as an ^{87}Sr optical lattice clock.

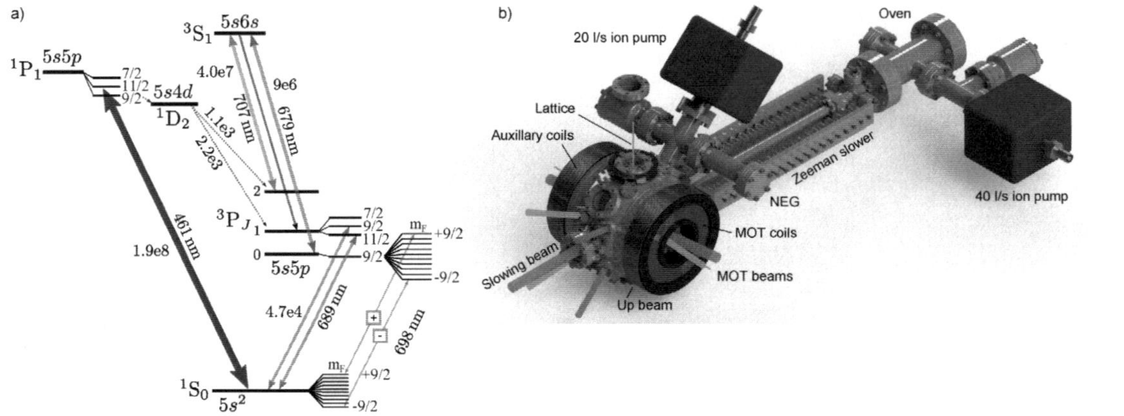

Figure 1. (a) The relevant energy levels of Sr for cooling, trapping, and probing the clock transition and (b) a rendering of the cold atom apparatus.

2. System overview

2.1. The cold atom apparatus

We adopt a similar approach to others in preparing cold samples of Sr [11, 12, 13, 4] using a 2-stage magneto-optical-trap (MOT) loaded from a Zeeman slowed atomic beam. Details of our oven source and novel transverse-field Zeeman slower, based on a 2D array of permanent magnets, can be found in ref. [14].

The first cooling stage uses the $\gamma = 30.2$ MHz broad $^1S_0 \leftrightarrow {}^1P_1$ cycling transition at 461 nm for efficient capture and cooling. For ^{87}Sr (^{88}Sr) a 'blue' MOT formed of 3 retro-reflected beams is loaded for 150–350 ms (60 ms) at saturation intensity of $s = 0.1$ radially and $s = 0.05$ axially with a quadrupole magnetic field gradient of 4 mT cm^{-1} and detuning of -40 MHz. The Zeeman slowing beam is then switched off, the MOT detuning is stepped to -35 MHz and the intensity ramped down to zero over 15 ms. This reduces the atomic temperature from around 5 mK to 2 mK and restores the MOT in the quadrupole field zero in preparation for transfer to the second stage MOT. A branching of 1 in 50,000 from the 5s5p $^1P_1$ to 5s4d $^1D_2$ state necessitates repumping lasers at 707 nm and 679 nm to clear out states 5s5p $^3P_2$ and 5s5p $^3P_0$ via 5s6s $^3S_1$ and return atoms to the cooling cycle via the 5s5p $^3P_1$ state (figure 1a).

The second cooling stage operates a 'red' MOT on the $\gamma = 7.5$ kHz narrow $^1S_0 \leftrightarrow {}^3P_1$ intercombination line at 689 nm. The quadrupole field gradient is switched in ~ 1 ms to 0.2 mT cm^{-1} to accommodate the weak cooling line. For ^{87}Sr a stirring laser operating on the $^1S_0|F = 9/2\rangle \leftrightarrow {}^3P_1|F = 9/2\rangle$ line is simultaneously overlapped with all MOT beams to redistribute population amongst m_F sublevels for efficient trapping. The red MOT is initially operated in 'broad-band' mode by frequency modulating the 689 nm MOT (stirring) light at 40 kHz with peak-to-peak deviation 4 MHz (2 MHz) to address the Doppler broadened atomic absorption spectrum. This generates approximately 100 (50) comb lines, with the nearest detuned by -400 kHz (< -100 kHz), and $s \simeq 4$ ($s \simeq 3$) per line in a 1 cm $1/e^2$ diameter beam. Around 10^5 atoms are collected and cooled to 10 μK during this 60–150 ms phase. The modulation is then turned off, the intensity jumped to $s \simeq 45$ ($s \simeq 45$), and the detuning stepped closer to resonance to form a narrow-band MOT. The cloud is cooled and compressed by ramping the intensity to $s \simeq 8$ while reducing the detuning towards resonance over a period of 60 ms. The intensity is stepped to $s = 0.8$ for a final stage of cooling in a low intensity narrow-band MOT reaching temperatures of ~ 1 μK after 75 ms. In this final stage an additional 'up' beam is needed ($s \simeq 30$) to counter gravity which is otherwise compromised by the geometry of the trapping beams.

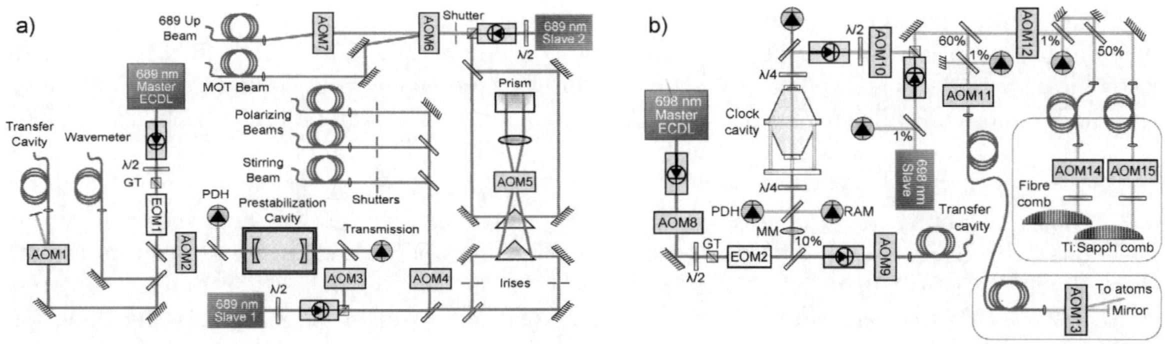

Figure 2. (a) 689 nm 2nd-stage cooling laser system and (b) the 698 nm clock laser system. PDH is Pound Drever Hall, GT is Glan Taylor polariser, MM is mode-matching lens, AOM (EOM) is acousto (electro) optic modulator, and ECDL is extended cavity diode laser.

Atoms are next loaded into a vertical one-dimensional optical lattice trap at 813 nm with $1/e^2$ radius of 45 μm. The lattice is formed by a retro-reflected beam with linear polarisation parallel to the quantisation field direction and up to 1 W from an M-Squared SolTiS Ti:Sapphire laser. The trap is tuned to near the magic frequency at 368554.5 GHz [15] and has a $1/e$ lifetime of 8 s. A spin polarising pulse of circularly polarised 689 nm stirring light, frequency modulated to cover a 2 MHz spectrum, is then applied to transfer population into $m_F = \pm 9/2$ stretched states in preparation for clock interrogation. During loading typical trap depths are $75\,E_{\rm r}$ ($\simeq$ 12 μK), where $E_{\rm r}$ is the photon recoil associated with the lattice frequency. To ensure a low atomic temperature a brief evaporation stage is carried out by ramping the lattice depth to 3 μK over 20 ms where it remains for a duration of 20 ms before ramping back to the final depth for clock probing. The clock transition is probed along the vertical axis of the optical lattice in the direction of tight confinement within the Lamb-Dicke regime. The clock laser has $1/e^2$ beam radius of around 200 μm and is linearly polarised along the quantisation field axis. The power of each cooling and trapping beam is actively stabilised at the fibre output following a polarising cube by feedback to the RF drive power of an acousto-optic modulator (AOM). Polarising cubes are positioned on entry to each fibre to maintain good alignment with the PM axis and lessen the effect of polarisation fluctuations due to thermal effects in the birefringent TeO_2 AOM crystal.

The quadrupole MOT fields are provided by a pair of 9×8 turn coils of 4 mm square copper Kapton-insulated wire with a 2.5 mm circular bore through which cooling water flows, and produce a field gradient 50 $\mu\mathrm{T\,cm^{-1}A^{-1}}$. Each coil is potted with Stycast into a Delrin former which provides a 2 mm thick insulation layer between the coil and chamber. Auxiliary coils are wound on to the outer of the former and provide a better approximation to a Helmholtz configuration for quantisation fields and mixing fields for magnetically induced spectroscopy [16] in ^{88}Sr, and produce 125 $\mu\mathrm{TA^{-1}}$ with long-term instability < 20 nT. MOT (auxiliary) coil currents are adjusted by FET (opamp) actuation in an actively stabilised loop using a Hall (flux gate) sensor. A bank of transient-voltage-suppressing diodes is used to clamp the switching voltage high to increase the rate of energy transfer from the MOT coils. The temperature inhomogeneity of the vacuum chamber is maintained at 0.3 K and monitored by 11 Pt-100 sensors calibrated to ± 10 mK.

2.2. Laser apparatus

2.2.1. 461 nm and repumping. Around 350 mW of 461 nm light is produced with a commercial frequency-doubled extended cavity diode laser (ECDL) plus tapered amplifier system with bow-tie cavity enhanced second-harmonic generation using a $KNbO_3$ crystal. The light is split and frequency shifted by AOMs to perform Zeeman slowing, MOT, and absorption and fluorescence imaging probe duties. Repumping is provided by two grating-based ECDLs at 679 nm and 707 nm which are frequency stabilised to a commercial wavemeter via piezo actuation on the cavity length. We modulate the current of the 707 nm diode and individually address all five hyperfine levels of 5s5p $^3P_2$ via separate orders of two AOMs, enabling blue MOT $1/e$ lifetimes > 1 s.

2.2.2. 689 nm cooling, stirring, and spin-polisaring. The 689 nm laser system shown in figure 2a is based on a home-built Littrow-configuration master ECDL seeding two injection-locked slave diode lasers (Opnext HL6750MG). Slave 1 is the source of the light addressing the $F = 9/2 \rightarrow 9/2$ 'stirring' and spin-polarising transition, while slave 2 sources the light for the main $f = 9/2 \rightarrow 11/2$ MOT transition. We narrow the linewidth of the master laser by a Pound Drever Hall (PDH) lock to a 60,000-finesse prestabilization cavity (also used to filter the light for injecting the slaves), and we provide long-term frequency stability by a PDH lock to the transfer cavity actuating on the RF drive frequency of AOM2 and 3. The accuracy of the transfer cavity lock is enhanced by implementation of an AOM-based residual-amplitude modulation (RAM) servo (figure 3 and section 2.3.1). The frequency of slave 2 can be selected to address either the ^{88}Sr or ^{87}Sr MOT transition by opening or closing the irises before AOM5.

2.2.3. 698 nm clock laser. The clock laser system shown in figure 2b consists of a home-built Littrow-configuration grating-based ECDL (non-AR-coated Opnext HL7001MG diode) PDH locked to a 7.75 cm long vertical ULE optical cavity [17] whose transmission is used to inject a slave diode laser in a similar scheme to [18]. The optical cavity has a measured finesse of 420,000 and bandwidth of 4.6 kHz which provides a roughly 23 dB suppression of the transmitted locking servo sidebands at 1 MHz. The cavity temperature is stabilised (< 1 mK) close to its minimum of thermal expansion at 25.7 °C by peltier control of a polished aluminium cylindrical shield isolated within a steel vacuum chamber at 10^{-7} mbar. Care is taken to avoid residual amplitude modulation by exluding polarising optics in the path between the free-space EOM output and optical cavity input mirror. A Glan-Taylor polarising optic is placed immediately infront of the EOM to ensure a high degree of linear polarisation along the crystal axis. To avoid problematic parasitic etalons and optical feedback to the laser, optical isolation is ensured by insertion of Faraday isolators, multiple beam splitters, and AOM frequency shifting. The complete optical setup is mounted on a two tier 60×50 cm active-vibration-isolation platform inside a 12 mm thick acrylic box.

To extend the locked period of the master laser a low bandwidth servo of the cavity transmission actuating on the laser current is implemented. When the laser is near a mode-hop we observe an increase in cavity transmission due to a change in the modulation character of the diode. Using this change as a discriminator we ensure the mode of the laser is well-tracked.

Light from the clock slave laser is distributed to the atoms and two femtosecond frequency combs labs by path length stabilised fibre links actuated by AOM11/14/15. At the remote end of the atom link AOM13 is used to control the clock light frequency stepping, phase, and amplitude. To eliminate thermally driven phase chirps during switching of the AOM RF drive power, the compensated path is extended to include AOM13. Here, the zero order beam is used in preference to the first order to avoid transient servo capture effects.

2.3. Transfer cavity scheme

Using a single tunable in-vacuum optical cavity whose length is stabilised to the 698 nm clock laser we transfer the stability of the clock to all cooling and trapping lasers, degraded only by the quality of the individual laser locks to the cavity and the cavity environment. This removes the need for atomic spectroscopy cells and multiple optical cavities that must be separately characterised. The scheme, shown in figure 3, implements a low bandwidth PDH lock at each wavelength with separation by optical filtering (by ensuring sufficiently different PDH modulation frequencies the number of detectors and optical filters could be reduced). AOM9 is updated to track the atomic reference such that the clock cavity drift is removed. The lattice frequency is easily controlled to within < 100 kHz as required for 10^{-18} clock uncertainty [19].

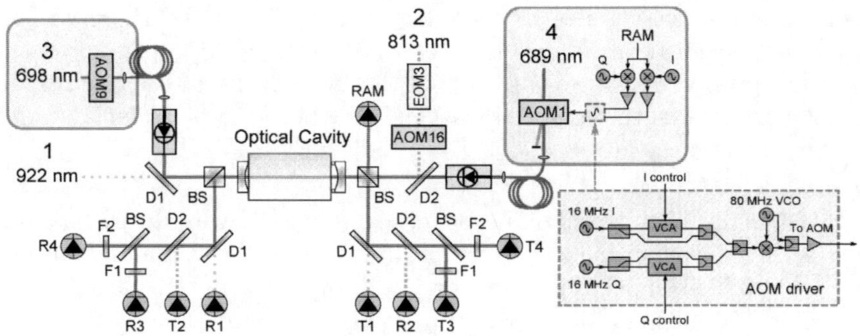

Figure 3. The transfer cavity locking scheme. Transmission and reflection photodiodes are labeled T and R and numbered according to wavelength. BS is a 50:50 beam splitter, D1 is a long pass filter above 900 nm, D2 is a short pass filter below 800 nm, and F1 and F2 are bandpass filters for 698 nm and 689 nm respectively.

2.3.1. AOM residual amplitude modulation servo. To allow resolution of the 900 kHz cavity linewidth to the kHz level, the 689 nm light incident on the transfer cavity is picked off onto a separate RAM photodetector and both the in-quadrature (Q) and the in-phase (I) components of RAM are mixed down to DC. Using voltage-controlled attenuator setups that can pass positive- or negative-amplitude RF signals, we can flexibly apply I and Q amplitude modulation to the RF drive carrier of AOM1 (figure 3), diverting a modulated amount of light away from the zero-order beam. We note that the optical fibre after the AOM is an essential component—it spatially filters the beam, homogenising the AM applied by the AOM. Closing the loop, the AM applied to the AOM will exactly compensate the RAM from the EOM (EOM1 figure 2), preventing offsets in the PDH lock. We observe good RAM suppression from typically around −50–60 dBm to below −80 dBm on our photodetector with a lock bandwidth of a few kHz. We acknowledge a similar system recently reported in ref [20].

2.3.2. 689 nm frequency correction by lattice loading position servo. The stability transfered to the 689 nm laser was at times unsatisfactory, evidenced by deviations in the atom loading position in the optical lattice. Since the red MOT sags under gravity, its vertical position is proportional to the magnetic field and MOT laser detuning. Given that the magnetic field is sufficiently stable, we implement a position servo on the centre of mass of the lattice-trapped atom cloud by feeding corrections to AOM9 in preference to those from the clock. The typical vertical extent of the atomic cloud is about 200 μm and is 1:1 imaged by a CCD array with pixel size of 16 μm. For our loading conditions we obtain a frequency sensitivity of approximately 7 kHz per pixel with sub pixel control resolution. The residual frequency difference between this method and clock corrections is shown in figure 4e.

2.4. Automated locking algorithms

We implement automated injection routines on both 689 nm slave lasers using a commercial wavemeter and knowledge of the laser current injection window. When a slave unlocks, the current is jumped by $\pm 1/2$ the injection window. If the slave injects on either of these two test currents, detected by an appropriate reading on the wavemeter, the new operating current will be reset to this value, else the injection window is increased and the process is repeated. Once injected, a similar process is used periodically to ensure the operating current is well centred in the injection window.

The wavemeter is also used to stabilise and reacquire the lock of both repump lasers. In the event of a laser mode-hop that is not recovered by the action of the PI servo feeding back to the laser piezo, a lock acquisition routine is entered. By ramping the piezo and monitoring the laser frequency, an appropriate step is applied to the laser current to move the mode-hop-free window of the ECDL towards the required wavelength where the piezo lock is reinstated.

2.5. Experimental control

An experimental control system based on bespoke, scalable FPGA hardware has been developed. Scripts which describe the experimental sequence are written in Python and compiled using the Labscript software suite [21], which has been adapted for use with our hardware. This includes a stackable 8-channel, 16-bit DAC board with a 2 μs update period and 24 times 40 MHz digital outputs, accessed via a breakout header board with galvanic isolation. Each DAC board contains an FPGA and 1 Mbit of memory per channel for storing data that is to be clocked out during the experimental shot. A pseudo-clock architecture allows efficient data compression allowing more complex shots to be written and executed. Data acquisition and clock servo operation is by a separate computer based system.

3. Performance

3.1. Clock spectroscopy

We perform high resolution Rabi spectroscopy of the clock transition in the Lamb-Dicke regime ($\eta \simeq 0.24$) with motion resolved from the carrier. Spectroscopy of the first-order motional sidebands in an unpolarised sample, shown in figure 4a, is used to determine the apparent axial (T_a) and radial (T_r) temperatures [22] and the axial trap frequency for characterisation of lattice shifts. Fourier limited line shapes are achieved for Rabi pulses approaching 100 ms in spin-polarised samples, limited by acoustic noise coupled to the clock laser reference cavity in the region of 2 Hz (see figure 4d).

3.2. Stability and reliability

To operate the clock we run two servos locked to transitions between the ground and excited $m_F = \pm 9/2$ stretched states and compute the average frequency $f^{87} = (f^- + f^+)/2$. The stretched states are split by $\simeq 1$ kHz in a 100 μT quantisation field. State populations are read out by normalised shelving detection using a single fluorescence imaging probe at 461 nm. We achieve an instability of $3.2 \times 10^{-15}/\sqrt{\tau}$, where τ is the averaging time, for the non-interleaved clock, measured by an interleaved self-comparison of two f^{87} clock servos with 80 ms Rabi time (see figure 4c). An instability of 2×10^{-17} was reached after 10^5 s of continuous averaging in interleaved operation.

In a recent 25 day measurement campaign the system recorded 502 hours of ‘good’ frequency data, corresponding to an uptime of 84 %. The relocking algorithms described in section 2.4 were implemented for the last 16 days of the campaign and aided an achieved uptime of 94 % for this period. The transfer fidelity of the 698 nm to 689 nm transfer cavity scheme was assessed over a 70 hour period with reference to the 689 nm frequency servo described in section 2.3.2 (figure 4e). We observe a peak deviation of 10 kHz, corresponding to a PDH lock deviation of approximately 1 %. An increased low frequency servo gain or increased cavity finesse should yield an improved performance with sub 1 kHz stability transfer.

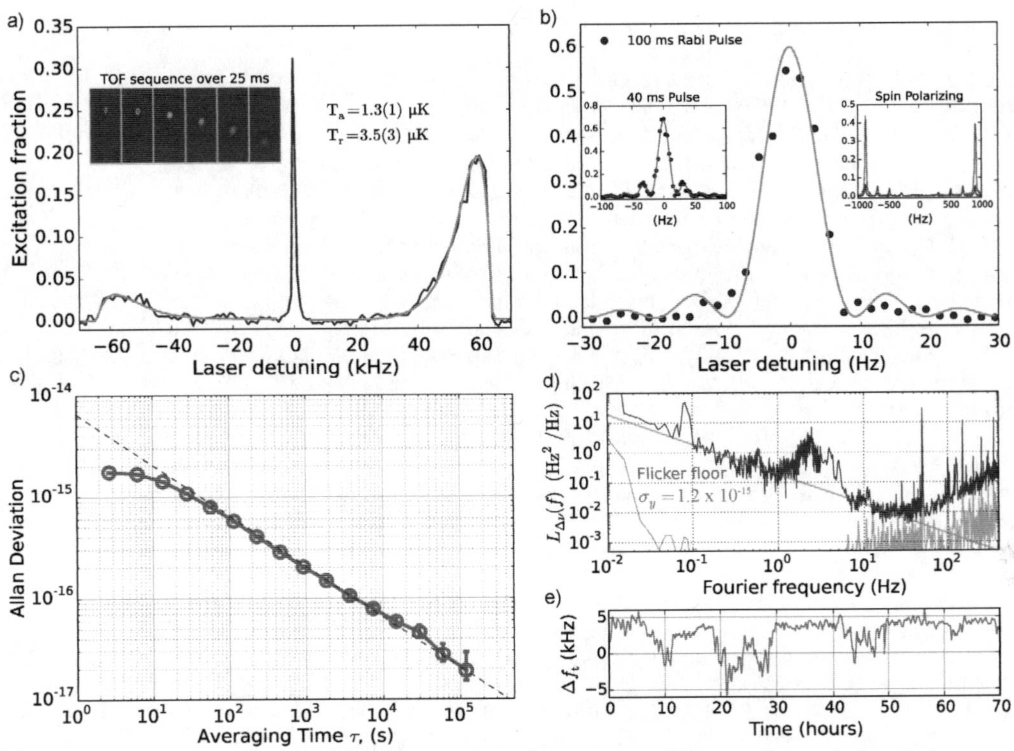

Figure 4. (a) Overdriven clock spectroscopy showing first order motional sidebands of lattice-trapped cold Sr and (inset) time-of-flight sequence for 1 μK temperature atom cloud released from the red MOT. (b) 100 ms (main) and 40 ms (inset left) Rabi excitation of the spin-polarised clock transition and (inset right) spectroscopy of m_F states with and without spin-polarisation. (c) Fractional frequency instability of an interleaved clock self-comparison. (d) Power spectral density of the clock laser frequency noise measured by comparison to an ultra-stable laser at 1064 nm via an optical frequency comb. The bump at 80 mHz is from a temperature oscillation present on the 1064 nm laser system. (e) Frequency corrections made by the lattice loading position servo compared to the clock referenced transfer cavity stabilisation.

4. Conclusion

We have described a reliable apparatus for a Sr optical lattice clock centred around a transfer cavity laser stabilisation scheme that enables long periods of operation with low maintenance and high accuracy. The additional servo loops and automatic recapture routines described in this paper enable continuous, reliable clock operation with little human input. Such a system is well placed to contribute to the realisation of an optical time-scale.

Acknowledgments

The authors thank Yuri Ovchinnikov, Marco Menchetti, Ross Williams, and Matt Earnshaw for contributions to the setup. This work was funded by the National Measurement System Electromagnetics and Time programme.

References

[1] Baillard X, et al 2008 *EPJD* **48** 11–17
[2] Katori H 2002 Spectroscopy of Strontium Atoms in the Lamb-Dicke Confinement *Frequency Standards and Metrology* ed Gill P pp 323–330

[3] Campbell G K, et al 2008 *Metrologia* **45** 539
[4] Falke St, et al 2011 *Metrologia* **48** 399
[5] Hong F L, et al 2009 *Opt. Lett.* **34** 692–694
[6] Nicholson T L, et al 2015 *Nature Communications* **6** 6896
[7] Poli N, Wang F Y, Tarallo M G, Alberti A, Prevedelli M and Tino G M 2011 *Phys. Rev. Lett.* **106**(3) 038501
[8] Meiser D, Ye J, Carlson D R and Holland M J 2009 *Phys. Rev. Lett.* **102**(16) 163601
[9] Daley A J, Boyd M M, Ye J and Zoller P 2008 *Phys. Rev. Lett.* **101**(17) 170504
[10] Kotochigova S, Zelevinsky T and Ye J 2009 *Phys. Rev. A* **79**(1) 012504
[11] Katori H, Ido T, Isoya Y and Kuwata-Gonokami M 1999 *Phys. Rev. Lett.* **82**(6) 1116–1119
[12] Loftus T H, Ido T, Boyd M M, Ludlow A D and Ye J 2004 *Phys. Rev. A* **70**(6) 063413
[13] Sorrentino F, Ferrari G, Poli N, Drullinger R and Tino G M 2006 *Mod. Phys. Lett. B* **20** 1287–1320
[14] Hill I R, Ovchinnikov Y B, Bridge E M, Curtis E A and Gill P 2014 *J. Phys. B* **47** 075006
[15] Katori H 2001 *Proceedings of the 6th Symposium in Frequency Standards and Metrology* 323–330
[16] Taichenachev A V, Yudin V I, Oates C W, Hoyt C W, Barber Z W and Hollberg L 2006 *Phys. Rev. Lett.* **96**(8) 083001
[17] Ludlow A D, Huang X, Notcutt M, Zanon-Willette T, Foreman S M, Boyd M M, Blatt S and Ye J 2007 *Opt. Lett.* **32** 641–3
[18] Sterr U, Legero T, Kessler T, Schnatz H, Grosche G, Terra O and Riehle F 2009 Ultrastable lasers: new developments and applications vol 7431 pp 74310A–74310A–14
[19] Katori H, Ovsiannikov V D, Marmo S I and Palchikov V G 2015 *Phys. Rev. A* **91**(5) 052503
[20] Hall J L, Zhang W and Ye J 2015 Accurate Removal of RAM from FM Laser Beams *2015 Joint Conference of the IEEE International Frequency Control Symposium & the European Frequency and Time Forum* (IEEE) pp 713–716
[21] Starkey P T, Billington C J, Johnstone S P, Jasperse M, Helmerson K, Turner L D and Anderson R P 2013 *Rev. Sci. Instrum.* **84** 085111
[22] Blatt S, Thomsen J W, Campbell G K, Ludlow A D, Swallows M D, Martin M J, Boyd M M and Ye J 2009 *Phys. Rev. A* **80**(5) 052703

Journal of Physics: Conference Series **723** (2016) 012020 doi:10.1088/1742-6596/723/1/012020

A transportable optical lattice clock

Stefan Vogt, Sebastian Häfner, Jacopo Grotti, Silvio Koller, Ali Al-Masoudi, Uwe Sterr and Christian Lisdat

Physikalisch-Technische Bundesanstalt, Bundesallee 100, D-38116 Braunschweig, Germany

E-mail: `stefan.vogt@ptb.de`

Abstract. We present the experimental setup and first results of PTB's transportable ^{87}Sr clock. It consists of a physics package, several compact laser breadboards, and a transportable high finesse cavity for the clock laser. A comparison of the transportable system with our stationary optical lattice clock yields an instability of $2.2 \times 10^{-15}\ \sqrt{s/\tau}$ for the transportable clock. The current fractional uncertainty of 1×10^{-15} is still limited by the not yet fully evaluated light shift from the free running optical lattice laser operated near the magic wavelength. We are currently improving our transportable system to reach an uncertainty at or below the 10^{-17} level, which will finaly be limited by the uncertainty in blackbody radiation shift correction.

1. Introduction

The performance of optical lattice clocks is now exceeding the Cs clock performance by two orders of magnitude in accuracy and stability. The best lattice clocks reach an uncertainty in the low 10^{-18} range [1, 2] and achieve instabilities of below $2 \times 10^{-16}\sqrt{s/\tau}$ [3]. This excellent performance offers new prospects for both fundamental science and applications that include the search for variations of fundamental constants, tests of general relativity on a higher level of accuracy, and relativistic geodesy. However, optical clocks that show a fractional instability of 10^{-15} or even 10^{-16} at one second cannot efficiently be compared by the traditional radio-frequency methods that are at best limited at 10^{-13} at one second. In order to exploit the full potential of optical clocks for these scientific applications, new methods for frequency comparisons beyond the current limitations of radio-frequency comparisons have to be developed. Therefore transportable frequency standards or stabilized optical fiber links are necessary. In addition, transportable optical frequency standards in combination with optical fiber links allow for relativistic geodesy (height measurement) [4, 5] at the centimetre level between far-separated points. This method to determine the geopotential is complementary to the state-of-the-art methods of geodesy, since it directly measures differences in the local gravitational potential, while established methods use leveling together with local measurements of the earth acceleration (the spatial derivative of the potential) or a combination of a detailed gravity model and GNSS satellite data. Therefore relativistic geodesy will eventually provide a useful verification and even improvement of geodetic models.

In this paper we describe an optical lattice clock setup based on neutral strontium atoms. The experimental setup is made robust and compact to allow for a transportation to another laboratory or even a remote location using an air-conditioned trailer. The setup was not optimized for extreme power efficiency or small volume but to provide ultimately an optical

Journal of Physics: Conference Series **723** (2016) 012020 doi:10.1088/1742-6596/723/1/012020

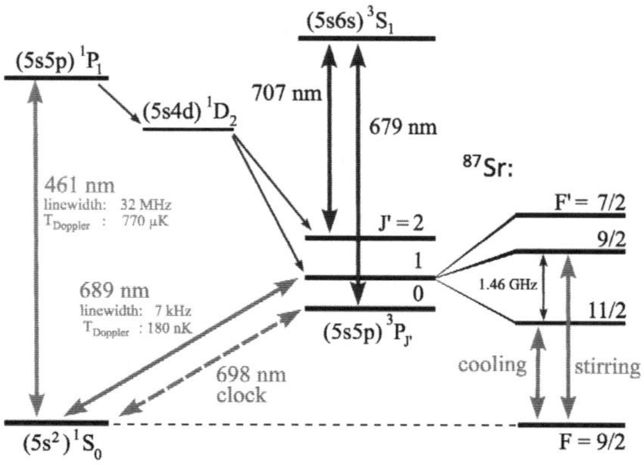

Figure 1. Partial level scheme of strontium. The lattice laser is operating at the Stark-shift-cancellation wavelength near 813 nm. (color online)

Figure 2. Top view of the physics package. The dimensions of the setup are 1200 mm × 900 mm. (color online)

lattice clock with metrologically competitive characteristics, i.e. uncertainties below 10^{-17} after 10^4 seconds of averaging.

First measurements with the transportable setup have been performed including a comparison with the stationary strontium clock at PTB [6, 3]. The current uncertainty budget has been evaluated leading to an uncertainty of 1.3×10^{-15}. We show that with the completion of the setup that is currently underway, the uncertainty of the system will be improved even below 10^{-17}.

2. Experimental setup

2.1. Laser Modules

For laser cooling of neutral ^{87}Sr atoms, trapping in an optical lattice, and interrogation of the 698 nm $^1S_0$–$^3P_0$ clock transition in total lasers at six different wavelengths are required (see Fig. 1). The lasers are based on commercial extended cavity diode lasers (ECDLs). For atom preparation, a two-stage Magneto-Optical Trap (MOT) sequence is used. In the first stage the 461 nm $^1S_0$–$^1P_1$ transition in the singlet system is used for Zeeman slowing and to efficiently trap the atoms in a MOT. In this stage two repump lasers (679 nm and 707 nm) are used to depopulate the long lived dark $^3P_2$ state, that is populated by a decay channel from the $^1P_1$ state. In the second MOT stage operating on the narrow intercombination line ($^1S_0$–$^3P_1$, 689 nm) the atoms are further cooled down to the μK level.

The lasers are integrated on breadboards that are connected via polarization-maintaining fibers to the physics package. Laser distribution, frequency shifting and switching are done on these compact aluminum breadboards (size: 450 mm × 300 mm) with half inch optics, acousto-optical modulators (AOMs) and mechanical shutters to prevent ac Stark frequency shifts from residual cooling light during clock interrogation. The 461 nm blue light is generated by second-harmonic generation in a resonant doubling cavity.

For second-stage cooling of ^{87}Sr, we use two lasers at 689 nm, called cooling- and stirring laser, which address the 1.46 GHz separated hyperfine transitions $^1S_0$–$^3P_1$, $F = 9/2$ – $F' = 11/2$ and $F = 9/2$ – $F' = 9/2$ [7]. The power requirements of about 10 mW per laser can be met with an ECDL without need of an additional injection-locked laser diode. The lattice light is provided by a commercial tapered amplifier (TA) system with an output power of up to 2 W.

For frequency stabilization the lasers are locked to individual 10-cm-long ultra low expansion glas (ULE) cavities. For the 461-nm cooling laser a cavity housed in a simple vacuum chamber with one passive heat shield is sufficient. The temperature of the vacuum chamber is stabilized by thermoelectric coolers (TEC). Changes of the cavity resonance frequency due to residual temperature fluctuations and day-to-day drifts are negligible compared to the linewidth of the cooling transition of about 32 MHz. Because of the considerably narrower linewidth of the red cooling transition ($\approx$ 7 kHz) more care has been taken to stabilize the temperature of the reference resonator for the second-stage cooling lasers. Here, a two-stage temperature stabilization has been applied that actively controls the outside vacuum chamber as well as a heat shield within the vacuum by TECs. The lattice laser will soon be stabilized to a similar cavity setup, which however had not been finalized for the first clock measurements. The finesse of the resonator for the 689 nm cooling is high ($F = 240\,000$) to relax the requirements on the laser stabilization electronics, while the other two cavities have a finesse of around 1000. The frequency stability of the 689 nm cooling laser is transferred to the stirring laser via a phase lock loop with a frequency offset. We do not stabilize the frequency of the repump lasers, because a frequency modulation of several 100 MHz that has to be applied to cover the hyperfine structure of ^{87}Sr leads to relaxed requirements for the frequency stability of the lasers. If necessary the lasers could be locked to a wavelength meter with an accuracy of 200 MHz in the future.

The clock laser system was previously described [8] while the transportable reference cavity was replaced by an improved design [9]. The resonator consists of a 12 cm long ULE spacer with fused silica mirrors. The resonator is rigidly mounted via a wire-bar mounting system, which is attached to Invar pins that are glued into holes in the spacer. The mounting of the cavity is optimized for low vibration sensitivity by applying all retaining forces only normal to the respective symmetry planes and by a kinematic mount, effectively reducing the degrees of freedom within the mounting structure to avoid overdetermination [10, 11]. The sensitivity of the cavity resonance frequency to accelerations in three orthogonal directions is $\kappa = (0.3, 4.3, 11.6) \times 10^{-10}$/g. We attribute the relatively high sensitivity along the spacer axis to imperfections in the realization of the mounting. Using an active vibration isolation platform (Table Stable TS-150), the thermal noise floor of the resonator setup of 4×10^{-16} is almost reached for timescales of 1 s up to a few 10 s (Fig. 6), as we could show by comparison to a non-transportable clock laser with an instability of below 1×10^{-16} [12].

2.2. Physics Package

As the frequency shift due to blackbody radiation of the environment is a critical factor in strontium lattice clocks [13, 2], special care has been taken to provide a good temperature homogeneity across the vacuum chamber surrounding the interrogated atoms. This includes an efficient water cooling of the coils that are used to generate the magnetic field for magneto-optical trapping and the design of a Zeeman slower using permanent magnets.

The main part of the physics package is the vacuum chamber, which is mounted on a table with the size of 1200 mm $\times$ 900 mm (see Fig. 2). This chamber is separated into two parts: the oven chamber where strontium atoms are evaporated from an atomic oven at about 480 °C, and the main chamber where laser cooling, trapping and the interrogation are performed. Both parts are connected by a 50 cm long Zeeman slower tube. In between the oven region and the Zeeman slower tube, a 2-mm-diameter pinhole is used to shield the MOT region from the blackbody radiation of the oven. It also enables differential pumping between the oven and the MOT chamber to maintain a pressure ratio of about one order of magnitude during operation of the apparatus. In the oven chamber a rotation shutter is used to fully block the radiation from the hot oven during the clock interrogation. The switching of the shutter with an servo motor outside the vacuum takes less than 35 ms, so that switching during one experimental cycle is possible. The surfaces closest to the atom position are the windows of re-entrant flanges with

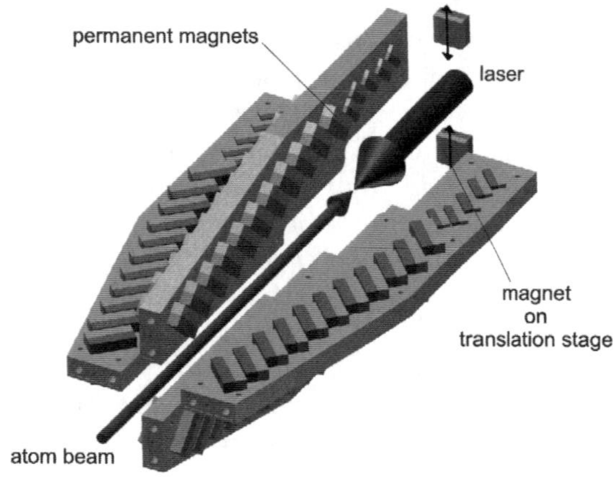

Figure 3. Sketch of the mechanical setup for the Zeeman slower with B-field along the axis. (color online)

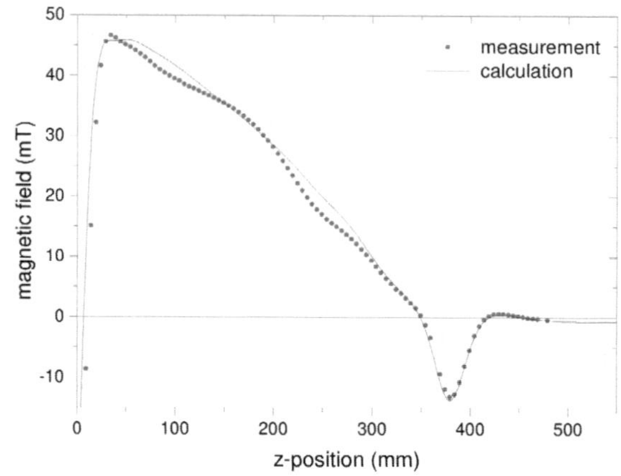

Figure 4. Longitudinal magnetic field component along the slowing axis (z). (color online)

a separation of 54 mm. Those are coated on the inside with a conducting ITO layer to avoid patch charges that would lead to a dc Stark effect. A residual electric charge of 100 mV would lead to a fractional frequency shift in the order of 10^{-19}. Both parts of the chamber are pumped with 40 litres/second ion getter pumps. During operation, the pressure in the main part of the chamber is 2×10^{-9} mbar, leading to a lattice lifetime of about one second. This lifetime is long enough for clock operation with typical interrogation times of about 150 ms limited by the coherence time of the clock laser.

2.3. Zeeman slower with permanent magnets

While direct loading atoms from the background using 2D-MOT – 3D-MOT techniques has been proven to be very useful to cool alkaline atoms, Zeeman slowing [14] is a robust method to provide high flux of slow atoms from elements with low vapor pressure. To avoid the power dissipation and thus heating by the solenoid permanent magnets can be used; an approach that has been chosen by several groups [15, 16, 17, 18] recently.

However, most of these Zeeman slowers produce a magnetic field that is oriented perpendicularly to the atomic beam. The main disadvantage of this transverse field configuration is that only half of the power of the slowing beam has the correct polarization for slowing the atoms – the other half is wasted. The design of a magnet-based Zeeman slower with a magnetic field parallel to the atomic beam is more challenging and more magnetic material has to be used [19]. For this setup we designed and built a Zeeman slower with a longitudinal field, because the ineffective use of blue laser power in the transverse case would be a major drawback especially in a transportable setup.

For our Zeeman slower, we have tried to match the magnetic field of an existing solenoid-based slower, which shows an excellent performance in our stationary strontium clock. The maximum capture velocity of the setup is adapted to an oven temperature of 380 °C to 450 °C.

To get the desired longitudinal field standard sized NdFeB magnet blocks[1] are arranged symmetrically around the beam axis such that the transverse components cancel out (Fig. 3). A ring of four magnets is needed to provide the desired field. The distance of each quadruple of

[1] Two types of magnets are used, remanence $B_r = 1.35$ T (1.3 T) and dimensions 30 mm × 30 mm × 15 mm (20 mm × 20 mm × 10 mm).

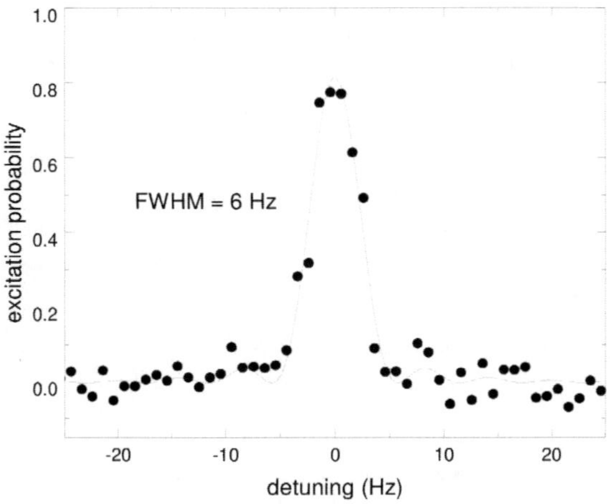

Figure 5. Nearly Fourier-limited linespectrum of the $^1S_0$–$^3P_0$ clock transition in a spin polarized sample of ^{87}Sr atoms interrogated with a probe time of 170 ms. (color online)

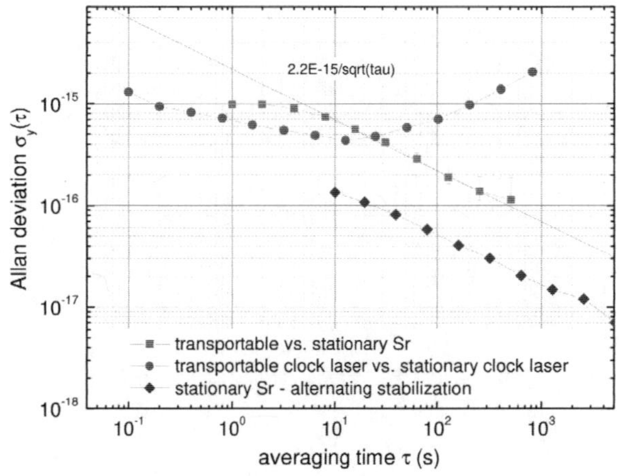

Figure 6. Allan deviation for a 20 minute comparison between stationary and transportable strontium lattice clock (squares). For comparison we show the instability of the reference cavity (circles) [12] and the instability of the stationary strontium clock (diamonds) [3]. (color online)

magnets to the axis can be used for fine tuning of the field near their z-position. An iterative algorithm had to be applied to fit all magnet positions to produce the design field. We have been systematically varying different parameters such as the total number of magnet blocks distributed equidistant along the length of the slower and the size of magnets for each ring. The magnets have been tilted by 45° with respect to the optical axis, to reduce unwanted stray-fields near the MOT. The final design was realized by tightly fitting the magnets into a rigid aluminum holder. With this fixed mount, we were able to get very close to the first (positive) part of the field without further adjustment of the individual magnet blocks as shown in Fig. 4.

To experimentally optimize the field at the output of the Zeeman slower, we firstly adjust the position of the two last magnets that form the part of the magnetic field where the atoms leave the slowing region. These are mounted on translation stages and can be radially moved to tune the magnetic field and hence the minimum velocity of the atoms interacting with the slowing beam, which is a critical part for MOT loading. The second parameter is the detuning of the slowing laser to the atomic resonance. A smaller detuning leads to a reduced capture velocity from the thermal beam, but makes the system more robust against rapid changes of the magnetic field along the slowing axis. Due to such imperfections atoms could otherwise leave the slowing process at a high velocity and would not be captured in the MOT.

We characterized the performance of our slower by measuring the velocity distribution through the Doppler profile on the 461-nm transition of the slowed atomic beam and the loading rate of the MOT. We find that 5×10^8 atoms (^{88}Sr) per second can be captured in the MOT. The velocity distribution shows that atoms can be slowed from about 450 m/s down to 70 m/s without significant loss at intermediate velocities. The loading time needed for regular clock operation with ^{87}Sr is 300 ms, which is similar to the performance of the solenoid Zeeman slower in our stationary experiment.

effect	**correction** (10^{-16})	**uncertainty** (10^{-16})
lattice ac Stark	0	13.0
BBR ac Stark	48.8	0.07
BBR ac Stark oven	0	1.1
2nd order Zeeman	1.2	0.12
line pulling	0	0.07
servo error	0	1.5
ac Stark probe	0	0.05
unstabilized fiber	0	3.0
cold collisions	0	1.0
background gas collisions	0	0.2
total	50.0	13.5

Table 1. Uncertainty budget of the transportable strontium clock. Currently it is dominated by the not yet fully evaluated Stark shift from the free running lattice laser.

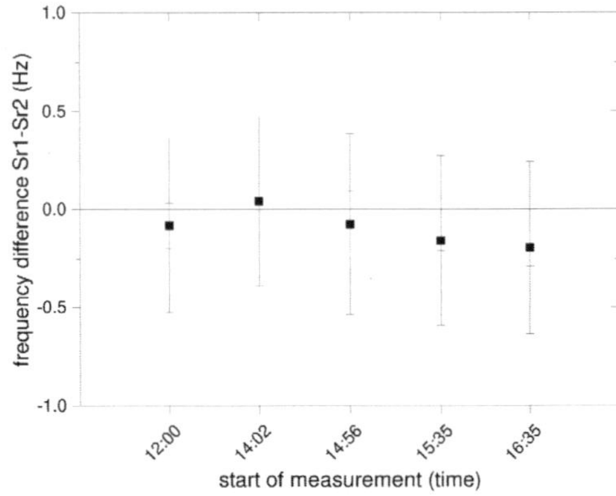

Figure 7. Frequency difference between stationary (Sr1) and transportable (Sr2) strontium clock from typically 1200 s long comparisons. The blue error bars indicate the statistical uncertainty while the black bars indicate the total uncertainty. (color online)

3. Clock comparison and uncertainty budget

We compared the frequency of our transportable clock to our stationary strontium lattice clock [6, 3], while both systems were locked to the clock transition in ^{87}Sr using the average of the two stretched m_F states. For this comparison two completely independent clock lasers were used and the interrogation cycles of the clocks were not synchronized. The stationary clock has shown an estimated instability of $5 \times 10^{-16}\ \sqrt{\mathrm{s}/\tau}$ in an interleaved self comparison (Fig. 6) which is an upper bound for the instability under normal operation. The measured instability of $2.2 \times 10^{-15}\ \sqrt{\mathrm{s}/\tau}$ for the comparison, also shown in Fig. 6, is thus dominated by the performance of the transportable clock. The instability of the transportable clock is limited by the transportable clock laser, which allows us to resolve a 6 Hz Fourier-limited linewidth for the clock transition (Fig. 5). This result was achieved in a fully independent fashion, i.e. without using techniques like correlated interrogation. Therefore this instability can be delivered by the transportable clock in future measurements without relying on the availibility of additional equipment.

The frequencies of the two clocks agree within the uncertainty of the measurement (Fig. 7). The stationary clock has an uncertainty of 3×10^{-17}. This does not contribute significantly to the uncertainty of the measurements. The contributions to the uncertainty of the transportable clock was evaluated in Table 1.

3.1. Lattice light shift

Even though the frequency of the lattice laser was not stabilized during the first measurement campaign, a preliminary uncertainty budget for the transportable clock was evaluated (Fig. 1). The frequency of the lattice laser was monitored via a beatnote with the lattice laser of the stationary strontium system. By comparison with this system, the lattice frequency of the transportable system was known to be near the Stark-shift-cancellation wavelength for ^{87}Sr ($m_F = 9/2$). It is necessary to measure the exact Stark-shift cancellation wavelength for each clock separately, because parameters like the angle between magnetic field and polarization or spurious spectral components of the lattice laser influence the the Stark-shift cancellation wavelength significantly. During the time we were running a comparison measurement (about 20 minutes), the lattice laser frequency was stable to within 50 MHz, corresponding to an uncertainty contribution of 2.3×10^{-16}. With this lattice laser frequency we tried to determine the magic wavelength by operating the clock at 5 lattice depths between about 70 E_{recoil} and 140 E_{recoil}. However, because of inconsistencies in the data that might be due to the application of a TA laser system for the optical lattice [20], we can only obtain a frequency uncertainty due to the unknown Stark-shift cancellation wavelength of 1.3×10^{-15}.

It is known that spurious spectral components in the lattice lasers emission can compromise the measurement of the Stark-shift cancellation wavelength [20]. Thus we use a volume Bragg grating with a linewidth of 0.1 nm for spectral filtering and a short optical fiber between grating and AOM for spatial mode filtering and spatial superposition of remaining spectral components. Further investigations are urgently required to decide whether this setup will be able to support a more precise determination and cancellation of the lattice light shift.

3.2. Blackbody radiation shift

The blackbody radiation shift has contributions from two sources. One stems from the blackbody radiation of the hot (480 °C) oven. This radiation can be easily blocked by the mechanical shutter in the oven chamber. With the shutter this relatively high uncertainty contribution of 1.1×10^{-16} will be reduced to below the 10^{-18} level. The second part of the blackbody radiation shift is coming from incomplete knowledge of the characteristic temperature of the vacuum chamber. We measure the temperature of the chamber with eight Pt100 sensors placed at critical places around the chamber. During clock operation we find a temperature difference of 250 mK between the hottest and the coldest point. Taking this difference as the width of are rectangular probability distribution for the true representative temperature [6], we obtain an uncertainty contribution of 7×10^{-18}, where the uncertainty of the shift coefficient does not play a significant role [13, 1].

3.3. Other effects

We have performed a preliminary investigation of the other frequency shifting effects that could be relevant at an uncertainty level of 10^{-16} (as shown in Tab. 1). As the measurement uncertainty was limited by the residual lattice ac Stark shift, no special care was taken to minimize these effects. This will be the next step after the characterization of the lattice light shifts. However, even without special optimization most of the effects are at the low 10^{-16} level and can be lowered by one order of magnitude or even more, as has been shown in our stationary lattice clock experiment [6], which employs identical techniques in all critical parts. At the moment the largest uncertainty contributions are due to an unstabilized part of the optical path, which is connecting the clock laser and the physics package, and the servo error, from the non-optimized gain settings for the stabilization to the atomic resonance. We give a comparatively large number for the optical path-length stabilization, because we have observed frequency shifts of this size due to AOM heating in the past. Furthermore, the frequency shift due to cold collisions was only calculated from an estimated density, based on a roughly known atom number. This results in a very conservative uncertainty for the collisional shift of 1×10^{-16}.

4. Summary and Conclusion

In this paper we have presented the first evaluation of an autonomous transportable optical clock based on fermionic strontium using a fully transportable clock laser. The clock showed an excellent instability of $2.2 \times 10^{-15} \sqrt{s/\tau}$ in a comparison with a more stable laboratory system, and an agreement between the two clocks within the uncertainty of the transportable clock. This uncertainty was limited to 1.3×10^{-15} due to the not yet fully evaluated lattice light shift. With the modification that are mentioned in the text, we expect that the uncertainty can be further improved the near future. Especially the proximity of a well evaluated strontium lattice clock will simplify the evaluation of the transportable clock and make it well suited for measurements in relativistic geodesy that are planned in the near future. In a few years from now, when optical fiber links will be available between various metrology labs throughout Europe, the transportable clock can enable relativistic geodesy with high accuracy along the links.

5. Acknowledgments

This work was performed within the framework of the Centre of Quantum Engineering and Space-Time Research (QUEST). We acknowledge funding from the German Research Foundation DFG within the collaborative research center CRC 1128 geo-Q and research training group RTG 1729, from the Marie-Curie Action ITN "FACT" and from the ITOC and QESOCAS projects in the European Metrology Research Programme EMRP. The EMRP is jointly funded by the EMRP participating countries within EURAMET and the European Union.

References

[1] Nicholson T L *et al* 2015 Nature Com. **6** 6896
[2] Ushijima I *et al* 2015 Nature Photonics **9** 185
[3] Al-Masoudi A *et al* 2015 Phys. Rev. A **92** 063814
[4] Bjerhammar A *et al* 1985 Bulletin Géodésique **59** 207
[5] Verhaar B J *et al* 1987 Phys. Rev. A **35** 3825
[6] Falke S *et al* 2014 New J. Phys. **16** 073023
[7] Mukaiyama T *et al* 2003 Phys. Rev. Lett. **90** 113002
[8] Vogt S *et al* 2011 Appl. Phys. B **104** 741
[9] Häfner S 2015 Ultrastabile Lasersysteme für Weltraum und Boden-Anwendungen Ph.D. thesis Leibniz Universität Hannover (Online: http://edok01.tib.uni-hannover.de/edoks/e01dh16/84568972X.pdf)
[10] Sterr U 2011, Frequenzstabilisierungsvorrichtung German patent DE 10 2011 015 489
[11] Chen Q F *et al* 2014 Rev. Sci. Instrum. **85** 113107
[12] Häfner S *et al* 2015 Opt. Lett. **40** 2112
[13] Middelmann T *et al* 2012 **109** 263004
[14] Phillips W D and Metcalf H 1982 Phys. Rev. Lett. **48**, 596
[15] Ovchinnikov Y B 2007 Opt. Commun. **276** 261
[16] Ovchinnikov Y B 2008 Eur. Phys. J. Special Topics **163** 95
[17] Chen Z *et al* 2011 Phys. Rev. Lett. **106** 133601
[18] Hill I R *et al* 2012 in *European Frequency and Time Forum (EFTF)* (IEEE) pp. 545–549.
[19] Ovchinnikov Y B 2011 Opt. Commun. **285** 1175
[20] Le Targat R *et al* 2012 *European Frequency and Time Forum (EFTF)* (IEEE) Contribution to European Frequency and Time Forum 2012, Gothenburg, Sweden

Journal of Physics: Conference Series **723** (2016) 012021 doi:10.1088/1742-6596/723/1/012021

The NIM Sr Optical Lattice Clock

Y Lin[1], Q Wang[1,2], Y Li[1,2], F Meng[1,3], B Lin[1,2], E Zang[1], Z Sun[1], F Fang[1], T Li[1] and Z Fang[1]

[1]Division of Time and Frequency, National Institute of Metrology, Beijing, 100029, China
[2]Department of Precision Instrument, Tsinghua University, Beijing 100084, China
[3]School of Electronics Engineering and Computer Science, Peking University, Beijing 100871, China

linyige@nim.ac.cn

Abstract. A ^{87}Sr optical lattice clock is built at the National Institute of Metrology (NIM) of China. The atoms undergo two stages of laser cooling before being loaded into a horizontal optical lattice at the magic wavelength of 813 nm. After being interrogated by a narrow linewidth 698 nm clock laser pulse, the normalized excitation rate is measured to get the frequency error, which is then used to lock the clock laser to the ultra-narrow $^1S_0$-$^3P_0$ clock transition. The total systematic uncertainty of the clock is evaluated to be 2.3×10^{-16}, and the absolute frequency of the clock is measured to be 429 228 004 229 873.7(1.4) Hz with reference to the NIM5 cesium fountain.

1. Introduction
In recent years, the accuracy of optical clocks have surpassed the best cesium fountain clocks [1, 2, 3, 4]. There are extensive discussions on the future redefinition of the SI (International System of Units) second using optical atomic frequency standards [5, 6, 7]. Therefore, before the redefinition it is always desired to have more laboratories worldwide to research on optical clocks and make absolute frequency measurements independently to demonstrate their agreement and consistency. Here we report our work on building a ^{87}Sr optical lattice clock and its evaluation result.

2. Experimental setup
The experimental details of our Sr clock can be found in ref [8, 9, 10, 11], here a brief introduction is given. Figure 1 shows the schematic of our experimental setup. The Sr atom vapor coming out of an effusive oven is firstly collimated by a hot micro-channel nozzle and then enters a transverse cooling cube to be further collimated by 2-dimensional (2D) optical molasses. After that, the atomic beam passes through a 30 cm long spin-flip type Zeeman slower [8] and finally enters the magneto-optical trap (MOT) chamber. The viewport opposite to the Sr oven is extended ~50 cm away from the MOT chamber by a piece of vacuum tube and heated up to more than 110 °C to prevent it from the deposition of Sr atoms. The long vacuum tube ensures that the temperature increase of the MOT chamber induced by the heated window is insignificant and the solid angle of the heated window seen by the atoms is reduced.

Two stages of laser cooling are used to trap ^{87}Sr atoms. The first stage laser cooling is operated on the $^1S_0$-$^1P_1$ transition at 461 nm with a natural linewidth of 32 MHz [8]. The 461 nm laser output is divided into five beams, including absolute frequency lock, Zeeman slowing, 2D molasses cooling, MOT cooling and probe beams. All these five laser beams are frequency shifted by different acousto-optic modulators (AOMs) to meet specified frequency detuning requirements. These AOMs are also used as fast light switches in order to run the experiments according to a specified time sequence. At the same time, mechanical shutters are employed in the light paths to shut the light completely off. Two re-pump lasers at 679 nm and 707 nm are necessary for improving the cooling efficiency and for the transition probability measurement [9].

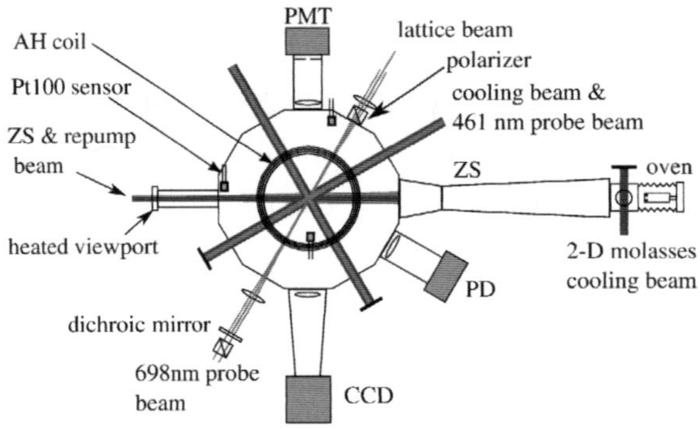

Figure 1. (color online) The schematic of the experimental setup. AH: anti-Helmholtz, ZS: Zeeman slower, PD: photodiode, PMT: photomultiplier tube

After the first stage laser cooling, the temperature of the atoms is ~3 mK. The atoms are then transferred to the second stage MOT utilizing the $^1S_0$-$^3P_1$ transition at 689 nm with a natural linewidth of 7.5 kHz. The master laser for the second stage laser cooling is an external-cavity diode laser (ECDL). The frequency of this 689 nm ECDL is locked to a reference cavity (finesse ~ 10,000) installed in a temperature controlled vacuum chamber through Pound-Drever-Hall (PDH) method. The output of the master laser is amplified by two slave lasers, one with injection lock and the other with offset-frequency phase lock. The output of the two slave lasers are delivered to the MOT by two pieces of single mode polarization maintaining (SMPM) fiber. Two AOMs are used for broad-band frequency modulation of these two slave lasers to improve the transfer efficiency between the first stage and second stage MOT [9]. At the beginning of the transfer from the blue MOT to the red MOT , the 461 nm cooling laser is switched off and the 689 nm laser is switched on. The field gradient is rapidly dropped from 50 Gauss/cm to 3 Gauss/cm and then ramped up to 10 Gauss/cm in 100 ms. The 689 nm laser frequency is wideband modulated (deviation 3 MHz, modulation frequency 30 kHz) by the AOMs during the dropping down and the ramping up of the field gradient. The frequency modulation of the lasers are switched off at the end of the field ramp. After another 55 ms of narrow line laser cooling, the ^{87}Sr atoms can be cooled down to a temperature as low as 3 μK.

During the whole cooling process, the 813 nm lattice laser is switched on and overlapped with the atom cloud. The lattice laser is a commercial Ti: sapphire laser (Coherent MBR110 pumped by a Verdi-V10). The frequency of the lattice laser is locked to a homemade F-P cavity. During the evaluation and frequency measurement of the Sr clock, the absolute frequency of the lattice laser is measured by the optical frequency comb once per day. The lattice beam (~300 mW) is focused to the center of the MOT and then retro-reflected by a dichroic mirror to form a standing wave optical lattice. The dichroic mirror has high reflectivity at 813 nm and high transparency at 698 nm. The optical lattice is set to be horizontally oriented [9]. The lattice beam waist is ~ 42 μm and the trap depth is 29 μK. After the second stage laser cooling, more than 10^4 atoms are trapped and confined in the Lamb-Dicke regime in the longitudinal direction of the 1-dimensional optical lattice.

Our 698 nm master laser is an ECDL locked to a high finesse reference cavity with the PDH method to narrow its linewidth. Both the cavity spacer and the mirror substrates are made of ultra-low expansion (ULE) glass, which set the thermal noise limited stability to $\sim 1\times10^{-15}$ at 1 s [12]. The stability of our 698 nm laser is measured to be 3×10^{-15} @ 1 s from the beat note of two similar laser systems [10]. Since then, some modifications focusing on reducing the residual amplitude modulation noise of the electro-optic modulator have been made, and the stability of the laser has been further improved. The output of the master 698 nm ECDL is frequency shifted by a double-pass 240 MHz AOM and is used to injection lock an F-P laser diode to amplify its power. The AOM driven by a direct-digital synthesizer (DDS) is used to compensate the drift of the cavity and bridge the gap between the cavity mode and the clock transition frequency. At the MOT, the 698 nm light is precisely co-aligned with the 813 nm lattice laser from the opposite direction using the dichroic mirror mentioned above.

Rabi excitation is used to probe the $^1S_0$-$^3P_0$ clock transition. A normalized technique for minimizing the effect of the shot to shot atom number fluctuation is used to measure the transition probability [9]. When the clock laser frequency is scanned over a span of more than 200 kHz, sideband resolved spectra of the clock transition are acquired, as shown in Figure 2. After the second stage laser cooling, the atoms are spin-polarized by being optically pumped to $|F=9/2, m_F=+9/2\rangle$ and $|F=9/2, m_F=-9/2\rangle$ stretched states alternatively using either σ+ or σ- polarization of the 689nm laser at $^1S_0(F=9/2)$-$^3P_1(F=9/2)$ transition frequency [11]. The 698 nm probe laser is co-aligned with the lattice laser, and their polarizations are aligned with the bias magnetic field (~ 56 μT), which is used to resolve the two stretched states. When atoms are optically pumped to a single spin sub-state, the Rabi transition linewidth can be as narrow as 3 Hz with a 698 nm probe pulse width of 320 ms, as shown in the inset of Figure 2. The actual atomic lock uses 80 ms probe pulse width (the Fourier-limited Rabi linewidth is ~10 Hz) to make the locking more reliable. Two independent digital servos are used to lock the clock laser time-multiplexed to the two stretched states. The average of the two digital locks, which takes 4 clock cycles, gives the center frequency of the Sr clock. A complete single clock cycle including two stages of laser cooling, spin polarizing, Rabi excitation and transition probability measurement takes 1018 ms.

3. Systematic shifts evaluation

Recently, the first systematic shifts evaluation of NIM's Sr clock has been made [13]. The evaluation of the systematic uncertainties is performed using the self-comparison method [14]. Two independent atomic servos that share the same physical apparatus are compared in a time-interleaved way. The systematic effects and their corrections with uncertainties are listed in Table 1.

Table 1. Frequency corrections and uncertainties of the Sr clock and of its absolute frequency measurement.

Contributor	Correction (10^{-16})	Uncertainty (10^{-16})
Lattice Stark	17.2	2.2
BBR Stark	49.7	0.7
2nd order Zeeman	1.8	0.1
Density	12.0	0.3
Clock laser Stark	0	0.1
Line pulling	0	0.1
Sr total	**80.7**	**2.3**
statistical		13
Gravitational	-50.6	1.1
Fountain calibration		31
Fiber transfer	0	< 1
Measurement total		**34**

The dominant correction comes from the black-body radiation (BBR) induced frequency shift. The BBR shift and its correction factors have been studied in depth by many other groups [3, 15, 16]. In our clock, the ambient temperature is recorded continuously during the experiments. Three thin film PT100 sensing resistors are installed onto the MOT chamber. One is embedded in a screw hole in the stainless steel MOT chamber. The second one is stuck to the flange which connects the extended hot Zeeman slowing beam window and the MOT chamber. The last one is stuck to the top MOT window. During the frequency measurements, the temperature difference between these sensors is less than 0.3 °C, and the fluctuation of the temperature is less than 0.5 °C. The ambient temperature 22(1) °C gives the BBR shift correction $49.7(0.7)\times10^{-16}$.

The second dominant frequency shift and uncertainty comes from the lattice AC Stark shift. In order to measure the "magic wavelength", the lattice laser frequency is changed by locking to different modes of the reference cavity. At one lattice frequency, the lattice trap depth is modulated with a depth difference as large as 70 Er (Er is the lattice photon recoil energy) between two atomic servos to measure the shift sensitivity. Because the modulation of the lattice depth also changes the atomic density, the frequency shift caused by this atomic density change needs to be corrected. The zero shift frequency is determined to be 368 554 672(44) MHz. During the absolute frequency measurements, the lattice laser frequency is locked to the nearest cavity mode which is ~370(3) MHz away from the zero shift frequency and measured by the optical frequency comb. The lattice trap depth during the measurements is 176(5) Er. The correction of the lattice AC Stark shift is $17.2(2.2)\times10^{-16}$.

The density shift is measured with self-comparison method by alternating the atomic density. The density of the atoms trapped in the lattice is changed by applying different loading times from 200 ms to 800 ms in the first stage laser cooling. The stability of a typical differential measurement is shown (red squares) in Figure 3. Due to the high atom density and relatively high atom temperature (~4 μK in the longitudinal direction of the lattice and ~10 μK in the radial direction), we observed a large density dependent frequency shift. The correction for running the clock at high density is $12.0(0.3)\times10^{-16}$.

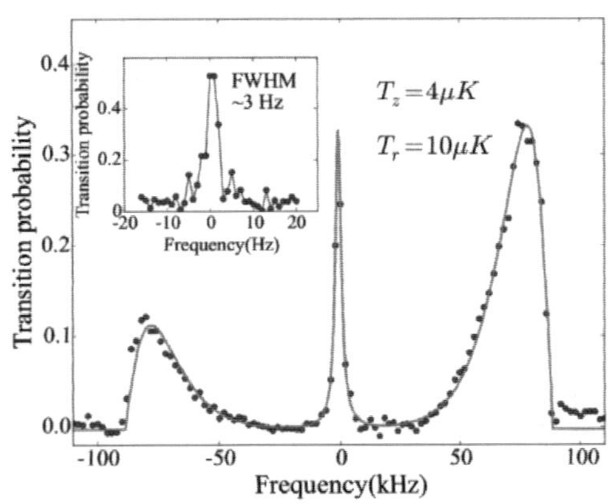

Figure 2. (color online) Sideband resolved spectra of the $^1S_0$-$^3P_0$ transition. The red solid line is a fit to the experimental data, which gives the atoms' temperature. The inset is a high resolution spectrum (full width at half maximum ~3 Hz) of a single Zeeman component of the $^1S_0$-$^3P_0$ transition with a Rabi excitation pulse width of 320 ms.

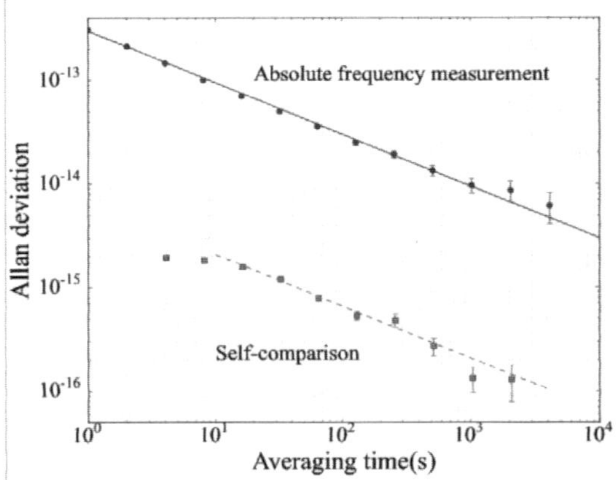

Figure 3. (color online) The measurement stability. The red solid squares show the Allan deviation (ADEV) of a typical self-comparison measurement of the density shift. The ADEV fits to (red dashed line) $6.6\times10^{-15}/\sqrt{\tau}$. The blue circles show the ADEV of a typical absolute frequency measurement which fits to (blue solid line) $3.0\times10^{-13}/\sqrt{\tau}$.

The clock laser is alternatively locked to the m_F=+9/2 and m_F=-9/2 components of the clock transition at a small bias magnetic field, and the first order Zeeman shift is canceled out when using the average of the two locks as the clock frequency. The frequency difference between these two locks gives the estimation of the magnetic field experienced by the atoms. With the coefficient in ref [2], the second order Zeeman shift is estimated to be $1.8(0.1)\times10^{-16}$.

The clock laser power is several nW and the Stark shift caused by the clock laser is very small (less than 0.1×10^{-16}) using the coefficient value in ref [17]. With the transition linewidth of 10 Hz and the split between two adjacent m_F states of 62 Hz, less than 10% of atoms remain in the other states after spin-polarization contribute less than 0.1×10^{-16} uncertainty by line pulling effect. DC Stark shift may occur when electric charges trapped in the coating of the MOT chamber fused silica viewports [18]. The charges can be removed effectively by shining UV light on the viewports [18, 19]. An UV lamp is used to treat the viewports, and we found no significant shift. Other effects, that contributed very small shifts and uncertainties considering our present total uncertainty level, are omitted. The total systematic uncertainty of NIM's Sr clock is 2.3×10^{-16}.

4. Absolute frequency measurement

The system setup for the absolute frequency measurement is shown in Figure 4. In order to trace the absolute frequency of the NIM's Sr clock to NIM5 cesium fountain, the frequency of the flywheel H-Maser is delivered to the Sr laboratory (located at a different campus about 50 km apart) with a fiber noise cancellation system. The uncertainty introduced by the fiber transfer system is less than 1×10^{-16} [20]. The gravitational shift, according to Einstein's general theory of relativity, is calculated to be $50.6(1.1)\times10^{-16}$ with the Sr clock altitude of 46.4 (1.0) m.

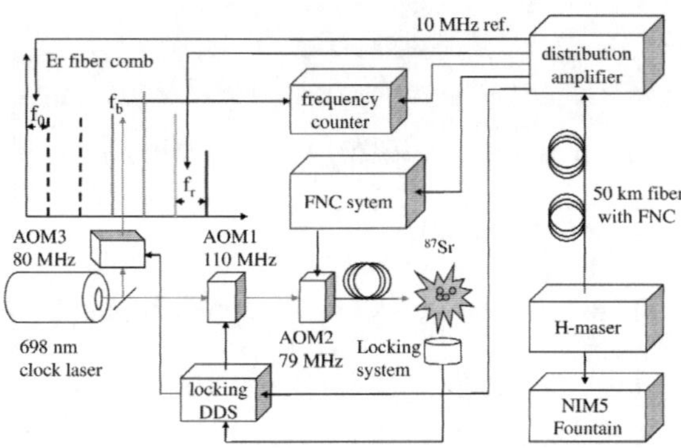

Figure 4. (color online) The schematic of the absolute frequency measurement. AOM: acousto-optic modulator. FNC: fiber noise cancellation.

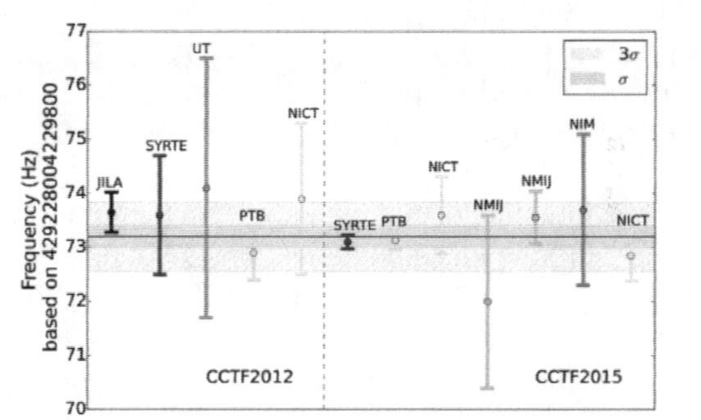

Figure 5. (color online) Sr clock frequency reported to CCL-CCTF FSWG. The black solid line is the 2015 recommended value. σ is the standard uncertainty of the recommended value.

A total of 49,113 s effective measurement data was acquired with 5 measurements. The Allan deviation of one of the measurements is shown (blue circles) in Figure 3. The statistical uncertainty of the total measurements is 1.3×10^{-15}. As shown in Table 1, the calibration uncertainty of NIM5 during the course of the measurements is 3.1×10^{-15}. The absolute frequency of the Sr clock is 429 228 004 229 873.7(1.4) Hz. This measurement result has been published in ref [13] and reported to the Consultative Committee for Length and the Consultative Committee for Time and Frequency Frequency Standards Working Group (CCL-CCTF FSWG) of the International Committee for Weights and Measures (CIPM) and agreed with the new recommended value of the Sr clock (shown in Figure 5 as black solid line), which is calculated by the weighted mean of all the effective measurements up-to-date [21, 22, 23, 24, 25, 26].

5. Conclusion

An optical lattice clock based on ^{87}Sr has been built and evaluated at NIM. The total systematic uncertainty of the clock is 2.3×10^{-16}. The absolute frequency of the clock has been traced to NIM5 cesium fountain with a measurement uncertainty of 3.4×10^{-15}. A new clock laser based on a long reference cavity is currently under development to further improve the stability of this Sr clock. We are planning to build the second Sr lattice clock in the near future.

6. Acknowledgment

The research is supported by the National Natural Science Foundation of China under Grant 91336212 and 91436104. The authors would like to thank W Chen, A Zhang, Y Gao and K Liang for their kind help.

References

[1] Chou C, Hume D, Koelemeij J, Wineland D and Rosenband T 2010 *Phys. Rev. Lett.* **104** 070802
[2] Bloom B, Nicholson T, Williams J, Campbell S, Bishof M, Zhang X, Zhang W, Bromley S and Ye J 2014 *Nature* **506** 71
[3] Nicholson T *et al.* 2015 *Nat. Commun.* **6** 6896
[4] Ushijima I, Takamoto M, Das M, Ohkubo T and Katori H 2015 *Nat. Photon.* **9** 185
[5] Gill P 2011 *Phil. Trans. R. Soc. A* **369** 4109
[6] Parker T 2012 *Review of Scientific Instruments* **83** 021102
[7] Riehle F 2015 *arXiv:1501.02068*
[8] Wang S, Wang Q, Lin Y, Wang M, Lin B, Zang E, Li T and Fang Z 2009 *Chin. Phys. Lett.* **26** 093202
[9] Lin Y *et al.* 2013 *Chin. Phys. Lett.* **30** 014206
[10] Li Y *et al.* 2014 *Chin. Phys. Lett.* **31** 024207
[11] Wang Q, Lin Y, Li Y, Lin B, Meng F, Zang E, Li T and Fang Z 2014 *Chin. Phys. Lett.* **31** 123201
[12] Numata K, Kemery A and Camp J 2004 *Phys. Rev. Lett.* **93** 250602
[13] Lin Y *et al. Chin. Phys. Lett.* **32** 090601
[14] Nicholson T, Martin M, Williams J, Bloom B, Bishof M, Swallows M, Campbell S and Ye J 2012 *Phys. Rev. Lett.* **109**
[15] Yudin V, Taichenachev A, Okhapkin M, Bagayev S, Tamm C, Peik E, Huntemann N, Mehlstäubler T and Riehle F 2011 *Phys. Rev. Lett.* **107** 030801
[16] Middelmann T, Falke S, Lisdat C and Sterr U 2012 *Phys. Rev. Lett.* **109** 263004
[17] Baillard X, Fouché M, Le Targat , Westergaard P, Lecallier A, Le Coq, Rovera G, Bize S and Lemonde P 2007 *Opt. Lett.* **32** 1812
[18] Lodewyck J, Zawada M, Lorini L, Gurov M and Lemonde P 2012 *IEEE Trans. Ultra. Freq. Cont.* **59** 411
[19] Pollack S, Turner M, Schlamminger S, Hagedorn C and Gundlach J 2010 *Phys. Rev. D. Particles and fields* **81** 02110
[20] Wang B, Gao C, Chen W, Miao J, Zhu X, Bai Y, Zhang J, Feng Y, Li T and Wang L 2012 *Scientific Reports* **2** 556
[21] Campbell G *et al.* 2008 *Metrologia* **45** 539
[22] Hong F *et al.* 2009 *Opt. Lett.* **34** 692
[23] Yamaguchi A, Shiga N, Nagano S, Li Y, Ishijima H, Hachisu H, Kumagai M and Ido T 2012 *Appl. Phys. Express* **5** 022701
[24] Le Targat *et al.* 2013 *Nat. Commun.* **4** 2109
[25] Falke S *et al.* 2014 *New J. Phys.* **16** 073023
[26] Akamatsu D, Inaba H, Hosaka K, Yasuda M, Onae A, Suzuyama T, Amemiya M and Hong F 2014 *Appl. Phys. Express* **7** 012401

Recent progress of neutral mercury lattice clock in SIOM

R C Zhao[1,2,3], X H Fu[1,2,3], K K Liu[1,2,3], W Gou[1,2,3], J F Sun[1,2], Z Xu[1,2,4], and Y Z Wang[1,2,4]

[1] Key Laboratory of Quantum Optics, Chinese Academy of Sciences, Shanghai 201800, China

[2] Shanghai Institute of Optics and Fine Mechanics, Chinese Academy of Sciences, Shanghai 201800, China

[3] University of Chinese Academy of Sciences, Beijing 100049, China

[4] Center of Cold atom physics, Chinese Academy of Sciences, Shanghai 201800, China

E-mail: xuzhen@siom.ac.cn

Abstract. Neutral mercury atom is one of good candidates of optical lattice clock. Due to its large atomic number, mercury atom is insensitive to black body radiation, which is the severe limitation for the development of optical clocks. However, the challenge of neutral mercury lattice clock is the requirement of high power deep-UV lasers, especially for both the cooling laser and the lattice laser. Here, we report the recent progress of neutral mercury lattice clock in SIOM, including the development for laser cooling of mercury atom and the cooling laser system with fiber laser amplifier. We have realized the magneto-optical trap of mercury atoms and measured the parameters of cold mercury atoms. A home-made external cavity diode laser works as a seed laser for a room temperature 1014.8 nm fiber laser amplifier. A new efficient frequency-doubling cavity from 1015 nm to 507 nm has been developed.

1. Introduction

Mercury is the heaviest stable atom that has been laser cooled. Its large nuclear charge number (Z=80) is a great advantage for precision measurement of some physical constants, such as permanent electric dipole moment and fine structure constant [1, 2]. With the improved performance of ultra-stable lasers and development of optical frequency comb, optical clocks advanced very quickly. The instability and uncertainty of optical lattice clock has gone to $2\times10^{-16}/\sqrt{\tau}$ and 10^{-18} level respectively, which has already surpassed microwave fountain clocks [3-6]. While Sr and Yb optical lattice clock mainly limited by blackbody radiation (BBR) shift, neutral mercury atom is a good candidate for optical lattice clock due to its low sensitivity of BBR at room temperature [7-10], which is no necessary to take a complex system to control the temperature uncertainty or operate under the cryogenic environment. However, the laser source for neutral mercury lattice clock is tough challenging because of their deep-UV region. Especially for cooling and lattice laser, high UV power is required. Here, we report the recent progress of laser cooling for neutral mercury atom and the development of a new laser cooling system in Shanghai Institute of Optics and Fine Mechanics (SIOM).

2. Laser cooling of neutral mercury atoms

An experimental setup of magneto-optical trap (MOT) was established for laser cooling of neutral mercury atoms. It is composed of vacuum system, cooling laser system, imaging system and time sequence control system. We can produce and measure cold mercury atoms in this setup.

2.1. Vacuum system

The vacuum system is consisted of two chambers, a source chamber and a science chamber, which are connected by a gate valve [11]. Source chamber contains a cooled mercury source, which have three main parts: the mercury cup, a heat sink and a multi-stage thermoelectric cooler (TEC). The mercury cup is cooled by the TEC, and the TEC is glued to the water cooled copper heat sink by thermally conductive low outgassing epoxy adhesive (H77, EPO-TEK). For daily operation, the mercury cup cools down to -70 °C to achieve a high vacuum with 2×10^{-9} Torr. For MOT operation, the temperature is kept at about -40 °C, corresponding to vacuum pressure of about 2×10^{-8} Torr.

The science chamber is a spherical octagonal cavity and it has eight indium-sealed UV fused silica (UVFS) windows with anti-reflection coatings at 254 nm. Six windows are for MOT beam access, and two are for detection and monitoring. The whole system is pumped only by an ion pump and the vacuum pressure can be lower than 10^{-9} Torr finally. A pair of MOT coils is mounted on the science chamber. After the compensation of the residual magnetic field by three pairs of Helmholtz coils, the total residual magnetic field is less than 10 mG in the MOT region.

2.2. Cooling laser system and optical configuration

A commercial frequency quadrupled semiconductor laser (Toptica, TA-FHG-1030) works as the cooling laser, which is locked on the cooling transition by sub-Doppler frequency modulation (FM) spectroscopy [12]. The commercial laser can maximally generate output power of 36 mW after two stages of frequency doubling with LBO and BBO crystals. The UV output laser is split into two beams by a beam splitter. About 10% of the UV laser is used to FM spectroscopy and the rest UV laser goes to the MOT chamber.

Because of the limitation of bandwidth in the servo loop of the frequency doubling, a feed-forward (FF) method is adopted to improve the frequency tunability of the UV laser system. FF module also reduces the laser frequency noise, when the laser frequency is locked on atom spectrum. The enhancement of the tunability and stability of UV laser gives a lot of advantages in the experiments of laser spectroscopy and laser cooling of mercury atom.

The sub-Doppler frequency modulation (FM) spectroscopy is adopted in our experiment. Almost all of the transitions have been found in saturated absorption spectroscopy (SAS) except ^{196}Hg because of its low natural abundance (~0.6%). With a high speed servo controller (New Focus, LB1005), the UV laser can be locked on the FM signals. Figure 1 shows the FM signals and saturated absorption signals for 8 transitions.

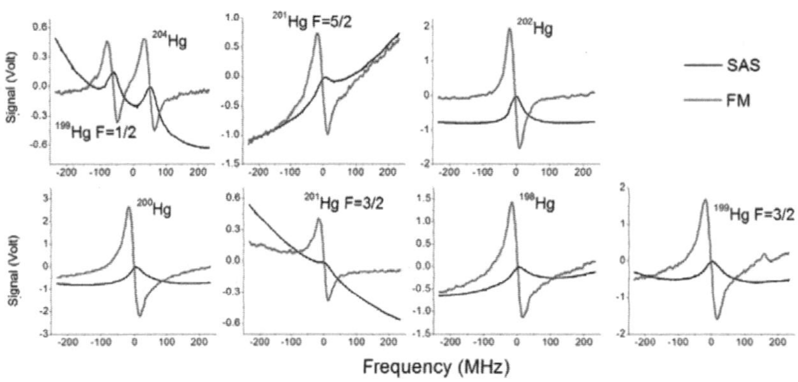

Figure 1. FM spectroscopy for all observed transitions.

Because of the shortage of the UV laser power, we adopt a folded beam configuration with a single shot to replace the normal configuration with six cooling beams for MOT [11]. The cooling laser power is about 10~20 mW and the beam diameter is magnified to 10 mm in the MOT. The MOT signal is detected by an electron multiplying charge coupled device (EMCCD, Andor, iXon3 885DU) and a photomultiplier (Hamamatsu, H9305-1) through 4f imaging system with fluorescence method. The magnification is 0.5 and the overall photon collection efficiency is about 0.0033.

2.3. Measurement and optimization of cold mercury atom in MOT

We have clearly observed the atom signals for all six rich abundance isotopes and measured the atom numbers by the fluorescence imaging of the cold mercury atoms. Background vapor pressure is an important factor in experiment, which could be changed by the Hg source temperature through TEC current. We measured the atom number of ^{202}Hg at different Hg source temperature as shown in figure 2. We choose -45 °C as work temperature. The capture rate and stable atom number in MOT is strongly influenced by laser detuning. The relation of the atom number in MOT and the laser detuning is measured for ^{202}Hg atom, as shown in figure 3. Maximal atom number is about 1.7×10^6 at about -10 MHz detuning. Due to the limited power of the UV laser, we can't measure the relation of the atom number and the intensity of cooling laser. Similarly, the maximal atom number is about 1.5×10^6 for ^{199}Hg atom.

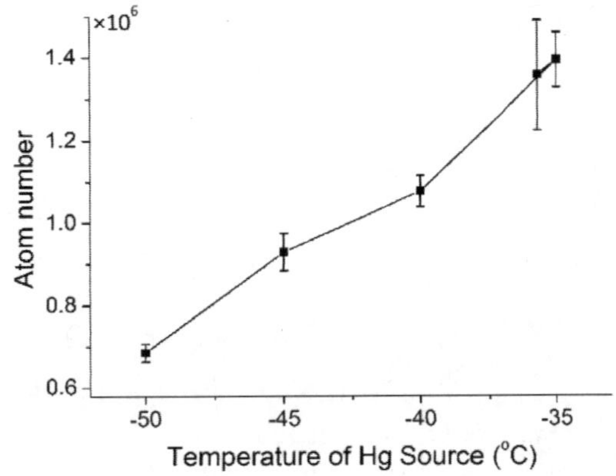

Figure 2. Dependence of atom number and temperature of mercury cup for ^{202}Hg atom.

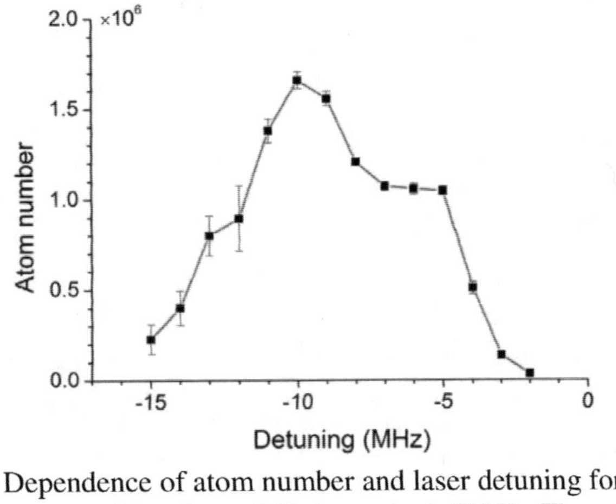

Figure 3. Dependence of atom number and laser detuning for ^{202}Hg atom.

The temperature was measured by time of flight method (TOF). The minimum flight time is 7 ms which is limited by the action of mechanical shutter. Figure 4 shows the TOF pictures at different flight time (left) and fitting result (right). Because of the theory of Doppler cooling, the lowest temperature had to be achieved in small detuning, which conflicted with the maximization of atom number [13]. So we decrease the laser detuning at the end of MOT loading. The lowest temperature is 170 μK for ^{202}Hg and 50 μK for ^{199}Hg.

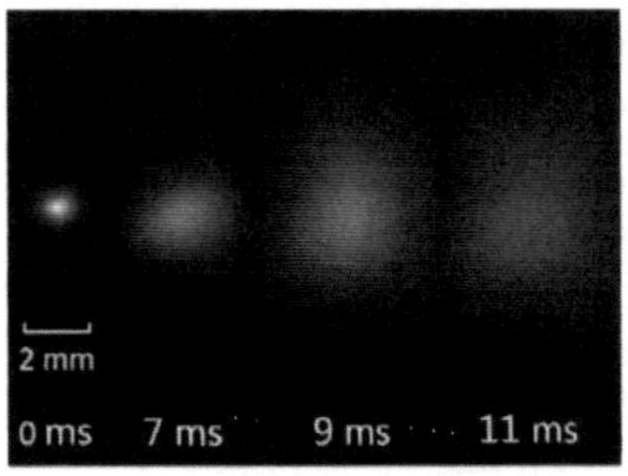

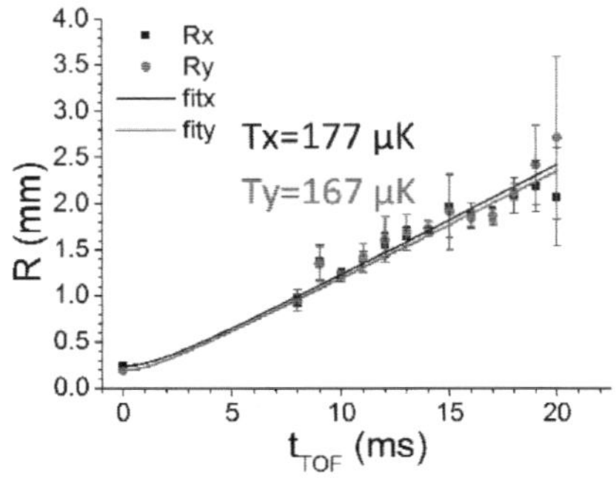

Figure 4. Temperature measurement for ^{202}Hg. Atom images at different TOF time (left) and temperature calculation by fitting result (right).

3. The development of cooling laser system

As mentioned above, to establish a neutral mercury optical lattice clock, the most challenging thing is to generate a high power deep UV laser at 253.7 nm for laser cooling of neutral mercury atom. In our previous work, we developed a 1014.8 nm fiber laser amplifier with short polarization-maintaining Yb-doped single mode fiber, which can work at room temperature [14]. But only 75 mW UV laser is generated at 4 W input IR laser power with two commercial doubling cavities. Here we present our recent development of cooling laser system.

3.1. The design of cooling laser system

The schematic diagram of our cooling laser system is shown in figure 5. A home-made external cavity diode laser (ECDL) works as seed laser for the 1014.8 nm fiber laser amplifier. After a single stage of fiber laser amplifier, the fundamental power can reach 7 W. The first doubling cavity with LBO crystal was completely demonstrated, but the second doubling cavity with BBO crystal is under constructed.

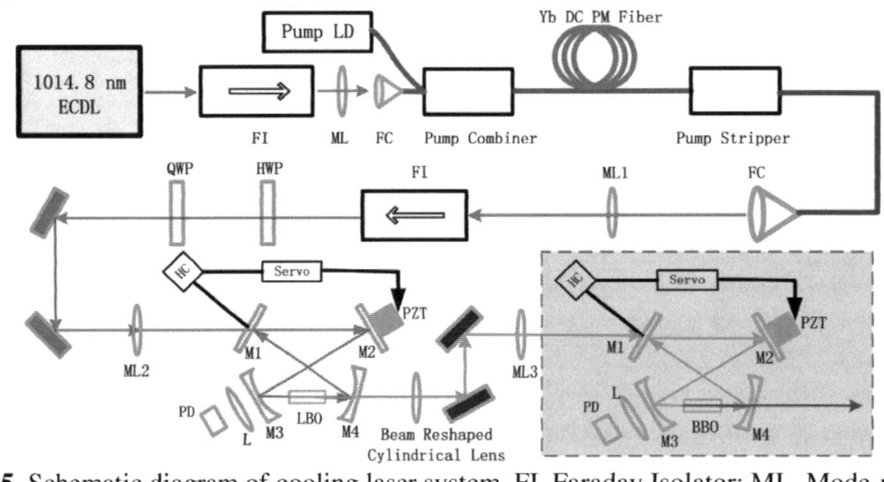

Figure 5. Schematic diagram of cooling laser system. FI, Faraday Isolator; ML, Mode-matching lens; FC, Fiber collimator; Yb DC PM Fiber, Yb Double Clad PM Fiber; HC, Hansch-Couillaud method; PZT, piezoelectric transducer; LBO, lithium triborate; L, lens; PD, photodiode.

3.2. *The home-made ECDL*

A home-made external cavity diode laser at 1014.8 nm is made as seed laser. The ECDL adopted Littrow configuration with an AR-coated laser diode (EYP-RWE-1060, Eagleyard) and 1200 lines/mm

Journal of Physics: Conference Series **723** (2016) 012022 doi:10.1088/1742-6596/723/1/012022

grating with 30% diffraction efficiency at 1015 nm. The output power is about 45 mW, and the typical line width is about one hundred of kHz. Its ASE suppression ratio is about 55 dB which even better than our used commercial laser (DL pro, Toptica), as shown in figure 6. After an optical isolator, the laser is coupled into a PM single mode fiber (PM 980) with laser power of 20 mW, which is sufficient for the 1014.8 nm fiber laser amplifier.

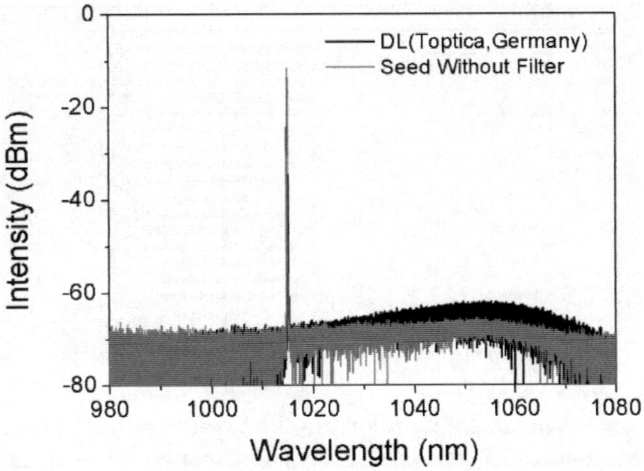

Figure 6. The comparison of ASE between the home-made ECDL (red) and the commercial ECDL (DL pro, Toptica) (black). The ASE of the home-made ECDL is about 10 dBm lower than the commercial ECDL at 1050 nm.

3.3. The 1014.8 nm fiber laser amplifier at room temperature

The fiber laser amplifier is same as Ref. [14], but the power of seed laser is much lower. When it operated at high output power, self-excitation will existed. Due to this reason, we insert a bandpass filter (FF01-935/170-25, Semrock) to purify the spectrum around 1030 nm. Now the maximal output power is 7 W, which limited by the current supply of pump laser, and the corresponding ASE suppression ratio is about 45 dB, as shown in figure 7.

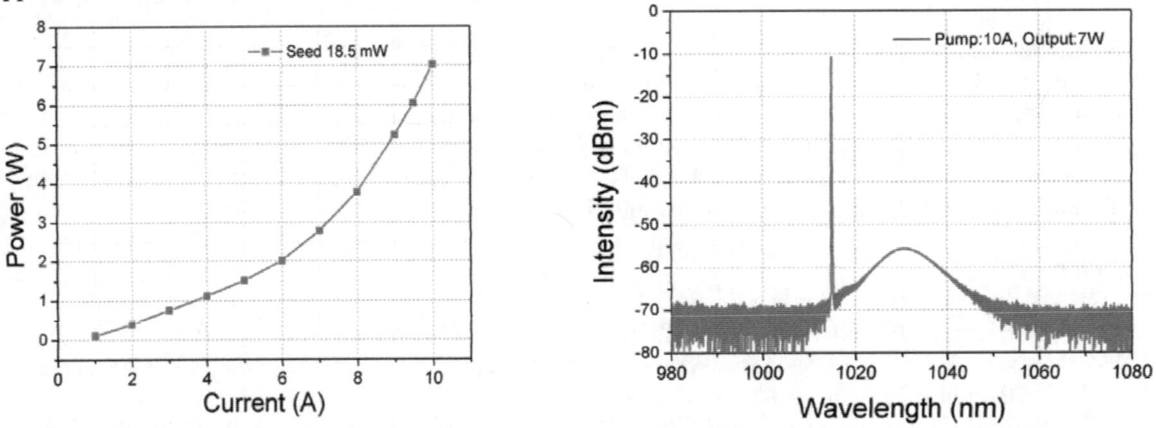

Figure 7. The output features of the 1014.8 nm fiber laser amplifier. The left figure is the relation of the output power and current of pump diode, and the right figure is the spectrum measurement at 7 W output power.

3.4. The first doubling cavity with LBO crystal

The high conversion efficiency and high stability from IR fundamental laser (1014.8 nm) to green laser (507.4 nm) is crucial to generate high power cooling laser in UV. A bow-tie ring cavity is adopted with a 20 mm length AR-coated LBO crystal, as shown in figure 5. The beam waist of IR

TEM_{00} mode is about 45μm at the center of LBO crystal, and the transmissivity of input coupler is about 3% around 1014.8 nm. Two lenses are used to match the mode of laser beam from fiber collimator and the fundamental mode of doubling cavity with high efficient of about 98%. The doubling cavity is locked by Hansch-Couillaud method [15]. The maximal output power of green laser is about 3.0 W at 4.9 W input power of IR laser with conversion efficiency of 61%, as shown in figure 8. The performance of this doubling cavity is well agreed with our design and calculation.

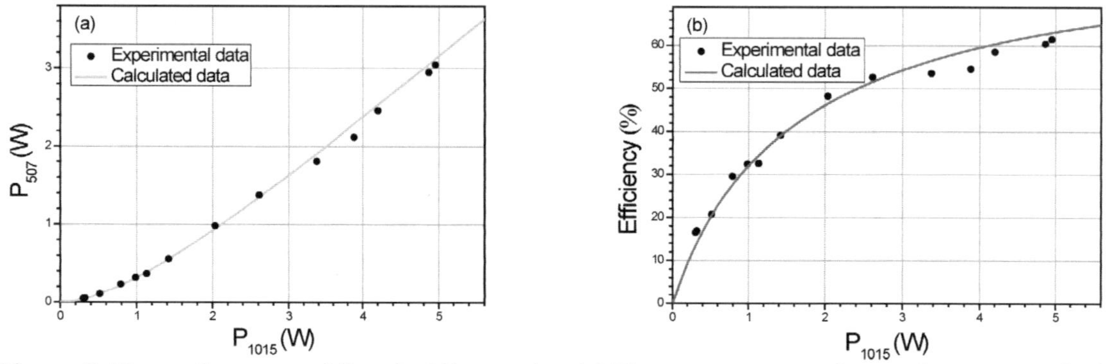

Figure 8. The performance of first doubling cavity. (a) The output power of green laser versus the input power of IR laser. (b) The conversion efficiency versus the input power of IR laser. The green (red) line is given by theoretical calculation.

4. Conclusions

In conclusion, we established an experimental system for laser cooling of neutral mercury atom, and successfully trapped six mercury isotopes in the magneto-optical trap. The atom number is about 1.7×10^6 for ^{202}Hg and 1.5×10^6 for ^{199}Hg atom. The temperature is about 170 μK for ^{202}Hg and 50 μK for ^{199}Hg. To enhance the laser power, a 1014.8 nm fiber laser amplifier was developed, which can work at room temperature. With our new develop of cooling laser system, the maximal output power at 507 nm is about 3 W. These works laid a good foundation to realize the neutral mercury lattice clock.

5. Acknowledgement

We thank Hongli Liu, Shiqi Yin, Guojian Shen, Damin Xu and Haibin Wu for their previous contributions on this system. The authors acknowledge the partial financial support from the National Natural Science Foundation of China under Grant No. 91436105, the National Basic Research Program of China under Grand No.2011CB921504, and the Research Project of Shanghai Science and Technology Commission under Grant. No. 09DJ1400700.

References

[1] Walther T 2007 *J. Mod. Optic.* **54** 2523-32
[2] Flambaum V V and Tedesco A F 2006 *Phys. Rev. C* **73** 055501
[3] Bloom B J, Nicholson T L, Williams J R, Campbell S L, Bishof M, Zhang X, Zhang W, Bromley S L and Ye J 2014 *Nature* **506** 71-5
[4] Ushijima I, Takamoto M, Das M, Ohkubo T and Katori H 2015 *Nat. Photonics* **9** 185-9
[5] McFerran J J, Yi L, Mejri S, Di Manno S, Zhang W, Guéna J, Le Coq Y and Bize S 2012 *Phys. Rev. Lett.* **108** 183004
[6] McFerran J J, Yi L, Mejri S, Zhang W, Di Manno S, Abgrall M, Guéna J, Le Coq Y and Bize S 2014 *Phys. Rev. A* **89** 043432
[7] Hachisu H, Miyagishi K, Porsev S G, Derevianko A, Ovsiannikov V D, Pal'chikov V G, Takamoto M and Katori H 2008 *Phys. Rev. Lett.* **100** 053001
[8] Petersen M, Chicireanu R, Dawkins S T, Magalhães D V, Mandache C, Le Coq Y, Clairon A

and Bize S 2008 *Phys. Rev. Lett.* **101** 183004
[9] Yamanaka K, Ohmae N, Ushijima I, Takamoto M and Katori H 2015 *Phys. Rev. Lett.* **114** 230801
[10] Gibble K 2009 *Phys. Rev. Lett.* **103** 113202
[11] Liu H L, Yin S Q, Liu K K, Qian J, Xu Z, Hong T and Wang Y Z 2013 *Chin. Phys. B* **22** 043701
[12] Liu K K, Liu H L, Zhao R C, Fu X H, Xu Z and Wang Y Z 2014 *Chin. J. Lasers* **41** 1202004
[13] McFerran J J, Yi L, Mejri S and Bize S 2010 *Opt. Lett.* **35** 3078-80
[14] Hu J, Zhang L, Liu H, Liu K, Xu Z and Feng Y 2013 *Opt. Express* **21** 30958-63
[15] Hansch T W and Couillaud B 1980 *Opt. Comun.* **35** 441-4

Journal of Physics: Conference Series **723** (2016) 012023 doi:10.1088/1742-6596/723/1/012023

Quantum projection noise limited stability of a $^{88}Sr^{+}$ atomic clock

B Jian[1,2], P Dubé[1] and A A Madej[1,2]

[1] National Research Council of Canada, Ottawa, Ontario K1A 0R6, Canada
[2] Department of Physics and Astronomy, York University, Toronto, Ontario M3J 1P3, Canada

E-mail: Bin.Jian@nrc-cnrc.gc.ca

Abstract. The evaluated accuracy of a single trapped $^{88}Sr^{+}$ ion clock referenced to the 5s $^2S_{1/2}$ – 4d $^2D_{5/2}$ transition at 445 THz at the National Research Council of Canada has reached 1.2×10^{-17} over recent years. On the other hand, the stability of an atomic clock determines how long the signals from two similar clocks have to be compared to reach a given level of uncertainty. Here, we report on the improvement of the stability of NRC's $^{88}Sr^{+}$ single ion clock by reducing the Allan deviation from 1×10^{-14} to 3×10^{-15} at 1 second averaging time. This is done by the implementation of a clear out laser that transfers the ion from the metastable state to the ground state at each cycle, followed by a state-preparation step that transfers the ion to the desired ground state magnetic sublevel of the probed transition.

1. Introduction

Optical atomic clocks based on single trapped ions have witnessed a dramatic development during the last few years [1-5]. At the National Research Council of Canada (NRC), we have developed an atomic clock referenced to the 5s $^2S_{1/2}$ – 4d $^2D_{5/2}$ transition of $^{88}Sr^{+}$ at 445 THz [3-5]. The evaluated fractional frequency uncertainty of the clock has reached 1.2×10^{-17} by using several methods to reduce uncertainties. For example, the electric quadrupole shift and the tensor Stark shift are cancelled by probing several pairs of Zeeman components [6]. The micromotion shifts are reduced to the 10^{-19} level by minimizing ion micromotion and by using the accurately known value of the differential scalar polarizability of the clock transition, $\Delta\alpha_0$. This allows one to tune the trap frequency such that the time dilation and scalar Stark shifts cancel each other by a factor of about 200 [5]. Furthermore, the accurate knowledge of $\Delta\alpha_0$ has also reduced the uncertainty associated with the blackbody radiation shift [5].

It is crucial to optimize the clocks' stability for the comparison of two similar clocks with such a low uncertainty level. Single-clock stabilities of $2\sim3 \times 10^{-16}/\sqrt{\tau}$ have been demonstrated with optical lattice clocks [7, 8]. However, the best stability of optical clocks based on a single trapped ion is about one order of magnitude larger due to the fact that the single atomic particle offers a significant smaller single-to-noise ratio to steer the local oscillator frequency [1, 9]. In this summary, we report on our recent improvement of the stability of our $^{88}Sr^{+}$ clock by reducing the Allan deviation from 1×10^{-14} to 3×10^{-15} at 1 second averaging time, comparable to the quantum projection noise limit for a 100 ms pulse [10]. The pulse sequence, state-preparation step, interrogation method, and stability measurement results are presented.

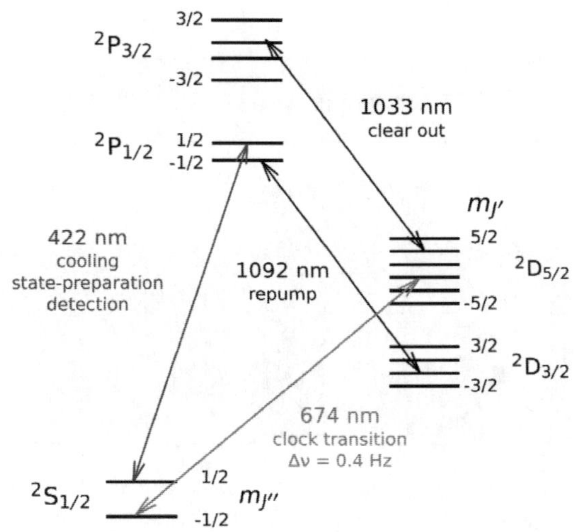

Figure 1. (Colour online) $^{88}Sr^+$ partial energy-level diagram.

2. Experimental methods

2.1. The $^{88}Sr^+$ ion clock

Figure 1 shows the relevant energy levels of the $^{88}Sr^+$ ion. The $^{88}Sr^+$ ion is cooled using a frequency-stabilized diode laser at 422 nm that is red detuned by approximately a half-linewidth from the $5s\ ^2S_{1/2} - 5p\ ^2P_{1/2}$ line center. A repump laser at 1092 nm prevents the ion from decaying to the metastable $^2D_{3/2}$ state [11]. The average ion kinetic temperature observed reaches ~2 mK. The $S - D$ clock transition of the $^{88}Sr^+$ ion, having a natural linewidth of 0.4 Hz, is probed by an ultra-stable laser at 674 nm. A small magnetic field of 1~2 μT is applied to split the clock transition into ten resolved Zeeman components and to define the quantization axis for optical pumping. A "clear out" laser at 1033 nm is applied to return the ion from the $^2D_{5/2}$ metastable state to the ground state after excitation and before the next interrogation pulse. Usually, three symmetric pairs of Zeeman components are probed to determine cancellation of the first-order Zeeman shift, the electric-quadrupole shift, and other tensor shifts [4, 6]. The NRC $^{88}Sr^+$ reference ion is confined in an end-cap trap which is operated with a voltage amplitude of 212 V at a frequency of 14.408 MHz to reduce the micromotion shifts.

2.2. Pulse sequence and state-preparation

Figure 2 shows the pulse sequence used in the experiment. The total cycle time, T_c, is the sum of the probe pulse length, T_p, and the dead time, T_d. During the cooling pulse, the ion state is detected by monitoring the fluorescence at S_1. The clear out laser then transfers the ion from the metastable state $^2D_{5/2}$ to the ground state with an efficiency of essentially 100% even if the ion is found in the ground state at S_1. The fluorescence is monitored again at the end of the cooling stage, S_2, to verify that the clear out action was successful and that the ion was cooled for several ms.

State-preparation has been implemented in our current setup by optical pumping using circularly polarized 422 nm light. This assures the probe laser will interact with the ion placed in the ground state sub-level on every probe cycle since the $^{88}Sr^+$ ion has two ground state magnetic sublevels. A longitudinal Pockels cell followed by a quarter-wave plate provides a voltage-controlled wave plate on the 422 nm beam path. The voltage applied determines whether the polarization at the ion is linear, σ^+ or σ^-. State-preparation starts 2 ms before the 422 nm beam is turned off and lasts 2 ms longer to ensure that only circular polarized light interacts with the ion. Figure 3 shows the effect of using state-preparation with σ^+ light on the $S - D$ Zeeman spectrum with the ion found in the $m_{J''} = +1/2$ sublevel 99% of the time.

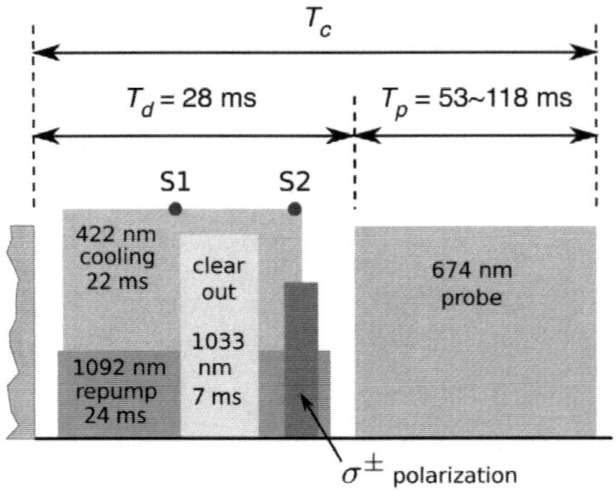

Figure 2. (Colour online) Pulse sequence for one interrogation cycle. The 422 nm polarization is switched from linear to circular (±) for a duration of 4 ms, and starts about 2 ms before the end of the pulse [10].

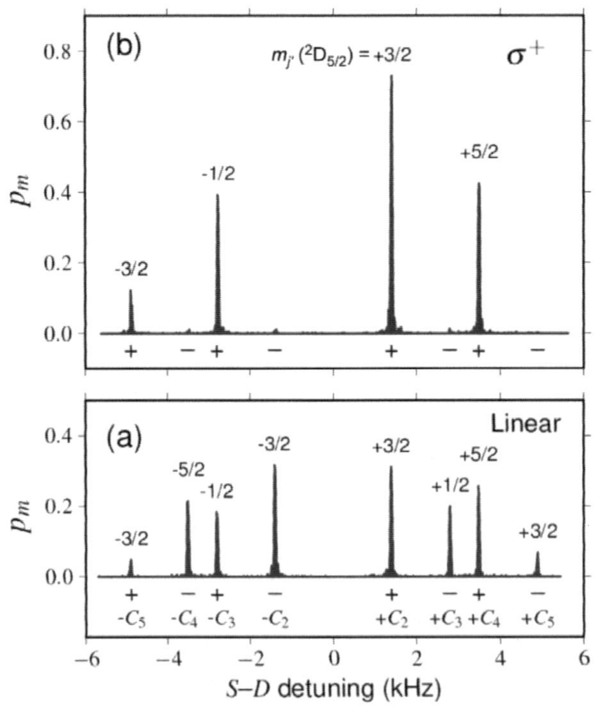

Figure 3. (Colour online) Zeeman-resolved spectra of the $S-D$ transition of $^{88}Sr^+$ showing transition probability to the $^2D_{5/2}$ state, p_m, as a function of 674 nm laser detuning for the 422 nm light linearly (a) and σ^+ polarized during the last 2 ms of the pulse (b) [10].

2.3. Lock method

Experimentally, the line centers of the Zeeman components of the $S-D$ clock transition are probed at frequency offsets of $\pm\delta$ from line center to determine updated center frequencies [12-14]. The center frequency of the clock transition is the average of two symmetric Zeeman components (there is no first-order magnetic insensitive transition in the $^{88}Sr^+$ clock transition spectrum). This sequence is repeated for an averaging time τ to reduce the statistical uncertainty according to [10]

$$\sigma_y(\tau) = \frac{G}{\nu_0}\sqrt{\left(\frac{1-p_x}{p_x}\right)\frac{T_c}{\tau}} \tag{1}$$

where σ_y is the standard Allan deviation, G is a gain parameter, p_x is the probability that the ion is in the metastable state after interaction with a probe pulse which is either blue or red detuned by δ with respect to the line center, and T_c is the cycle time defined earlier.

The lock is usually operated by measuring the frequencies of six Zeeman components from three symmetric pairs that probe all the sublevels of the $^2D_{5/2}$ metastable state. This is because other important shifts can be canceled by averaging the frequencies of Zeeman pairs that connect to all of the $^2D_{5/2}$ state sublevels. These shifts include the linear Zeeman shift, electric quadrupole shift, and tensor Stark shift [4, 6].

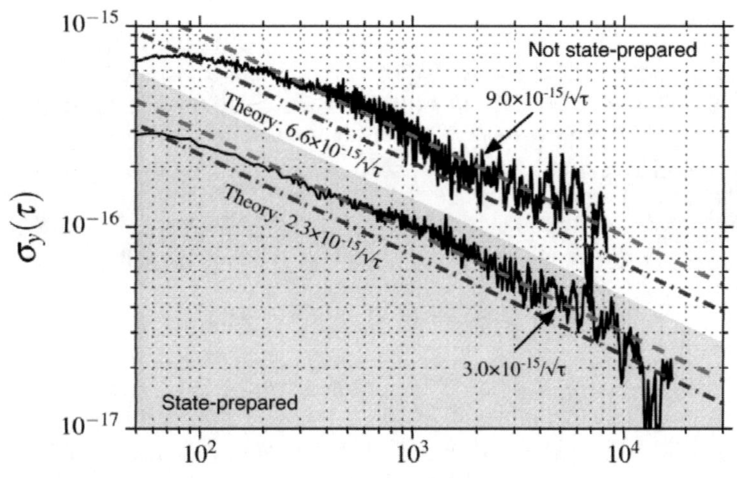

Figure 4. (Colour online) Single-clock Allan deviation of $^{88}Sr^+$ ion as a function of averaging time, τ, with and without state preparation [10].

3. Results

Figure 4 shows the single-clock Allan deviations as a function of averaging time. The result without state-preparation was obtained by comparing two $^{88}Sr^+$ ion optical frequency standards (an end-cap trap and a rf Paul trap with evaluated uncertainties of 1.2×10^{-17} and $\sim 1 \times 10^{-16}$, respectively). The observed stability of $9 \times 10^{-15}/\sqrt{\tau}$ is in reasonable agreement with the calculated optimal stability without state preparation of $6.6 \times 10^{-15}/\sqrt{\tau}$.

For the result with the state-preparation, the stability measurements have been obtained by using self-comparisons in the end-cap trap. The probe laser frequency is locked to the C_2, C_3, and C_4 symmetric pairs as shown in Figure 3. Allan deviations can be calculated from all possible combinations, C_2 vs C_3, C_2 vs C_4, and C_3 vs C_4. The three Allan deviations from each run are averaged and the result divided by $\sqrt{6}$ to obtain the single-clock stability. The observed stability of $3 \times 10^{-15}/\sqrt{\tau}$ is in good agreement with the calculated quantum projection noise (QPN) limit of $2.3 \times 10^{-15}/\sqrt{\tau}$ with a probe pulse length of 118 ms [10]. This is one of the lowest instabilities reported for a single trapped ion clock. Further improvement of the instability will necessitate better control of the probe laser frequency noise and of the background magnetic field noise.

4. Conclusions and outlook

We have presented improvements of the instability of our $^{88}Sr^+$ ion clock which now approaches the quantum projection noise limit. The improvements are made possible by introducing a clear out laser returning the ion back to the ground state after it is excited from the metastable state as well as an optical pumping step to put the ion into the desired ground state magnetic sublevel of the probed transition. The instability obtained is $3 \times 10^{-15}/\sqrt{\tau}$ which reduces the averaging time by an order of magnitude when averaging down to a certain accuracy compared to $1 \times 10^{-14}/\sqrt{\tau}$ when these two steps were not implemented.

The current accuracy of the NRC $^{88}Sr^+$ single ion clock is mainly limited by the uncertainty from the blackbody thermal field. In order to push down the uncertainty of the ion clock to the low 10^{-18} level, evaluation of the thermal field of the ion trap has been started recently. A test end-cap ion trap will be constructed and run under the normal operating conditions. The trap thermal field will be imaged by an IR camera to determine the temperatures of the ion trap parts. Combined with measurements of the emissivities of the various ion trap parts, the effective black body radiation field temperature at the ion will be determined by comparing the temperature measurement results with a finite element simulation [15].

A study of the gravitational red shift is planned by comparing the clock transition frequencies of $^{88}Sr^+$ ions trapped in rf Paul and end-cap traps [16]. A vertical translation stage with a displacement range of ~3 meters has been constructed. One of our ion clocks, the rf Paul trap, will be mounted on this translation stage and the comparison of the frequency between the two clocks will be able to precisely determine the relative gravitational red shift.

References

[1] Chou C W, Hume D B, Koelemeij J C J, Wineland D J and Rosenband T 2010 *Phys. Rev. Lett.* **104** 070802
[2] Huntemann N, Okhapkin M, Lipphardt B, Weyers S, Tamm Chr and Peik E 2012 *Phys. Rev. Lett.* **108** 090801
[3] Madej A A, Dubé P, Zhou Z, Bernard J E and Gertsvolf M 2012 *Phys. Rev. Lett.* **109** 203002
[4] Dubé P, Madej A A, Zhou Z and Bernard J E 2013 *Phys. Rev.* A **87** 023806
[5] Dubé P, Madej A A, Tibbo M and Bernard J E 2014 *Phys. Rev. Lett.* **112** 173002
[6] Dubé P, Madej A A, Bernard J E, Marmet L, Boulanger J-S and Cundy S 2005 *Phys. Rev. Lett.* **95** 033001
[7] Hinkley N, Sherman J A, Phillips N B, Schioppo M, Lemke N D, Beloy K, Pizzocaro M, Oates C W and Ludlow A D 2013 *Science* **341** 1215
[8] Nicholson T L, Campbell S L, Hutson R B, Marti G E, Bloom B J, McNally R L, Zhang W, Barrett M D, Safronova M S, Strouse G F, Tew W L and Ye J 2015 *Nat. Commun.* **6** 6896
[9] Rosenband T, Hume D B, Schmidt P O, Chow C W, Brusch A, Lorini L, Oskay W H, Drullinger R E, Fortier T M, Stalnaker J E, Diddams S A, Swann W C, Newbury N R, Itano W M, Wineland D J and Bergquist J C 2008 *Science* **319** 1808
[10] Dubé P, Madej A A, Shiner A and Jian B 2015 *Phys. Rev.* A **92** 042119
[11] Fordell T, Lindvall T, Dubé P, Madej, Wallin A E and Merimaa M 2015 *Opt. Lett.* **40** 1822
[12] Bernard J E, Marmet L and Madej A A 1998 *Opt. Commun.* **150** 170
[13] Riis E and Sinclair A G 2004 *J. Phys.* B: *At. Mol. Opt. Phys.* **37** 4719
[14] Peik E, Schneider T and Tamm Chr 2006 *J. Phys.* B: *At. Mol. Opt. Phys.* **39** 145
[15] Doležal M 2015 private communication
[16] Chou C W, Hume D B, Rosenband T and Wineland D J 2010 *Science* **329** 1630

Trapped ion $^{88}Sr^+$ optical clock systematic uncertainties – AC Stark shift determination

GP Barwood, G Huang, SA King, HA Klein and P Gill

National Physical Laboratory, Teddington Middlesex TW11 0LW U.K.

E-mail: geoffrey.barwood@npl.co.uk

Abstract. A recent comparison between two trapped-ion $^{88}Sr^+$ optical clocks at the UK National Physical Laboratory demonstrated agreement to 4 parts in 10^{17}. One of the uncertainty contributions to the optical clock absolute frequency arises from the blackbody radiation shift which in turn depends on uncertainty in the knowledge of the differential polarisability between the two clocks states. Whilst a recent NRC measurement has determined the DC differential polarisability to high accuracy, there has been no experimental verification to date of the dynamic correction to the DC Stark shift. We report a measurement of the scalar AC Stark shift at 1064 nm with measurements planned at other wavelengths. Our preliminary result using a fibre laser at 1064 nm agrees with calculated values to within ~3%.

1. Introduction

Since the seventh symposium, there has been strong progress in the development of optical frequency standards based on both cold trapped atoms and ions. Both projected and measured clock reproducibilities are now in the low parts in 10^{17} to few parts in 10^{18} for a number of atom and ion systems and details are given elsewhere in these proceedings. One of the optical frequency standards under development at the UK National Physical Laboratory (NPL) is based on the 5s $^2S_{1/2}$ – 4d $^2D_{5/2}$ electric quadrupole transition in $^{88}Sr^+$ at 674 nm. A partial term scheme, showing the cooling lasers and clock transition is shown in figure 1. This transition is also being studied at the National Research Council (NRC) in Canada and at the VTT Technical Research Centre of Finland. At NPL, we have recently compared two optical frequency standards based on $^{88}Sr^+$ at 674 nm and shown agreement to within 4 parts in 10^{17} [1]. This measurement confirmed experimentally the projected uncertainty by the NRC [2]. Other NPL work has centred on improvements in real-time monitoring and control of the micromotion [3] and a study on the effect of magnetic field noise on clock frequency stability [4]. A key advantage of the $^{88}Sr^+$ system is the ability to operate the RF trap at a frequency where the second order Doppler and Stark shifts due to the micromotion cancel [5]. Reference [5] also includes a measurement of the DC differential polarisability that had previously been a limiting factor in the uncertainty of the $^{88}Sr^+$ optical standard. However, in order to obtain the blackbody shift coefficient from the DC differential polarisability, a small "dynamic correction" [6] is applied. Although this correction is small, there has been, to date, no direct experimental confirmation of the AC Stark shift theory used to estimate this correction. In these proceedings, we present our first measurements of the Stark shift at 1064 nm, using a 100-mW fibre laser and compare our result with the calculated shift.

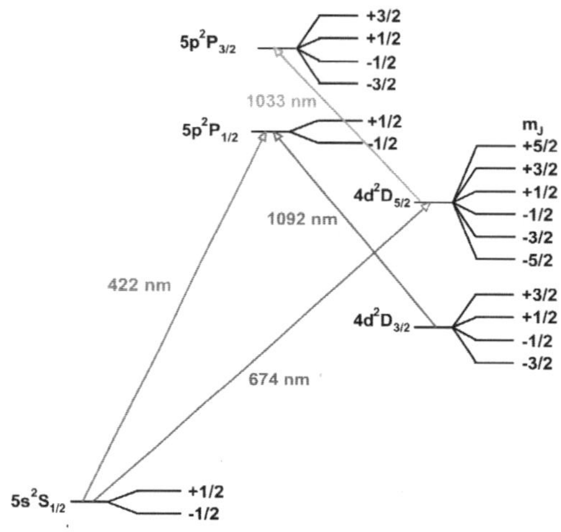

Figure 1. Partial term scheme for $^{88}Sr^{+}$, showing the wavelengths required to cool the ion, together with the 674 nm clock transition.

Figure 2 shows an overall schematic of our laser system, together with one of our two traps. Cooling light at 422 nm is provided by a frequency doubled 844-nm extended cavity laser; distributed feedback lasers provide the clear-out sources at 1033 nm and 1092 nm. Recently, VTT have demonstrated that these lasers can be replaced with amplified spontaneous emission sources [7], offering significant simplification. Full details of our laser system are given elsewhere [1, 3, 4]. Stark shift measurements are undertaken via two trap comparison with an additional laser directed into one of our two traps (figure 2). A computer-controlled shutter is used to accumulate data on the frequency shift with this laser shuttered “on” and “off”.

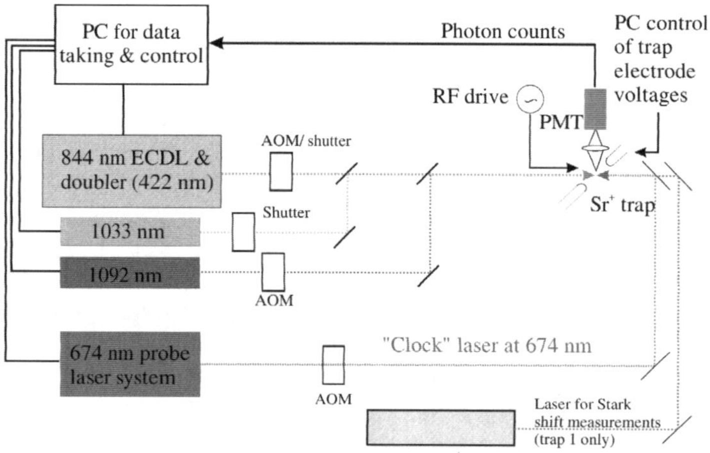

Figure 2. Overall schematic for cooling and probing $^{88}Sr^{+}$, with an additional laser for AC Stark shift measurements

2. Stark shift theory and measurement

The theory of Stark shifts is well-documented and tensor, vector and scalar Stark shifts can all be observed under different circumstances [8]. Both the tensor Stark shift and quadrupole shift [8, 9] vary as the square of the magnetic quantum number ($\propto m^2$) and the angle between the trap axis and the magnetic field. Interrogating and averaging over three specific pairs of Zeeman components [9, 10] eliminates both the quadrupole shift and the tensor Stark shift. A vector Stark shift [8] depends linearly on m and can be observed if some of the 674 nm light is circularly polarised. However, this is eliminated by averaging over pairs of Zeeman components symmetrically placed around line centre; this is the same routine that eliminates the first order Zeeman shift. To determine the blackbody shift, we mainly need to consider the scalar Stark shift, since blackbody radiation is normally isotropic. Where heat sources are present [12], there could be a non-isotropic electric field giving a tensor Stark shift but this is averaged out by measuring on different Zeeman components as already discussed. We therefore consider here only the scalar Stark shift.

From [6], the scalar Stark shift $\Delta\omega_{AC}$ in an electric field E for the level with total angular momentum J as a function of frequency ω is given by:

$$\Delta\omega_{AC} = \frac{E^2}{3\hbar^2(2J+1)}\sum_{J'}\frac{\omega_{JJ'}\left|<J\parallel p\parallel J'>\right|^2}{\omega_{JJ'}^2-\omega^2} \quad (1)$$

Here, $\left|<J\parallel p\parallel J'>\right|^2$ is a reduced dipole (p) matrix element and $\omega_{JJ'}$ is the frequency between the states J and J' listed in tables 1 and 2 of [6] with $\hbar$ the reduced Planck's constant. Equation 1 gives the shift for the $^2S_{1/2}$ ($J = \frac{1}{2}$) and $^2D_{5/2}$ ($J = \frac{5}{2}$) levels; the shift of the 674-nm clock transition is the difference between these values. For a Stark shift resulting from a laser with power P and $1/e$ amplitude radius w_0, the Stark shift can be calculated from the following relationship between the rms electric field E and the output power at the centre (i.e. point of maximum intensity) of a TEM_{00} Gaussian beam:

$$E^2 = \frac{2P}{\pi c\varepsilon_0 w_0^2} \quad (2)$$

Substituting equation 2 into equation 1 gives values for the power shift of the $^2S_{1/2}$ and $^2D_{5/2}$ levels as a function of optical frequency ω. The difference between these values is the power shift of the 674-nm clock transition, calculated in figure 3 for $w_0 = 100$ μm. This figure indicates the frequencies of the 422-nm cooling transition and 1033-nm clear-out laser shown in figure 1. The wavelength at 408 nm is the 5s $^2S_{1/2}$ – 5p $^2P_{3/2}$ transition.

From measurements at different wavelengths in the region ~370 nm to ~1550 nm, we plan to experimentally verify some of the oscillator strengths of equation (1). In the initial work reported here, we observe the scalar Stark shift using a 100-mW 1064-nm fibre laser; this shift is predominantly determined by the reduced matrix element (oscillator strength) of the 4d $^2D_{5/2}$ – 5p $^2P_{3/2}$ transition at 1033 nm. Other laser wavelengths where measurements are planned are 423 nm (close to the cooling laser frequency) and 406 nm (close to the 5s $^2S_{1/2}$ – 5p $^2P_{3/2}$ transition). An additional measurement is planned in the near ultra-violet at ~370 nm. Although equation 1 contains a large number of unknowns, most of the terms from transitions in the ultra-violet can be approximated as a DC term for wavelengths longer than 400 nm. Including the transitions at 1033 nm, 422 nm and 408 nm together with one term for the closely spaced 4d $^2D_{5/2} \rightarrow$ 4f $^2F_{5/2,\,7/2}$ uv doublet limits the number of unknowns to five. A least squares fitting routine will then be used to extract the required oscillator strengths.

Of additional interest is the determination of two "magic wavelengths", where the AC scalar Stark shift is zero; these wavelengths are calculated to be at 1545.4 nm and 417.0 nm. Both wavelengths are readily accessible; the first generated by a telecoms laser and the second either using a direct output from a blue diode or a frequency doubled 834 nm laser. The 417-nm wavelength is the equivalent to that recently measured in $^{40}Ca^+$ [13]. Although ion "magic wavelengths" do not have the same key significance as they do for atoms [14] they provide two additional linear relationships between the oscillator strengths independent of laser power and beam size. Uncertainty in the measurement of these two parameters is a significant contributor to the overall AC Stark shift uncertainty.

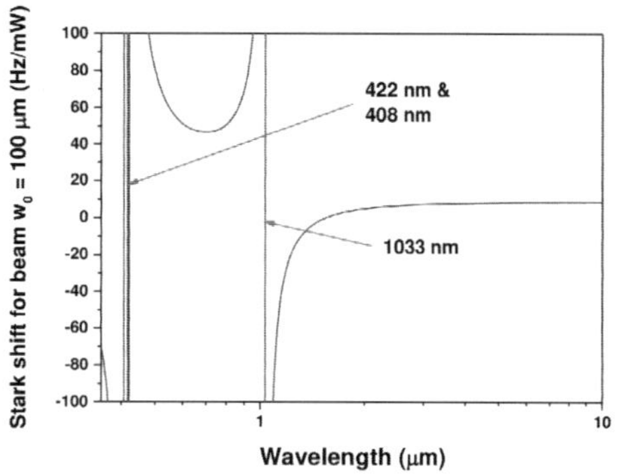

Figure 3. Calculated AC scalar Stark shifts over the wavelength region 350 nm to 10 μm for the $^{88}Sr^+$ 674 nm optical clock transition.

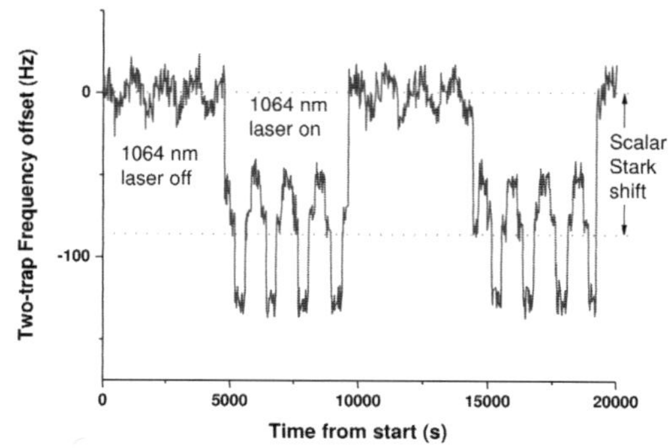

Figure 4. Stark shift observed at 1064 nm by two-trap comparison; both scalar and tensor Stark shifts are observed (see text).

Figure 4 shows our observation of the Stark shift using a 100-mW fibre laser tuned to 1064.534 nm with a $1/e$ amplitude radius (= w_0) of 600 μm. Two-trap comparison measurements are shown with the fibre laser being toggled on or off every ~80 minutes. During both the laser "on" and "off" periods, the 674-nm laser is cycled every ~3 minutes between Δm = 0, ±1 and ±2 Zeeman components. Each Zeeman component has a different quadrupole and tensor Stark shift; this accounts for the sudden frequency changes as the Zeeman component is changed. The size of these frequency steps changes when the 1064-nm laser is toggled between being on and off. This change arises from the AC tensor Stark shift at this wavelength. In order to measure the scalar Stark shift, we average the trap offset over the three Zeeman components during the periods when the 1064-nm laser is shuttered on and off (the offset with the laser shuttered off is consistent with zero).

This measurement was repeated at different 1064-nm fibre laser input beam alignments and power levels. At each power level, measurements were made at different input beam alignments, both horizontally and vertically. Since the laser output is via a single mode polarisation maintaining fibre, a spatially clean and circularly symmetric beam profile is expected; this was verified experimentally. The results for different beam alignments were fitted to a Gaussian profile to determine the frequency shift corresponding to the maximum beam intensity. For our most recent measurements, this process of beam centring has been automated by replacing the final beam-steering mirror with a type fitted with computer-controlled piezoelectric steering. The beam size was measured by placing a mirror before the trap and using a knife edge at a distance corresponding to the ion position. The accuracy of the beam-size measurement (2.5% in w_0^2; k = 1) was the largest uncertainty contribution. The frequency shift was measured at nine different powers between ~5 mW and ~18 mW and the results were fitted using linear regression. In addition to the statistical uncertainty of the regression fit, the uncertainty budget (table 1) includes contributions from uncertainties in the power including the vacuum window transmission, pointing stability and beam size determination. The result is a measurement of the scalar Stark shift which we determine to be -5.4(2) Hz/mW for w_0 = 600 μm. Using equations 1 and 2, we calculate the expected shift as -5.51 Hz/mW, consistent with the observed result to within experimental uncertainty.

Table 1. Uncertainty budget for the Stark shift measurement at 1064 nm.

Parameter	**Relative uncertainty (k = 1)**
Beam area ($\propto w_0^2$)	2.5%
Power measurement uncertainty	1.9%
Beam pointing stability	1.6%
Regression fit	0.7%
TOTAL	3.6%

In order to use equation 1 to calculate the blackbody shift, we require the expression for the blackbody mean square electric field ξ^2 per unit bandwidth (i.e. in units of $(\mathrm{V/m})^2(\mathrm{radian/s})^{-1}$):

$$\xi^2 = \frac{\hbar}{\pi^2 \varepsilon_0 c^3} \frac{\omega^3}{\exp(\hbar\omega / kT) - 1} \tag{3}$$

Integrating equation 3 over all frequencies ω gives the familiar result for the total blackbody radiation field $\langle E^2 \rangle = (832\ \mathrm{V/m})^2$ at T = 300 K. To obtain the blackbody shift for the $^2S_{1/2}$ and $^2D_{5/2}$ levels, we need to substitute equation 3 into equation 1 and integrate to obtain equation 4:

$$\Delta\omega_{BB} = \frac{1}{3\hbar\pi^2\varepsilon_0 c^3 (2J+1)} \sum_{J'} \int_0^\infty \frac{\omega_{JJ'}\omega^3 \left| < J \| p \| J' > \right|^2}{\left(\omega_{JJ'}^2 - \omega^2\right)\left(\exp\left(\frac{\hbar\omega}{kT}\right) - 1\right)} d\omega \tag{4}$$

We then sum over all the listed [6] transitions from the $^2S_{1/2}$ and $^2D_{5/2}$ levels, as indicated in equation 4. This summation also needs to include the final frequency-independent corrections listed in tables 1 and 2 of [6] to the polarisability (for example, the polarisability contribution from the electrons in the ionic core). This calculation leads to the published result for the blackbody shift, including the "dynamic correction". Once our AC Stark shift measurements at other wavelengths are complete, equation 4 could be used to update the value for the blackbody shift by replacing some of the matrix elements with their experimentally determined values.

3. Conclusion

We have presented the results of our first AC Stark shift measurements on the $^{88}Sr^+$ 674-nm optical clock transition, using a 1064-nm fibre laser. Together with future measurements planned at wavelengths close to transitions at 422 nm and 408 nm, we expect to be able to determine experimentally some of the matrix elements required to calculate the blackbody shift (equation 4). The measurements are limited by the uncertainty of the beam size (w_0) determination; the total shift varies with the inverse square of this parameter. Of particular interest is the determination of the two accessible "magic wavelengths" for $^{88}Sr^+$ at 1545.4 nm and 417.0 nm where the AC Stark shift is calculated to be zero. These wavelengths will provide two linear relationships between the matrix elements that are independent of laser power and beam-size. The uncertainties in the measurement of these two parameters limit the accuracy of the measurement reported here. Once these measurements have been completed, a least squares fit will allow determination of the matrix elements for transitions at wavelengths longer than 400 nm; these transitions contribute most to the variation of the AC Stark shift with frequency.

Acknowledgments

This work was funded by the UK Department for Business, Innovation and Skills as part of the National Measurement System Electromagnetics and Time programme and by the European Metrology Research Programme (EMRP). The EMRP is jointly funded by the EMRP participating countries within EURAMET and the European Union.

References

[1] Barwood G P, Huang G, Klein H A, Johnson L A M, King S, Margolis H S, Szymaniec K and Gill P 2014 *Phys. Rev. A* **89**, 050501(R)
[2] Dubé P, Madej A A, Zhou Z and Bernard J E 2013 *Phys. Rev. A* **87**, 023806
[3] Barwood G P, Huang G, Klein H A and Gill P 2015 *Meas. Sci. Technol.* **26** 075203
[4] Barwood G P, Huang G, King S A, Klein H A and Gill P 2015 *J. Phys. B: At. Mol. Opt. Phys.* **48** 035401
[5] Dubé P, Madej A A, Tibbo M and Bernard J E 2014 *Phys. Rev. Lett.* **112**, 173002
[6] Jiang D, Arora1 B, Safronova M S and Clark C W 2009 *J. Phys. B: At. Mol. Opt. Phys.* **42** 154020
[7] Fordell T, Lindvall T, Dubé P, Madej A A, Wallin A E and Merimaa M 2015 *Opt. Lett.* **40** 1822
[8] Auzinsh A, Budker D and Rochester S M 2010 "*Optically Polarized Atoms: Understanding light-atom interactions*" (Oxford University Press) p 329
[9] Itano W M 2000 *J. Res. Natl. Inst. Stand. Technol.* **105**, 829
[10] Dubé P, Madej A A, Bernard J E, Marmet L, Boulanger J-S and Cundy S 2005 *Phys. Rev. Lett.* **95**, 033001
[11] Margolis H S, Barwood G P, Huang G, Klein H A, Lea S N, Szymaniec K and Gill P 2004 *Science* **306** 1355
[12] Flambaum V V, Porsev S G and Safronova M S 2016 *Phys. Rev. A* **93**, 022508
[13] Liu P-L, Huang Y, Bian W, Shao H, Guan H, Tang Y-B, Li C-B, Mitroy J and Gao K-L 2015 *Phys. Rev. Lett.* **114**, 223001
[14] Katori H and Takamoto M, Pal'chikov V G and Ovsiannikov V D 2003 *Phys. Rev. Lett.* **91**, 173005

Compact Yb^+ optical atomic clock project: design principle and current status

Clément Lacroûte, Maël Souidi, Pierre-Yves Bourgeois, Jacques Millo, Khaldoun Saleh, Emmanuel Bigler, Rodolphe Boudot, Vincent Giordano and Yann Kersalé

FEMTO-ST, TF Dpt., UMR 6174 CNRS-ENSMM-UFC-UTBM, 26 ch de l'Epitaphe, 25030 Besançon, France

E-mail: `clement.lacroute@femto-st.fr`

Abstract. We present the design of a compact optical clock based on the $^2S_{1/2} \rightarrow ^2 D_{3/2}$ 435.5 nm transition in $^{171}Yb^+$. The ion trap will be based on a micro-fabricated circuit, with surface electrodes generating a trapping potential to localize a single Yb ion a few hundred μm from the electrodes. We present our trap design as well as simulations of the resulting trapping pseudo-potential. We also present a compact, multi-channel wavelength meter that will permit the frequency stabilization of the cooling, repumping and clear-out lasers at 369.5 nm, 935.2 nm and 638.6 nm needed to cool the ion. We use this wavelength meter to characterize and stabilize the frequency of extended cavity diode lasers at 369.5 nm and 638.6 nm.

1. Introduction

Optical atomic clocks have now surpassed atomic fountain clocks by two orders of magnitude both in terms of fractional frequency stability and systematic uncertainty [1, 2]. They have proven to be useful for frequency metrology, for fundamental physics tests [3, 4], and their resolution make them competitive sensors for relativistic geodesy [5, 6]. While theoretical proposals point out new ways to further improve their performances [7, 8, 9], and a new defintion of the SI second by an optical transition is being discussed [10], interest starts to arise for field measurements and transportable atomic clocks [11].

Several proposals for fundamental physics tests using space optical clocks have been written, and prototypes of such clocks are being developed [12, 13]. Time distribution through GNSS would benefit both from optical clocks in the satellites and in the ground stations [14]. Additionally, one of the most immediate benefits of a transportable optical clock would be centimetric resolution for relativistic geodesy surveys [15].

In this perspective, we are currently designing a compact, single-ion atomic clock based on the 435.5 nm transition in trapped $^{171}Yb^+$, with a target frequency stability one order of magnitude below hydrogen masers in a comparable volume, *e.g.* $10^{-14}\tau^{-1/2}$ in about 500 L. Yb^+ can be manipulated with diode lasers, and the 435.5 nm transition can easily be adressed by second-harmonic generation, which can be integrated in a compact optical setup. The clock will be based on a surface electrodes (SE) trap [16]. SE traps are now widely used in quantum information processing (QIP) experiments. They allow for a compact physics package, benefit from high-end microfabrication techniques, and allow for the integration of optical elements [17].

They are known to be hindered by the so-called "anomalous heating" [18], due to the proximity of the trapping electrodes that can radiate patch potentials [19], but a way to circumvent this phenomenon using ion-bombardment of the electrodes was recently found [20, 21], making such traps compatible with a metrological setup.

In this article, we present the general design of the clock, with an emphasis on the SE trap and the laser frequency stabilization of extended cavity diode lasers using a multi-channel, fibered wavelength meter.

2. Surface electrode trap ion clock: general principle

2.1. Clock design

We aim to build a compact atomic clock with a total volume of order 500 L, enabled by the use of a surface electrode ion trap combined with a compact vacuum chamber analogous to those used in "atom chips" experiments [22]. The trap will close a cubic glass cell connected to a hybrid ion/getter pump, see a preliminary schematic illustration figure 1. Such vacuum chamber designs have been used in compact (< 1 m^3) Bose-Einstein condensation setups [23]. The pressure obtained in such setups is in the low 10^{-10} mbars range, in principle compatible with Ytterbium ion trapping and cooling. To our knowledge, such "atom chips" cells have never been tested in ion trapping setups, and we will adapt and test such a design with Yb^+ ions.

The cooling and ionization lasers will be stabilized using a compact, multi-channel wavelength meter (see section 3). The clock laser will be generated by frequency doubling of an extended cavity diode laser (ECDL) at 871 nm. Second harmonic generation will be obtained with a fibered non-linear optical waveguide, avoiding the use of a resonant cavity. It will be pre-stabilized near the electric quadrupole transition at 435.5 nm using an ultra-stable optical cavity; we are developping an ultra-compact ULE Fabry-Perot cavity at telecom wavelength, with a projected total volume of about 54 L for the ultra-stable laser system. We will develop a similar cavity at 871 nm, in order to reduce the ultra-stable clock laser volume as much as possible.

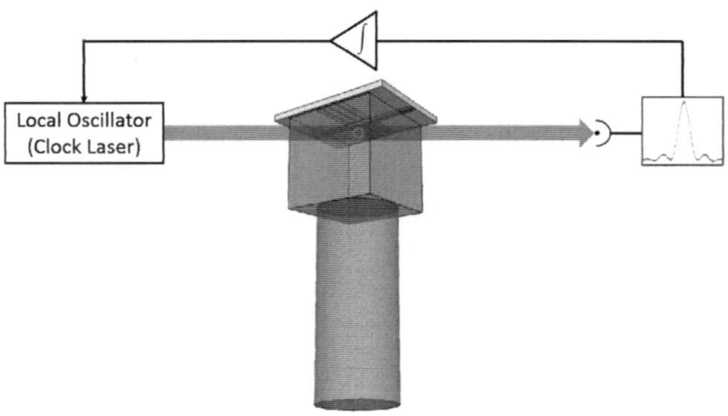

Figure 1. Schematic view of the experiment. A surface electrode trap confines an Yb^+ ion 600 μm below the chip surface. The chip closes a cubic (1 inch)^3 glass cell evacuated to ultra-high vacuum. The clock laser probes the clock transition and is locked to the atomic signal.

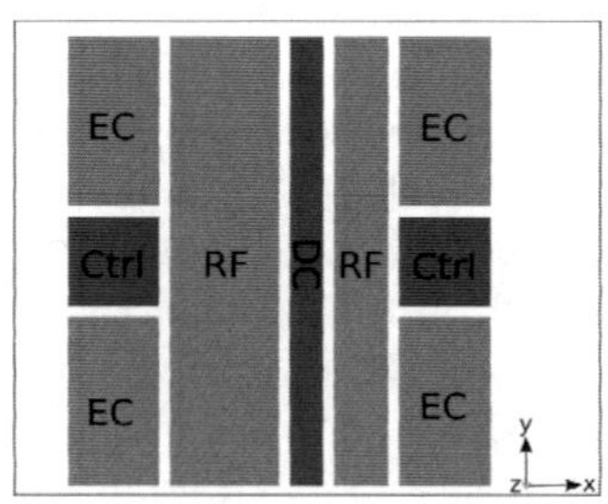

Figure 2. Schematic of the five-wire trap design. RF: radio-frequency electrodes. Ctrl: DC control electrodes. DC: central control electrodes. EC: DC endcap electrodes.

2.2. Trap design

In most metrological experiments with ions, 3D designs are used for the trapping electrodes. Similar trapping potentials can be obtained using surface electrodes traps. The main difference comes from the plane shaped electrodes which are all in the same plane and generate the confinement field above the surface of the trap. The pseudo-potential generated by surface electrodes can be computed analytically [24, 25, 16] and the field can be approximated as a quadrupolar one just near the center of the trap. Due to low electrodes radio-frequency (RF) breakdown voltages, typically close to 300 V, SE traps are usually weaker than traditional Paul traps.

Our system is a five wire RF SE trap illustrated in figure 2. It consists of two RF electrodes separated by a DC control electrode and surrounded by two segmented DC electrodes (the "end caps"). Our two RF electrodes are asymmetric, one of the electrodes being wider than the other (by a factor of two) in order to rotate the principal axes of the trap compared to the surface. The resulting tilt is about 8° with regard to the chip frame. This will allow us to laser cool all the motional axes of our ion even when the laser beam is parallel to the electrodes.

We aim to maximise the ion trapping distance from the electrodes, d, as the so-called "anomalous heating" which affects SE traps scales with d^{-4} [26]. In order to obtain at least 500 μm between the ion and the surface of the trap, the two RF electrodes are respectively 600 μm and 1200 μm wide (the distance between the ion and the trap surface is related to the electrodes width and spacing, see for instance [16]). The central DC control electrode is 600 μm wide, and the length of these three electrodes is about 5 mm. These dimensions locate the center of the saddle point at 580 μm above the surface of the trap and rotates the principal radial axes of motion. At this distance, we can expect a normalized electric field noise power spectral density $\omega S_E(\omega) \approx 4\times10^{-6}$ V^2/m^2 [20]. The resulting heating rate for a trapping frequency of 100 kHz is of order 2200 phonons/s $\approx$ 20 mK/s. Even though this is rather high, the corresponding temperature rise during a probe pulse of 50 ms is about 1 mK, corresponding to an additional second-order Doppler shift of about 3×10^{-19}. Moreover, this heating rate can be reduced by a factor of up to 100 using surface cleaning by an Ar beam [20], which would make this effect completely negligible.

We also want a trap depth higher than 200 meV and a radial stability parameter $q_r < 0.3$. Following simulations of our trap geometry based on [24], we decided to apply an RF voltage of 250 V at 5.5 MHz. That leads us to a theoretical depth of 222 meV and a radial stability parameter $q_r = 0.27$. By computing the curvature of the pseudo-potential in the center of the trap we find an oscillating motion frequency of 500 kHz in the radial plane and 97 kHz along the longitudinal axis. Figure 3 shows finite-element method (FEM) simulations of the pseudo-potential.

3. Characterization and wavelength stabilization of extended cavity diode lasers using a Fizeau wavelength meter

Ultra-violet (UV) laser diodes are still difficult to realize, and only few wavelengths are available below 400 nm. The 369.52 nm cooling transition in Yb^+ is therefore often obtained by frequency doubling a 740 nm extended cavity diode laser, which can be stabilized to a Fabry Perot transfer cavity [27], an iodine cell, or a wavelength meter [27, 28]. More recently, diodes emitting directly around 370 nm near room-temperature have become available, and such diodes can either be stabilized to Yb^+ using a discharge lamp [29], or to a wavelength meter, as demonstrated here.

The transitions involved in $^{171}Yb^+$ cooling have linewidth γ of order 10 MHz. The integrated laser frequency fluctuations must be well below γ on the laser-cooling timescale, which is about a few ten ms. This translates in a fractional frequency stability well below 10^{-9} at short term for the wavelengths which are considered here (369.5 nm and 638 nm). For a clock operated on a daily basis, the lasers should also have negligible drift rates (below 1 MHz/day) to stay on

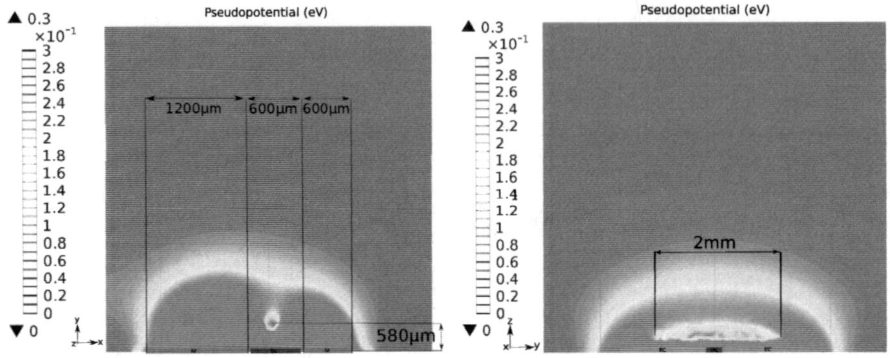

Figure 3. FEM simulations of the pseudopotential generated by our set of electrodes. *Left*: the ion will be localized at 580 μm above the trap surface. The axis of minimal energy is tilted compared to the vertical axis. This effect is due to the asymmetry of the RF electrodes widths. *Right*: pseudopotential, showing the trap in the longitudinal direction.

resonance on the longer term.

3.1. Wavelength meter performance

The wavelength meter we use for laser frequency stabilization is a Standard WS7 wavelength meter from HighFinesse, with a wavelength measurement range from 350 nm to 1120 nm. An optional multichannel fiber-switch unit can be used to measure up to eight lasers and an optional proportional integral (PI) regulator can be used to perform a frequency-lock of all these lasers. We plan to use the 4-channel version of the switch to simultaneously measure and stabilize the cooling, clear-out and ionization lasers at 369.5 nm, 638 nm and 398 nm. For the measurements presented here, only one channel is used.

We have characterized the wavelength meter in terms of fractional frequency stability using the experimental setup described in [30]. We calculated the Allan deviation of the measured wavelength (771 nm) when using an ultra-stable, frequency-doubled, 1542 nm laser as an input. The ultra-stable laser exhibits a fractional frequency instability of about 2×10^{-15} at 1s and a drift rate of 6 kHz/day [31], far below the wavelength meter instability. The resulting Allan deviation is therefore the wavelength meter fractional frequency stability. We obtain a fractional frequency instability of $2\times10^{-10}\ \tau^{-1/2}$ from 0.15 s to 10 s and a long-term frequency drift of $3\times10^{-13}\ \tau$ (about 10 MHz/day) at 771 nm. The resulting relative frequency instability is shown in figures 4 and 5, scaled to 369.52 nm and 638.61 nm, assuming that the resolution is the same at all wavelengths (dashed black line).

3.2. Frequency control with the wavelengthmeter

We have first calculated the Allan deviation of a free-running ECDL wavelength at 369.5 nm, as shown in figure 4 (pink triangles). The measured fractional frequency stability is below 10^{-9} for τ between 0.1 and 10 s, which is typical for ECDLs [32]. The red squares show the locked laser in-loop fractional frequency stability as measured by the wavelength meter when using the wavelength meter proprietary software. The dashed line indicates the best performance achievable by the locked laser, assuming that the wavelength meter resolution is the same at 771 nm and 369.52 nm. It is also important to note that the ECDL and the wavelength meter were not characterized at the same time, nor in the same room, which most probably affects the results. Assessing the ECDL locked performance would necessitate an independent reading of its wavelength, which we cannot perform at the moment.

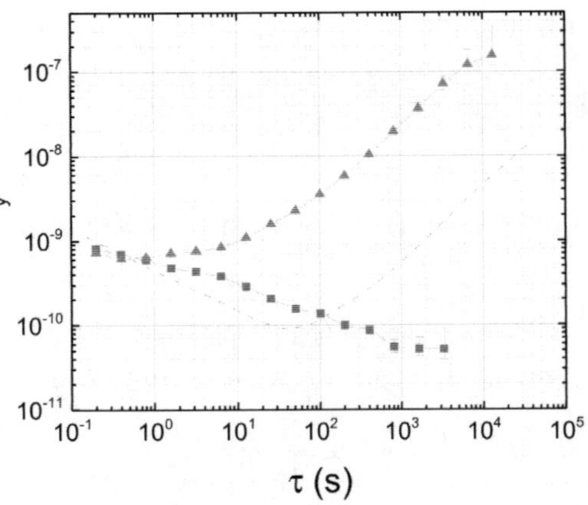

Figure 4. Characterization and frequency stabilization of the 369.52 nm cooling laser by the wavelength meter (WM). *Dashed line:* stability of the wavelength meter, scaled to 369.52 nm. *Pink triangles:* free-running Extended Cavity Diode Laser (ECDL) at 369.52 nm. *Red squares:* ECDL at 369.52 nm locked to the WM using the proprietary software, in-loop signal.

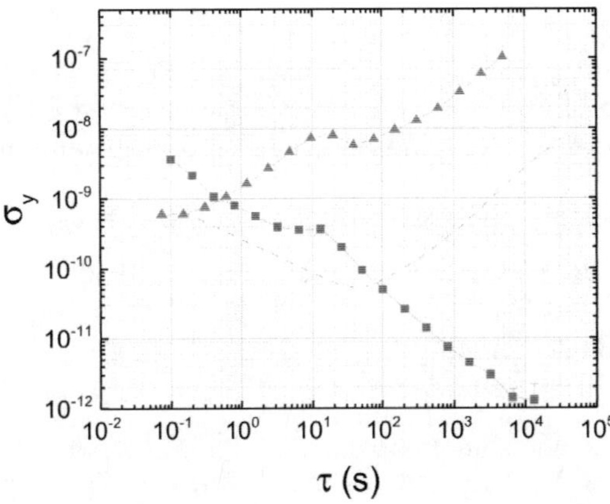

Figure 5. Characterization and frequency stabilization of the 638.61 nm cooling laser by the wavelength meter (WM). *Dashed line:* stability of the wavelength meter, scaled to 638.61 nm. *Pink triangles:* free-running Extended Cavity Diode Laser at 638.1 nm. *Red squares:* ECDL at 638.61 nm locked to the WM using our own digital PI loop, in-loop signal, outliers removed.

Nevertheless, it is fair to state that the best possible performance achievable by the locked laser is about 5×10^{-10} at 1 s, with a 30-fold improvement over its frequency drift.

The ion clock operation will require a complex control system, with inter-dependent parameters that will need to be optimized. We will use homebuilt digital electronics whenever possible, both for compacity and for the high degree of control allowed by multiple input-multiple output (MIMO) control systems. In that spirit, we have developped our own frequency control loop using the wavelength meter. The wavelength read by the wavelength meter is transferred via ethernet to a digital PI controller including a digital to analog converter and a FPGA. The correction signal acts on the piezoelectric transducer of the ECDL. Figure 5 shows preliminary results obtained when locking the clear-out laser at 638.61 nm. The free-running laser shows excess noise around $\tau = 10$ s, probably due to temperature fluctuations. Even though this noise is not fully rejected, the resulting in-loop stability is below 10^{-9} for $\tau > 0.5$ s. We suspect that the short-term excess frequency noise stems from communication timing jitter, as the data transits from the wavemeter server computer to the PI board. Once this problem is solved and with fully optimized gain parameters, stability below 3×10^{-10} from 1 to 1000 s should be achievable.

It is worth noting that the drift rate of about 10 MHz/day only allows autonomous operation on the day scale. This drift should therefore be compensated either numerically, or by using a slower servo loop based on the trapped ions fluorescence signal.

4. Conclusion

In conclusion, we have presented the status of our compact Yb^+ optical clock project. A testbed SE trap has been designed. The ion will be trapped at a distance greater than 500 μm in order to reduce the effect of the so-called anomalous heating. Transverse trapping greater than 500 kHz will ensure operation in the Lamb-Dicke regime.

The compact clock will of course incorporate a compact optical bench. The cooling lasers frequency stabilization will be performed using a multichannel, fibered wavelength meter. We have measured the wavelength meter frequency stability, and used it to characterize and lock the cooling and clear-out ECDLs at 369.52 nm and 638.61 nm. The performances are compatible with ion cooling and detection requirements.

This work is funded by the Agence Nationale de la Recherche (ANR) (ANR-14-CE26-0031-01-MITICC); ANR Programme d'Investissement d'Avenir (PIA) (First-TF network, Oscillator-IMP project); and the Région Franche Comté.

References

[1] Bloom B J, Nicholson T L, Williams J R, Campbell S L, Bishof M, Zhang X, Zhang W, Bromley S L and Ye J 2014 *Nature* **506** 71–75
[2] Ushijima I, Takamoto M, Das M, Ohkubo T and Katori H 2015 *Nat. Phot.* **9** 185–189
[3] Huntemann N, Lipphardt B, Tamm C, Gerginov V, Weyers S and Peik E 2014 *Phys. Rev. Lett.* **113** 210802
[4] Godun R, Nisbet-Jones P, Jones J, King S, Johnson L, Margolis H, Szymaniec K, Lea S, Bongs K and Gill P 2014 *Phys. Rev. Lett.* **113** 210801
[5] Hinkley N, Sherman J A, Phillips N B, Schioppo M, Lemke N D, Beloy K, Pizzocaro M, Oates C W and Ludlow A D 2013 *Science* **341** 1215–1218
[6] Margolis H S 2013 *Science* **341** 1185–1186
[7] Yudin V I, Taichenachev A V, Oates C W, Barber Z W, Lemke N D, Ludlow A D, Sterr U, Lisdat C and Riehle F 2010 *Phys. Rev. A* **82** 011804
[8] Derevianko A, Dzuba V A and Flambaum V V 2012 *Phys. Rev. Lett.* **109** 180801
[9] Zanon-Willette T, Yudin V I and Taichenachev A V 2015 *Phys. Rev. A* **92** 023416
[10] Riehle F 2015 *Comptes Rendus Physique* **16** 506–515
[11] Margolis H 2014 *Nat. Phys.* **10** 82–83
[12] Schiller S *et al* 2008 *Exp. Astro.* **23** 573–610
[13] Bongs K, Singh Y, Smith L, He W, Kock O, wierad D, Hughes J, Schiller S, Alighanbari S, Origlia S, Vogt S, Sterr U, Lisdat C, Le Targat R, Lodewyck J, Holleville D, Venon B, Bize S, Barwood G P, Gill P, Hill I R, Ovchinnikov Y B, Poli N, Tino G M, Stuhler J and Kaenders W 2015 *Comptes Rendus Physique* **16** 553–564
[14] Moudrak A, Klein H and Eissfeller B 2008 *InsideGNSS* **Sept./Oct.** 45–50
[15] Delva P and Lodewyck J 2013 *Acta Futura* **7** 67–78
[16] Chiaverini J, Blakestad R B, Britton J, Jost J D, Langer C, Leibfried D, Ozeri R and Wineland D J 2005 *Quant. Inf. and Comp.* **5** 419–439
[17] Eltony A M, Gangloff D, Shi M, Bylinskii A, Vuleti V and Chuang I L 2015 *arXiv:1502.05739 [physics, physics:quant-ph]* ArXiv: 1502.05739
[18] Turchette Q A, Kielpinski, King B E, Leibfried D, Meekhof D M, Myatt C J, Rowe M A, Sackett C A, Wood C S, Itano W M, Monroe C and Wineland D J 2000 *Phys. Rev. A* **61** 063418
[19] Dubessy R, Coudreau T and Guidoni L 2009 *Phys. Rev. A* **80** 031402
[20] Hite D A, Colombe Y, Wilson A C, Brown K R, Warring U, Jrdens R, Jost J D, McKay K S, Pappas D P, Leibfried D and Wineland D J 2012 *Phys. Rev. Lett.* **109** 103001
[21] McKay K S, Hite D A, Colombe Y, Jrdens R, Wilson A C, Slichter D H, Allcock D T C, Leibfried D, Wineland D J and Pappas D P 2014 *arXiv:1406.1778 [physics, physics:quant-ph]* ArXiv: 1406.1778
[22] Du S, Squires M B, Imai Y, Czaia L, Saravanan R A, Bright V, Reichel J, Hänsch T W and Anderson D Z 2004 *Phys. Rev. A* **70** 053606
[23] Farkas D M, Hudek K M, Salim E A, Segal S R, Squires M B and Anderson D Z 2010 *App. Phys. Lett.* **96** 093102
[24] House M G 2008 *Phys. Rev. A* **78** 033402
[25] Wesenberg J H 2008 *Phys. Rev. A* **78** 063410
[26] Deslauriers L, Olmschenk S, Stick D, Hensinger W K, Sterk J and Monroe C 2006 *Phys. Rev. Lett.* **97** 103007
[27] McLoughlin J J, Nizamani A H, Siverns J D, Sterling R C, Hughes M D, Lekitsch B, Stein B, Weidt S and Hensinger W K 2011 *Phys. Rev. A* **83** 013406
[28] Pyka K, Herschbach N, Keller J and Mehlstäubler T E 2013 *App. Phys. B* 1–11
[29] Lee M W, Jarratt M C, Marciniak C and Biercuk M J 2014 *Opt. Expr.* **22** 7210
[30] Saleh K, Millo J, Didier A, Kersalé Y and Lacroûte C 2015 *Appl. Opt.* **54** 9446–9449
[31] Didier A, Millo J, Grop S, Dubois B, Bigler E, Rubiola E, Lacroûte C and Kersalé Y 2015 *Appl. Opt.* **54** 3682–3686
[32] Rovera G D, Santarelli G and Clairon A 1994 *Rev. of Sci. Instr.* **65** 1502–1505

Journal of Physics: Conference Series **723** (2016) 012026 doi:10.1088/1742-6596/723/1/012026

Recent progress on the $^{27}Al^+$ ion optical clock

Z T Xu, W H Yuan, X Y Zeng, H Che, X H Shi, K Deng, J Zhang and Z H Lu

MOE Key Laboratory of Fundamental Physical Quantities Measurement, School of Physics, Huazhong University of Science and Technology, Wuhan 430074, China

zehuanglu@hust.edu.cn

Abstract. An aluminium ion optical clock is under development at Huazhong University of Science and Technology. The $^{25}Mg^+$ ion is chosen as logic ion to sympathetically cool an Al^+ ion and to detect its states. The $^{25}Mg^+$ ion is cooled to the motional ground state through Raman sideband cooling as the first step for quantum logic spectroscopy. Ultra-stable lasers for the interrogation of the clock transition are developed. The instability of the laser beat frequency is 1.2×10^{-15} at 1 s, which is close to the thermal noise limit of the reference cavity.

1. Introduction

The $^1S_0$-$^3P_0$ transition in Al^+ at 267.4 nm, which has a narrow natural linewidth of 8 mHz, has long been recognized as a good potential clock transition [1]. Its electric quadrupole shift is zero (J=0), therefore the transition frequency is not influenced by the gradient of the electric field in the ion trap. Furthermore, the blackbody radiation contribution to the fractional frequency uncertainty of the Al^+ clock is as low as 4×10^{-19} [2] at room temperature (300 K), which is the smallest among atomic species currently under consideration for optical clocks.

Since the Doppler cooling for Al^+ ions of 167 nm cannot be experimentally realized currently, the method of quantum logic spectroscopy (QLS) [3] is proposed to cool the ion and detect the clock transition. We choose $^{25}Mg^+$ as logic ion. Cooling the logic ion to ground state [4] is the first step to use QLS. In this paper we will report our work on Raman sideband cooling.

On the other hand, to detect the narrow clock transition we need an ultra-stable clock laser. We lock the laser to a very stable Fabry-Perot (FP) cavity with the Pound-Drever-Hall (PDH) technique [5]. There are many perturbation sources that affect the length of the cavity. Thermal noise limit is one of the fundamental noise sources due to the Brownian motion of the spacer, mirror substrate, and mirror coating [6]. Temperature fluctuations also affect the cavity length, so a very low coefficient of thermal expansion (CTE) material and a very good temperature control are needed. We investigate the noise contributions due to temperature fluctuations, optical power fluctuations and residual amplitude modulation to identify the limiting factors. We achieve two sub-Hz linewidth ultra-stable lasers. The instability of the lasers' beat frequency is 1.2×10^{-15} at 1 s, which is approaching the thermal noise limit of the two reference cavities.

2. Ion trap system

We use a linear Paul trap [7] to trap aluminium and magnesium ions. The linear trap consists of four blade electrodes and two end-cap tip electrodes. All the electrodes are made of titanium to reduce the influence of magnetic field. A helical resonator with a Q value of 200 is used to amplify the RF voltage to several hundred Volts, and its resonance frequency is about 23.8 MHz. The Q-factor is defined by the amount of power stored in the resonator divided by the energy losses per cycle. The fluorescence of ions is collected by an imaging system into a PMT. The fluorescence collection efficiency is about 0.2 %. Mg and Al atoms are photo-ionized by a 285 nm laser and a 395 nm laser, respectively.

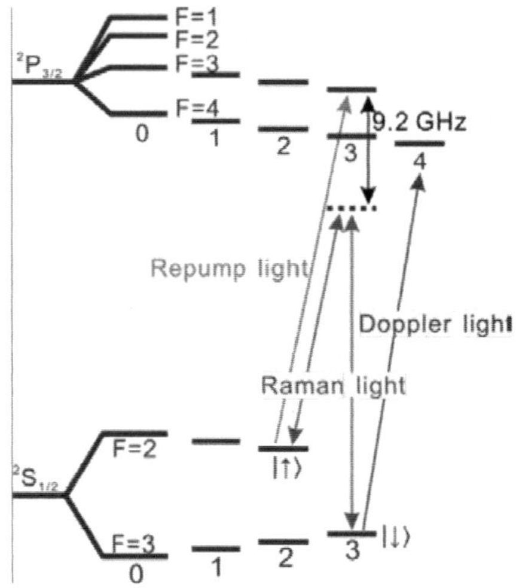

Figure 1. Partial energy levels of $^{25}Mg^{+}$ ion and used relevant lasers.

In our setup, two 280 nm frequency-quadruple laser systems are used for Raman sideband cooling [8]. One laser provides the two Raman beams and repumping light with the help of a 9.2 GHz EOM. Doppler cooling light is produced by a second 280 nm laser. So that during the detection process the background scattering rate can be reduced to 1 kHz. The relevant laser beams used are shown in figure 1. All the lasers are independently controlled with AOMs through RF switches. These RF switches are controlled by TTL signals generated by an NI 7851R FPGA board.

To determine the transition frequency between $|F = 2, m = 2\rangle$, i.e. $|\uparrow\rangle$, and $|F = 3, m = 2\rangle$ states which will be used to repump $|\uparrow\rangle$ state back to $|F = 3, m = 3\rangle$, i.e. $|\downarrow\rangle$, state during Raman sideband cooling, we scan the spectrum of the transition. We first pump the ion to the $|\uparrow\rangle$ state, and then apply a microwave pulse followed by Doppler cooling light to detect the states of the Mg ion. The resonance frequency is 1.777830 GHz at about 5×10^{-4} T magnetic field, and the Rabi frequency is 20 kHz. This Rabi frequency is still not high enough when performing quantum logic operation. Currently the antenna for the microwave is a homemade $\lambda/4$ antenna made by copper wire and mounted outside the vacuum chamber. We are considering two possible ways to increase the Rabi frequency. One is to shorten the distance between the antenna and the ion. Another is to impedance match the antenna with the microwave source.

The Raman spectrum is obtained by the following time sequence (shown in Fig. 2). First, we simultaneously switch on Doppler cooling and repumping light for 300 μs. Then we apply the Raman beams on the ion for 10 μs. During Raman sideband cooling, Raman interaction will repeat for 15 times. In between, repumping light will open for 2 μs followed by a 20 μs microwave pulse. In order to make sure the ion is completely repumped back to the ground state, this repumping process is repeated three times. The Raman sideband cooling is followed by scanning the Raman spectroscopy. Finally Doppler cooling light is applied for 50 μs to detect the states of $^{25}Mg^{+}$. The whole time sequence takes about 4 ms. To obtain the transition probability, we repeat the cycle 2000 times to compile statistics for each point on the sideband spectroscopy. In our current sequence, the microwave interaction is the most time-consuming process.

Theoretically, the temperature of the ion after Doppler cooling is expected to reach the Doppler limit. The actual secular motion frequencies are ω_x=2.2 MHz, ω_y=2.4 MHz, and ω_z=1.5 MHz, respectively. For an Mg^{+} ion at 1.5 MHz axial mode frequency, this temperature corresponds to an average vibration quantum number of 10. The vibration quantum number can be measured by measuring the ratio between the 1st red sideband (RSB) and the 1st blue sideband (BSB) of the Raman transition. Typically, in our system the average quantum number of an Mg ion after Doppler cooling is 16. This may be caused by the laser detuning during Doppler cooling. A half-linewidth red detuning gives the smallest temperature, but we use a near resonant laser to implement Doppler cooling. This is because we use the same laser for both Doppler cooling and detection. In order to obtain high enough fluorescence rates we have to use a small detuning. The Rabi frequencies of transitions corresponded to the 1st and the 2nd order RSB vary with vibration quantum numbers. At certain points the Rabi frequency will be near zero. Therefore the ion could not be cooled to the ground state using only the 1st order RSB cooling or the 2nd order RSB cooling. To avoid this problem, the 2nd order RSB are applied 15 times followed by 15 times 1st order RSB, and the process is repeated for two times. Figure 3 shows the sidebands under different Raman sideband cooling sequences. At last the population of n=0 is about 93 % in figure 3.

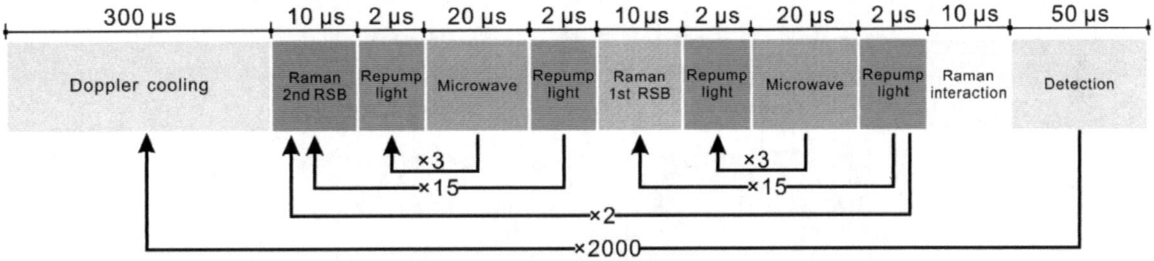

Figure 2. The time sequence for Raman sideband cooling.

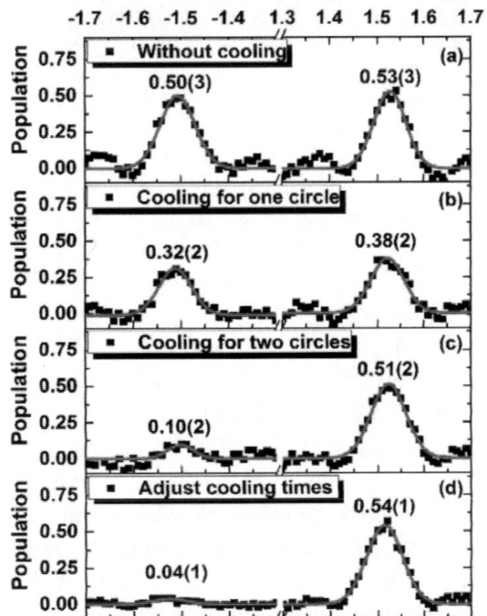

Figure 3. Measurement of sidebands under different Raman sideband cooling sequences. (a) the 1st BSB and RSB without Raman sideband cooling, n≈16(2); (b) the 1st BSB and RSB when cooled with 15 times 1st RSB and 15 times 2nd RSB, n=5.33(33); (c) the 1st BSB and RSB when cooled with twice 15 times 1st RSB and 15 times 2nd RSB, n=0.24(5); (d) the 1st BSB and RSB when the last 1st RSB

cooling times is changed from 15 times to 30 times, n=0.08(1), and the population of n=0 is about 93%.

After Raman sideband cooling, we try to sympathetically cool $^{27}Al^+$. Here we use $^{26}Mg^+$ to sympathetically cool $^{27}Al^+$ for demonstration purpose. At first, a single $^{26}Mg^+$ is loaded. Then the Al oven is fired and the 395 nm ionization laser is switched on for several minutes. One can easily know whether $^{27}Al^+$ is loaded by looking at the ion image on EMCCD, since the $^{26}Mg^+$ ion will shift a little bit on the image plane due to the trapped dark $^{27}Al^+$ ion. To confirm that the dark ion is $^{27}Al^+$, mass spectroscopy experiments have been carried out. We first measure the secular motion frequency of two $^{26}Mg^+$ ions, and then measure the secular motion frequency of a $^{26}Mg^+/^{27}Al^+$ pair. As shown in Fig. 4, the secular motion frequency of $^{26}Mg^+/^{26}Mg^+$ and $^{26}Mg^+/^{27}Al^+$ are 434.4 (7) kHz and 430.2(4) kHz, respectively. The results agree with the theoretical calculation, 434 kHz and 430 kHz, respectively.

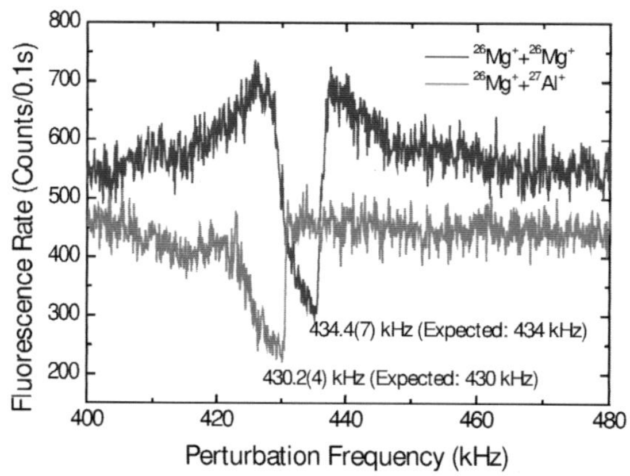

Figure 4. Secular motion frequency of Mg-Mg ion pair and Mg-Al ion pair.

3. Ultra-stable laser system

The $^{27}Al^+$ optical clock requires an ultra-narrow linewidth laser as a clock laser. To develop the first generation clock laser, two 1070 nm diode lasers from Toptica are phase stabilized to two independent 10 cm long all ULE cavities through PDH locking technique. The stabilized diode laser is quadrupled to 267.4 nm as the clock laser.

The thermal noise limit of one reference cavity which works at TEM_{00} mode is estimated to be 7.7×10^{-16} @ 1s, dominated by the thermal noise of the mirror substrates material. The linewidth of TEM_{00} modes of these two cavities are measured to be 13.6 kHz and 4.5 kHz, respectively. As the cavity length variation directly broadens the locked laser linewidth, the cavities are sealed in two vacuum chambers with pressures less than 1×10^{-6} Pa. The vacuum chambers are temperature controlled at the measured ULE cavity's zero crossing temperature T_0 of 36 °C and the temperature fluctuation is about 1 mK in few hours' time. In addition, the vacuum chambers are placed on active vibration isolation (AVI) stages and the whole systems are enclosed with a heavy box made of lead and iron plate to reduce the effect of acoustic noise and air disturbance. The ULE cavities are placed on four viton bars on their notched surfaces where the mounting positions are optimized to minimize the sensitivity to vibration through finite-element analysis (FEA) with Comsol Multiphysics.

The EOMs used in the PDH locking scheme are temperature controlled to reduce the residual amplitude modulation (RAM), and modulation depths are about 1.08 to reach a maximum slope for the PDH error signal. The error signals are sent to locking circuits (FALC, Toptica) that feed back to the current and piezo control of the diode lasers. We obtain a 2 MHz current feedback bandwidth and a 2 kHz piezo feedback bandwidth. As shown in figure 5, a beat note with a linewidth of 0.88 Hz is obtained between the two locked lasers, indicating an absolute linewidth of 0.6 Hz for one single laser. The Allan deviation of the beat note is 1.2×10^{-15} at 1 s which is close to the thermal noise limit contributions of the two cavities.

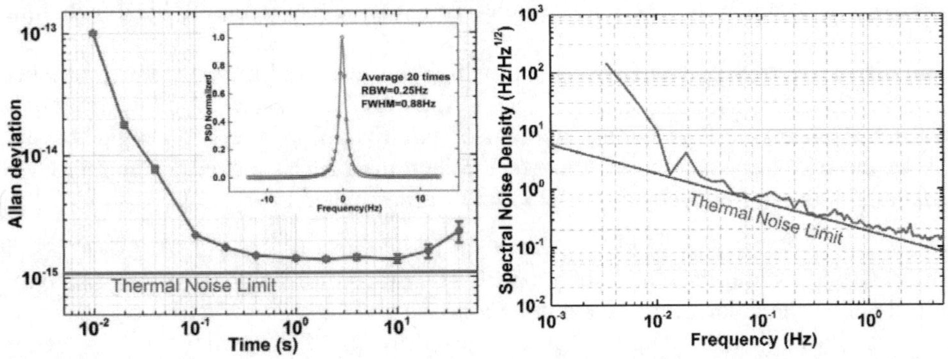

Figure 5. The Allan deviation and spectral noise density of the beatnote between lasers.

In order to develop better ultrastable clock lasers that can reach an instability of 10^{-16} or even 10^{-17}, noise contributions due to temperature fluctuations, optical power fluctuations and residual amplitude modulation are analysed to identify the limiting factors.

For the noise contribution due to temperature fluctuations, the thermal time constant τ is measured to be about 1 day by changing the temperature of the vacuum chamber with a step of 1 K and use an optical frequency comb to monitor the frequency of the locked laser. The temperature fluctuation of the cavity can be written by the following formula:

$$\mathrm{T_{in}}(\omega) = \frac{\mathrm{T_{out}}(\omega)}{\sqrt{1+(\omega\tau)^2}}. \tag{1}$$

$\mathrm{T_{out}}(\omega)$ is the vacuum chamber temperature in the frequency domain, and $\mathrm{T_{in}}(\omega)$ represents the ULE cavity temperature. The slope of the CTE β near the zero crossing temperature is measured to be $-1.2\times10^{-9}/\mathrm{K}^2$ by measuring the relationship between the cavity temperature and the locked laser frequency. The influence of temperature variation of the cavity on the Allan deviation $\sigma_{in}(\tau)$ can be estimated by Eq. 2.

$$\sigma_y(\tau) = \beta(\mathrm{T}-\mathrm{T_0})\sigma_{in}(\tau). \tag{2}$$

For conservative estimation, $\mathrm{T}-\mathrm{T_0}$ is chosen to be 1 K, which is larger than the real temperature difference between the zero crossing temperature $\mathrm{T_0}$ and the controlled temperature. As a result, $\sigma_y(\tau)$ is lower than the thermal noise limit of the 10 cm cavity at 1 s to 1000 s. In order to reduce the noise contributed from temperature fluctuations, larger thermal time constant of the vacuum chamber and better temperature control at $\mathrm{T_0}$ will help.

The power circulating in the cavity is very strong while the laser is resonant with the cavity. The total power in the cavity can be estimated to be

$$P_{\mathrm{cavity}} = P_{\mathrm{in}}\mathrm{F}/\pi\ , \tag{3}$$

where F is the cavity finesse, P_{in} is the laser power coupled into the cavity. In our case the measured finesse is about 300,000 for one of the cavities and P_{in} is about 10 μW, therefor the power circulating in the cavity P_{cavity} is near 1 W. The laser beam will induce a local heating on the mirror coatings and substrates that causes the cavity length change, and will result in the fluctuation of the locked laser frequency. In order to evaluate the influence of power fluctuations, we first measure the frequency sensitivity to power k by modulating the coupling power and measure the laser frequency change. When the the coupling power changes by 5 μW, the laser frequency changes 35.5 Hz, which corresponds to a sensitivity of 7.1 Hz/μW. After that we measure the laser power in front of the cavity when the laser is locked to the cavity and get the influence of the power fluctuation $\sigma_{P_{\mathrm{in}}}(\tau)$ on the Allan deviation

$$\sigma_y(\tau) = kP_{\mathrm{in}}\sigma_{P_{\mathrm{in}}}(\tau)/\nu\ . \tag{4}$$

Here, ν = 280.7 THz is the laser frequency. The Allan deviation contributed by the laser power fluctuation is lower than 10^{-15} during 1 s to 100 s. By active stabilization, it can be improved to the 10^{-17} level.

RAM is one of the dominant noises in ultrastable laser systems based on the PDH technique. In the ideal case, the error signal should be zero when the laser is on resonance with the cavity. But RAM introduces an offset voltage in the error signal which biases the locking point from the cavity transmission peak and fluctuates with time. The RAM can be caused by many effects, such as temperature fluctuation of the EOM, misalignment of polarization between the input beam and the principal axis of the EOM crystal, and etalon effect. When the laser is locked to the reference cavity, the RAM induced fractional frequency instability can be represented as [9]:

$$\sigma_y(\tau) = \sigma_{RAM}(\tau)\frac{\Gamma}{\nu}, \tag{5}$$

where Γ is the cavity linewidth. We implement an EOM temperature stabilization to reduce the RAM. We also carefully align the polarization of the injection beam to the axis of the crystal and suppress the etalon effect. It is shown that the Allan deviation caused by RAM is smaller than 10^{-15} from 1 s to 1000 s. It can be further reduced by active feedback control to reach the 10^{-17} level.

4. Conclusion and Outlook

We demonstrate Raman sideband cooling the $^{25}Mg^+$ ion to the motional ground state and sub-Hz linewidth ultra-stable lasers for Al^+ optical frequency standard. The achieved stability of the lasers is approaching the thermal noise limits of the cavities. We also investigate the noise contributions due to temperature fluctuations, optical power fluctuations and RAM to identify the limiting factors.

For our second generation clock laser, the diode lasers will be locked to 30 cm long ULE cavities with fused silica mirrors to reach an instability of 1×10^{-16} at 1 s. One of the cavities is a commercial product with a sandwich structure, with an additional ULE ring optically contacted to the back surface of the fused silica mirrors [10]. The second cavity is a homemade one with a re-entrant structure which has a wide tuning range of the zero crossing temperature of CTE to room temperature [11].

For the third generation clock laser, the laser will be locked to a 6 cm long cryogenic cavity with sapphire mirrors to reach an expected instability of 2×10^{-17} at 1 s. The designed square shape cavity will be maintained at 4 K where the CTE of cavity (4×10^{-11} /K) and the slope of CTE (3×10^{-11} /K^2) are both very small. A pulse tube cryostat for housing the cavity is currently under development.

References

[1] Yu N, Dehmelt H and Nagourney W 1992 *Proc. Natl. Acad. Sci. USA* **89** 7289
[2] Safronova M S, Kozlov M G and Clark C W 2011 *Phys. Rev. Lett.* **107** 143006
[3] Schmidt P, Rosenband T, Langer C, Itano W, Bergquist J and Wineland D 2005 *Science* **309** 749
[4] Monroe C, Meekhof D, King B, Jefferts S, Itano W and Wineland D 1995 *Phys. Rev. Lett.* **75** 4011
[5] Drever R W P, Hall J L, Kowalski F V, Hough J, Ford G M, Munley A J and Ward H 1983 *Appl. Phys. B* **31** 97
[6] Numata K, Kemery A and Camp J 2004 *Phys. Rev. Lett.* **93** 250602.
[7] Deng K, Che H, Lan Y, Ge Y P, Xu Z T, Yuan W H, Zhang J and Lu Z H 2015 *J Appl. Phys.* **118** 113106
[8] Monroe C, Meekhof D M, King B E, Jefferts SR, Itano W M and Wineland D J 1995 *Phys. Rev. A* **75** 4011
[9] Martin M, Benko C, Hall J, Ye J, Hagemann C, Legero T, Sterr U, Riehle F, Cole G and Aspelmeyer M 2014 *Opt. Lett.* **39** 1980
[10] Legero T, Kessler T and Sterr U 2010 *J. Opt. Soc. Am. B* **27** 914
[11] Zhang J, Luo Y, Ouyang B, Deng K, Lu Z H and Luo J 2013 *Eur. Phys. J. D* **67** 46

Evaluation of trap-induced systematic frequency shifts for a multi-ion optical clock at the 10^{-19} level

J Keller, T Burgermeister, D Kalincev, J Kiethe and T E Mehlstäubler

Physikalisch-Technische Bundesanstalt, Bundesallee 100, D-38116 Braunschweig, Germany

E-mail: `tanja.mehlstaeubler@ptb.de`

Abstract. In order to improve the short-term stability of trapped-ion optical clocks, we are developing a frequency standard based on $^{115}In^+$ / $^{172}Yb^+$ Coulomb crystals. For this purpose, we have developed scalable segmented Paul traps which allow a high level of control for multiple ion ensembles. In this article, we detail on our recent results regarding the reduction of the leading sources of frequency uncertainty introduced by the ion trap: 2nd-order Doppler shifts due to micromotion and the heating of secular motion, as well as the black-body radiation shift due to warming of the trap. We show that the fractional frequency uncertainty due to each of these effects can be reduced to well below 10^{-19}.

1. Introduction

Laser-cooled, trapped ions are among the most successful references for optical frequency standards, achieving fractional frequency uncertainties below 10^{-17} [1, 2]. However, current implementations of optical ion clocks are based on a single reference ion, which results in an intrinsically low signal-to-noise ratio. Averaging times of days to weeks are necessary for resolving the transition frequency well enough to benefit from these low systematic uncertainties. This impediment to applications can be overcome by interrogating an ensemble of ions simultaneously. To this end, we are developing a multi-ion clock based on ensembles of linear Coulomb crystals in a segmented Paul trap [3]. Our approach is suitable for any configuration in which the electric quadrupole shift from the field gradients within the crystals can be controlled [4] or is intrinsically low, such as in the $^1S_0 \leftrightarrow {}^3P_0$ lines in group 13 ions (e.g. In^+, Al^+), as well as the $^2S_{1/2} \leftrightarrow {}^2F_{7/2}$ E3 transition in Yb^+.

The challenge to maintain low systematic frequency uncertainties in such a clock based on a strongly coupled many-body system is addressed in two ways: The development of scalable ion traps with low heating rates, low on-axis micromotion and a well-known thermal environment, and experiments to gain an improved understanding of the Coulomb crystal dynamics. In the current implementation, we use $^{115}In^+$ ions, sympathetically cooled with $^{172}Yb^+$ (see Fig. 1). As detailed in [3], indium is chosen for its high mass, low sensitivity to electric [5] and magnetic [3,6] fields, and the facilitated scalability through direct state detection and second-stage cooling on the $^1S_0 \leftrightarrow {}^3P_1$ transition.

In this article, we present results concerning the reduction of three sources of uncertainty to well below 1×10^{-18}, which are also among the limiting effects in current single-ion clocks. Section 2 discusses the 2nd-order Doppler shift due to excess micromotion. In section 3, we

Figure 1. Example of a mixed-species Coulomb crystal in the prototype trap, consisting of 6 $^{115}In^+$ ions (dark, positions indicated by circles) and 6 fluorescing $^{172}Yb^+$ ions (from [7]).

present a measurement of external heating rates in a prototype trap [7]. The AC Stark shift due to black-body radiation is discussed in section 4, and the results are summarized in section 5.

2. Excess micromotion

Micromotion consists of fast oscillations driven by the rf field in a Paul trap when the ion position deviates from the electric field node. It can affect the clock transition frequency through both a 2nd-order Doppler and Stark shift, the latter of which is negligible in In^+ and Al^+ due to the low differential static polarizability of the clock transitions [5]. Precise knowledge of rf field amplitudes at the ion positions is therefore required, in particular for the storage of extended crystals. We investigated three techniques for micromotion determination in terms of their resolution [8]. In this section, we describe the two central results of this study: the explanation of an offset at a level of 10^{-19} seen in resolved-sideband measurements at Doppler temperature and the experimental verification of a newly derived model for the quantitative application of the photon-correlation technique in the commonly encountered regime of comparable linewidth and drive frequency. The (wrongful) application of the only previously published model [9] under these conditions would lead to an underestimation of the micromotion-induced frequency shift by an order of magnitude and a 1st-order sensitivity to laser detuning. With this new understanding, we were able to show that both techniques allow uncertainties in the 2nd-order Doppler shift due to excess micromotion (EMM) well below 10^{-19}.

2.1. Resolved sideband method

The resolved sideband method uses a transition which is narrow compared to the sideband spectrum produced by the 1st-order Doppler shift due to micromotion. By comparison of the excitation strength on the carrier (Ω_0) and micromotion sideband transitions (Ω_1), a modulation index $\beta \approx 2\Omega_1/\Omega_0$ is determined, which relates to the 2nd-order Doppler shift as

$$\frac{\Delta\nu_{\mathrm{D2}}}{\nu_0} = -\left(\frac{\Omega_{\mathrm{rf}}\lambda\beta}{4\pi c}\right)^2 , \tag{1}$$

where Ω_{rf} denotes the trap drive frequency, c the speed of light and λ the transition wavelength. While in the low temperature limit, the evaluation of the method is straightforward and the resolution is ultimately limited by the available laser power and lowest resolvable sideband strength, the uncertainty in measurements at Doppler temperature can be limited by a temperature-dependent offset. This signal arises from the nonlinearity of $\exp(i\vec{k}(\vec{x}_{\mathrm{sec}} + \vec{x}_{\mathrm{imm}}))$, which mixes the phase modulation due to secular motion $x_{\mathrm{sec}} \propto \cos(\omega_{\mathrm{sec}}t)$ and the induced intrinsic micromotion $x_{\mathrm{imm}} \propto \cos(\omega_{\mathrm{sec}}t)\cos(\Omega_{\mathrm{rf}}t)$, as illustrated in Fig. 2a. Apart from a geometrical factor ≤ 1, the expected Rabi frequency for a Fock state with n phonons is

$$\frac{\Omega_1}{\Omega_0} = \underbrace{\frac{q}{4}\eta^2(2n+1) + \mathcal{O}\left(\eta^4\right)}_{\text{intrinsic micromotion}} \underbrace{-i\left\langle n\left|e^{ikx}\right|n\right\rangle\left(\frac{\beta}{2} + \mathcal{O}\left(q^3\right)\right)}_{\text{excess micromotion}} , \tag{2}$$

where q is the Mathieu-q parameter of the rf confinement (see, e.g. [10]) and η is the Lamb-Dicke parameter (e.g. [11]). For a thermal state, the observed signals correspond to a thermal average

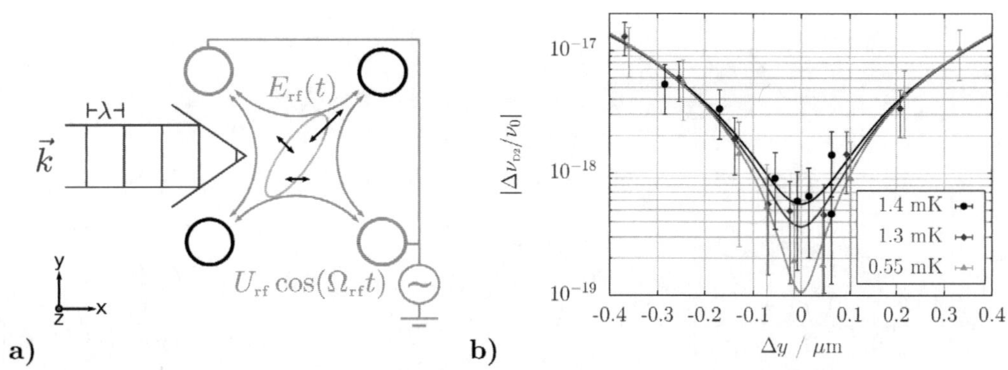

Figure 2. Sampling of intrinsic micromotion in resolved sideband measurements. **a)** Along its secular motion trajectory (grey line), the ion undergoes intrinsic micromotion (black arrows). The correlated components of the two along $\vec{k}$ produce a temperature-dependent sideband at $\Omega_{\rm rf}$. **b)** Experimental observation of the offset at a secular frequency of $\omega_{\rm sec} \approx 2\pi \times 500\,\rm kHz$ for different temperatures on the $^2S_{1/2} \leftrightarrow {}^2D_{5/2}$ transition in a single $^{172}Yb^+$ ion. No sideband excitation is observed at the minimum when the ion is cooled to the motional ground state.

method	$\Delta\nu_{\rm D2}/\nu_0$ resolution	limited by
sideband ($T_{\rm Doppler}$)	1.6×10^{-19}	sampled IMM
sideband (ground state)	3×10^{-20}	achievable Rabi frequency, decoherence
photon-correlation	3.5×10^{-21}	photon shot noise

Table 1. Achieved resolution for different micromotion compensation methods.

of Rabi oscillations, with the minimal excitation corresponding to minimized EMM due to the 90° phase difference of the two terms in (2). Figure 2b shows an experimental observation of the temperature dependence.

2.2. Photon-correlation method

The photon-correlation method employs a transition with linewidth Γ comparable to or wider than $\Omega_{\rm rf}$. While the evaluation had been described in the limit of $\Omega_{\rm rf} \ll \Gamma$ by Berkeland et al. [9], to our knowledge no published model existed for the nowadays more common regime where $\Gamma \approx \Omega_{\rm rf}$. Figure 3 shows an experimental verification of our newly derived model for this case by observing the signal dependence on laser detuning (**a**) and comparison to resolved-sideband measurements when a single $^{172}Yb^+$ ion is radially displaced from the rf node (**b**). While the analytic expression as given in [8, Eq. 18] assumes $\beta \ll 1$, this restriction can be removed by taking higher-order sidebands of the 1st-order Doppler shift into account. As the method uses only the fluorescence during Doppler cooling, it is in principle applicable without pausing clock interrogation, which is advantageous in terms of servo instability [12]. Table 1 summarizes the resolution limits achievable in our experimental setup for the photon-correlation and sideband methods.

Since rf confinement is only used in the radial plane of a linear trap, the electric field ideally vanishes along the trap axis. In real traps, however, segmentation and machining tolerances will produce a finite axial field component, which limits the range in which clock ions can be stored without deteriorating the frequency uncertainty. If multiple ions are to be stored for spectroscopy, this field needs to be minimized by design [3]. Figure 4 shows a 3D micromotion measurement in the prototype trap, in which a single ion was shifted along the axis, without

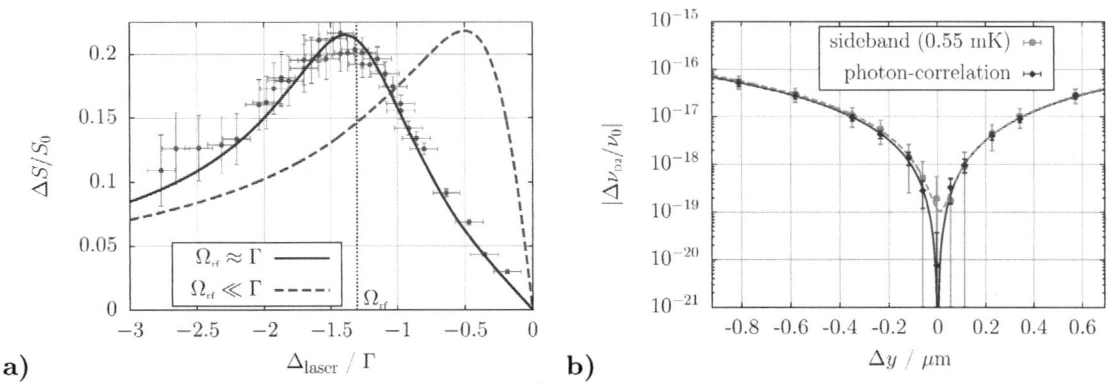

Figure 3. a) Measured photon-correlation signal as a function of laser detuning (filled circles) for a fixed micromotion amplitude ($\beta = 0.085$). The observed quantity is the fluorescence modulation amplitude ΔS, normalized by the mean fluorescence rate S_0. The solid curve shows our model for the experimental parameters $\Gamma = 2\pi \times 19.6\,\text{MHz}$ and $\Omega_{\text{rf}} = 2\pi \times 25.42\,\text{MHz}$ (indicated by the vertical line). The dashed curve shows the result derived by Berkeland et al. [9] in the limit $\Omega_{\text{rf}} \ll \Gamma$. The models agree under that condition. **b)** Quantitative comparison of photon-correlation and resolved-sideband measurements when a single $^{172}\text{Yb}^+$ ion is radially displaced from the rf node. The residual mismatch of the fitted minima corresponds to a fractional 2nd-order Doppler shift of 3×10^{-21}.

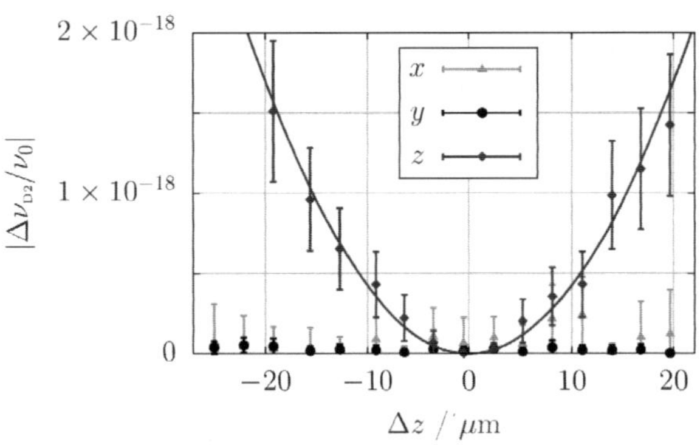

Figure 4. 3D micromotion measurement as a function of the axial ion position using the photon-correlation technique. A single ion is shifted axially through a segment of the prototype trap by varying the axial dc voltages, without adjustment of the radial compensation voltages.

adjustment of the radial compensation voltages.

3. Secular motion, external heating rates

Thermally excited secular motion also contributes to the 2nd-order Doppler shift. If static contributions to the confining potential can be neglected, intrinsic micromotion doubles the resulting 2nd-order Doppler shift for the directions of rf confinement, leading to a total shift of $\Delta\nu_{\text{D2}}/\nu_0 = -5k_BT/(2mc^2)$ in a linear trap. For an $^{115}\text{In}^+$ ion sympathetically cooled to the Doppler limit of Yb^+ at 0.5 mK, this shift amounts to -1.0×10^{-18}. A low temperature uncertainty during clock interrogation is therefore important to prevent this contribution from limiting the frequency inaccuracy. Since the different motional modes in extended crystals exhibit varying sympathetic cooling efficiency [13], low external heating rates are favorable in order to ensure that the steady-state temperatures are close to the Doppler limit.

External electric fields fluctuating at the secular frequency can resonantly transfer energy to the motion of the ions [14]. The traps used in our setup are designed to limit this effect by keeping a large distance of $\approx 0.7\,\text{mm}$ between the ions and trap electrodes and by onboard

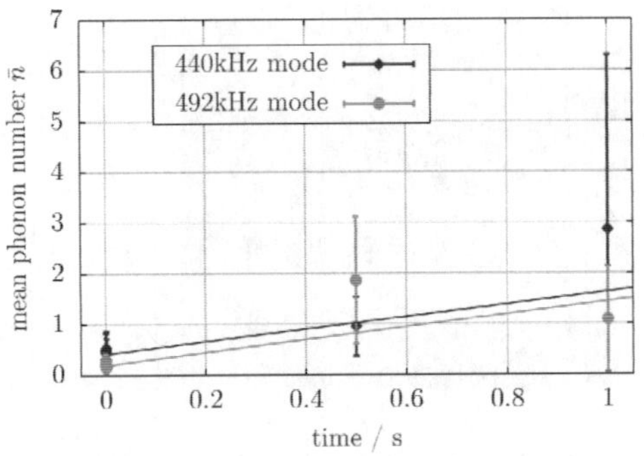

Figure 5. Radial heating rate measurement in the prototype trap. The circles and diamonds show the excitation of the radial modes for a single ion as a function of time without cooling. From the fits, we determine the heating rates as $(1.2 \pm 0.4)\,\mathrm{s}^{-1}$ (440 kHz mode) and $(1.3 \pm 0.5)\,\mathrm{s}^{-1}$ (492 kHz mode).

lowpass filtering of the DC voltages with a cutoff frequency of $\approx 100\,\mathrm{Hz}$ [7]. To determine the heating rates experimentally, we cool a single $^{172}\mathrm{Yb}^+$ ion close to the motional ground state of both radial modes at secular frequencies of $\omega_1 = 2\pi \times 440\,\mathrm{kHz}$ and $\omega_2 = 2\pi \times 492\,\mathrm{kHz}$, via the $^2\mathrm{S}_{1/2} \leftrightarrow {}^2\mathrm{D}_{5/2}$ transition at 411 nm. During cooling, the lifetime of the excited state is quenched using the 1650 nm $^2\mathrm{D}_{5/2} \leftrightarrow {}^2\mathrm{P}_{3/2}$ transition. After a varied amount of time, the mean phonon number in each mode is determined by comparison of 1st-order red and blue secular sidebands [15]. The result is shown in Fig. 5: The linear fits determine similar heating rates, the higher of which is $\dot{\bar{n}} = (1.3 \pm 0.5)\,\mathrm{s}^{-1}$, corresponding to an electric field power spectral density of $S_E(492\,\mathrm{kHz}) = (1.8 \pm 0.7) \times 10^{-14}\,(\mathrm{V/m})^2/\mathrm{Hz}$. In the absence of cooling, the energy transferred to the ion increases the 2nd-order Doppler shift due to each mode as

$$\left| \frac{\partial}{\partial t} \left(\frac{\Delta\nu_{\mathrm{D2}}}{\nu_0} \right) \right| = \frac{\dot{\bar{n}}\hbar\omega_{\mathrm{sec}}}{mc^2} , \tag{3}$$

where ω_{sec} denotes the secular frequency of the respective mode and m the mass of the ion. This shift amounts to $(2.5 \pm 1.0) \times 10^{-20}\,\mathrm{s}^{-1}$ and $(1.1 \pm 0.4) \times 10^{-19}\,\mathrm{s}^{-1}$ for a clock based on $^{115}\mathrm{In}^+$ and $^{27}\mathrm{Al}^+$, respectively.

4. Black-body radiation due to trap heating

The AC Stark shift due to black-body radiation (BBR) amounts to 1.36×10^{-17} for In^+ at 300 K [3, 5]. In Paul traps, a particular concern is the warming of the trap structure due to the applied rf voltages on the order of several hundred volts. In order to reduce the corresponding uncertainty, tests have been performed on a new trap design based on AlN wafers, shown in Fig. 6a. Besides the use of materials with a high heat conductivity and low dielectric losses to limit the temperature increase in operation, two onboard Pt100 sensors are included for in-situ temperature measurements. The effective thermal environment experienced by the ions is deduced from a finite-element model, the parameters of which are adjusted using measurements on a test trap [16].

Trap temperature measurements are performed with a calibrated IR camera and thermistors. Figure 6b shows a thermal image of the ceramics test trap when an rf amplitude of 1 kV at $\Omega_{\mathrm{rf}} = 2\pi \times 15.4\,\mathrm{MHz}$ is applied, which would yield trap frequencies of $\omega = 2\pi \times 1.0\,\mathrm{MHz}$ ($\omega = 2\pi \times 1.52\,\mathrm{MHz}$) for $^{172}\mathrm{Yb}^+$ ($^{115}\mathrm{In}^+$). FEM simulations with experimentally adjusted material parameters predict an effective temperature increase at the ion position of $\Delta T_{\mathrm{ion}} = 1.0\,\mathrm{K}$, which corresponds to an AC Stark shift of $\Delta\nu_{\mathrm{BBR,In}}/\nu_0 \approx 2 \times 10^{-19}$. Using the calibrated Pt100 sensors, the uncertainty of the temperature rise seen by the ions is reduced to 0.1 K, which corresponds to a BBR shift uncertainty of 2×10^{-20} [16]. The total uncertainty of the BBR shift in In^+ is then limited by the theoretical prediction of the differential polarizability [5]. This can be reduced with an experimental determination, e.g. by controlled heating of the trap via the Pt100 sensors during interrogation.

a) 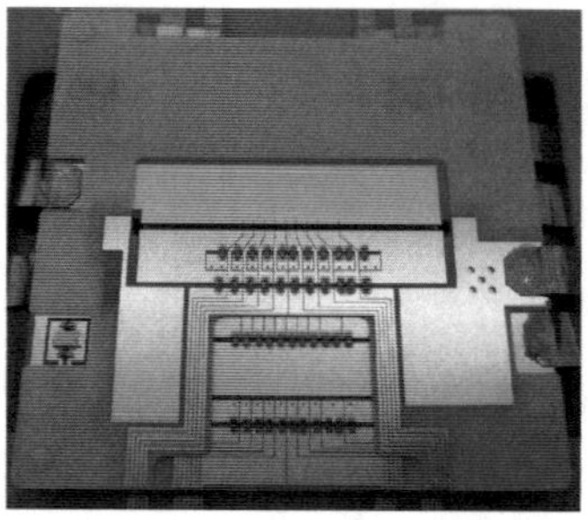 b) 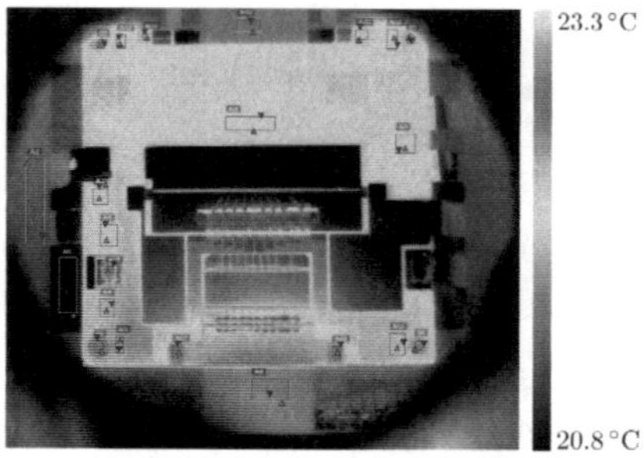

Figure 6. a) Image of the next-generation trap made from a stack of 4 AlN wafers. Ions are trapped in the central slit. On the bottom left, one of the temperature sensors can be seen. **b)** Thermal image showing the temperature increase at $U_{\rm rf} = 1\,\rm kV$, $\Omega_{\rm rf} = 2\pi \times 15.4\,\rm MHz$ (color online).

5. Summary

This article presents our recent results concerning the reduction of the major trap-related sources of frequency uncertainty in current ion clocks. Our analysis of micromotion measurement techniques has shown that the corresponding 2nd-order Doppler shift can be determined with an uncertainty well below 10^{-19}, independent of the ion mass, with both the resolved-sideband and photon-correlation methods [8] (see Table 1). The heating rates of less than 2 phonons per second measured in a prototype trap facilitate sympathetic cooling in extended mixed-species crystals and give rise to a 2nd-order Doppler shift increase of $|\partial/\partial t(\Delta\nu_{\rm D2}/\nu_0)| = (2.5 \pm 1.0) \times 10^{-20}\,\rm s^{-1}$ in $\rm In^+$ if cooling is paused during interrogation. In the next-generation AlN wafer trap, onboard temperature sensors, in combination with an experimentally verified thermal model, allow a precise knowledge of the black-body spectrum at the ion position due to trap warming. The residual temperature uncertainty corresponds to a fractional uncertainty of 2×10^{-20} in the BBR AC Stark shift [16].

References

[1] Chou C W, Hume D B, Koelemeij J C J, Wineland D J and Rosenband T 2010 *Phys. Rev. Lett.* **104** 070802
[2] Huntemann N 2014 *High-Accuracy Optical Clock Based on the Octupole Transition in* $^{171}Yb^+$ Ph.D. thesis Leibniz Universität Hannover
[3] Herschbach N, Pyka K, Keller J and Mehlstäubler T E 2012 *Appl. Phys. B* **107** 891
[4] Arnold K, Hajiyev E, Paez E, Lee C H, Barrett M D and Bollinger J 2015 *Phys. Rev. A* **92** 032108
[5] Safronova M S, Kozlov M G and Clark C W 2011 *Phys. Rev. Lett.* **107** 143006
[6] Becker T, Zanthier J V, Nevsky A Y, Schwedes C, Skvortsov M N, Walther H and Peik E 2001 *Phys. Rev. A* **63** 051802
[7] Pyka K, Herschbach N, Keller J and Mehlstäubler T E 2014 *Appl. Phys. B* **114** 231
[8] Keller J, Partner H L, Burgermeister T and Mehlstäubler T E 2015 *J. Appl. Phys.* **118** 104501
[9] Berkeland D J, Miller J D, Bergquist J C, Itano W M and Wineland D J 1998 *J. Appl. Phys.* **83** 5025
[10] Major F G, Gheorge V N and Werth G 2005 *Charged Particle Traps* (Springer-Verlag, Heidelberg)
[11] Wineland D J, Monroe C, Itano W M, Leibfried D, King B E and Meekhof D M 1998 *J. Res. Natl. Inst. Stand. Tech.* **103**
[12] Barwood G P, Huang G, King S A, Klein H A and Gill P 2015 *J. Phys. B: At. Mol. Opt. Phys.* **48** 035401
[13] Kielpinski D, King B E, Myatt C J, Sackett C A, Turchette Q A, Itano W M, Monroe C, Wineland D J and Zurek W H 2000 *Phys. Rev. A* **61** 032310
[14] Brownnutt M, Kumph M, Rabl P and Blatt R 2015 *Rev. Mod. Phys.* **87** 1419
[15] Turchette Q A, Kielpinski D, King B E, Leibfried D, Meekhof D M, Myatt C J, Rowe M A, Sackett C A, Wood C S, Itano W M, Monroe C and Wineland D J 2000 *Phys. Rev. A* **61** 063418
[16] Doležal M, Balling P, Nisbet-Jones P B R, King S A, Jones J M, Klein H A, Gill P, Lindvall T, Wallin A E, Merimaa M, Tamm C, Sanner C, Huntemann N, Scharnhorst N, Leroux I D, Schmidt P O, Burgermeister T, Mehlstäubler T E and Peik E 2015 *Metrologia* **52** 842

Different ways to active optical frequency standards

Duo Pan, Xiaobo Xue, Xiaogang Zhang, and Jingbiao Chen*
State Key Laboratory of Advanced Optical Communication System and Network, School of Electronics Engineering and Computer Science, Peking University, Beijing 100871, China

*E-mail: jbchen@pku.edu.cn

Abstract. Active optical frequency standard, or active optical clock, is a new concept of optical frequency standard, where a weak feedback with phase coherence information in optical bad-cavity limitation is formed, and the continuous self-sustained coherent stimulated emission between two atomic transition levels with population inversion is realized. Through ten years of both theoretical and experimental exploration, the narrow linewidth and suppression of cavity pulling effect of active optical frequency standard have been initially proved. In this paper, after a simple review, we will mainly present the most recent experimental progresses of active optical frequency standards in Peking University, including 4-level cesium active optical frequency standards and active Faraday optical frequency standards. The future development of active optical frequency standards is also discussed.

1. Introduction

Recently great progress has been made in optical clocks, and the 10^{-18} level accuracy and stability has been experimentally realized [1-3]. However, for optical frequency standards working at passive regime, a local oscillator with excellent short-term stability is necessary, where a probing laser is usually prestabilized to a super-cavity with the Pound–Drever–Hall (PDH) technique. The linewidth of the local oscillator is currently the main limitation of optical clocks' stability. However, due to the inevitable Brownian cavity-length noise, to further improve the stability of optical clocks is still a great challenge.

The concept of active optical frequency standard was conceived ten years ago [4,5], where a weak feedback with phase coherence information in optical bad-cavity limitation is formed, and the continuous self-sustained coherent stimulated emission between two atomic transition levels with population inversion is realized. The mechanism of active optical frequency standard makes the output frequency insensitive to the cavity length noise, thus has a potential much narrower linewidth than that of local oscillators in passive optical clocks. Meanwhile, since the output frequency is determined by the atomic transition, it can be applied as quantum frequency standards directly. The mechanism of active optical frequency standards is similar to that of Hydrogen masers, which has a superior short-term stability among microwave atomic clocks. Theoretically, the active optical frequency standard has the potential frequency stability as Hydrogen masers in microwave regime, thus has the advantage of much improved fractional stability with an operating frequency exceed the oscillators in the microwave domain by about five orders of magnitude. We would like to point out that, there is no analog of active Faraday optical clock in microwave regime.

Since the concept of active optical frequency standard was proposed, some preliminary considerations [4-6] summarized and presented in the 7th Symposium on Frequency Standards and Metrology [6]

have triggered wider interests on different ways [8-19] to the creation of active optical frequency standards with unique techniques and elegant ideas[8-12]. Different ways with neutral atoms at 2-level, 3-level, 4-level energy structures with thermal and laser cooled and trapped configurations [7-9], Raman laser [10], sequential coupling [11] and moving optical lattice configuration [12] have been investigated recently. To make continuous-wave active optical frequency standards with cold atoms, one also can apply sequential coupling technique [11,12], synchronization between two ensembles of 3-level atoms [12,13], or cold atomic beam. The bad-cavity Rb Raman laser has been investigated with very beautiful results [10]. The effect of interaction between atoms on frequency stability and accuracy of active optical clock has been calculated theoretically[14], which also shows that active optical frequency standards have the characteristic of suppressing the cavity pulling effect, and the output linewidth can break quantum noise limitation. Active ^{88}Sr optical clock at 7.6 kHz natural linewidth 689 nm transition with cold atomic beam is under developing in Niels Bohr Institute [20], and trapped 1D lattice atoms in JILA near the crossover from good-cavity to bad-cavity regime in quasi steady-state lasing shows a prospective linewidth of 6.0 kHz [21].
In this paper, we will mainly present the most recent experimental progresses of active optical frequency standards in Peking University, including 4-level Cesium active optical frequency standards with measured 380 Hz linewidth [15-18] and active Faraday optical frequency standards with measured 281 Hz linewidth [19]. The future development of active frequency standards is also discussed.

2. Four-level active optical frequency standards

2.1. Lasing of cesium four-level active optical frequency standards
The way of 4-level active optical frequency standard has been proposed to avoid the sensitivity of light shift due to pumping laser. Recently, in this way, the Cs four-level active optical frequency standards have made a progress experimentally [15,16]. With the 459 nm or 455 nm pumping laser applied to the atoms, the population inversion between $7^2S_{1/2}$ and $6^2P_{3/2}$ levels is performed. The continuous self-sustained stimulated emission at 1469.9 nm is realized with the whole system operating in deep optical bad-cavity regime, of which the ratio between cavity bandwidth and gain bandwidth is 55.08.
As the main characteristic of the cesium four-level active optical frequency standard, the suppression of cavity pulling effect is examined using a Fabry-Perot interferometer (FPI). It turns out that when the cavity mode frequency is detuned 281.6 MHz away, the 1469.9 nm output light frequency shifts only 6.8 MHz. Therefore the cavity pulling effect was reduced to 2.4% compared with ordinary laser. This effect can reduce the influence of environmental vibrations on active optical frequency standards, and in consequence the inevitable thermal Brownian-motion noise of cavities can be effectively suppressed in active optical clock. This experimental result of strongly suppressed cavity pulling effect clearly demonstrates that active optical clock owns potential narrower linewidth than that of super-cavity stabilized laser [22,23].
The linewidth of the 1469.9 nm output light is measured by the heterodyne signal between two independent setups. The results show that the output linewidth of each experimental setup is 380 Hz, which is much narrower than that of ordinary laser based on alkali atoms considering there is no any vibration isolation measure. Similarly, we have also observed the continuous self-sustained 1359 nm ($7^2S_{1/2}$ - $6^2P_{1/2}$) stimulation emission of the Cs four-level active optical frequency standard [17].

2.2. Good-bad cavity dual-wavelength active optical frequency standard
Even though the cavity pulling effect has been suppressed, the residual perturbation of the cavity still affects the frequency stability greatly. To stabilize the main cavity length, we propose a scheme of He-Ne 633 nm and Cs 1359 nm good-bad cavity dual-wavelength active optical frequency standard [18]. In the scheme, the 633 nm and 1359 nm stimulated emissions share a common cavity, and the 633 nm laser is designed to work in good-cavity laser regime, i.e., the cavity mode linewidth for 633 nm is designed to be much narrower than the He-Ne 1.5 GHz laser gain bandwidth. At the same time, the

1359 nm stimulated emission still works in bad-cavity regime. The 633 nm laser is subsequently frequency stabilized to a super-cavity with PDH technique [22,23], hence the main cavity of the good-bad cavity active optical frequency standard is stabilized. The stability of 1359 nm bad-cavity output frequency is then expected to be improved by at least 1 order of magnitude than that of the 633 nm PDH stabilized laser because of the suppression of cavity pulling effect. The quantum limited linewidth of 1359 nm laser of the good-bad cavity active optical clock is estimated to be 72.5 mHz, and it can be reduced to mHz when a large waist mode cavity is applied to increase the effective atoms. The frequency stability and accuracy of 1359 nm output signal we finally obtain with thermal Cs atoms will be worse than its theoretical value, because of the existence of classic noises. So that the utilization of laser cooled and trapped atoms is necessary to avoid the disadvantageous influence from thermal motion of atoms. For alkali atoms like Cs, Rb and K, the available diode lasers and techniques of laser cooling and trapping will be helpful in the way of active optical frequency standards based on cold alkali atoms.

3. Active Faraday optical frequency standards

In all above mentioned ways [4-18] to the active optical frequency standards, the quantum reference of optical frequency standard and the stimulated optical emission of gain medium are both from the same atomic transition of the same atoms. This requirement strongly limits available quantum system. The active Faraday optical frequency standard, as demonstrated by a recent experiment [19], spatially separates the quantum reference of frequency standard and the stimulated emission of gain medium. In the active Faraday optical frequency standard [19,24], a narrow linewidth Faraday atomic filter, which can be a very general 2-level atomic transition, is used as quantum reference of optical frequency standard, while the stimulated emission of gain medium can be provided by Ti: sapphire and dye, besides semiconductor diode materials [19,24]. This unique way of active optical frequency standard opens the door for various atomic transitions and alternative gain media to optical clocks. In this way, the Faraday effect found 170 years ago, starts to play a quantum reference role in modern optical clocks.

In the experimental scheme demonstrated in [19,24], an ultranarrow bandwidth Faraday atomic filter based on cesium atoms acts as a quantum frequency selector, of which the bandwidth is 25 MHz and the transmission is 18%. The filter is then placed inside the extended cavity of a laser diode with anti-reflection coating, the 852 nm ($6P_{1/2}$-$6S_{1/2}$) stimulated emission of active Faraday optical frequency standard is realized by adjusting the extended cavity mirror. A maximum output power of 75 μW is obtained. Suppression of cavity pulling effect in active Faraday optical frequency standard is measured by beating the output light with pumping laser. When changing the extended cavity frequency within 350 MHz, the centre frequency of the beat signal varied within a range of 40 MHz. Therefore the cavity pulling coefficient is reduced to 11% compared with common lasers.

The output light frequency linewidth is measured by the heterodyne signal between two independent setups to be 281 Hz [19,24,25]. The result is 19000 times smaller than the natural linewidth of the Cs 852 nm transition line and two orders of magnitude better than frequency-stabilized lasers in good-cavity regime with atomic filter or interference filter working as frequency selectors. In the next step, we will apply the mechanism of active Faraday optical clock to measure the Cs 852 nm transition frequency to an expected accuracy of Hz level, even though its natural linewidth is as wide as 5.2 MHz. In the view of laser spectroscopy, this unique mechanism of active Faraday optical clock provides us an active laser spectroscopy based on the stimulated emission within bad-cavity regime for many general optical transitions with natural linewidth as wide as several megahertz.

To further improve the frequency stability of active Faraday optical frequency standards, utilization of laser cooled and trapped atom in magic wavelength optical lattice is a promising solution [25]. One can use the 689 nm transition of strontium atoms trapped by magic wavelength optical lattice, which has a natural linewidth of 7.6 kHz. With the existence of homogeneous magnetic field with appropriate intensity, we can realize an ultra-narrow linewidth Faraday optical filter at 689 nm transition. If one uses semiconductor diode or Ti: Sapphire as gain medium, by weak feedback of laser resonator at bad-cavity regime, one can realize Faraday active optical clock based on cold atoms [25]. The relevant research is on-going. According to reference [26], the 689 nm transition of strontium atoms trapped in magic lattice in hollow-core photonic crystal fibres can realize an optical thickness of 2.5, so it is possible to achieve a rotation angle near to $\pi/2$ in an experiment scheme shown in figure 1. Considering the high symmetry of m=1 and m=-1 Zeeman sublevels, the influence of magnetic field to the Faraday active optical frequency standard can be controlled within required region.

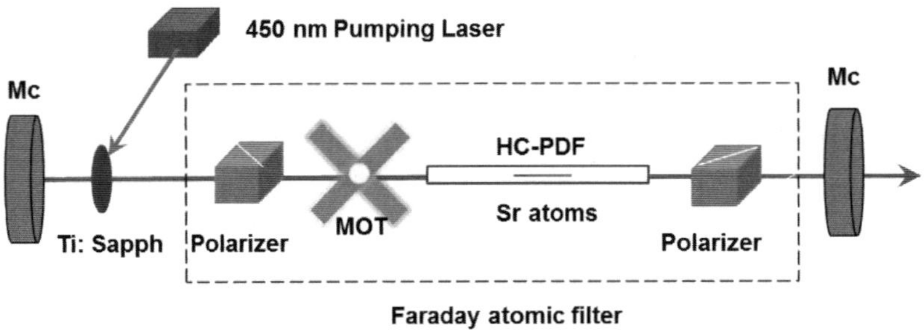

Figure 1. (Colour online) The proposed experimental scheme of active Faraday optical clock based on Sr atoms trapped in magic optical lattice in hollow-core photonic crystal fibres with Ti:Sapphire laser gain operating at bad-cavity regime.

4. Conclusion

In this paper we mainly present the most recent experimental progresses of active optical frequency standards in Peking University, including 4-level cesium active optical frequency standards and active Faraday optical frequency standards in bad-cavity regime [25]. Under the present experimental conditions, noise of cavity length is still the main limitation of frequency standard measured linewidth. For the future progress to improve the frequency stability, we need to stabilize the cavity length through the PDH technique combined with the dual-wavelength good-bad cavity method [17,18]. Since the thermal motion of atoms brings many influences caused by classic effects, it needs further investigation to realize active optical frequency standards based on laser cooled atoms and ions. We believe the ^{88}Sr active optical clocks at 689 nm 7.6 kHz transition at cold atomic beam configuration [20], trapped 1D lattice configuration [21], or Faraday configuration [19,25] are very promising for continuous self-sustained operation of Sr active optical clock in the near future. A blue Cs magneto-optical trap operating at 455 nm transition is under development in Peking University to investigate the 4-level Cesium active optical frequency standards and active Faraday optical frequency standards with cold Cs atoms.

Acknowledgments

This work is supported by International Science & Technology Cooperation Program of China under No. 2010DFR10900, and NSF of China under No.91436210.

References

[1] Bloom B J, Nicholson T L, Williams J R, et al 2014 An optical lattice clock with accuracy and stability at the 10^{-18} level *Nature*, **506** p 71
[2] Hinkley N, Sherman J A, Phillips N B, et al 2013 An Atomic Clock with 10^{-18} instability *Science*, **341** p 1215
[3] Nicholson T L, Campbell1 S L, Hutson R B et al 2015 Systematic evaluation of an atomic clock at 2×10^{-18} total uncertainty, *Nat. Commun*, **6** p 6896

[4] Chen J and Chen X 2005 Optical lattice laser Proceedings, *Proc. of 2005 IEEE Int. Freq. Control Symp. (Vancouver)* p 608
[5] Chen J 2009 Active optical clock *Chin. Sci. Bull.*, **54 (3)** 348 (*Preprint* physics/0512096)
[6] Yu D and Chen J 2008 Laser theory with finite atom-field interacting time *Phys. Rev. A.* **78** 013846
[7] Chen J 2009 Active optical clocks *Proceedings of the 7th Symposium in Frequency Standards and Metrology (Pacific Grove)* (World Scientific Publishing Co. Pte. Ltd.) p 525
[8] Meiser D, Ye J, Carlson D R and Holland M J 2009 Prospects for a millihertz-linewidth laser *Phys. Rev. Lett.* **102** 163601
[9] Meiser D and Holland M J 2010 Steady-state superradiance with alkaline-earth-metal atoms *Phys. Rev. A.* **81** 033847
[10] Bohnet J G, Chen Z, Weiner J M, Meiser D, Holland M J and Thompson J K 2012 A steady-state superradiant laser with less than one intracavity photon *Nature* **484** 78
[11] Kazakov G A and Schumm T 2013 Active optical frequency standard using sequential coupling of atomic ensembles *Phys. Rev. A.* **87** 013821
[12] Kazakov G A and Schumm T 2015 Active optical frequency standards using cold atoms: perspectives and challenges *Preprint* physics/1503.03998v1
[13] Xu M, Tieri D A, Fine E C, Thompson J K and Holland M J 2014 Synchronization of Two Ensembles of Atoms *Phys. Rev. Lett.* **113** 154101
[14] Maier T, Kraemer S, Ostermann L and Ritsch H 2014 Superradiant clocks using a magic wavelength optical lattice *Optics Express* **22** 13269
[15] Xu Z, Zhuang W, Wang Y, Wang D, Zhang X, Xue X, Pan D and Chen J 2013 Lasing of Cesium four-level active optical clock *Proc. Joint IEEE Int. Freq. Control Symp. and European Freq. and Time Forum (Prague)* p 395
[16] Pan D, Xu Z, Xue X, Zhuang W and Chen J 2014 Lasing of cesium active optical clock with 459 nm laser pumping *Proc. of the Joint IEEE Int. Freq. Control Symp. & the European Freq. and Time Forum (Taipei),* p 19
[17] Xu Z, Pan D, Zhuang W and Chen J 2015 Experimental scheme of 633 nm and 1359 nm good-bad cavity dual-wavelength active optical frequency standard *Chin. Phys. Lett.* **32** 083201
[18] Xu Z, Pan D, Zhuang W and Chen J 2015 Dual-wavelength bad cavity laser as potential active optical frequency standard *Chin. Phys. Lett.* **32** 093201
[19] Zhuang W and Chen J 2014 Active Faraday optical frequency standard *Opt. Lett.* **39(**21) 6339
[20] Christensen B R, Schäffer S A and Thomsen J W *Private communication*
[21] Norcia M A and Thompson J K 2015 A Cold-Strontium Laser in the Superradiant Crossover Regime *Preprint* physcis/1510.06733
[22] Jiang Y Y, Ludlow A D, Lemke N D, Fox R W, Sherman J A, Ma L S and Qates C W 2011 Making optical atomic clocks more stable with 10^{-16}-level laser stabilization *Nature Photonics* **5** 158
[23] Kesseler T, Hagemann C, Grebing C, Legero T, Sterr U, Riehle F , Martin M J , Chen L and Ye J 2012 A sub-40-mHz-linewidth laser based on a silicon single-crystal optical cavity *Nature Photonics* **6**(10) 687
[24] Zhang X, Zhuang W and Chen J 2015 Active Optical Frequency Standard Based on Narrow Bandwidth Faraday Atomic Filter *Proc. of the 2015 Joint Conf. of the IEEE Int. Freq. Control Symp. & European Freq. and Time Forum (Denver)* p 618
[25] Pan D, Zhuang W, Xue X, Zhang X, Chen M, Xu Z and Chen J 2015 Ten Years of Active Optical Frequency Standards *Proc. of the 2015 Joint Conf. of the IEEE Int. Freq. Control Symp. & European Freq. and Time Forum (Denver)* p 363
[26] Okaba S, Takano T, Benabid F, Bradley T, Vincetti L, Maizelis Z, Yampol'skii V, Nori F and Katori H 2014 Lamb-Dicke spectroscopy of atoms in a hollow-core photonic crystal fibre *Nat Commun.* **5** 4096

Journal of Physics: Conference Series **723** (2016) 012029 doi:10.1088/1742-6596/723/1/012029

Design of an ultra-compact reference ULE cavity

Alexandre Didier[1], Jacques Millo[1], Clément Lacroûte[1], Morvan Ouisse[1], Jérôme Delporte[2], Vincent Giordano[1], Enrico Rubiola[1] and Yann Kersalé[1]

[1] FEMTO-ST, CNRS/UBFC/ENSMM/UFC/UTBM, Besançon, France
[2] CNES, Toulouse, France

E-mail: yann.kersale@femto-st.fr

Abstract. This article presents the design and the conception of an ultra-compact Fabry-Pérot cavity which will be used to develop an ultra-stable laser. The proposed cavity is composed of a 25 mm long ULE spacer with fused silica mirrors. It leads to an expected fractional frequency stability of 1.5×10^{-15} limited by the thermal noise. The chosen geometry leads to an acceleration relative sensitivity below 10^{-12} /(m/s^2) for all directions.

1. Introduction

Ultra-stable lasers are a key element in various applications ranging from optical frequency standards [1, 2], gravitational wave detection [3], tests of fundamental physics [4] and generation of ultra-pure microwave signals [5]. In frequency metrology, and in particular for optical atomic clocks, lasers are stabilized onto ultra-stable Fabry-Pérot cavities providing the relative frequency stability at short integration times in order to probe the atomic transition. The principal limitations of such lasers stems from Fabry-Pérot cavities length fluctuations principally due to thermal and vibration perturbations. Therefore a lot of efforts were dedicated to isolating cavities and making them insensitive to environmental perturbations. Thermal noise floors below 10^{-15} for 100 mm long cavities [6, 7, 8, 9] and of 1×10^{-16} or below for 400 mm long cavity [10] have been already reached. Recently, relative frequency instabilities below 10^{-16} have been obtained by the use of a 480 mm long cavity [11] or with a cryogenic silicon single crystal cavity [12]. In non-laboratory environment, for applications such as the generation of ultra low phase noise microwave for radar [13], geodesy [14] and local oscillator for space missions [15], the relative frequency instability requirements is reduced to the 10^{-15} level surpassing the best quartz crystal oscillator performances. For these fields of applications, a complete redesign of the cavity and its assembly is needed to make the cavity compatible with transport. Recently, several research groups have built cavities that are designed to be vibration-insensitive [16], robust [17] and transportable [15, 18].

In this work, we report the design of a reference cavity with calculated acceleration sensitivities below 10^{-12}/(m/s^2) for all directions. The horizontally oriented cavity is rigidly held with three compressive forces applied in the cavity middle plane (orthogonal to the optical axis). The cavity within a custom made vacuum chamber fits on a 600×300 mm^2 breadboard with a height of 300 mm. The freespace optical components used for the mode matching are directly attached onto this vacuum chamber.

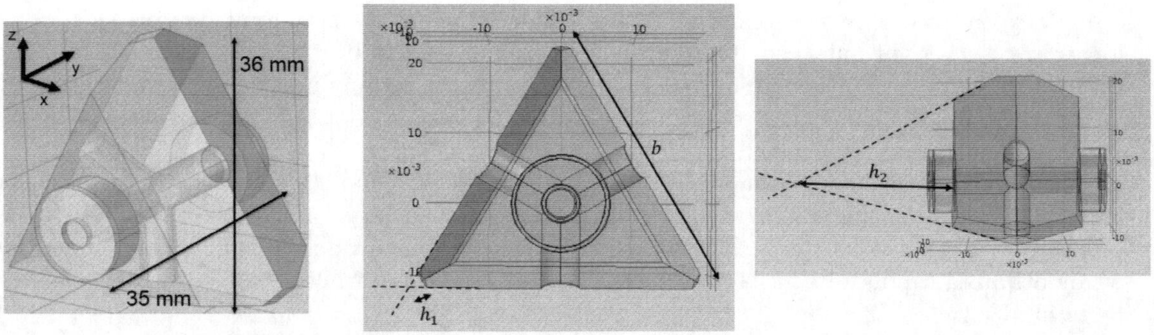

Figure 1. Left: Double tetrahedral spacer with SiO_2 substrates and ULE rings; Center: Cavity front view ; Right: Cavity side view

2. Cavity design and geometry

The optical length of the cavity is 25 mm corresponding to a free spectral range of 6 GHz with a plano-concave optical resonator configuration (radius of curvature of 240 mm). Fused silica mirror substrates are optically contacted to a spacer made of ultra-low expansion (ULE) glass. The overall coefficient of thermal expansion (CTE) is partially compensated by contacting rings in ULE on the back of the mirrors. Thank to the high mechanical quality factor of fused silica, the fractional frequency stability is limited at 1.5×10^{-15} by the thermal noise of the high dielectric reflective coatings for 1.54 μm laser wavelength [19]. It has to be mentioned that the thermal noise can be further reduced using crystalline coatings [20]. This cavity length results from a compromise between the compactness of the cavity and its thermal noise flicker limit. The full size of the final cavity geometry (including mirrors and compensation rings) fits in 35 mm x 36 mm x 41 mm (figure 1). Mechanical simulations using finite element modeling (FEM) have been performed in order to reduce the acceleration sensitivities of the cavity by optimizing its geometry. FEM is also used to dimension the ULE compensation rings to adjust the overall CTE.

2.1. Mechanical modeling

Several geometries have been used for ultra-stable Fabry-Pérot cavities, such as the cylinder [6, 8, 9] and the cube [16] with four holding points. Vertical cavities [7, 9] are held by a monolithically attached midplane ring resting on three Teflon rods or viton. For the spherical cavity, only two holding points on its diameter are used with a "magic angle" [22] where its length is insensitive to the forces used to support it.

We observe the same kind of deformations than those described in [9, 22]. Movements of the mirrors under accelerations can be approximated by a translation along the optical axis (y-axis) and tilts around the two other axis (x-axis or z-axis). For the design of vibration insensitive cavities it is important to keep geometric and stress symmetries. For advantageous geometries the translation do not lead to vibration sensitivity, only remain tilts of the mirrors which can be minimized by tuning the spacer parameters.

In the above-mentioned geometries, such as the cylinder, the measured acceleration sensitivity is always significantly larger for one acceleration direction. In order to further reduce the acceleration sensitivities and to homogenize them, the ends of the cylinder are cut with an angle, thus removing matter near the mirrors. Following this trend, we designed a spacer geometry based on two tetrahedral (see figure 1), allowing to have less matter near the mirrors.

One issue that could increase the acceleration sensitivities of the cavity is the inhomogeneity of the forces used to hold it. To overcome this problem, very careful position adjustments of the cavity on its supports are needed. With a non hyperstatic three-point holding system, the forces are always self balanced. Our cavity is therefore held in its middle plan on three points. The cavity being held away from the mirrors, the mirrors strains due to the static restraining forces are smaller. The sharp edges of the spacer increase also the stiffness of the spacer and decrease its tendency to bend at its ends.

The double tetrahedral shape of the cavity possesses enough symmetries so that the axial translation of the cavity theoretically vanishes. The biggest strain contribution to the acceleration sensitivities stems therefore from the tilts of the mirrors. The cavity dimensions were optimized through FEM in order to find spacer dimension where the the mirrors tilts could be minimized.

In this article, we focus on the most important parameters on the vibration sensitivity such as the main edge of the spacer b or the depth of cut of the cavity vertices h_1 creating the area for its support (see figure 1). The depth of cut of the end vertices of the cavity h_2, allowing the attachment of the mirrors on the spacer , have also an impact on the acceleration sensitivity. The diameter of the holes for light propagation and out-gassing decrease slightly the acceleration sensitivities. More importantly, having three holes with 120° axial symmetry instead of just one enables to have the same amplitude of acceleration sensitivity along the three directions.

The parameters were swept in the simulations and the relative length fluctuations of the cavity were extracted from the FEM analysis. We could check that the simulated axial translation of the mirrors is always below $10^{-12}/(\mathrm{m/s}^2)$ (see figure 2), which is at the level of the apparent uncertainty of our simulation. The acceleration sensitivities 1 mm away from the mirrors centers for each transverse direction gave us the contribution of the mirrors rotations to the acceleration sensitivities of the cavity, in the pessimistic case where the optical axis is shifted by 1 mm from the geometric axis of the cavity.

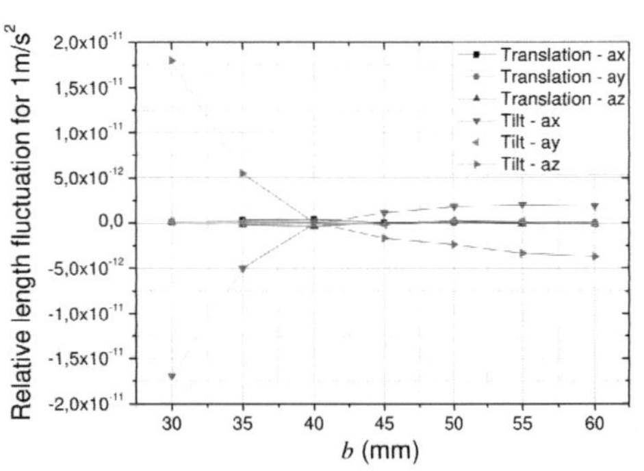

Figure 2. Relative cavity length fluctuation as a function of the main edge of the spacer for the three directions, when the other parameters are fixed ($h_1 = 2$ mm, $h_2 = 35$ mm).

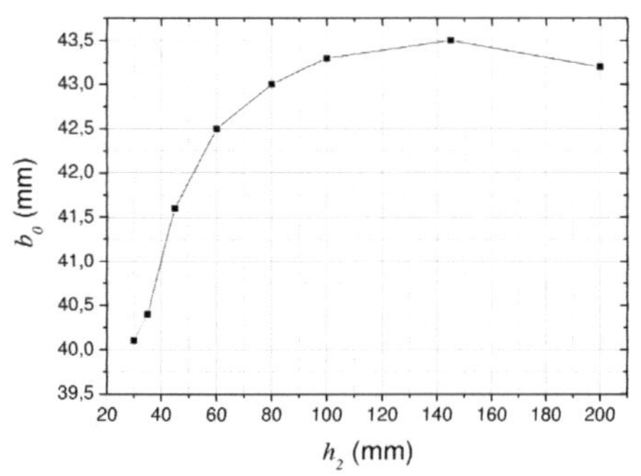

Figure 3. Specific edge length b_0 for several values of h_2 when other parameters are fixed ($h_1 = 2$ mm).

The figure 2 shows an example of vibration sensitivity function as the parameter b. The tilts due to two transverse accelerations (a_x and a_z) vanishes for the same value of the parameter $b_0 = 40$ mm. On figure 3, the values of b_0 is reported as function of the parameter h_2 for a constant value of $h_1 = 2$ mm. This mean that h_2 can be used to tune and minimize b_0. The cut of depth h_1 also affect the sensitivity due to tilt induced by transverse acceleration (see figure 4) and can be used to optimize the geometry. By tuning b, h_1 and h_2 parameters all together, one can therefore decrease the size of the cavity while maintaining very low acceleration sensitivities.

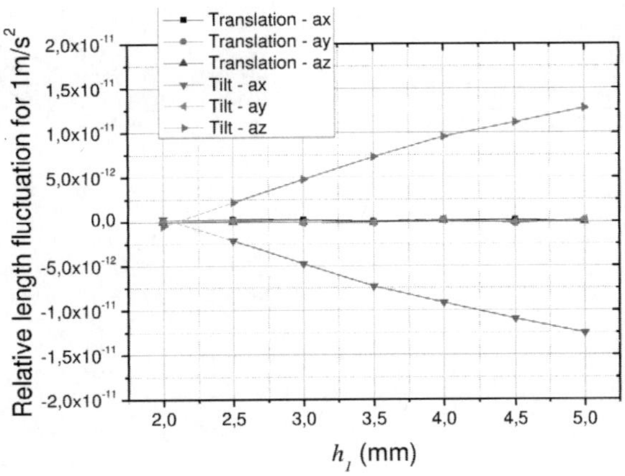

Figure 4. Relative cavity length fluctuation as a function of the depth of cut of the cavity vertices h_1 for the three directions, when the other parameters are fixed (h_2 = 35 mm, b = 41.5 mm).

The cut angle given by h_2 for a constant length is limited by the surface needed to attach the mirrors to the spacer. The cut depth h_1 has to be superior than 2 mm, in order to be able to hold the cavity on a large enough surface.

All sets of parameters have also an influence on the small offset depicted on figure 2 at the acceleration inversion point. This offset results from the lack of planar symmetry between the x and z-axis.

The best configuration is obtained for a main edge of the spacer b = 41.5 mm, a cut depth h_1 = 2 mm and a depth of cut of the end vertices of the cavity h_2 = 35 mm. With these values we can see that the mirrors relative length fluctuations are below $10^{-12}/(\mathrm{m/s^2})$ in all directions, even for a strong shift of the optical axis.

2.2. Thermal modeling

The shift of the cavity CTE turning point can be compensated by the use of ULE rings optically contacted to the back of the mirrors [21].

In the present case the mass added on the two ends of the cavity is not negligible compared to the mass of the cavity and influences the vibration sensitivity. Both thermal and mechanical simulations had therefore to be repeated alternatively updating compensation rings and spacer dimensions. Generally speaking, when the mass of the rings increase, the temperature of the overall CTE turning point increase but the vibration sensitivity also tends to rise. A compromise has therefore been made between the thermal and mechanical inversion points.

With further iteration of finite element analysis, ring dimensions of 12.7 mm outer diameter, 5 mm inner diameter and 1 mm thickness has been found to shift the cavity CTE zero crossing near 11°C for the cavity dimensions mentioned in the previous section.

3. Experimental setup

A custom vacuum chamber, with a cubic shape (figure 5) was designed in order to reduce the total volume of the system.

Supporting a transportable cavity by three points only in its middle plan is unsafe. In order to increase the contact surface between the cavity and the holding system, and to add a mechanical

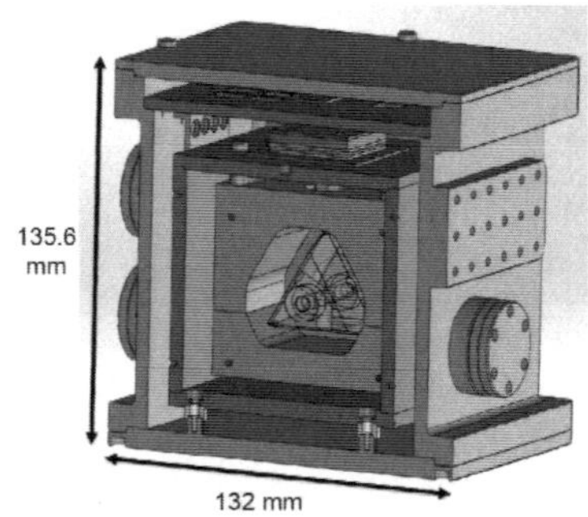

Figure 5. Cavity vacuum chamber.

Figure 6. Cavity mounted in it stainless steel support.

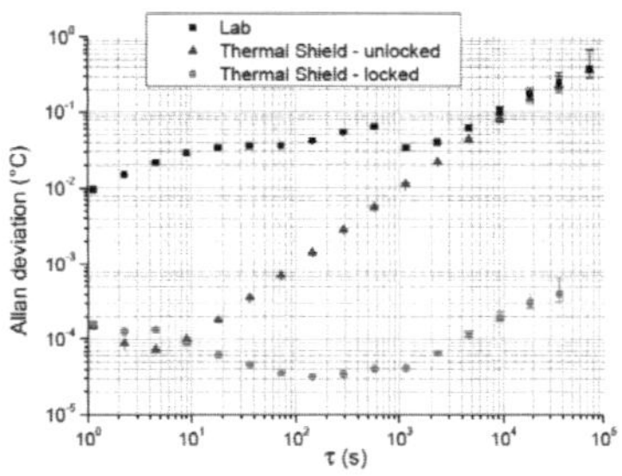

Figure 7. Thermal isolation between the lab and the copper shield.

Figure 8. Experimental set-up on a 600 × 300 mm^2 breadboard.

damping and a good grip, two 2 mm diameter and 1 mm thick Viton pads are placed at the bottom of the enclosure with a re-entrant cutoff of 0.5 mm. A Viton sphere, on the cavity top vertices, of 3.75 mm diameter is used to transfer the force exerted by a home made stud to the cavity (figure 6). The centering of the cavity is managed by eyesight and the positioning errors on each support contact are estimated to be under 0.5 mm. This system has already been able to stand a few g acceleration manually applied in all directions without noticeable sliding movements of the cavity.

To reduce the room temperature influences on the cavity temperature a set of three or four thermal shields are typically used. For a compact and transportable cavity this number of shield has to be considerably reduced. In that way our first thermal shield is the cavity enclosure made in stainless steel in order to get a high thermal capacitance. A second shield in copper will then reduce the thermal radiation from the external environment to the cavity and allow homogeneous temperature around the cavity. To reduce the thermal link between the vacuum chamber and the copper thermal shield, insulating materials (*eg* Teflon and glass reinforced polymer) are used for the posts and the screws holding the thermal shield, in order to provide the needed thermal resistance of 400 K/W per post so that the thermal conduction is reduced to the level of the thermal radiation.

An attenuation of 95 dB is obtained at 1 Hz between the room temperature and the cavity. The thermal time constant was calculated to be around 5 days. Figure 7 shows the experimental result of the copper shield temperature instability. A 40 dB gain is obtained between the room temperature fluctuations and the copper shield. When this shield is temperature controlled the temperature instability is about 0.1 mK for integration times from 1 s to 10^4 s, compatible with the 10^{-15} level of relative frequency instability for an offset of 1 mK from the CTE inversion.

A 600×300 mm^2 breadboard is used to support the vacuum chamber containing the cavity. The optics for the cavity mode matching are attached on the vacuum chamber, where custom designed M3 breadboards have been machined. All the fibered optical components used for the laser frequency Pound-Drever-Hall locking are also placed on the breadboard (not seen on figure 8).

4. Conclusion

We reported on the design of an ultra-compact ULE cavity with a thermal noise of 1.5×10^{-15} which fits in a 2.5 liters vacuum chamber. FEM simulations predict acceleration sensitivities below $10^{-12}/(\mathrm{m/s^2})$ in all directions. The stabilized laser total volume will be of the order of 50 liters. Experimental relative frequency stability and acceleration sensitivities are under investigations.

Acknowledgments

The work has been realized in the frame of the ANR project "Projets d'Investissements d'Avenir (PIA) Equipex Oscillator-Imp" and by CNES. The authors would like to thank the Council of the Région de Franche-Comté for its support to the PIA and to the Labex FIRST-TF for fundings.

References

[1] Nicholson T L, Campbell S L, Hutson R B, Marti G E, Bloom B J, McNally R L, Zhang W, Barrett M D, Safronova M S, Strouse G F, Tew W L and Ye J 2015 *Nature communications* **6** 6896
[2] Ushijima I, Takamoto M, Das M, Ohkubo T and Katori H 2015 *Nature Photonics* **9** 185
[3] Willke B, Danzmann K, Frede M, King P, Kracht D, Kwee P, Puncken O, Savage R L, Schulz B, Seifert F, Veltkamp C, Wagner S, Webels P and Winkelmann L 2008 *Class. Quantum Grav.* **25** 114040
[4] Herrmann S, Senger A, Mohle K, Nagel M, Kovalchuk E and Peters A 2009 *Phys. Rev. D* **80** 105011.
[5] Fortier T M, Kirchner M S, Quinlan F, Taylor J, Bergquist J C, Rosenband T, Lemke N, Ludlow A D, Jiang Y, Oates C W and Diddams S A 2011 *Nature Photonics* **5** 425
[6] Nazarova T, Riehle F and Sterr U 2006 *Appl. Phys. B* **83** 531
[7] Ludlow A D, Huang X, Notcutt M, Zanon-Willette T, Foreman S M, Boyd M M, Blatt S and Ye J 2007 *Optics Letters* **32** 641
[8] Webster S A, Oxborrow M, Pugla S, Millo J and Gill P 2008 *Physical Review A* **77** 033847
[9] Millo J, Magalhães D V, Mandache C, Le Coq Y, English E M L, Westergaard P G, Lodewyck J, Bize S, Lemonde P and Santarelli G 2009 *Physical Review A* **79** 053829
[10] Swallows M D, Martin J M, Bishof M, Benko C, Lin Y, Blatt S, Rey AM and Ye J 2012 *IEEE Transactions on Ultrasonics, Ferroelectrics, and Frequency Control* **59** 416
[11] Häfner S, Falke S, Grebing C, Vogt S, Legero T, Merimaa M, Lisdat C and Sterr U 2015 *Optics Letters* **40** 2112
[12] Kessler T, Hagemann C, Grebing C, Legero T, Sterr U, Riehle F, Martin M J, Chen L and Ye J 2012 *Nature Photonics* **6** 687
[13] Didier A, Millo J, Grop S, Dubois B, Bigler E, Rubiola E, Lacroûte C and Kersalé Y 2015 *Applied Optics* **54** 3682
[14] Chou C W, Hume D B, Rosenband T and Wineland D J 2010 *Science* **329** 1630
[15] Argence B, Prevost E, Lévèque T, Le Goff R, Bize S, Lemonde P and Santarelli G 2012 *Optics Express* **20** 25409
[16] Webster S and Gill P 2011 *Optics Letters* **36** 3572
[17] Vogt S, Lisdat C, Legero T, Sterr U, Ernsting I, Nevsky A and Schiller S 2011 *Appl. Phys. B* **104** 741
[18] Leibrandt D R, Thorpe M J, Bergquist J C and Rosenband T 2011 *Optics Express* **19** 10278
[19] Numata K, Kemery A and Camp J 2004 *Phys. Rev. Lett.* **93** 250602
[20] Cole G D, Zhang W, Martin M J, Ye J and Aspelmeyer M 2013 *Nature Photonics* **7** 644
[21] Legero T, Kessler T and Sterr U 2012 *J. Opt. Soc. Am. B* **27** 914
[22] Leibrandt D R, Thorpe M J, Notcutt M, Drullinger R E, Rosenband T and Bergquist J C 2011 *Optics Express* **19** 3471

The Autonomous Cryocooled Sapphire Oscillator: A Reference for Frequency Stability and Phase Noise Measurements

V Giordano[1], S Grop[2], C Fluhr[1], B Dubois[3], Y Kersalé[1], E Rubiola[1]

[1] Dept TF, FEMTO-ST Institute, CNRS UMR 6174, Besançon, France
[2] Alemnis GmbH, Feuerwerkerstrasse 39, 3602 Thun, Switzerland
[3] FEMTO Engineering, 32 avenue de l'Observatoire 25000 Besançon France

E-mail: `giordano@femto-st.fr`

Abstract. The Cryogenic Sapphire Oscillator (CSO) is the microwave oscillator which feature the highest short-term stability. Our best units exhibit Allan deviation $\sigma_y(\tau)$ of 4.5×10^{-16} at 1 s, $\approx 1.5\times10^{-16}$ at 100 s $\leq \tau \leq$ 5,000 s (floor), and $\leq 5\times10^{-15}$ at one day. The use of a Pulse-Tube cryocooler enables full two year operation with virtually no maintenance.

Starting with a short history of the CSO in our lab, we go through the architecture and we provide more details about the resonator, the cryostat, the oscillator loop, and the servo electronics.

We implemented three similar oscillators, which enable the evaluation of each with the three-cornered hat method, and provide the potential for Allan deviation measurements at parts of 10^{-17} level. One of our CSOs (ULISS) is transportable, and goes with a small customized truck. The unique feature of ULISS is that its $\sigma_y(\tau)$ can be validated at destination by measuring before and after the roundtrip. To this extent, ULISS can be regarded as a traveling standard of *frequency stability*.

The CSOs are a part of the Oscillator IMP project, a platform dedicated to the measurement of noise and short-term stability of oscillators and devices in the whole radio spectrum (from MHz to THz), including microwave photonics. The scope spans from routine measurements to the research on new oscillators, components, and measurement methods.

1. Introduction

The Cryogenic Sapphire Oscillator (CSO) is the microwave oscillator which features the highest short-term stability. The Allan deviation (ADEV) $\sigma_y(\tau)$ is below 1×10^{-15} for integration time $\tau = 1$–10^4 s. Such stability enables the operation of laser-cooled microwave clocks at the quantum limit [1, 2]. The CSO proved to be useful at improving the resolution in Very Long Baseline Interferometry (VLBI) and in the deep-space netwoks for space-vehicle ranging and Doppler tracking [3, 4, 5, 6, 7, 8]. And it can enhance the calibration capability of Metrological Institutes, and help for the qualification of high performances clocks and oscillators [9].

2. History

Research on microwave sapphire Whispering Gallery mode (WG) oscillators started in our lab in 1995 with room-temperature and 77 K resonators. We developed several techniques to compensate for the high temperature sensitivity of the sapphire (70 ppm/K at 300 K) [10, 11]. At

Journal of Physics: Conference Series **723** (2016) 012030 doi:10.1088/1742-6596/723/1/012030

the end of the 90s, CNES (the French space agency) and BNM (the French metrology agency, now LNE) encouraged us to develop a 4 K CSO for ground applications. We got help from the FSM lab at the University of the Western Australia (UWA) in a rather informal collaboration, still going on. Our first resonator was tested at UWA, and in 2001 we started the first liquid-He cooled resonator in Besançon [12]. In 2004 our open-cavity resonator achieved $\sigma_y(\tau) \leq 7\times10^{-15}$ at short term, and $< 2\times10^{-14}$ at one day [13, 14, 15].

Figure 1. Liquid helium cryostats of the early CSOs at the Femto-ST Institute (2000).

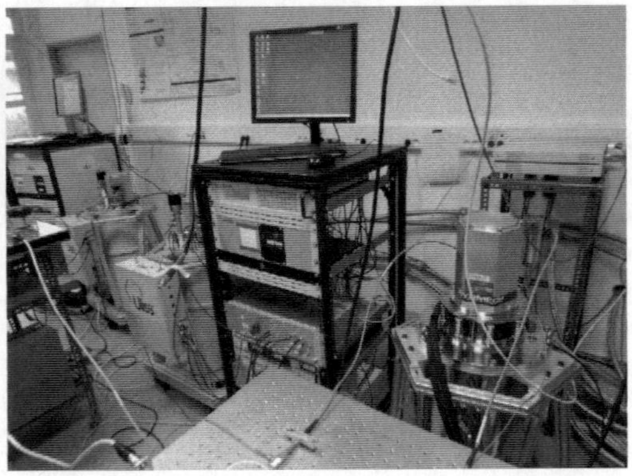

Figure 2. 2014: The three CSOs of the Oscillator IMP platform.

The ELISA project, funded by the European Space Agency (ESA) started in 2007, with National Physical Laboratory (UK) and TimeTech GmbH (DE) as subcontractors. We demonstrated the first state-of-the-art frequency stability with an autonomous cryocooler. ELISA features $\sigma_y(\tau) \leq 3\times10^{-15}$ for $\tau = 1{-}10^4$ s [4, 5, 16, 17]. It was later moved to the ESA Deep Space Antenna Station DSA-3 in Malargüe, Argentina, where it is operational since.

In 2011, we designed a new CSO, codenamed ULISS, which can be transported with a small truck. In two years ULISS visited several European sites, traveling more than 10,000 km [18]. A set of three CSOs is now operational at our site. So, measuring ULISS before and after traveling, we can validate stability and spectral purity at destination. These CSOs have been used for example to check on the the stability of a laser stabilized on a ULE cavity, and to run with a cold atom fountain at SYRTE.

3. CSO Technology at FEMTO-ST

3.1. Architecture and Design

Frequency stability relies on the resonator, which is a cylinder of low-defect Al_2O_3 monocrystal. Al_2O_3 is chosen for lowest dielectric loss and good mechanical properties. The Whispering Gallery (WG) mode resonance provides the strongest confinement of energy inside the resonator, and in turn the highest quality factor Q. The unloaded Q of a 30–50 mm diameter, 10–30 mm high resonator exceeds 10^9 at 10 GHz and at liquid-He temperature. Industrial Al_2O_3 crystals show a useful turning point at 5–8 K due to impurities, detailed later.

The oscillator (Figure 3) is a Pound-Galani scheme [19], in turn derived from the Sulzer quartz oscillator [20]. In the Pound scheme [21], a voltage-controlled oscillator is frequency locked to the resonator using the reflected microwave to get the error signal. The Pound-Galani scheme differs in that the same resonator is also used as the reference of the locked oscillator. In our implementation, the resonator power is stabilized by a second control. Two tunnel diodes near the resonator input detect forward and reflected power.

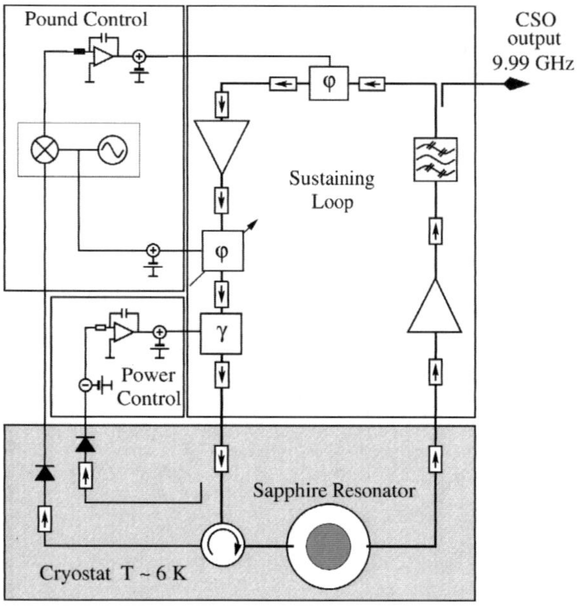

Figure 3. Oscillator main loop, with Pound and power servos.

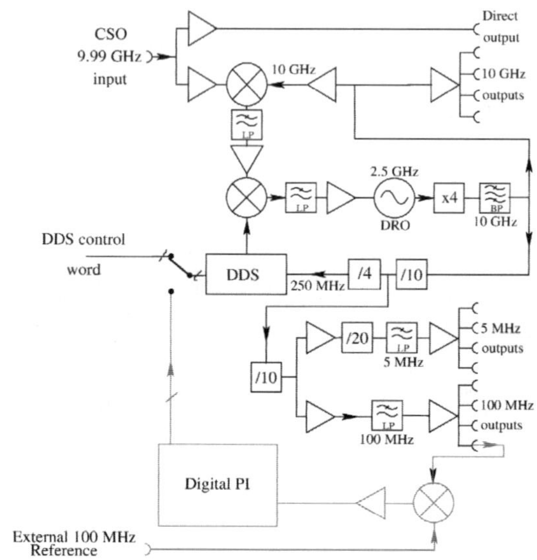

Figure 4. Low noise frequency synthesis for the generation of round frequencies: 10 GHz, 100 MHz and 5 MHz.

The complete machine includes a low-noise synthesizer which delivers 5–10–100 MHz and 10 GHz signals (Figure 4). The synthesizer can be frequency locked to a 100 MHz input, intended to discipline the CSO to a H maser for long term stability, letting the core oscillator unperturbed.

3.2. Resonator

As a general rule, the highest Q is expected from pure crystals with the lowest defect density. The problem with pure Al_2O_3 is that the the sensitivity of the resonant frequency to temperature is too high even at 4 K and monotonic. A turning point, necessary for stability, is provided by impurities. Paramagnetic ions as Mo^{3+}, Ti^{3+} Cr^{3+} substitute Al^{+3} and induce a small magnetic susceptibility χ. In the appropriate conditions, the thermal variation of χ compensates for the natural sensitivity of the sapphire, and the resonator sensitivity nulls at the turning-point temperature T_0. The value of T_0 is specific to each crystal, as it depends on the concentration of dopants. However, a comfortable turning point just above 4 K has been found in most crystals grown with different methods, namely, Heat Exhange (HEMEX), Kyropoulos and Czochralski, and from different manufacturers. This is a fortunate outcome of technology. HEMEX grade sapphire proved to be the best choice for stability, as first demonstrated the UWA [22]. In [23] we demonstrate that crystals grown with the Kyropoulos or the Czochralski method can be used for short-term stability of 1×10^{-15} or better. A sketch of the resonator assembly, mounted in a copper cavity is shown in the figure 5. In our most implementations, the resonator has diameter of 54 mm and height of 30 mm, and it is operated in the $WGH_{15,0,0}$ mode, close to 10 GHz. Machining tolerance limits the accuracy to 5×10^{-5}, i.e., ± 5 MHz at 10 GHz. Hence we design the resonator for $\nu_0 = 10\,\text{GHz} - \delta_{\text{rf}}$, with $\delta_{\text{rf}} = 10 \pm 5$ MHz. In this way ν_0 is always lower than 10 GHz, and δ_{rf} is compensated by a DDS with a resolution $< 10^{-15}$. With $\delta_{\text{rf}} \lesssim 20$ MHz, the DDS stability and noise do not impact on the signal purity. Interestingly, machining tolerance guarantees the statistical independence of oscillators in a room by preventing microwave-leakage coupling. In fact, the frequency of two nominally equal resonators differ by 10^4–10^5 times the

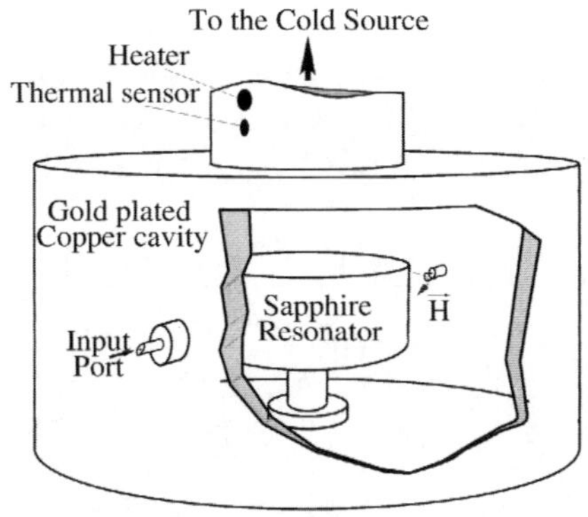

Figure 5. Sapphire resonator inside its Au-plated copper cavity. WGH-mode oriented loops provide external coupling (transverse magnetic field).

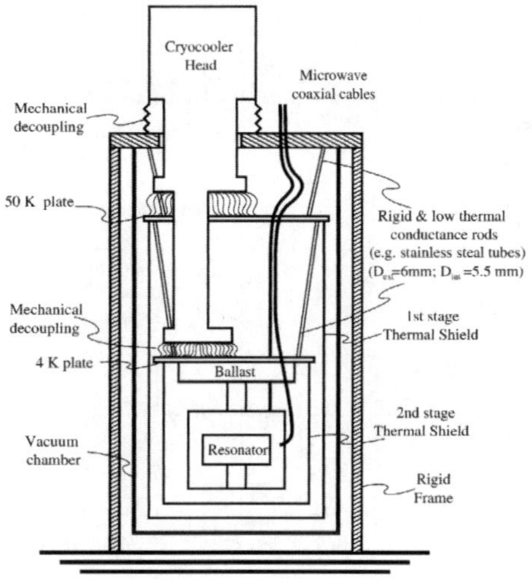

Figure 6. Scheme of the cryostat.

bandwidth.

3.3. Cryostat

The early CSOs were implemented in large liquid Helium Dewars (Figure 1). This setup is almost free from mechanical vibes, and provides good temperature uniformity. The main problem is refilling liquid He every 3 weeks. Since ELISA, we opted for a Pulse-Tube (PT) cryocooler because it can operate two years with no maintenance. Figure 6 sketches the principle our cryostats inspired from [24]. The 4 K plate that holds the resonator is rigidly coupled to the cryostat top flange by rigid rods. Conversely it is mechanically decoupled from the PT 2nd stage by means of a flexible thermal link made with OFHC copper braids. We tried some configurations using stainless steel tubes, fiber or even mylar for the rigid rods. The copper braids are made with 50 μm diameter OFHC copper wires. The number of wires and their length differ from one CSO to one other. The design criterium was to maintain the total equivalent thermal resistance below 10 K/W. Eventually we found the design of the mechanical suspension is not critical. For all the tested configurations a fractional frequency stability better than 1×10^{-15} at short-term can be obtained providing the other parts of the instrument are correctly tuned. We should notice that the typical displacement on the PT 2nd stage is *only* 24 μm peak-to-peak [25]. A reduction of the vertical vibrations to less than 1 μm rms is easily obtained by choosing the ratio between the stiffness of the rigid rods and those of the copper braids. A ratio of 100 can be easily obtained. The link's thermal resistance, combined with a ballast mass, low-passes the temperature fluctuations and enables stabilization better than witnin 100 μK.

3.4. Oscillator Electronics

The sustaining amplifier, based on commercial components at room temperature, compensates for the ≈ 30 dB loss of the resonator. Two low-noise amplifiers provide 54 dB gain. A Voltage Controlled Attenuator (VCA) controls of the power in closed loop. Two Voltage Controlled Phase Shifters (VCPS) are used for the Pound servo, for phase modulation and for frequency control. The first one is used to modulate the phase of the injected signal and the control signal is injected in the second one.

3.5. Pound Control

Thermal and mechanical fluctuations modulate the phase of the oscillator-loop path. This induces frequency noise through the Leeson effect [26], and in our case limits the stability to some 10^{-14} at $\tau = 1$ s. Random walk and drift makes the instability even worse at longer τ depending on the environment. These fluctuations are fixed with a Pound servo.

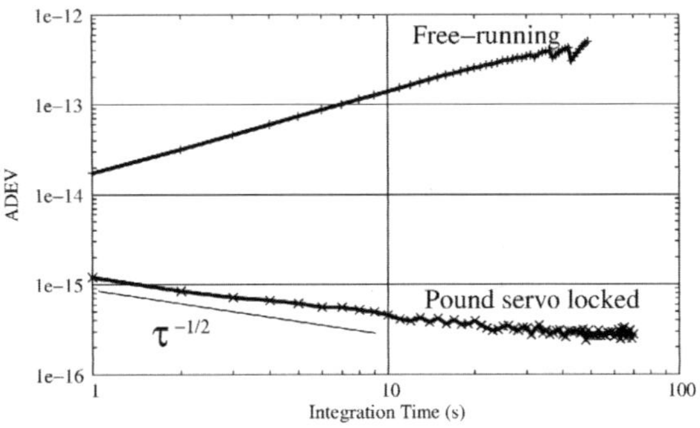

Figure 7. Allan standard deviation for a free and Pound-locked CSO.

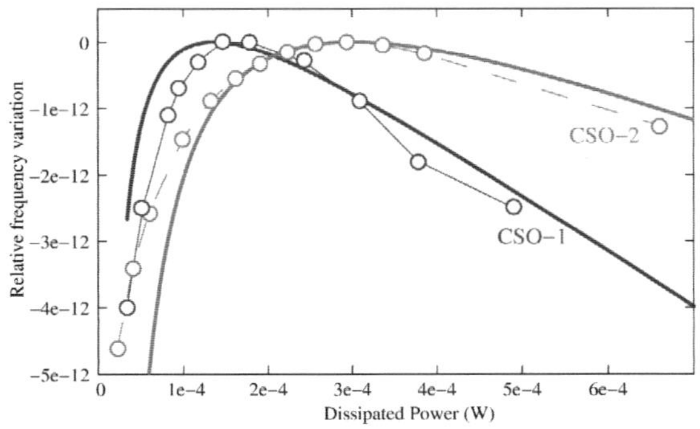

Figure 8. Power sensitivity of two resonators presenting different Q-factors.

A phase modulation at $\nu_m = 70$ kHz (typically) is applied at the input of the resonator, which has a bandwidth of ≈ 10 Hz. The tunnel diode DT1 beats the reflected carrier with the modulation sidebands. The DT1 signal output, down-converted from ν_m to dc, provides the error signal which is integrated and applied to the second VCPS, which controls ν_0. The control bandwidth is of ≈ 1 kHz. The slope of the Pound discriminator is $D \approx 1{\times}10^{-3}$ V/Hz (typical). The noise measured at the output of DT1 at ν_m is $e_n \approx 10$ nV/$\sqrt{\text{Hz}}$. Figure 7 shows the Allan deviation of a CSO in free-running and Pound-locked conditions. Here, $D \approx 7{\times}10^{-4}$ V/Hz. The expected stability is

$$\sigma_y(\tau) = \frac{e_n}{\sqrt{2}D\nu_0}\tau^{-1/2} \approx 1{\times}10^{-15}\tau^{-1/2}. \quad (1)$$

With a given noise in the loop, increasing D results in improved stability. However, below $1{\times}10^{-15}$ other noise phenomena on the resonator become dominant.

3.6. Power Control

The CSO sensitivity to microwave power results from thermal effects and from the radiation pressure. The latter shows up at high Q. Beyond 300 μW, the power shift is of the order of $1{\times}10^{-8}$/W. At lower power, the power shift is affected by the paramagnetic impurities for modes laying nearby or in the Electron Spin Resonance (ESR) bandwidth. A sweet point P_0 has been demonstrated at 10 GHz, where the effect of ESR saturation of Cr^{3+} ions compensates for the radiation pressure, and the overall power shift nulls [27] (Figure 8). Of course, this relaxes the specs for power stabilization.

4. Current Status

We have implemented three CSOs, which are a part of the Oscillator IMP project (Figure 2). They were assembled successively since 2012. They are in principle equal, but for the resonator and minors details. Two of them (CSO-1 and CSO-2) are equiped with a HEMEX crystal, the third one uses a Kyropoulos crystal. The main characteristics of each resonator are given in

Table 1. Resonator characteristics.

	CSO-1	CSO-2	CSO-3
Frequency ν_0 (GHz)	9.988	9.995	9.987
Material	HEMEX	HEMEX	Kyropoulos
Loaded Q-factor Q_L	1×10^9	350×10^8	400×10^8
Input coupling coeff. β_1	1	1	0.92
Turnover temperature T_0	6.238 K	5.766 K	6.265 K
Injected power	100 μW	300 μW	70 μW

table 1. Only CSO-1 is fully optimized, CSO-2 and CSO-3 need an additional adjustment of the resonator coupling.

The individual stabilities are measured with the three-cornered hat method. The CSO output signals are mixed to obtain the three beatnotes: $\nu_{12} = 7$ MHz, $\nu_{13} = 0.9$ MHz and $\nu_{23} = 7.9$ MHz. These beatnotes are analysed using a multi-channels K&K-FXE SCR digital phase recorder [28]. The three channels work in parallel thereby the data acquisitions are synchronous. All datas are processed using the frequency average mode with one second averaging time (equivalent to a Λ windowing for $\tau = 1$ s). The Allan standard deviations (ADEV) are computed for the different integration times τ by grouping the 1-s data. The hypothesis of statistical independence is supported by the spectral separation of the core oscillators, and also by the independent temperature controls.

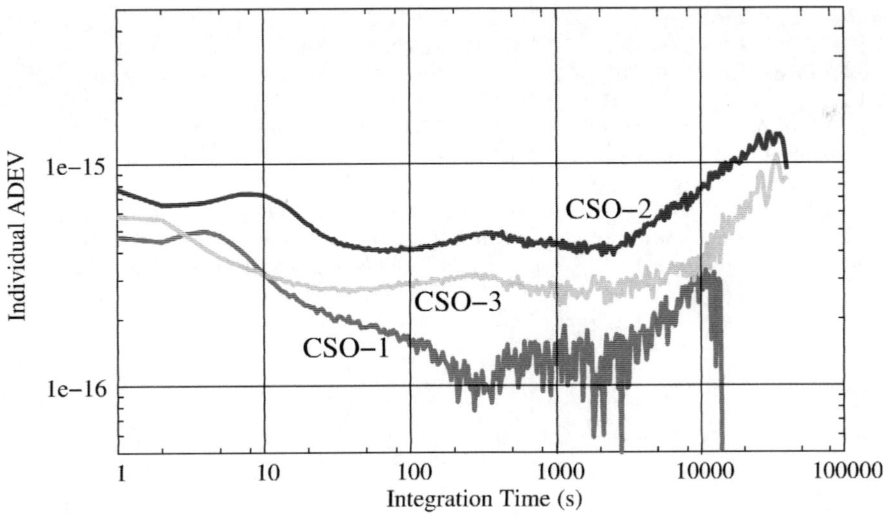

Figure 9. Individual ADEV from the 3-cornered-hat method.

At $1\,\mathrm{s} \leq \tau \lesssim 50\,\mathrm{s}$, the individual ADEVs roll off, but they do not follow the expected $\tau^{-1/2}$ law. The white FM noise due to the Pound discriminator is hidden under a bump at 10 s, most likely due to the 4-K temperature control. At longer τ, ADEV hits a floor, which is $\sigma_y(\tau) \approx 1.5\times10^{-16}$ at $100\,\mathrm{s} \leq \tau \leq 5{,}000\,\mathrm{s}$ for the best oscillator. The small bumps at 400 s match the fluctuation due to the on-off air conditioning system. In this region, there is some correlation between the units. Anyway, we are in the process of installing a new PI control which stabilizes room temperature and humidity.

Acknowledgments

This work is supported by the French ANR Programme d'Investissement d'Avenir (Oscillator IMP, First-TF and Refimeve+ projects in progress at the TF Departments of Femto-ST Institute and UTINAM). The authors would also like to thank the Council of the Région de Franche-Comté for its support to our projects and the FEDER (Fonds Européen de Développement Economique et Régional) for funding one CSO.

References

[1] Santarelli G et al., 1999 *Phys. Rev. Lett.* **82** (23) 4619–22
[2] Takamizawa A et al. 2014 *IEEE Transact. UFFC* **61** (9) 1463–69
[3] Dick G J, Santiago D G and Wang R T 1995 *IEEE Transact. UFFC* **42** (5) 812–19
[4] Grop S et al. 2010 *Rev. Sci. Instrum.* **81** (2) 025102
[5] Grop S et al. 2011 *Electronics Letters* **47** (6) 386–88
[6] Nand N, Hartnett J, Ivanov E and Santarelli G 2011 *IEEE Transact. MTT* **59** (11) 2978–86
[7] Doeleman S et al. 2011 *Pub. of the Astronomical Soc. of the Pacific* **123** (903) 582–95
[8] Rioja M, Dodson R, Asaki Y, Hartnett J and Tingay S 2012 *The Astronomical J.* **144** (4) 121
[9] Dolgovskiy V et al 2014 *Appl. Phys. B* **116** (3) 593–601
[10] Kersalé Y, Vallet O, Vives S, Meunier C and Giordano V 2001 *Electronics Lett.* **37** (23) 1392–93
[11] Kersalé Y, Vives S, Meunier C and Giordano V 2003 *IEEE Transact. UFFC* **50** (12) 1662–66
[12] Bourgeois P Y, Kersalé Y, Bazin N, Chaubet M and Giordano V 2003 *Electronics Lett.* **39** (10) 780
[13] Bourgeois P Y et al. 2004 *Electronics Lett.* **40** (10) 605
[14] Bourgeois P Y et al. 2004 *IEEE Transact. UFFC* **51** (10) 1232–39
[15] Giordano V et al. 2005 *Europ. Phys. J.: Appl. Phys.* **32** 133–41
[16] Grop S et al. 2010 *Electronics Lett.* **46** (6) 420–22
[17] Grop S et al. 2011 *IEEE Transact. UFFC* **58** (8) 1694–97
[18] Giordano V et al. 2012 *Rev. of Sci. Instrum.* **83** (8) 085113
[19] Galani Z et al. 1984 *IEEE Transact. MTT* **32** (12) 1556–65
[20] Sulzer P 1955 *Proc. IRE* **43** (6) 701–07
[21] Pound R 1946 *Rev. Sci. Instrum.* **17** (11) 490–505
[22] Luiten A N, Mann A G and Blair D G 1993 *Electronics Lett.* **29** (10) 879–81
[23] Giordano V, Fluhr C, Grop S and Dubois B 2016 *IEEE Trans. on MTT* **64** (1) 78–85
[24] Ikushima Y et al. 2008 *Cryogenics* **48** (9) 406–412
[25] Wang C and Gifford P E 2002 *AIP Conference Proceedings* **613** (1) 641–48
[26] Rubiola E 2008 *Phase noise and frequency stability in oscillators* (Cambridge: Cambridge University Press)
[27] Giordano V, Grop S, Bourgeois P Y, Kersalé Y and Rubiola E 2014 *J. Appl. Phys.* **116** (5) 054901
[28] Kramer G and Klische W *Proc. of the 2001 IEEE Int. Frequency Control Symposium* 144–51

A second generation of low thermal noise cryogenic silicon resonators

D G Matei[1,3], T Legero[1], Ch Grebing[1], S Häfner[1], Ch Lisdat[1], R Weyrich[1], W Zhang[2], L Sonderhouse[2], J M Robinson[2], F Riehle[1], J Ye[2] and U Sterr[1]

[1]Physikalisch-Technische Bundesanstalt (PTB), Bundesallee 100, 38116 Braunschweig, Germany
[2]JILA, National Institute of Standards and Technology and University of Colorado, 440 UCB, Boulder, CO 80309-0440, USA

[3]Corresponding author: dan.matei@ptb.de

Abstract. We have set up an improved vertically mounted silicon cavity operating at the zero-crossing temperature of the coefficient of thermal expansion (CTE) near 123 K with estimated thermal noise limited instability of 4×10^{-17} in the modified Allan deviation. Owing to the anisotropic elasticity of single-crystal silicon, the vertical acceleration sensitivity was minimized *in situ* by axially rotating the resonator with respect to the mounting frame. The control of the resonator temperature is greatly improved by using a combination of two thermal shields, monitoring with several temperature sensors, and employing low-thermal conductivity materials. The instability of the resonator stabilized laser was characterized by comparing with another low-noise system based on a 48 cm long room temperature cavity of PTB's strontium lattice clock, resulting in a modified Allan deviation of 7×10^{-17} at 100 s.

1. Introduction

Even today's most stable lasers only achieve a coherence time of a few seconds, which are still at least an order of magnitude away from the coherence time of the clock transitions in neutral atoms or ions that are employed in the best optical clocks [1,2]. Thus, at present, the resolved linewidths of these clock transitions are mostly limited by the frequency fluctuations of the interrogating laser. In addition, in a discontinuous interrogation of the clock transition, frequency fluctuations further decrease the clock stability via the Dick effect [3].

While promising results have been obtained from alternative laser stabilization methods like spectral hole burning [4] or active clocks [5], the longest coherence times are provided by laser systems stabilized to optical cavities. Thermal noise fundamentally limits the length stability of optical cavities and thus the frequency stability of ultrastable lasers that use cavities as a frequency reference [6]. To reduce the influence of thermal noise, various approaches have been followed, such as increasing the cavity length [7], lowering the temperature [8,9], or using spacer and coating materials with high mechanical quality factor [10].

Extending previous work based on PTB-JILA collaboration [8,11], we are setting up two nearly identical silicon cavities that operate at the zero-crossing temperature of the thermal expansion coefficient (CTE) near 123 K [12] with estimated thermal noise limited instability of mod $\sigma_y \approx 4\times10^{-17}$. Here we report on the first of the two improved cryogenic silicon resonators. For comparison, we employ another ultra-low thermal noise system at 698 nm, based on a 48 cm long ULE spacer with dielectric mirror coatings on fused silica substrates. The comparison includes a frequency comb and noise cancelled fiber links between the systems. In previous three-cornered-hat comparisons, the ULE system achieved an instability of $\sigma_y = 8\times10^{-17}$ [7].

Throughout this paper we use the modified Allan deviation for analysis, since the comparison between the 1.5 μm laser and the 698 nm laser is affected by high-frequency phase noise from the frequency comb [13] and from the limited servo bandwidth of the frequency lock to the cavity. In this case the standard Allan deviation depends on the measurement bandwidth while the modified Allan deviation does not show this ambiguity. In addition, its steeper slope clearly allows identifying low-frequency fluctuations [14].

2. Cavity setup and reduction of technical noise

The most important technical noise sources which degrade the laser performance are seismic and acoustic vibrations, pressure and temperature fluctuations and residual amplitude modulation (RAM). To reduce these influences we re-engineered the first generation silicon resonator system [8]. As a first step we significantly increased the cavity finesse of the TEM_{00} mode to 480 000 by replacing the contaminated mirrors. This provides better tolerance against RAM and electronic offsets in the Pound-Drever-Hall (PDH) [15] servo setup. The additional improvements to the cavity support structure, the vacuum, and the temperature control are presented in detail in the following.

2.1. Silicon cavity mounting

As in the previous setup [8], a double-cone shaped single-crystal silicon cavity with a length of 21 cm is used with its vertical optical axis parallel to the <111> direction of the silicon crystal. The cavity is placed on an optimized tripod mount as shown in figure 1a. A support ring made of silicon decouples the cavity from the thermal expansion of the base plate. The ring is connected by three polyether ether ketone (PEEK) rods to the base plate and supports the spacer on three small PEEK posts.

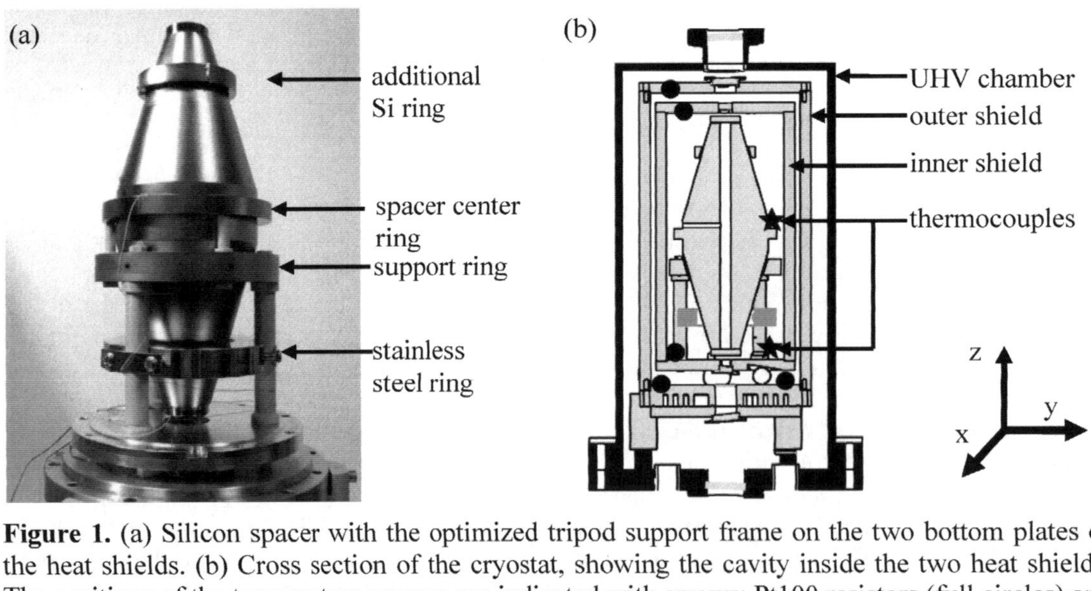

Figure 1. (a) Silicon spacer with the optimized tripod support frame on the two bottom plates of the heat shields. (b) Cross section of the cryostat, showing the cavity inside the two heat shields. The positions of the temperature sensors are indicated with arrows: Pt100 resistors (full circles) and the thermocouple junction on the spacer (star); (Colour online).

Any mechanical resonances in the support structure enhance vibrations acting on the cavity. As a result of replacing the polytetrafluoroethylene (PTFE) support structure used in [8] with PEEK and adding a stainless steel clamping ring, the lowest FEM-simulated mechanical resonance frequency is now moved from 15 Hz to above 70 Hz at room temperature, away from the noisy region in the vibration spectrum (figure 2b).

The vibration sensitivity was measured by applying sinusoidal accelerations along the *x*, *y* or *z* direction and observing the corresponding optical frequency change with respect to a stable reference laser. With the improved support frame we found at 5 Hz excitation frequency horizontal sensitivities of $k_x = 2.5(12)\times10^{-12}$ (m/s²)$^{-1}$ and $k_y = 0.7(6)\times10^{-12}$ (m/s²)$^{-1}$. The residual sensitivity to horizontal accelerations can be attributed to related bending of the spacer in combination with an offset of the cavity mode from the symmetry axis [16]. Also asymmetries in the elastic properties of the tripod lead to radial forces on the spacer that will affect the cavity length.

The anisotropic structure of silicon with a 3-fold rotational symmetry around the <111> direction allows a sine-like variation of the vertical sensitivity of $\pm\ 1.5\times10^{-11}$ (m/s²)$^{-1}$ simply by rotating the spacer azimuthally with respect to the tripod support. The vertical acceleration sensitivity can thus be minimized experimentally. In addition, the base line of the sensitivity can be shifted by the position of the spacer's center ring or by adding some weight on top of the spacer. We see a clear difference between the FEM predictions and the experimental results (triangles in figure 2a) and could not experimentally reach a zero crossing in the vertical vibration sensitivity. Therefore we attached a small silicon ring on top of the spacer (figure 1a) in order to shift the sensitivity curve to a zero crossing. This allows us to optimize the azimuthal angle φ and obtain a residual vertical acceleration sensitivity of $k_z = 0.4(2)\times10^{-12}$ (m/s²)$^{-1}$. We note the important benefit of being able to minimize the vibration sensitivity via *in situ* adjustment of φ after the manufacturing process has already been completed for the cavity.

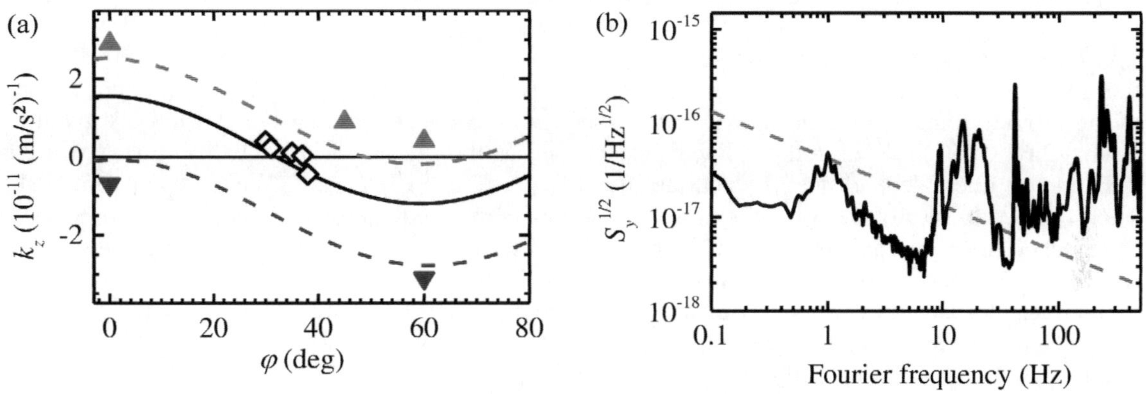

Figure 2. (a) Measured (symbols) and simulated (lines) vertical acceleration sensitivity k_z as a function of the azimuthal rotation angle φ. Upward triangles: the spacer in normal position (support plane Δz = 2.8 mm below the transverse symmetry plane), downward triangles: spacer vertically flipped (Δz = 7.5 mm below); open diamonds: spacer in normal position with an additional silicon ring. The error bars are approximated by the size of the symbols. The lines indicate the values predicted by finite element method (FEM). (b) Calculated frequency noise induced by the seismic perturbations assuming uncorrelated vibrations along *x*, *y*, *z* and expected thermal noise (dashed line); (Colour online).

The seismic spectrum during cryostat operation at 123K was measured close to the cavity along three directions and converted to fractional frequency noise spectral density (figure 2b). The corresponding contribution to the frequency instability of the laser system is thus expected to be well below the thermal noise level for averaging times above 0.1 s, as will be discussed in Section 3.

2.2. *Temperature stabilization*

The impact of temperature fluctuations on the cavity length depends on the operational offset ΔT from the zero-crossing temperature and the slope of the CTE at this point. To reduce cavity temperature fluctuations, we use a system of two concentric heat shields for the silicon spacer, separated by low-thermal-conductivity materials. The outer shield is actively cooled with cold nitrogen gas evaporated from a Dewar. The passive inner shield acts as an additional low pass filter for reducing residual temperature fluctuations. For improved thermal control and reliable operation precisely at the zero-crossing CTE, the temperature is monitored at several points across the setup (figure 1b): the temperatures at the top and bottom of the heat shields are measured with Pt100 sensors. The temperature of the spacer relative to the inner heat shield is measured with a pair of miniature copper-constantan thermocouples attached to the spacer and to the inner heat shield. This setup allows us to reliably stabilize the cavity temperature to within less than 50 mK of the zero-CTE temperature.

The thermal model of the system described in [11] was extended by including direct heat transfer from the room temperature environment to the inner shield and to the spacer to account for conduction through the sensor wires, and for thermal radiation entering the optical ports. The corresponding coefficients were determined from observed temperature offsets between outer shield, inner shield and spacer to be 82(3) μW/K and 3.1(1) μW/K, respectively. From the temperature evolution recorded while cooling the system to 123 K, we calculate time constants of 6.4 days for the transfer from the outer to the inner shield, and of 3.4 days for the transfer from the inner shield to the spacer.

Temperature data recorded for 10 days was used to evaluate the thermal stability of the spacer. As the expected short-time temperature fluctuations are well below the instrument resolution of 0.1 mK, we applied the thermal model to calculate the temperature fluctuations at the inner shield and cavity from the measured temperature of the outer shield and of the vacuum chamber (figure 3). Assuming a conservative offset value of ΔT = 50 mK from the CTE zero-crossing temperature, with a resulting CTE: $\alpha \approx 8.5\times10^{-10}$ K^{-1}, we calculate the corresponding length fluctuations of the cavity and find that they do not limit the performance of the system for averaging times up to 1000 s.

2.3. *Vacuum pressure stability*

Variations in the residual gas pressure inside the cavity induce frequency variations through changes of the refractive index. The refractivity of nitrogen at atmospheric pressure and 30 °C is given by $n - 1 = 2.65\times10^{-4}$ [17]. Assuming a linear dependence of the refractive index with pressure, we can estimate the linear dependence of fractional frequency as 2.65×10^{-7} mbar^{-1}. We have experimentally determined this sensitivity by switching off the ion pump and measuring the induced frequency dependence to be approximately 5×10^{-7} mbar^{-1}. The difference between the measured and estimated values is probably due to a different composition of the residual gas and the separation between vacuum gauge and cavity mode. But the difference is sufficiently small that we can confidently estimate the upper bound of the contribution of intracavity pressure fluctuations to the fractional frequency instability of the cavity.

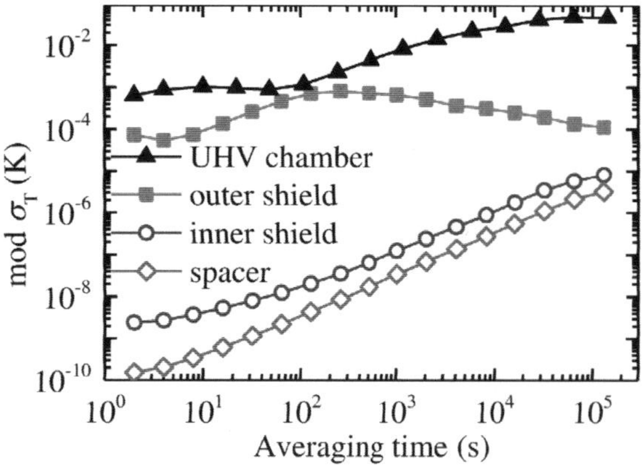

Figure 3. Temperature instability measured on the chamber and the outer shield (filled symbols) and calculated from our thermal model for the inner heat shield and the spacer (open symbols); (Colour online).

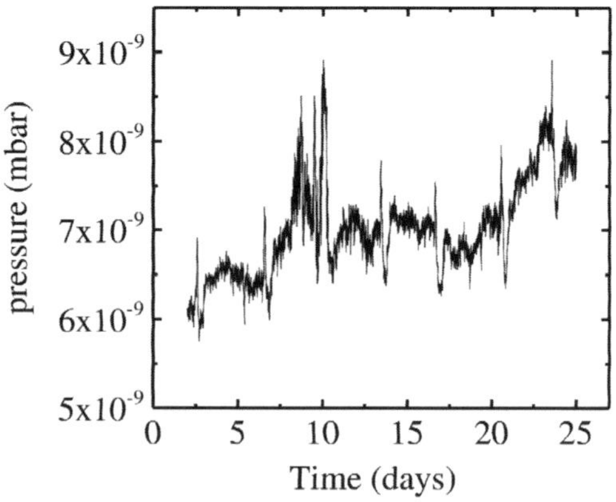

Figure 4. Vacuum pressure recorded for three weeks. The disturbances occurring every 3-4 days are generated by short-time temperature changes of the outer shield during the refilling of the Dewar with liquid nitrogen.

A base pressure in the 10^{-9} mbar range is maintained in the UHV chamber by an ion pump and the cryogenic pumping action of the cold surfaces. An improved vacuum system and ion pump avoid pressure bursts observed in the previous setup [18]. The absolute pressure recorded for three weeks (figure 4) was used to calculate the corresponding instability (figure 5). The increased estimates at lower averaging times could be attributed to noise in the measurement circuit and might be higher than the actual pressure fluctuations. The worsening instability for times above 1000 s could be related to fluctuations arising from daily temperature excursions in the laboratory.

2.4. Residual amplitude modulation

The optical setup is largely identical to the first generation setup described in [8]. A commercial fiber-laser is frequency stabilized to the cavity using the PDH locking technique. The necessary phase modulation is obtained with a Ti-diffused waveguide fiber-coupled electro-optical modulator (EOM) driven with a frequency of 8 MHz. Such EOMs generate relatively large RAM due to imperfect aligning of the polarization maintaining fibers to the principal axes of the birefringent crystal. Thus, the amplitude modulation was controlled by applying a bias voltage to the EOM [19]. The remaining RAM was measured with the PDH lock detector with the laser detuned from the cavity resonance. The observed RAM contribution is clearly below the expected thermal noise limit (figure 5).

3. Frequency stability

All the noise sources presented above contribute to frequency noise at levels well below the expected thermal noise limit for averaging times between 0.1 s to 1000 s, as shown in figure 5. At shorter averaging times, the laser is expected to be limited by vibration-induced noise. At the longer time scales, temperature or pressure instability is expected to limit the laser performance.

In order to measure the laser instability, we have compared the Si stabilized laser with the 698 nm clock laser [7] and with the Sr optical clock transition (figure 5). The frequency of the beat signal was counted with Λ-mode counters [14]. Compared to the modified Allan deviation that we mostly use in this paper the instability of the Sr laser is the regular Allan deviation obtained from a three-cornered-hat measurement. Note that for flicker frequency noise these two deviations are related by mod $\sigma_y \approx 0.82\ \sigma_y$ [20]. For averaging times below 100 s, the instability of the recorded beat signal is very close to be limited by that of the Sr clock laser, while for longer times larger fluctuations are still observed. They are most likely caused by parasitic etalons which will be minimized through optimizing the optical setup.

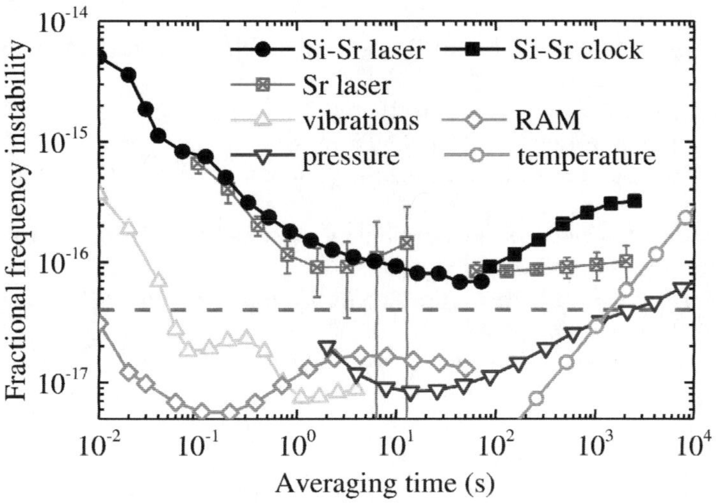

Figure 5. Instability between the silicon system and the Sr laser (full circles) and the Sr clock (full squares), and calculated instabilities due to the discussed noise sources. The instability of the Sr laser from [7] is plotted with crossed squares. Except for the Sr laser instability, which shows overlapping Allan deviation, all other curves show modified Allan deviations. The calculated thermal noise is shown with a dashed line (Colour online).

Journal of Physics: Conference Series **723** (2016) 012031 doi:10.1088/1742-6596/723/1/012031

4. Conclusion and outlook

We have characterized the performance of an updated cryogenic silicon optical resonator by determining the technical noise sources that could potentially contribute to the frequency instability of the laser. The measurements show that the thermal noise floor can be reached for averaging times from tenths of seconds to hours. A frequency comb assisted stability comparison with a state of the art room-temperature ULE resonator stabilized laser at 698 nm was mainly limited by the latter system. We are currently setting up a second cryogenic Si resonator for direct characterization and to push both systems towards their design instability.

Using the same silicon spacer, we can further decrease the thermal noise limit by using low thermal noise crystalline mirror coatings and by decreasing the resonator temperature. Even at 123 K the improved mirror coatings lead to an expected thermal noise limit of mod $\sigma_y \approx 8\times10^{-18}$. Cooling the current cavities to 4 K can reduce the estimated thermal noise to the same value. Combining both approaches has the potential to reduce the thermal noise limit by more than one order of magnitude to the low 10^{-18} level.

Acknowledgments

The cryogenic silicon cavity laser system employed in this work was developed jointly by the JILA Physics Frontier Center (NSF) and the National Institute of Standards and Technology (NIST), the Centre for Quantum Engineering and Space-Time Research (QUEST), and Physikalisch-Technische Bundesanstalt (PTB). We acknowledge funding from the DARPA QuASAR program, the European Community 7th Framework Programme (Grant No. 263500), and the European Metrology Research Programme (EMRP) under QESOCAS. The EMRP is jointly funded by the EMRP participating countries within EURAMET, and the European Union. J. Y. acknowledges support from the Humboldt foundation.

References

[1] Bloom B J, Nicholson T L, Williams J R, Campbell S L, Bishof M, Zhang X, Zhang W, Bromley S L and Ye J 2014 *Nature* **506** 71

[2] Huntemann N, Sanner C, Lipphardt B, Tamm C and Peik E 2016 *Phys. Rev. Lett.* **116**(6) 063001

[3] Dick G J 1988 Local oscillator induced instabilities in trapped ion frequency standards *Proc. 19th Annu. Precise Time and Time Interval Meeting, Redendo Beach, 1987* (Washington, DC: U.S. Naval Observatory) p 133

[4] Thorpe M J, Rippe L, Fortier T M, Kirchner M S and Rosenband T 2011 *Nature Photonics* **5** 688

[5] Norcia M A and Thompson J K 2016 *Phys. Rev. X* **6**(1) 011025

[6] Numata K, Kemery A and Camp J 2004 *Phys. Rev. Lett.* **93** 250602

[7] Häfner S, Falke S, Grebing C, Vogt S, Legero T, Merimaa M, Lisdat C and Sterr U 2015 *Opt. Lett.* **40** 2112–2115

[8] Kessler T, Hagemann C, Grebing C, Legero T, Sterr U, Riehle F, Martin M J, Chen L and Ye J 2012 *Nature Photonics* **6** 687

[9] Wiens E, Chen Q F, Ernsting I, Luckmann H, Rosowski U, Nevsky A and Schiller S 2014 *Opt. Lett.* **39** 3242 and Erratum: 2015 *Opt. Lett.* **40**, 68

[10] Cole G D, Zhang W, Martin M J, Ye J and Aspelmeyer M 2013 *Nature Photonics* **7** 644–650

[11] Hagemann C, Grebing C, Lisdat C, Falke S, Legero T, Sterr U, Riehle F, Martin M J and Ye J 2014 *Opt. Lett.* **39** 5102

[12] Middelmann T, Walkov A, Bartl G and Schödel R 2015 *Phys. Rev. B* **92**(17) 174113

[13] Hagemann C, Grebing C, Kessler T, Falke S, Lemke N, Lisdat C, Schnatz H, Riehle F and Sterr U 2013 *IEEE Trans. Instrum. Meas.* **62** 1556

[14] Benkler E, Lisdat C and Sterr U 2015 *Metrologia* **52** 565

[15] Drever R W P, Hall J L, Kowalski F V, Hough J, Ford G M, Munley A J and Ward H 1983 *Appl. Phys. B* **31** 97–105
[16] Millo J *et al.* 2009 *Phys. Rev. A* **79** 053829
[17] Egan P F, Stone J A, Hendricks J H, Ricker J E, Scace G E and Strouse G F 2015 *Opt. Lett.* **40** 3945
[18] Hagemann C 2013 *Ultra-Stable Laser Based on a Cryogenic Single-Crystal Silicon Cavity* Ph.D. thesis Fakultät für Mathematik und Physik der Gottfried Wilhelm Leibniz Universität Hannover
[19] Zhang W *et al.* 2014 *Opt. Lett.* **39** 1980
[20] Rubiola E 2005 *Rev. Sci. Instrum.* **76** 054703

Journal of Physics: Conference Series **723** (2016) 012032 doi:10.1088/1742-6596/723/1/012032

Autonomous cryogenic sapphire oscillators employing low vibration pulse-tube cryocoolers at NMIJ

Takeshi Ikegami[1], Ken-ichi Watabe[1], Shinya Yanagimachi[1], Akifumi Takamizawa[1] and John G. Hartnett[2]

[1] Radio-Frequency Standards Group, National Metrology Institute of Japan (NMIJ), AIST, Tsukuba, Ibaraki, Japan
[2] Institute of Photonics and Advanced Sensing (IPAS), and, the School of Physical Sciences, University of Adelaide, Adelaide, Australia

E-mail: t.ikegami@aist.go.jp

Abstract. Two liquid-helium-cooled cryogenic sapphire-resonator oscillators (CSOs), have been modified to operate using cryo-refrigerators and low-vibration cryostats. The Allan deviation of the first CSO was evaluated to be better than 2×10^{-15} for averaging times of 1 s to 30 000 s, which is better than that of the original liquid helium cooled CSO. The Allan deviation of the second CSO is better than 4×10^{-15} from 1 s to 6 000 s averaging time.

1. Introduction

Cryogenic sapphire oscillators (CSOs) are still the most frequency-stable oscillator operating in the microwave frequency region [1]. Their frequency stability, expressed as Allan standard deviation, reaches a few times 10^{-16} over averaging times of 1 s to 1 000 s. This short-term stability cannot be obtained in any commercial device, for example, in BVA OCXOs or hydrogen masers.

Recently, the cesium atomic-fountain primary frequency standards, when limited by a quantum projection noise, have a short-term frequency stability approaching a few times 10^{-14}. In order to avoid the influence of the Dick effect (the degradation of frequency stability due to dead time), a more stable local oscillator than obtainable in commercially available devices is required. Thus CSOs have been successfully implemented as local oscillators for such frequency standards [2], [3].

On the other hand, the development of optical atomic clocks is rapid and their frequency stability and uncertainty approaches 10^{-18}. In near future, the redefinition of the unit of time, the "second," is expected [4]. One purpose of optical clocks is the improvement of the international timescale, TAI (International Atomic Time). The uncertainty of TAI has been determined by a balance of the frequency stabilities between primary frequency standards (currently Cs atomic fountains), time comparison methods (GPS or two-way satellite time transfer), and local oscillators (LOs, currently, hydrogen masers), to the current value of about 3×10^{-16}. Because primary frequency standards (current Cs atomic fountains, and optical clocks in future) do not operate continuously over years, better LOs to replace the current hydrogen masers will be necessary. The LO should be stable enough to keep good time and phase, even when a primary standard stops operating and hence it cannot calibrate the LO during those periods. If stable LOs are successfully developed that can operate continuously without phase jumps over many years, they can be calibrated with the best optical clocks and can be used as LOs for timekeeping. Using an optical comb technology, a microwave signal, which is phase-coherent to the optical frequency of an optical clock, can be easily produced by frequency division. Such a microwave signal is easily compared to the microwave output of a CSO, and the CSO can be conditioned by an optical clock intermittently. Thus the CSO acts as a flywheel between conditioning cycles. Therefore, a CSO may also be considered as a candidate LO for timekeeping, which can possibly replace the current hydrogen masers in an era of optical clocks. Therefore, either a LO in the microwave or optical regimes will form a new LO after the re-definition of the second. The choice should be made on the better frequency stability, regardless of it being at optical or microwave frequencies. From a practical viewpoint, a LO in the microwave region might be convenient to produce electrical signals such as 10 MHz, which are most used commonly.

However, it has been difficult to use liquid-helium-cooled CSOs for any purpose that requires long-term continuous operation, due to the high maintenance cost of regular liquid helium refills and the phase jumps which occur on the output signal during refilling of the liquid cryogen, about every 3 weeks. In order to overcome these problems, a low vibration cryostat has been implemented using a pulse-tube-cryocooler [5], [6]. Cryogenic sapphire oscillators have been converted to use this technology (hereafter, designated 'cryoCSO') and as a result their frequency stability has been measured to be better than that of liquid-helium-cooled versions [6], [7]. At NMIJ, two liquid-helium-cooled CSOs, which had been operated since 2005, were converted using pulse-tube cryo-refrigerators and low vibration cryostats and their resulting performance evaluated.

2. NMIJ Cryogenic Sapphire Oscillators using liquid helium

We operated 2 cryogenic sapphire oscillators since 2005. The characteristics of our liquid-helium cooled CSOs is briefly reviewed below [8].

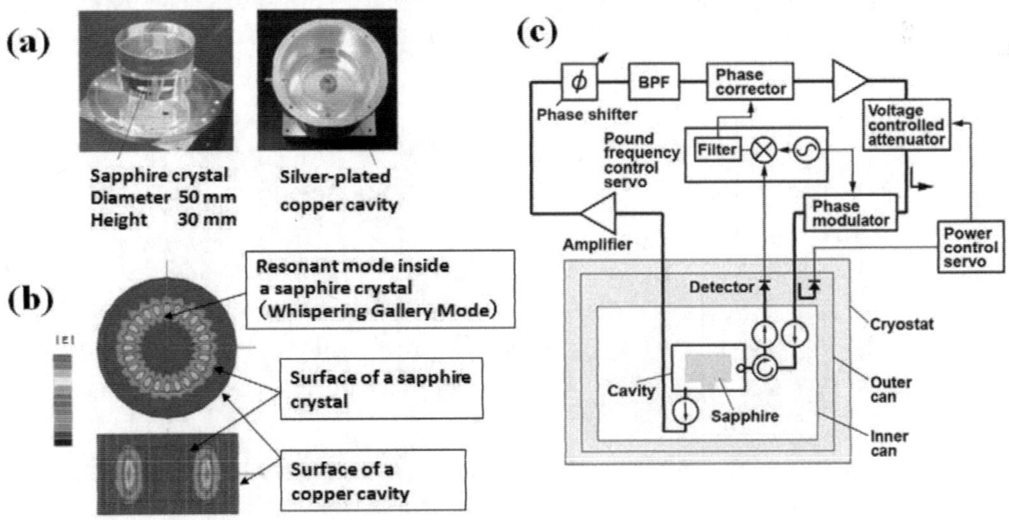

Figure 1. (colour online) (a) Sapphire crystal and silver-plated copper cavity, (b) the distribution of the electric field intensity inside a crystal ($WGH_{12,0,0}$ mode), (c) schematics of cryogenic sapphire oscillator.

Three high quality (HEMEX grade) sapphire crystals from Crystal Systems were cut and polished cylindrically with a height of 30 mm and a diameter of 50 mm (Fig.1(a)), and are used as electromagnetic resonators. The Whispering-Gallery (WG) mode (Fig.1(b)) is used, which has a high loaded quality factor of about 10^9 at a crystal temperature below 10 K. We used the $WGH_{15,0,0}$ mode, which has the highest quality factor. All crystals have a turning point at a temperature greater than the liquid helium temperature of 4.2 K, where the temperature dependence of the resonance frequency of the crystal is nulled to first order. The turning points are 6.1 K at 10.812 GHz (designated crystal1 hereafter, with a loaded quality factor $Q_L=5\times10^8$ at this temperature), 7.0 K at 10.810 GHz (crystal2, $Q_L=7\times10^8$), and 7.6 K at 10.811 GHz (crystal3, $Q_L=4\times10^8$). Each crystal is mounted inside a silver-plated copper cavity with 2 SMA connectorized coaxial probes for input and output as may be seen in Fig. 1(a). These are made of flexible coaxial cable where the inner conductor is bared for a straight probe (sensitive to the E-field) or looped and soldered onto the coaxial shielding for a loop probe (sensitive to the H-field).

As shown in Fig.1(c), the cavity is set inside a vacuum can, and sealed off for thermal isolation. This inner vacuum can was installed inside a second larger vacuum can. This outer vacuum can was then inserted into a liquid helium cryostat and cooled to 4.2K. The crystal was controlled to its turning point temperature with a precision of better than 1 mK. The resonator is used both as a high Q-factor filter and as a frequency discriminating element in an oscillator. The output signal is amplified, phase-corrected, and fed back to the input of the cavity to achieve oscillation. The oscillation frequency is controlled by the resonance frequency of the crystal so as to minimize the reflected power from the resonator using a Pound stabilization method. The reflected signal is detected, with a tunnel-diode microwave power detector located in the cryogenic environment, demodulated with a lock-in amplifier, and fed back to the phase corrector. Also, the oscillating signal power is stabilized using a second cryogenic microwave power detector. Initially we built two CSOs using crystal1 and crystal2 between 2003 and 2005, and evaluated their frequency stability. Their Allan deviation was measured to be less than 2×10^{-15} at the averaging time of 1 s to 1 000 s [8].

3. Modification of the liquid-helium-cooled CSOs using cryo-refrigeration

The first and second CSOs, which used liquid helium as a cryogen, were modified with pulse-tube cryocoolers and low-vibration cryostats. The changes were implemented in 2013 and 2015, respectively. Formerly, we had tried to cool the sapphire crystal by directly attaching its copper cavity to the second stage of a pulse-tube cryocooler, but that did not result in the needed performance [9].

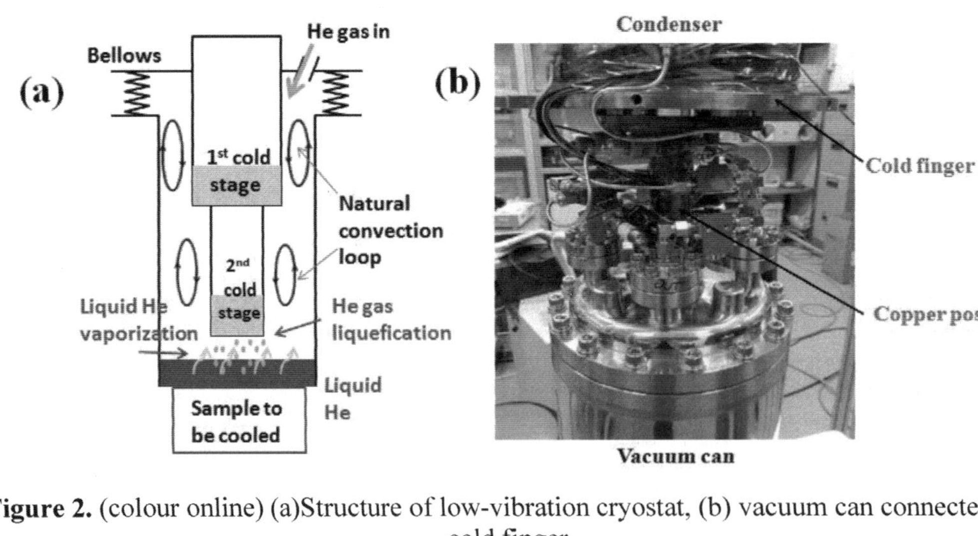

Figure 2. (colour online) (a)Structure of low-vibration cryostat, (b) vacuum can connected to a cold finger.

Fig. 2(a) is a schematic of the low-vibration cryostat and Fig. 2(b) is a photograph of the vacuum can attached to the cold finger of the new cryoCSO. We used a Cryomech PT407RM pulse-tube cryocooler. In this low-vibration cryostat, helium gas is liquefied around the second stage condenser of the cryocooler and stored in the liquid helium container below the second stage as shown in Fig. 2(a). Hereafter, this second stage is called the 'condenser'. Once a few liters of liquid helium have accumulated, the gas inlet is closed. After that the vaporized helium gas in the chamber is constantly re-liquefied by the condenser in a closed system. No additional external supply of helium gas is needed. Below the liquid-helium container, a copper 'cold plate' is attached, called the 'cold finger' hereafter. The inner vacuum can containing the silver-plated copper cavity is attached to the cold finger, via a copper shaft (called 'copper post'), which thermally connects the cold finger to the cavity. The liquid helium container and cold finger are mechanically separated from the cold stages of the cryocooler by a bellow supported by springs, and the vibrations of the cold stage are significantly attenuated. At the center of the copper post, a temperature sensor and a heater are attached. The heater controls the temperature of the copper post using a Lakeshore 340 temperature controller.

3.1. First cryoCSO (cryoCSO1)

For the first cryoCSO, we reused the inner vacuum can from our second liquid-helium cooled CSO (CSO2), whose heat capacity is very large and its structure is not optimized for the cryostat. The crystal2 was used for the cryoCSO1. The loaded Q-factor and coupling on the primary reflection port were re-measured to be Q_L=5×10^8 and β_1^*=0.51, respectively. The incident power on the resonator and modulation index were set to 100 μW and 0.2 rad, respectively.

However, the temperature of the condenser did not reach 4.2K and the liquefaction of helium gas did not occur (maybe due to too large a heat flow to the vacuum can). Fortuitously, the resonator turning point temperature is 7 K and that was achievable and good frequency stability was obtained. The operation of the cryostat using a similar helium heat exchange gas was proposed by G.J. Dick (JPL) [10]. However our intention was to operate at about 50 kPa (meaning below atmospheric pressure) so a partial vacuum develops in the chamber, providing some additional decoupling from room temperature. Another difference from Dick's idea is we use rigid springs to support the bellows against this partial vacuum. But in this implementation the pressure in the helium container remained around 1 atm.

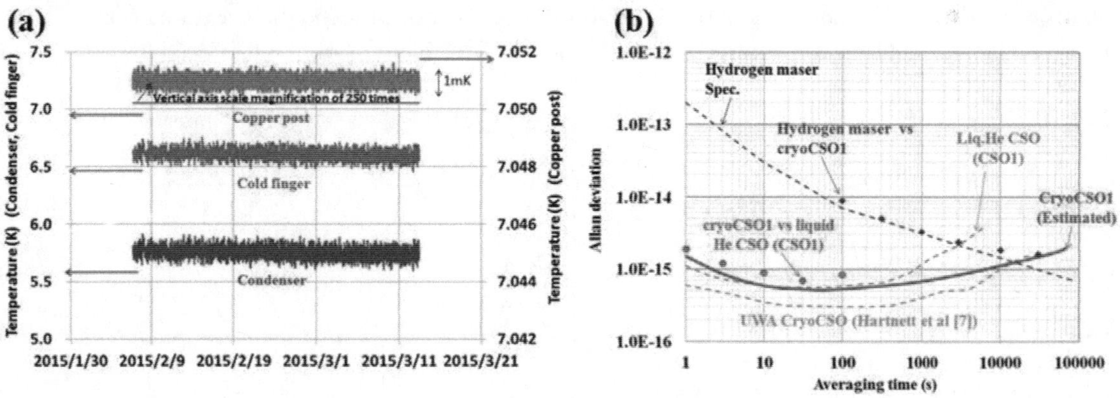

Figure 3. (colour online) (a) temperature stability of cryoCSO1, (b) Allan deviation of cryoCSO1.

As shown in Fig. 3(a), the temperature stability of the cold finger and the condenser were almost the same. The temperature stability at the copper post was not as good as one might have expected had the liquefaction occurred.

The frequency stability of cryoCSO1 was evaluated using our first liquid helium-cooled CSO (CSO1) kept as a short-term frequency stability reference, because it had been previously evaluated. The Allan deviation of the frequency stability for the single CSO1 is shown as a gray dotted line in Fig. 3(b) [8]. For the longer term frequency stability evaluation (τ > 1000 s), a hydrogen maser (HM) was used as a reference, whose specified Allan deviation is shown as a black dotted line in Fig. 3(b). The Allan deviation of the beat signal between cryoCSO1 and CSO1 for the short averaging times of 1 s to 100 s is shown as filled pink circles in Fig. 3(b). The Allan deviation of the beat signal is almost equal to $\sqrt{2}$ times of that of CSO1, which means that the frequency stability of cryoCSO1 is approximately that of CSO1 at averaging times between 1 s and 100 s. The Allan deviation of the beat signal between cryoCSO1 and the HM is shown as blue filled diamonds in Fig. 3(b). That result is dominated by that of the HM except the last 2 points at averaging times of 10 000 s and 30 000 s. From this we estimated the Allan deviation of the single cryoCSO1 as a solid red line in Fig. 3(c).

An Agilent 53132A frequency counter was used in an auto mode to measure the beat signal. For the averaging time from 1 s to 100 s, the gate time was set to each averaging time (for example, 10 s gate time for 10 s averaging time). For an averaging time greater than 100 s, the gate time was fixed to 100 s and the beat frequency was averaged to calculate the beat frequency for longer averaging time (for example, we averaged 3 of the 100 s gate time data to obtain the 300 s averaging time data).

As shown in Fig. 3(b), our short-term frequency stability was a factor of three times that of the best data of Hartnett et al. [7], but at longer averaging times our result was slightly better, possibly due to better environmental conditions.

3.2. *Second cryoCSO (cryoCSO2)*

In the second CSO, an inner vacuum can was custom-designed for use with the cryocooler and it successfully liquefied the helium gas. Crystal3 was implemented in cryoCSO2, which had not been used in either CSO1 or CSO2, before. The loaded Q-factor and coupling of this resonator were measured to be Q_L=4×10^8 and β_1^*=0.52, respectively. The power on the resonator and the modulation index were set to about 100 μW and 0.6 rad, respectively.

The temperature stability was much better in cryoCSO2 than in cryoCSO1 and was close to the resolution of the temperature controller (≤ 100 μK), as shown in the Fig. 4(a). The absolute value of the temperature of the cold finger is not accurate because an uncalibrated sensor was used. However, the temperature fluctuations are reasonably accurate. Because helium gas is liquefied around the second stage of the condenser and liquid helium is contained above the cold finger, the temperature of the cold finger is fixed at the boiling temperature of helium at the final pressure developed in that chamber. In this cryoCSO a pressure of about 50 kPa (i.e. below atmospheric pressure) was obtained, which also means that the temperature of the boiling liquid is about 3.5 K, a lot less than 4.2 K.

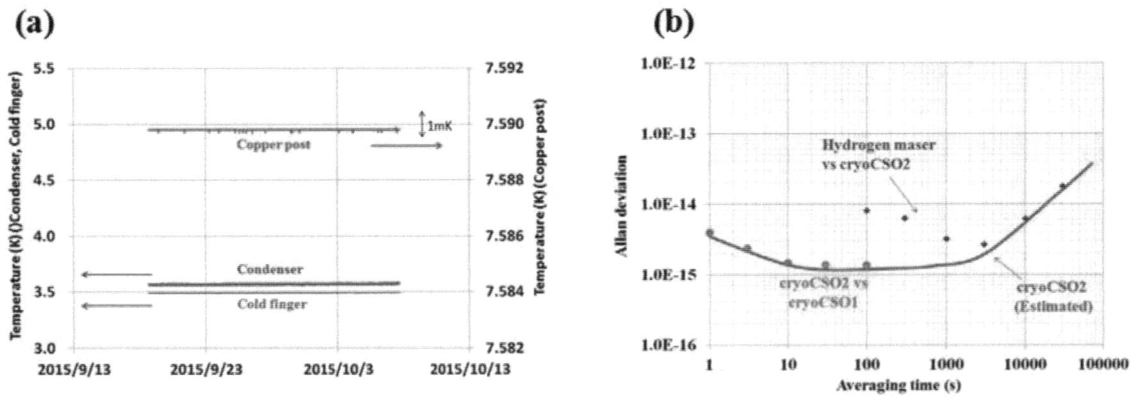

Figure 4. (colour online) (a) Temperature stability of cryoCSO2, (b) Allan deviation of cryoCSO2.

The Allan deviation of frequency fluctuations for cryoCSO2 was determined by a comparison with cryoCSO1 for the short-term stability, and against a hydrogen maser for the longer term. The Allan deviation of the beat signal between cryoCSO2 and cryoCSO1 is shown as filled pink circles in Fig 4(b). The stability of the beat between cryoCSO2 and the HM is shown as blue filled diamonds. The stability for the single cryoCSO2 is estimated as a solid red line in the figure. The frequency stability of the cryoCSO2 is worse than that of the cryoCSO1 by a factor of 2 in the short term. The reason for this at averaging time less than 1 000 s is most probably due to reduced unloaded Q-factor at the operational temperature equal to about half that for cryoCSO1. There remains a residual frequency drift (for averaging times > 1 000 s) because the data were taken very soon after the cryocooler of cryoCSO2 was switched on. The long-term stability improves as the drift rate exponentially decays over time [7].

4. Summary

We have now modified 2 conventional liquid-helium cooled CSOs by introducing low-vibration pulse-tube cryocoolers and low-vibration cryostats. The first cryoCSO is now used with our Cs atomic fountain frequency standard, NMIJ-F2, which has achieved a nearly quantum-projection-noise-limited frequency stability of 8×10^{-14} at 1 second [3]. The cryostat design achieves heat exchange between the cryocoolers cold stage and the cold finger by use of helium gas. In the first implemented cryoCSO, helium liquefaction did not occur. However, the frequency stability was better than 2×10^{-15} between 1s and 30 000 s and better than 1×10^{-15} between 3 s and 4 000 s.

We now have 2 units of cryocooler CSOs. Thus we plan to improve the short-term stability at 1s into 10^{-16} regime. We plan to optimize all control parameters and reduce technical noise sources where possible as well as operate the cryoCSOs continuously, without phase jumps, over many years. Finally, we plan to operate the cryoCSOs as practical local oscillators, which will be able to keep time, with a short-term performance better than our hydrogen masers.

Acknowledgments

This work was supported by JSPS KAKENHI Grant Number 26246038.

References

[1] Clayton C R, Ivanov E N, Hartnett J G, Stanwix P L and Tobar M E 2008 *Rev. Sci. Instrum.* **79** 051301

[2] Santarelli G, Laurent Ph, Lemonde P, Clairon A, Mann A G, Chang S, Luiten A N and Salomon C 1999 *Phys. Rev. Lett.* **82** 4619

[3] Takamizawa A, Yanagimachi S, Tanabe T, Hagimoto K, Hirano I, Watabe K, Ikegami T and Hartnett J G 2014 *IEEE Trans. Ultrason. Ferroelect. Freq. Control* **61** 1463

[4] Margolis H and Gill P 2005 *Phys. World* **May** 18

[5] Wang C and Hartnett J G 2010 *Cryogenics* **50** 336

[6] Grop S, Bourgeois P Y, Bazin N, Kersalé Y, Rubiola E, Langham C, Oxborrow M, Clapton D, Walker S, De Vincente J and Giordano V 2010 *Rev. Scientific Instrum.* **81** 025102

Grop S, Schäfer W, Bourgeois P Y, Kersalé Y,Oxborrow M, Rubiola E and Giordano V 2011 *IEEE Trans. Ultrason. Ferroelect. Freq. Control* **58** 1694

Giordano V, Grop S, Dubois B, Bourgeis P Y, Kersalé Y, Rubiola E, Haye G, Dolgovskiy V, Bucalovicy N, Di Domenico G, Schilt S, Chauvin J and Valat D 2012 *Rev. Scientific Instrum.* **83** 085113

[7] Hartnett J G, Nand N and Lu C 2012 *Appl. Phys. Lett.* **100** 183501

Hartnett J G, Nand N R, Wang C and Le Floch J-M IEEE 2010 *IEEE Trans. Ultrason. Ferroelect. Freq. Control* **57** 1034

Hartnett J G and Nand N R 2010 *IEEE Trans. Microw. Theory Techn.* **58** 3580

[8] Watabe K, Hartnett J G, Locke C R, Santarelli G, Yanagimachi S, Shimazaki T, Ikegami T and Ohshima S 2006 *Jpn. J. Appl. Phys.* **45** 9234

[9] Watabe K, Koga Y, Ohshima S, Ikegami T and Hartnett J G 2003 *Proc. 2003 IEEE Freq. Control Symp.* 388

[10] Dick G J 2002 *IEEE Freq. Control Symp. Tutorial*

Precision optical metrology with alkali-atom isoclinic points

Nathan Wells, Travis Driskell, and James Camparo[1]

Physical Sciences Laboratories, The Aerospace Corporation
PO Box 92957, Los Angeles, CA 90009, USA

E-mail: james.c.camparo@aero.org

Abstract. Vapour-phase spectroscopy rarely involves transitions between well-isolated atomic states. Routinely, the spectra comprise overlapped Doppler/pressure-broadened resonances, which leads to a "pulling" of the spectral peaks from their true atomic resonance frequencies. This pulling gives the absorption resonances a temperature sensitivity, which limits their utility for precision spectroscopy when sub-Doppler techniques are not viable. Here, we discuss the use of alkali isoclinic points as a solution to this problem.

1. Introduction

Though vapor-phase spectroscopy is ubiquitous in atomic physics, it rarely involves the text-book case of well-isolated quantum state transitions. More often, the spectra comprise overlapped Doppler/pressure-broadened resonances, with the overlap yielding spectral peaks that are "pulled" from their true atomic resonance frequencies. Further, as vapor temperature fluctuates the Doppler-broadened widths vary, which in turn leads to more or less spectral overlap, and as a consequence fluctuations in resonance pulling. Thus, the resonance peaks in a vapor phase spectrum rarely provide an accurate measure of the atom's intrinsic resonance frequency, and they are routinely sensitive to systematic temperature variations. Though sub-Doppler spectroscopy has long been recognized as a means for mitigating such problems, it is not viable under all conditions: sub-Doppler signals can be sensitive to microphonics, and sub-Doppler techniques are inoperative in regimes of collision broadening. Here, we discuss the use of alkali isoclinic points for high-precision and high-accuracy vapor-phase spectroscopy when sub-Doppler techniques fail.

According to the IUPAC Compendium of Chemical Terminology, an isoclinic point is "a wavelength, wavenumber, or frequency at which the first derivative of an absorption spectrum of a sample does not change upon a chemical reaction or physical change of the sample" [1]. If we take this to include changes in state variables (*e.g.*, temperature and pressure), then the D_1 spectra of $I = 3/2$ alkalis contain isoclinic points (where I is the nuclear spin) [2], and these correspond to the midpoint between their $n^2S_{1/2}(F_g=2) - m^2P_{1/2}(F_e=1,2)$ transitions as illustrated in figure 1. Specifically, given the equality of the absorption cross sections for the $|F_g=2\rangle \rightarrow |F_e=1,2\rangle$ transitions (*i.e.*, the "doublet"), the amplitudes of these two over-lapped absorption resonances are identical. Consequently, any state-variable that only influences the widths of the lines (*e.g.*, temperature via Doppler broadening or alkali pressure via collision broadening) will not affect the midpoint's frequency, ν_m: the zero-crossing of the absorption's first derivative at ν_m will not change with temperature or pressure. Alkali isoclinic points

[1] To whom any correspondence should be addressed.

can therefore be employed as state-variable independent atomic frequency references, either for laser stabilization, or as "markers" against which other (state-variable dependent) spectral features can be measured [3,4]. In the following sections we discuss some of these applications.

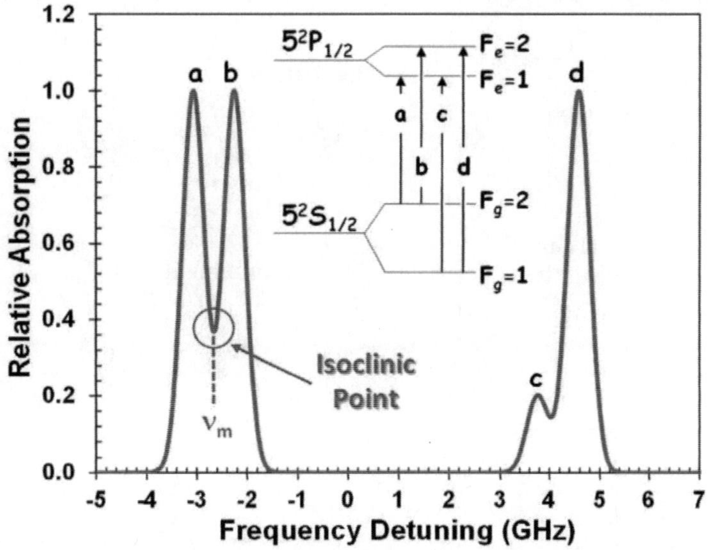

Figure 1: The D_1 absorption spectrum of ^{87}Rb at 795 nm. The mid-point between the 'a' and 'b' absorption resonances corresponds to an isoclinic point.

2. Present experimental system

2.1. Measurement system

Figure 2 shows our typical experimental arrangement: a VCSEL diode laser is split into two beams, each of which has a diameter of 0.4 cm, and is separately attenuated using a rotating polarizer. One beam passes through the reference cell, which contains a vapor of isotopically enriched ^{87}Rb without any buffer gas, maintained at 35 ± 0.01 °C. The other beam passes through the measurement cell, whose temperature T can be varied between 32.5 °C and 60 °C, and which routinely contains ^{87}Rb in the presence of a buffer gas. The cell diameters are 2.5 cm and their lengths are 4.4 cm, and we employ low laser powers to limit distortions of the spectra due to optical pumping (on the order of 10^2 nW).

The relative absorption cross section in both cells, $\sigma(\nu)$, is estimated using Beer's law: $N\sigma(\nu)L = \ln[I(\nu)/I_o(\nu)]$, where $I(\nu)$ is the transmitted light intensity and $I_o(\nu)$ is the light intensity reaching the photodetector in the absence of the vapor, each evaluated at laser frequency ν. $I_o(\nu)$ can be estimated by tuning the laser both above and below resonance, and fitting the off-resonance transmitted light intensity to a quadratic in ν.

2.2. Measuring resonance shifts

The midpoint point frequency in each cell (*i.e.*, ν_m in figure 1) is determined by taking a numerical derivative of the absorption cross section, with the isoclinic point in the reference cell providing a temperature independent frequency marker for the spectral features observed in the measurement cell. The frequency separation between the two cells' midpoint frequencies is a measure of the D_1 frequency shift for the Rb atoms in the measurement cell due to temperature and/or buffer gas pressure. Figure 3a shows the relative absorption cross sections in our reference and measurement cells for the case of 25 torr Kr in the measurement cell at a temperature of 42.5 °C. Figure 3b shows the numerical derivatives of the absorption cross sections, and highlights the ^{87}Rb atom's D_1 collision shift in this system. (Though the spectra look "clean" in figure 3a, taking the numerical derivative of the data makes the noise in the spectra manifest.)

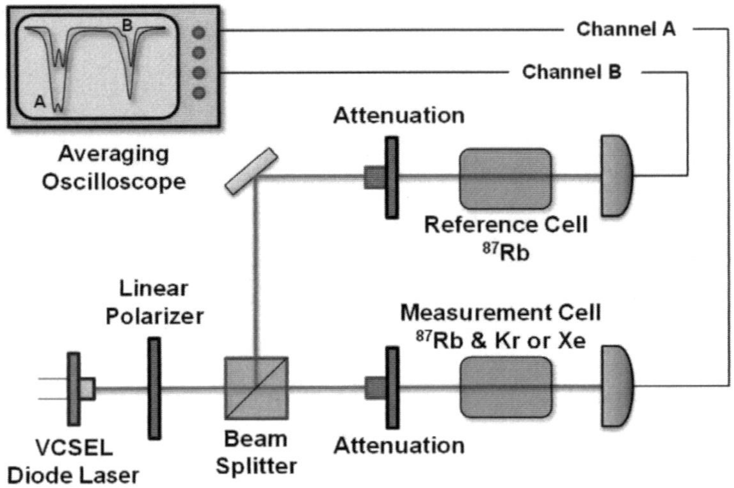

Figure 2: Experimental arrangement for our present isoclinic-point spectroscopy experiments.

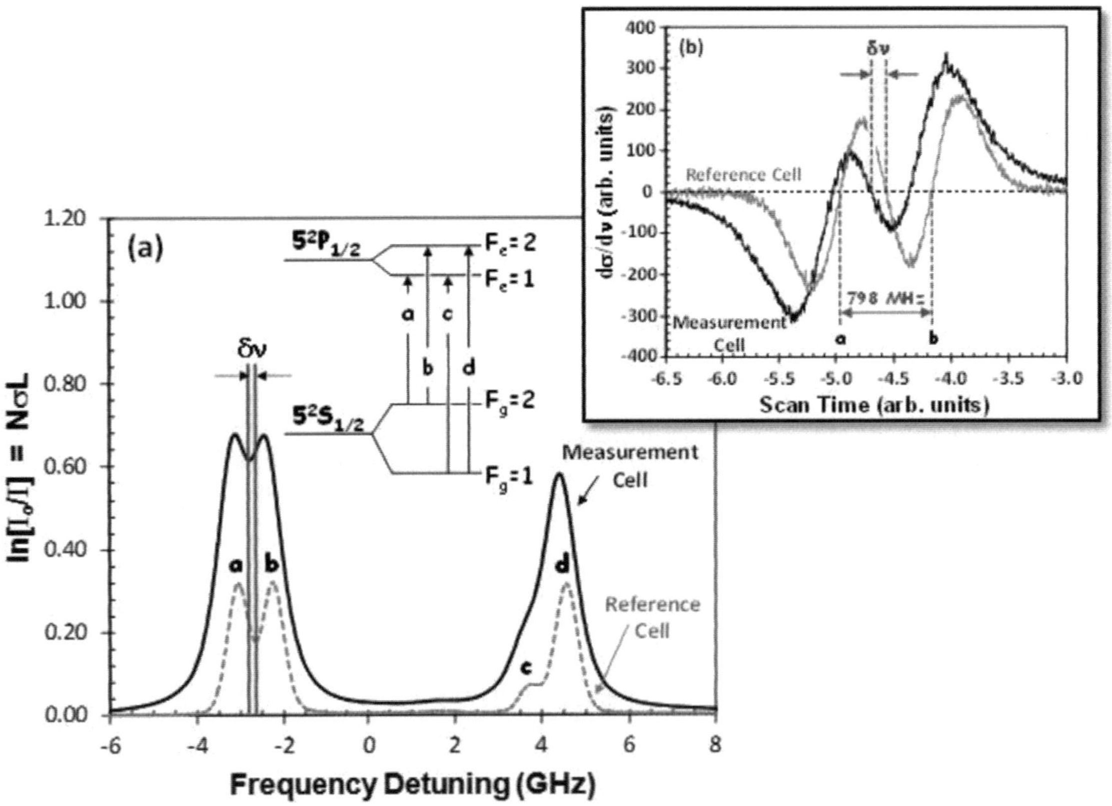

Figure 3: (a) The relative absorption cross section in our reference cell and our measurement cell at 42.5 °C with 25 torr Kr. (b) The numerical derivative of the absorption cross sections, indicating the collision shift of the ^{87}Rb D_1 line in the measurement cell. Abscissa values in "scan time" are calibrated to frequency using the known excited-state hyperfine splitting of ^{87}Rb (as observed in the reference cell) after taking account of Doppler broadening's pulling of the lineshape peaks.

3. Some recent isoclinic point applications

3.1. Accurate measurement of collision shifts

With the isoclinic point, we can make very accurate measurements of collision shifts. Recently, we compared the Rb/Kr and Rb/Xe D_1 collision shifts with theory [5], and as illustrated in figure 4 the experimental uncertainty in the collision shifts for these systems has been steadily decreasing over time. In particular, the most recent measurements by Rotondaro and Perram [6] and our group are in excellent agreement. In this regard, it is important to note that the results from our two groups were obtained by very distinct measurement techniques: Rotondaro and Perram employed a traditional spectroscopic technique to estimate the center-of-mass collision shift of the hyperfine absorption manifold, whereas we employ the ^{87}Rb isoclinic point. Thus, while unknown systematic effects might unduly influence our results or equivalently those Rotondaro & Perram, it is difficult to imagine a systematic effect that would influence both techniques in the same way and to the same extent.

The figure also shows that there is a slight, but statistically significant discrepancy between theory and experiment (*i.e.*, ~ 20%). The theories of both Rebane [7] and Kielkopf [8] overestimate the collision shifts. While Kielkopf's modeling of the interaction potential is more detailed than Rebane's (as Kielkopf considers a quadrupole, $1/R^8$, term in the potential [9]), Rebane's theory explicitly accounts for the fine and hyperfine structure of the alkali. Interestingly, both theories determine the magnitude of the van der Waals ground and excited-state coefficients of the interaction potential in a similar fashion, suggesting that there is a need for a closer theoretical examination of the noble-gas atom's interaction with the alkali $^2P_{1/2}$ state. (The alkali ground-state's interaction with the noble gases is arguably better understood than the excited-state interaction, suggesting the later as the more likely suspect for error.)

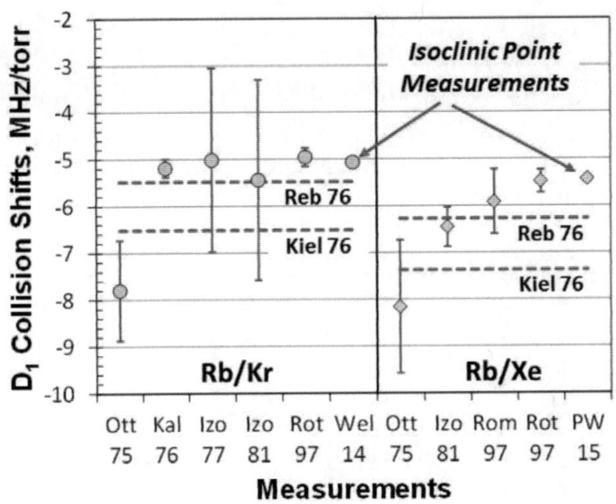

Figure 4: Measurements of the D_1 collision shift by Kr and Xe over time along with two theoretical estimates: Reb 76 ⇒ Rebane [7] and Kiel 76 ⇒ Kielkopf [8]. Abscissa values correspond to particular measurements; for example, Rot 97 ⇒ Rotondaro & Perram [6]. For the others, Ott 75 ⇒ [10], Kal 76 ⇒ [11], Izo 77 ⇒ [12], Wel 14 ⇒ [3], Rom 97 ⇒ [13], and PW 15 ⇒ [5].

3.2. The collision shift's energy dependence

The energy dependence of the collision cross section gives rise to a temperature dependence of the D_1 collision shift: $\delta\nu(T) = \delta\nu(T_o)\times(T/T_o)^{\kappa}$, where $\delta\nu(T)$ is the collision shift at temperature T, and T_o is a reference temperature. Though the temperature exponent, κ, provides unique information regarding the perturbative interaction [14], measurements of κ have proven difficult. In large part, this stems from attempting to measure small changes in the center-of-mass shift of a manifold of optical hyperfine transitions as the manifold shifts and broadens with temperature (*e.g.*, figure 1). One of the few unambiguous techniques relies on the isoclinic point [3].

Figure 5 shows a complete set of published experimental κ exponents for the Rb and Cs D_1 transitions: A ⇒ Romalis *et al.* [13], B ⇒ Klutz *et al.* [15], C ⇒ Kozlova *et al.* [16], D ⇒ Wells *et al.* [3], E ⇒ Wells *et al.* [5]; circles correspond to Rb, while diamonds correspond to Cs. The dashed (brown) and long-dash (blue) lines correspond to the only theoretical estimates of κ for Rb and Cs, respectively [8]. Barring the isoclinic point measurements, the general scatter of the experimental data among themselves, and between experiment and theory, is much larger than the quoted experimental uncertainties. This attests to the difficulty of κ exponent measurements using traditional spectroscopic techniques. Conversely, the Rb/Kr and Rb/Xe κ exponents based on isoclinic point measurements are not only consistent with one another (as one would expect for Rb interacting with the heavier noble gases), but agree very well with theory: $\kappa_{th} = 0.31$ [8].

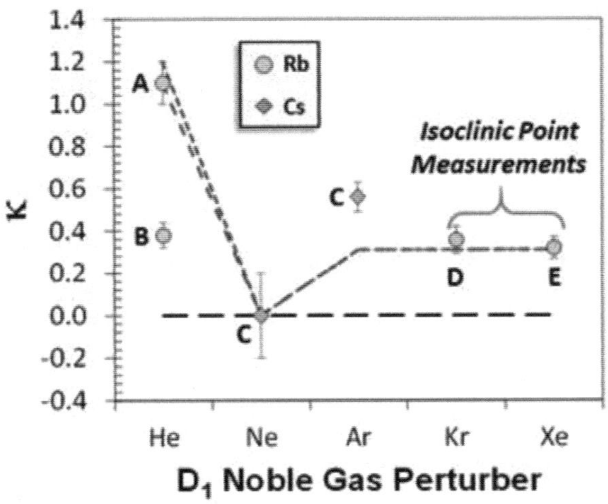

Figure 5: Complete set of published κ exponent values for the Rb and Cs D_1 collision shifts. The dashed lines correspond to theory, which overlap beyond Ne. Barring the isoclinic point measurements, the scatter in the data attests to the difficulty of κ measurements, even though κ provides fairly unique information regarding the perturbative interaction.

3.3. Understanding the detailed character of the isoclinic point

The isoclinic nature of the midpoint between the ‘a’ and ‘b’ transitions illustrated in figure 1 requires the equality of oscillator strengths for the $5^2S_{1/2}(F=2) \rightarrow 5^2P_{1/2}(F=1)$ and $5^2S_{1/2}(F=2) \rightarrow 5^2P_{1/2}(F=2)$ transitions. Any process that breaks that equality destroys the isoclinic nature of the midpoint, which has prompted research in our laboratory carefully examining ‘a’ and ‘b’ transition-amplitude asymmetry.

Figure 6: We measured the asymmetry parameter α (defined in the text) as a function of Xe pressure (circles). (The diamond corresponds to a cell that was mostly filled with Xe, but had a small amount of additional N_2: total pressure, $P_T = 18.0$ torr, $P_{Xe}/P_T = 0.98$.) The measured asymmetry agrees very well with a theory ascribing the asymmetry to mixing of the $P_{1/2}$ and $P_{3/2}$ oscillator strengths.

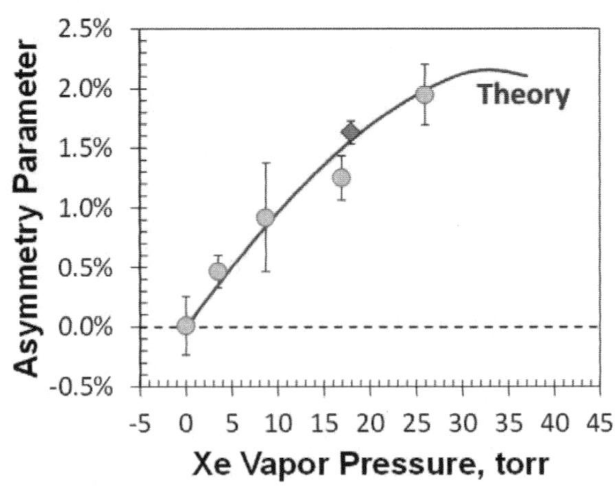

Recently, we found evidence that elastic $P_{1/2} \leftrightarrow P_{3/2}$ fine-structure mixing collisions add D_2 oscillator strength into the D_1 transition [17] causing an isoclinic point asymmetry. Our results are shown in figure 6, where the asymmetry parameter, α, is defined as the relative change in the ‘a’ and ‘b’ transition amplitudes (*i.e.*, S_a and S_b, respectively):

$$\alpha = \frac{2(S_a - S_b)}{S_a + S_b}. \qquad (1)$$

As the figure shows, this collision process has a small but noticeable effect on the equality of the 'a' and 'b' transition amplitudes, and our measurements are in very good agreement with a theory ascribing the asymmetry to $P_{1/2} \leftrightarrow P_{3/2}$ oscillator-strength mixing [17]. We note, however, that while the asymmetry is observable, it likely has little influence on the utility of the isoclinic point for most precision spectroscopy purposes [17].

4. Summary
Here, we have discussed the alkali isoclinic point, *i.e.*, the midpoint between the $n^2S_{1/2}(F=2) \rightarrow m^2P_{1/2}(F=1,2)$ transitions in (nuclear spin) I = 3/2 alkalis, and some recent isoclinic-point applications. In particular, alkali isoclinic points can be employed as state-variable independent atomic frequency references, which can be employed for laser stabilization, or as "markers" against which other (state-variable dependent) spectral features can be measured. To date, isoclinic points have been employed to make precision measurements of collision shifts and the energy dependence of those shifts [3,5]. In our laboratory, we are continuing to exploit isoclinic points in order to explore rubidium's interaction with diverse perturbers, notably collision shift and κ measurements for Rb interacting with He, Ne, Ar, and N_2. Additionally, we are working to develop a thermometry capability based on isoclinic points, which will (potentially) lead to the improved long-term stability of vapour-cell atomic clocks [4].

References
[1] IUPAC Compendium of Chemical Terminology, 2nd ed. (the "Gold Book"). http://goldbook.iupac.org.
[2] Wells N P and Camparo J C 2010 ^{87}Rb D_1 isoclinic point *Phys. Rev. A* **82** 062505.
[3] Wells N P, Driskell T U and Camparo J C 2014 Kr-collision shift of the Rb D_1 transition: The isoclinic piont and precision optical spectroscopy *Phys. Rev. A* 89 052516.
[4] Wells N P, Driskell T U and Camparo J C 2014 Stabilizing vapor-cell temperature: The isoclinic point "atomic thermometer" *Proc. 28th Eur. Freq. and Time Forum* (The Printing House: Stoughton, WI) pp. 548-53.
[5] Wells N P, Driskell T U and Camparo J C 2015 Rubidium D_1 collision shift by the heavy noble gases *Phys. Rev. A* **92** 042509.
[6] Rotondaro M D and Perram G P 1997 Collisional broadening and shift of the rubidium D_1 and D_2 lines ($5^2S_{1/2} \rightarrow 5^2P_{1/2}$, $5^2P^{3/2}$) by rare gases, H_2, D_2, N_2, CH_4 and CF_4 *J. Quant. Spectrosc. Radiat. Transfer* **57** 497-507.
[7] Rebane V N 1976 Calculation of the broadening and shift of hyperfine components of rubidium and cesium D_1 lines in collisions with atoms of heavy inert gases *Opt. Spectrosc.* **41** 526-28.
[8] Kielkopf J F 1976 Predicted alkali collision broadening by noble gases based on semiempirical potentials *J. Phys. B: Atom. Molec. Phys.* **9** L547-50.
[9] Kielkopf J F 1974 Semiempirical potentials for alkali-noble gas interactions *J. Chem. Phys.* **61** 4733-39.
[10] Ottinger Ch, Scheps R, York G W and Gallagher 1975 A Broadening of the Rb resonance lines by the noble gases *Phys. Rev. A* **11** 1815-28.
[11] Kaliteevskii N I, Rish O M and Chaika M P 1976 Measurement of the broadening and shift of the D_1 line of rubidium 87 due to collision with atomic krypton *Opt. Spectrosc.* **41** 293-5.
[12] Izotava S L, Kantserov A I and Frish M S 1977 Measurement of the shift and broadening of STS components of a rubidium line in the presence of krypton *Opt. Spectrosc.* **42** 120-2.
[13] Romalis M V, Miron E and Cates G D 1997 Pressure broadening of Rb D_1 and D_2 lines by ^{3}He, ^{4}He, N_2, and Xe: Line cores and near wings *Phys. Rev. A* **56** 4569-78.
[14] Peach G 1981 Theory of the pressure broadening and shift of spectral lines *Adv. Phys.* **30** 367-474.

[15] Kluttz K A, Averett T D and Wolin B A 2013 Pressure broadening and frequency shift of the D_1 and D_2 lines of Rb and K in the presence of ^{3}He and N_2 *Phys. Rev. A* **87** 032516.
[16] Kozlova O, Guérandel S and de Clercq E 2011 Temperature and pressure shift of the Cs clock transition in the presence of buffer gases: Ne, N_2, and Ar *Phys. Rev. A* **83** 062714.
[17] Wells N P, Driskell T U and Camparo J C 2015 Fine-structure mixing in Rb-Xe elastic collisions and isoclinic point asymmetry *Phys. Rev. A* **92** 022505.

Rb-stabilized laser at 1572 nm for CO_2 monitoring

R Matthey, W Moreno, F Gruet, P Brochard, S Schilt and G Mileti

Laboratoire Temps-Fréquence, Institut de Physique,
Université de Neuchâtel, Neuchâtel, Switzerland

E-mail: renaud.matthey-de-lendroit@unine.ch

Abstract. We have developed a compact rubidium-stabilized laser system to serve as optical frequency reference in the 1.55-μm wavelength region, in particular for CO_2 monitoring at 1572 nm. The light of a fiber-pigtailed distributed feedback (DFB) laser emitting at 1560 nm is frequency-doubled and locked to a sub-Doppler rubidium transition at 780 nm using a 2-cm long vapor glass cell. Part of the DFB laser light is modulated with an electro-optical modulator enclosed in a Fabry-Perot cavity, generating an optical frequency comb with spectral coverage extending from 1540 nm to 1580 nm. A second slave DFB laser emitting at 1572 nm and offset-locked to one line of the frequency comb shows a relative frequency stability of $1\cdot10^{-11}$ at 1 s averaging time and $<4\cdot10^{-12}$ from 1 hour up to 3 days.

1. Introduction

Some applications require a frequency-stable laser source operating at a precise wavelength where no direct suitable molecular or atomic reference transition exists. A reason may be the weakness of the considered transition line-strength that would require long optical interaction length to ensure a sufficient absorption signal for laser stabilization. In other cases, the laser must precisely emit at a frequency lying on the wing of an absorption line, at a detuning of up to several GHz from the line center, or even out of any absorption line. An example of such application is the measurement of the concentration of carbon dioxide (CO_2) in the atmosphere based on the integrated-path differential-absorption lidar (IPDA) technique [1], as envisaged by space agencies [2] to identify on a global scale and more precisely the CO_2 exchanges between the atmosphere and the Earth's lands and oceans [3]. The constraints imposed on the lidar instrument for a CO_2 mission were defined by feasibility studies [3, 4]. For a transmitter laser emitting at 1572 nm, on the slope of a CO_2 absorption line, a frequency stability – in terms of Allan deviation – below 20 kHz ($<10^{-10}$ fractional frequency stability) is requested for integration times from 700 s up to 3 years of mission operation time [5].

A relative frequency stability of about $3\cdot10^{-11}$ was demonstrated by Numata *et al.* for integration time up to 1,000 s at a wavelength of 1572.34 nm [6]. In this case, a distributed feedback (DFB) laser diode was frequency-stabilized at the center of a Doppler-broadened CO_2 line using a multipass cell with a path-length of 17 m combined with a laser frequency modulation technique. In the implemented approach, additional lasers could be offset-locked up to ± 7 GHz away from the master laser. Besides, compact 780-nm Rb-based frequency-stabilized laser heads were previously developed in our premises for atomic clocks, using DFB lasers and evacuated Rb cells with an optical path-length of a few centimeters only. These laser heads show a relative frequency stability better than $1\cdot10^{-11}$ at all time-scales between 1 s and 1 day [7]. The main goal of the present work consisted in transferring the high frequency stability achievable with an Rb cell to an arbitrary wavelength in the 1.55-μm wavelength region, in particular around 1572 nm, for CO_2 monitoring purpose.

Journal of Physics: Conference Series **723** (2016) 012034 doi:10.1088/1742-6596/723/1/012034

The laser system is illustrated in Figure 1. A fiber-pigtailed DFB laser emitting at 1560 nm (master laser) is frequency-doubled and stabilized to the D_2 line of ^{87}Rb at 780 nm. A part of the master laser light is injected into an electro-optical modulator (EOM) enclosed in a Fabry-Perot cavity to generate an optical frequency comb (OFC). A second DFB laser emitting at 1572 nm (slave laser) is offset-locked to one line of the generated comb. The output of the system consists of a fraction of the stabilized slave laser light. The optical system is fully fibered.

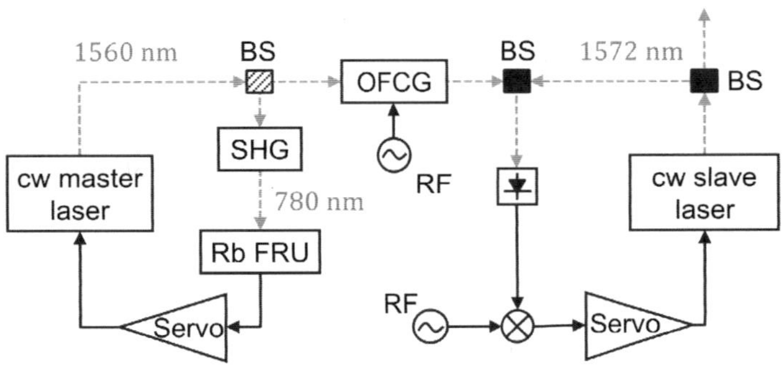

Figure 1. Schematics of the laser system. Dashed grey lines and black lines correspond to optical fibers and electrical links, respectively. BS: beam splitter (hatched one: see text); OFCG: optical frequency comb generator; Rb FRU: rubidium frequency reference unit; RF: radio frequency oscillator; SHG: second harmonic generator.

2. Experimental Setup

2.1. Frequency Doubling for Stabilization at 1560 nm

The master laser is a fiber-pigtailed DFB laser (Emcore) emitting up to 80 mW of optical power at 1560 nm. The laser has a full width at half-maximum (FWHM) linewidth of 1-2 MHz (at 4-ms observation time), determined using the concept of the β-separation line [8] from its measured frequency noise spectrum and also from the recorded beat note with two independent lasers. A fibered optical isolator is disposed at the output of the master laser to prevent unwanted feedback. A 50/50 polarization-maintaining fiber beam splitter (PM-BS) splits the light into two outputs; the first one is used for frequency stabilization and the second to generate the OFC.

The first output of the PM-BS is frequency-doubled using a second harmonic generator (SHG) module made of a fiber-pigtailed 5-cm-long periodically-poled lithium niobate (PPLN) waveguide. The optimal operation temperature and the efficiency of the module were determined to be 50°C and 300 %/W (including the fiber coupling losses), respectively [9]. The frequency-doubled light is then injected into a frequency reference unit (FRU) that has a similar design as the one integrated in the laser heads described by Gruet *et al.* [7]. There, it optically pumps the ^{87}Rb atoms (D_2 line, 780.24 nm) contained in a home-built evacuated cylindrical cell (10-mm diameter by 19-mm length). The cell is thermalized and mounted in a magnetically-shielded package. The light is retro-reflected in order to probe the Rb vapor in a sub-Doppler absorption scheme that provides narrow resonances (~15 MHz FWHM) for frequency-locking.

Frequency-stabilization of the master laser is achieved by modulating its injection current with a 50-kHz sinusoidal signal and applying synchronous demodulation of the detected light intensity at the first harmonic of the modulation frequency. The correction signal obtained after a proportional-integrator servo-controller is applied to the laser current in a feedback loop.

2.2. Optical Frequency Comb Generation

The frequency-stabilized master laser is used to seed an OFC generator (OptoComb). The injected light is phase-modulated by an EOM enclosed in a waveguide Fabry-Perot cavity with a finesse of 60 and a free spectral range of 2.5 GHz. The EOM is driven by a radio-frequency (RF) signal in resonance with the cavity at the forth harmonic of its free-spectral range, i.e., 10 GHz, to enhance the efficiency of the sideband generation process. Hence, a widely-spanning frequency comb is generated with a spectral coverage of 40 nm centred at the master laser frequency (Figure 2). The synthesizer generating the RF signal applied to the EOM is referenced to a hydrogen maser for long-term stability.

By denoting f_m and ν_M the RF and master laser frequency, respectively, the frequency of the *n*th-order optical comb line, ν_n, is given by $\nu_n = \nu_M + n \cdot f_m$, where *n* can take a positive or negative integer value. As the frequency stability of the RF signal considerably exceeds that of the master laser, the frequency of all comb modes follows the fluctuations of the master laser. The frequency dithering applied to stabilize the master laser is thus also transferred to each mode of the frequency comb, which gives the possibility to offset-lock a slave laser to one mode of the comb without requiring any further frequency modulation, e.g., of the slave laser.

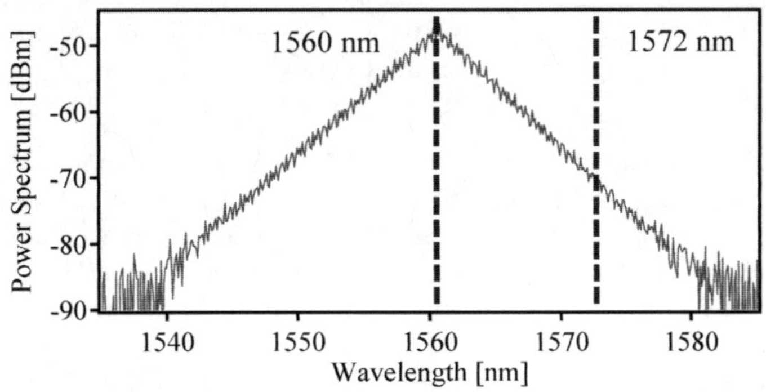

Figure 2. Spectral envelope of the optical frequency comb covering a span larger than 40 nm around the seed central line of 1560.48 nm, as obtained when driving the EOM at 10 GHz.

2.3. Offset-Locking for Stabilization at 1572 nm

For the slave laser, a linewidth of 1-2 MHz (FWHM) was determined in the same way as for the master laser. An optical isolator is disposed at the laser output and a 50/50 PM-BS divides the optical power into two parts. One part constitutes the optical output of the system, while the other is used to offset-lock the slave laser to one mode of the optical frequency comb using a similar method as proposed by Schilt *et al.* [10]. For this purpose, a fast photodiode detects the beat note between the slave laser and the nearest comb mode. After amplification, the beat signal is frequency down-converted by mixing with a local oscillator of frequency f_{LO} referenced to the same hydrogen maser. The resulting intermediate frequency signal is further filtered using the combination of a 1-MHz high-pass filter and an 11-MHz low-pass filter. Finally, an envelope detector measures the power of the transmitted RF signal, which, after demodulation at the master laser modulation frequency, provides a voltage error signal for an offset-locking stabilization loop [10]. The combination of the filters and envelope detector acts as a frequency discriminator and creates an absorption-like feature in the low-frequency electrical domain (see Figure 3). The frequency down-converted beat note between the slave laser and the comb line can thus be locked to the centre of this feature by generating a derivative-like signal using the dithering of the beat signal that results from the modulation of the master laser. The absence of dithering of the slave laser frequency is beneficial for the injection seeding process of the power oscillator of a lidar transmitter or to serve as an unmodulated optical frequency reference.

The comb line and the offset frequency are selected so that the slave laser emits at the desired frequency $\nu_S = \nu_M + n \cdot f_m \pm f_{LO}$. The sign of the offset frequency is determined from the observed change in the injection current of the locked slave laser in response to a variation of the offset frequency f_{LO}. The mode number *n* can be determined when varying the EOM drive frequency f_m or using an optical spectrum analyser.

3. Frequency Stability Results

The frequency stability of the laser system was evaluated at the main stages of the setup, corresponding to different wavelengths, using heterodyne detection with external optical frequency references. Hence, we were able to follow how the stability of the Rb reference is transferred to the slave laser by successively evaluating the master laser before (1560 nm) and after (780 nm) the frequency-doubling process, an optical comb mode (1557 nm) and at last the slave laser (1572 nm).

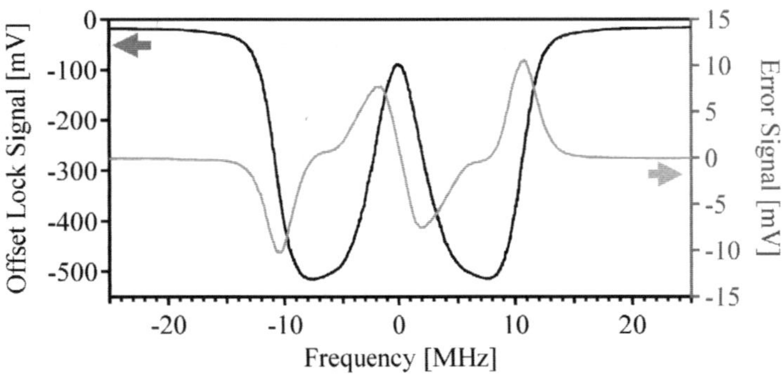

Figure 3. Black line: offset-locking signal measured at the output of the envelope detector. Grey line: error signal obtained after demodulation.

3.1. Master Laser

The frequency stability of the master laser was measured at 1560 nm and 780 nm. At 1560 nm, a beat note was performed between the master laser locked onto the Rb direct F2→F'3 transition (which is the most stable at long term due to its lower temperature sensitivity [8]) and a self-referenced commercial Er-fiber comb fully stabilized to an H-maser. At 780 nm, the frequency-doubled master laser was heterodyned with an external Rb-stabilized reference laser head locked onto the cross-over CO22/23 transition, which demonstrated relative frequency stability better than $1\cdot10^{-11}$ at all time-scales from 1 s up to 1 day [7]. The measurements of the master laser at 780 nm and 1560 nm were carried out simultaneously – although they did not always have the same overall duration – and for two different types of fiber splitters at the laser output (hatched splitter in Figure 1).

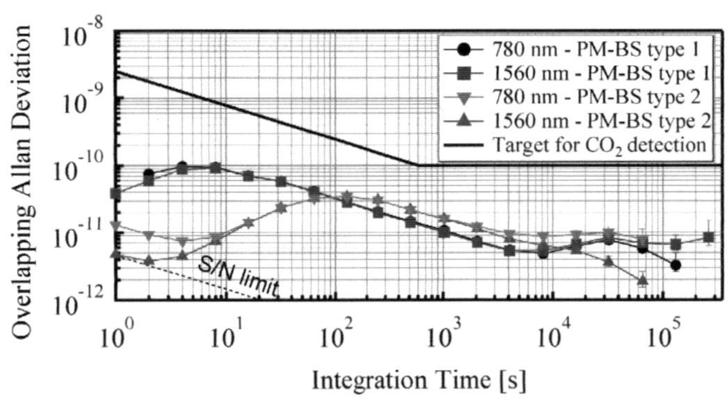

Figure 4. Fractional frequency stability of the beat note between the master laser (1560 nm) and the self-referenced comb (■, ▲), and between the frequency-doubled master laser (780 nm) and an external Rb-stabilized reference laser head (●, ▼). (■, ●), (▲, ▼): simultaneous measurement done with the first and second type of PM-BS, respectively.

As presented in Figure 4, a fractional frequency stability of $5.2\cdot10^{-12}$ at 1-s averaging time (corresponding to <1 kHz absolute instability at 1560 nm) is reached with the second type of fiber splitter. This value matches the estimated signal-to-noise limit at 1 s. Above 4,000 s, the long-term relative instability is always lower than $1\cdot10^{-11}$ (2 kHz absolute frequency at 1560 nm). All stability measurements performed at 780 nm were limited by the resolution of the frequency counter, which type and settings differed from the ones used at 1560 nm. This effect is responsible for the difference observed at short term between the results at 780 nm and 1560 nm. The long-term fractional frequency stability at 780 nm is also further limited slightly below $1\cdot10^{-11}$ by the reference laser head for integration times longer than 10^4 s.

The first type of fiber splitter induced spurious reflections and associated etalon fringes that degraded the short-term stability at the level of $1\cdot10^{-10}$ at 5 s. The replacement of this low-performing splitter with a better one (type 2 in Figure 4) drastically improved the short-term stability by more than an order of magnitude. With the second type of splitter a previously hidden instability bump reaching $3\cdot10^{-11}$ at around 100-s integration time is observed. It is caused by the presence of a fiber splice at the output of the frequency-doubling module that induces etalon interferences.

Journal of Physics: Conference Series **723** (2016) 012034 doi:10.1088/1742-6596/723/1/012034

3.2. Optical Frequency Comb

When the first type of splitter was installed to split the emission of the master laser, the frequency stability of the generated OFC was retrieved from the beat-note frequency measured between one line of the comb and a narrow-linewidth (<10 kHz) auxiliary semiconductor laser operating at 1557 nm. The laser was phase-locked to the self-referenced comb to ensure its long-term frequency stability. The low optical power per mode in the two combs prevented the direct beating of the generated OFC against the self-referenced comb. As shown in Figure 5, the stability of the OFC line at 1557 nm faithfully reproduces the master laser stability at 1560 nm – already presented in Figure 4 but replicated for comparison purpose. The non-concomitance of both measurements results in slight short- and mid-term differences due to modified temperature-induced etalon fringes in the master laser locking loop, and in variations in the long term due to the slowly fluctuating laboratory background temperature.

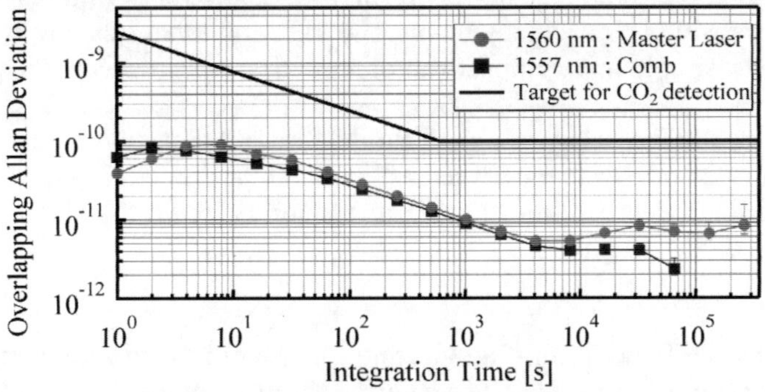

Figure 5. Fractional frequency stability of the beat note between one line of the OFC and a 1557-nm auxiliary laser, phase-locked to a self-referenced comb (■), and between the master laser and the self-referenced comb (●). Measurements were not carried out simultaneously.

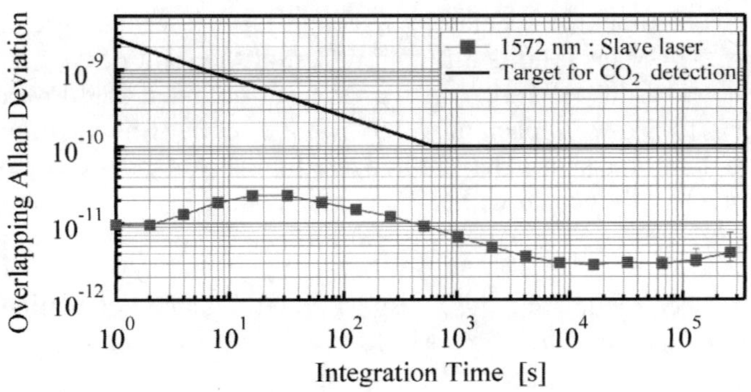

Figure 6. Fractional frequency stability of the beat note between the slave laser at 1572 nm and the self-referenced comb (■).

3.3. Slave Laser

In the same manner as applied to the master laser, the frequency stability of the slave laser was determined from the beat note with the self-referenced comb fully-stabilized to the H-maser. The offset frequency f_{LO} in the slave laser stabilization loop was set to 50 MHz and the drive frequency of the OFC generator was chosen so that the slave laser was lying at the center of the CO_2 R20 absorption line (190,743,066.5 MHz). To properly stabilize the slave laser, the modulation amplitude of the master laser, still locked to the direct transition F2→F'3 with the fiber splitter of the second type, was increased, implying a degradation of the short-term stability of the slave laser with respect to the master laser stability previously reported in Figure 4.

The obtained fractional frequency stability of the slave laser is displayed in Figure 6. It amounts to $1 \cdot 10^{-11}$ at 1 s and stays below $2.5 \cdot 10^{-11}$ for all timescales up to the maximum assessed integration time of 3 days, reaching $4 \cdot 10^{-12}$ on the long term. The target stability for atmospheric CO_2 monitoring is by far fulfilled. The measurement was performed under environmental temperature conditions that were different from those prevailing for the evaluation of the master laser stability (see Figure 4). The setup was less protected against rapid temperature changes, which made the temporal frequency fluctuations linked to etalon-like cavity more severe. This resulted in a shift of the instability bump in the Allan deviation towards shorter integration times (~30 s).

4. Discussion and Perspective

For the demonstration of the developed scheme, an H-maser was employed as a frequency reference for the OFC generator driving signal as well as for the offset-locking frequency. Both RFs can be referenced to a more compact and transportable frequency reference such as a commercially-available GPS-referenced Rb atomic clock that offers a sufficient stability compatible with the RF requirements of the system (fractional frequency stability of $1 \cdot 10^{-9}$). Depending on the desired output wavelength, the mixer and RF oscillator of the offset-locking loop might be omitted, locking the slave laser directly to the optical comb and fine-tuning its frequency with the OFC generator modulation frequency.

The accuracy and stability of the output frequency at 1572 nm are most of all dictated by the performance of the master laser as the RF sources used in the setup are much more stable. The accuracy and reproducibility of the master laser frequency, which depend on the selected reference sub-Doppler Rb line and on the frequency stabilization method, are presently being investigated. Our preliminary results suggest that for the direct F2→F'3 transition an accuracy below 100 kHz and a reproducibility of 20 kHz could be reached with a relatively simple and compact rubidium cell setup, in accordance with the results of Ye *et al.* [11].

At present, an imperfection in the fiber at the frequency-doubling module output that induces etalon-like cavities limits the frequency stability of our system between 10-s and 100-s integration time. The reduction of these etalon-like effects should decrease the short-term instability down to the estimated signal-to-noise limit of the master laser ($5.2 \cdot 10^{-12} \tau^{-1/2}$).

5. Conclusion

We have demonstrated that the stability of a DFB laser emitting at 1560 nm, stabilized to rubidium using frequency-doubling and sub-Doppler spectroscopy, can be transferred through the generation of an optical frequency comb and a frequency offset-locking scheme to a second DFB laser emitting at 1572 nm. The relative frequency stability of the master laser is $5.2 \cdot 10^{-12}$ at 1 s, at the limit of the signal-to-noise. At 1572 nm, a relative frequency stability of $1 \cdot 10^{-11}$ at 1 s has been obtained, and lower than $2.5 \cdot 10^{-11}$ for all assessed timescales (1 s to 3 days), reaching $4 \cdot 10^{-12}$ above 3,000 s. Such result fulfils the requirements for a CO_2 satellite lidar mission. The applied method is not restricted to 1572 nm and can be implemented for any wavelength located within the comb span.

6. Acknowledgement

This work was supported by the European Space Agency and the Swiss National Science Foundation (project number 156621). We thank our colleague Dr. Christoph Affolderbach and Dr. Andreas Fix (DLR, Germany) for helpful discussions.

References

[1] Ehret G, Kiemle C, Wirth M, Amediek A, Fix A and Houweling S 2008 *Appl. Phys. B* **90**, 593
[2] ESA 2008 *Report for Assessment*, SP-1313/1
[3] Bréon F M and Ciais P 2009 *Comput. Geosci.* **342** 412
[4] Caron J, Durand Y, Bezy J L and Meynart R 2009 *Proc. Lidar Technologies Techniques and Measurements for Atmospheric Remote Sensing V* (SPIE, 2009), paper 74790E
[5] Fix A, Matthey R, Amediek A, Ehret G, Gruet F, Kiemle C, Klein V, Mileti G, Pereira do Carmo J and Quatrevalet M 2014 *Proc. Int. Conf. on Space Optics (Tenerife)*, session 2C
[6] Numata K, Chen J, Wu S, Abshire J and Krainak M 2011 *Appl. Opt.* **50** 1047
[7] Gruet F, Pellaton M, Affolderbach C, Bandi T, Matthey R and Mileti G 2012 *Proc. of the Int. Conf. in Space Optics (Ajaccio),* paper 048
[8] Di Domenico G, Schilt S and Thomann P 2010 *Appl. Opt.* **49** 4801
[9] Matthey R, Gruet F, Schilt S and Mileti G 2015 *Opt. Lett.* **40** 2576
[10] Schilt S, Matthey R, Kauffmann D, Affolderbach C, Mileti G and Thévenaz L 2008 *Appl. Opt.* **47**, 4336
[11] Ye J, Swartz S, Jungner P and Hall J 1996 *Opt. Lett.* **21** 1280

Octave-spanning supercontinuum generation via microwave frequency multiplication

D C Cole[1, 2,*], K M Beha[1,3], S A Diddams[1] and S B Papp[1]

[1]Time and Frequency Division, National Institute of Standards and Technology, 325 Broadway, Boulder, CO 80305 USA
[2]Department of Physics, University of Colorado, Boulder, CO 80309, USA
[3]Current: Menlo Systems, Am Klopferspitz 19a, D-82152 Martinsried, Germany

[*]Corresponding author: daniel.cole@nist.gov

Abstract. We demonstrate a system based on telecom components for the generation of a coherent octave-spanning supercontinuum from a continuous-wave laser. The system utilizes direct multiplication of a 10 GHz signal derived from a commercial synthesizer to carve pulses from the laser, which are then iteratively chirped and compressed in two stages. After reducing the repetition rate of the resulting pulse train to 2.5 GHz using selective transmission through an electro-optic gate, propagation through highly-nonlinear fiber generates an octave-spanning supercontinuum spectrum. We discuss the impact of the noise of the modulation frequency on the coherence of the supercontinuum and discuss its mitigation. Close agreement between experiment and theory is shown throughout, and we use our ability to precisely model the experiment to propose an extension of the system to 20 GHz repetition rate.

The self-referenced optical frequency comb has revolutionized precision metrology and spectroscopy by providing a set of known, equidistant optical frequencies and linking these frequencies to microwave signals traceable to the SI second [1, 2]. Recently, the potential to use frequency combs for new applications and to expand their utility in present ones has driven efforts to develop combs with pulse repetition rates and mode spacings of 10 GHz and above. This is much higher than the typical 0.1-1 GHz repetition rates of most femtosecond mode-locked laser combs found in commercial production and in research labs. The applications that stand to benefit from higher repetition rates exploit individual access to widely-spaced comb modes and high power per mode, and include astronomical spectrograph calibration [3, 4], precision spectroscopy [5], optical arbitrary waveform generation [6], and optical communications [7, 8]. These applications, and in particular spectrograph calibration and precision spectroscopy, will benefit from measurement and stabilization of the comb's carrier-envelope offset frequency $f_{\rm CEO}$. Due to the properties of mode-locked lasers, it has proven challenging to build mode-locked laser combs operating above 10 GHz repetition rate for which $f_{\rm CEO}$ can be detected [9]. Thus, there is a need for new means to provide high repetition rate trains of short, high peak power pulses for generation of octave-spanning optical spectra and $f-2f$ interferometry.

One method for generation of high-repetition-rate pulse trains is the electro-optic modulation (EOM) comb scheme [10-14]. In this scheme, a microwave frequency input is multiplied through the generation of many intensity- and phase-modulation sidebands on a continuous-wave (CW) seed laser, yielding a pulse train in the time domain. This multiplication can be continued through subsequent nonlinear spectral broadening of the pulses. For the case of octave-spanning supercontinuum spectra, the multiplication factor is ultimately on the order of the ratio of the CW laser's optical frequency to the microwave modulation frequency – for the 10 GHz system discussed here the factor is 19340.

The EOM comb scheme for comb generation has advantages in the flexibility of the comb's repetition rate f_r, the fast and independent tunability of the repetition rate and seed frequency for any particular comb, the flatness of the initial comb's spectrum, and the compressibility of the comb's pulses. Recently, we demonstrated self-referencing of an EOM comb system with 10 GHz repetition rate and measured the frequency of the comb's CW seed laser to a stability beyond 5 parts in 10^{14} [15]. In this proceeding, we discuss in detail our 10 GHz EOM comb system and demonstrate that it may be precisely modelled with a side-by-side comparison of experimental and calculated spectra. We also discuss a unique challenge in using the EOM comb scheme to generate a coherent octave-spanning supercontinuum, which is the linearly increasing contribution of the frequency noise of the modulation tone from which f_r is derived to the frequency fluctuations of the optical comb modes. Finally, we propose a simplification and an extension of our 10 GHz system to 20 GHz repetition rate and beyond.

The experimental system we discuss here is shown in figure 1a. In order to generate the short optical pulses required for coherent supercontinuum generation in highly-nonlinear fiber (HNLF), our system consists of several stages. In the first stage, an intensity modulator carves 50 % duty cycle optical pulses from a fiber-coupled 1550 nm CW seed laser and the pulses are chirped using two $LiNbO_3$ phase modulators, then the pulses are temporally compressed in single-mode fiber (SMF). In the second stage, the pulses are amplified in an erbium-doped fiber amplifier (EDFA), chirped again via self-phase modulation (SPM) in 100 m of HNLF (see e.g. Ref. 16 for previous work using this technique), and temporally compressed using a spatial light modulator (SLM). In the final stage, the pulse repetition rate may be reduced with an electro-optic gate, after which the pulses are amplified and launched into hybrid HNLF for generation of an octave-spanning supercontinuum spectrum. Also included in our system in the second stage is an optical Fabry-Perot filter cavity for the purpose of reducing the impact of the frequency noise of the modulation tone; this is discussed in detail below.

To introduce the EOM comb system, we present a standard calculation of the comb's electric field following intensity and phase modulation for generation of the initial chirped pulse train. The seed laser's field $E_0 e^{i\omega_c t}$ is multiplied by factors $\frac{1}{2}\left[1+\exp\left(i\frac{\pi}{2}(1+\sin\omega_r t)\right)\right]$ and $\exp(i\beta_m \sin\omega_r t)$, which result respectively from 50 % duty-cycle intensity modulation and phase modulation with modulation index β_m at the angular frequency $\omega_r = 2\pi f_r$. Up to an overall optical phase shift and a shift in time, the resulting field is:

$$E = E_0 \times \cos\left[\frac{\pi}{2}\sin^2(\omega_r t/2)\right] \times \exp\left[i\omega_c t - i\left(\beta_m + \frac{\pi}{4}\right)\cos\omega_r t\right]. \quad (1)$$

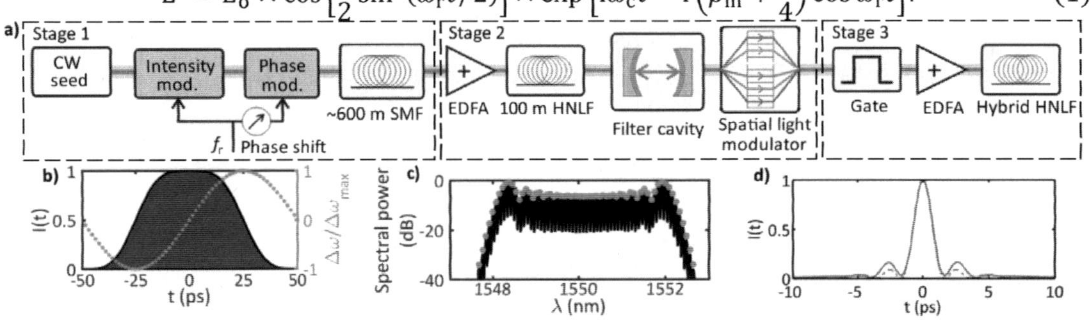

Figure 1 (colour online). (a) Experimental apparatus for generation of octave-spanning supercontinuum from a CW laser. (b) Relation between intensity profile (solid blue) and instantaneous carrier frequency (dotted green) of the initial chirped pulses generated by the EOM comb scheme after intensity and phase modulation. (c) A comparison of the experimental spectrum of the EOM comb (black) to a calculation of the spectrum (green circles). (d) Calculations of the temporal intensity profile of the EOM comb pulses after compression in 570 m SMF (solid blue) and for perfect compression to the Fourier-transform limit (dashed red). Full-width at half maximum (FWHM) of both traces is ~1.5 ps.

This can be understood as the product of a time-varying real amplitude and a phase factor from which the instantaneous carrier frequency $\omega_c + \Delta\omega(t)$ can be calculated: $\Delta\omega(t) = \omega_r\left(\beta_m + \frac{\pi}{4}\right)\sin\omega_r t$. The pulse intensity and the oscillating carrier frequency are plotted in figure 1b. The carrier frequency of the pulses increases in time, permitting pulse compression in standard anomalously dispersive SMF. Additionally, the range of instantaneous carrier frequencies $\omega_c + \Delta\omega(t)$ occurring during the flat top of the intensity profile shown in figure 1b is represented with equal weight in the optical spectrum of the pulse train, yielding a wide, flat central feature in the spectrum. The phase-modulation index sets the optical bandwidth. A comparison between the measured spectrum of our EOM comb and a calculation using $\beta_m = 15\pi/2$ is shown in figure 1c.

After generation of the EOM comb, we propagate the pulse train through ~600 m of SMF for pulse compression. Simulated results of this compression are shown in figure 1d, and indicate that compression to near the Fourier transform limit is possible. Intuitively, this is because components of the pulse occurring later in time have higher carrier frequency and propagate more quickly than components occurring earlier. The resulting ~1.5 ps duration of the pulses enables efficient generation of SPM chirp in the second stage of the experiment. After amplification to 400 mW average power, the pulses are launched into 100 m of low-dispersion HNLF. The specified dispersion values at 1550 nm are $D = -0.04$ ps/nm · km and $D' = 0.01$ ps/nm^2 · km, and SPM dominates the pulse-evolution dynamics [17]. The SPM chirp increases the 4 nm initial bandwidth of the pulses to ~40 nm, in good agreement with a numerical simulation, as shown in figure 2a. We perform the simulation using the non-linear Schrödinger equation (NLSE) including 3rd-order dispersion, taking as initial conditions the calculated intensity profile of the EOM comb pulses after compression in 570 m SMF (blue trace in figure 1d). The dispersion values for the HNLF used in the simulation are $D = -0.04$ ps/nm · km and $D' = 0.003$ ps/nm^2 · km, close to the values specified by the manufacturer.

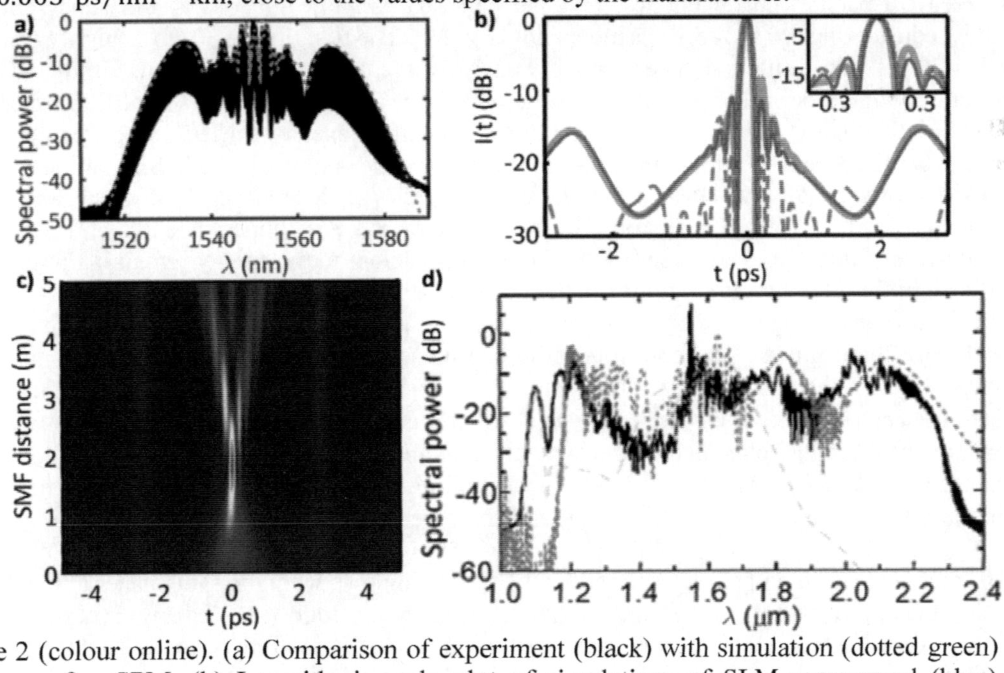

Figure 2 (colour online). (a) Comparison of experiment (black) with simulation (dotted green) for the spectrum after SPM. (b) Logarithmic-scale plot of simulations of SLM-compressed (blue), SMF-compressed (thick green) and transform-limited (dashed red) intensity profiles corresponding to the simulated spectrum in (a). Asymmetry in the SMF-compressed pulse is due to higher order dispersion. Inset: the same, plotted to emphasize the asymmetry. (c) Simulated compression of the SPM-chirped pulses during propagation in SMF. (d) Experimental octave-spanning supercontinuum generated by the EOM comb system (black), plotted with simulated spectra resulting from propagation in the first, 30 cm section (long-dashed teal) and the second, 7.7 m section (short-dashed red) of the hybrid HNLF.

After generation of the SPM chirp, the pulses are transmitted through a Fabry-Perot cavity for suppression of optical frequency fluctuations, as we will discuss below. We then pass the pulse train through an SLM, which provides direct control of each comb mode's optical phase. Using the SLM, we iteratively adjust the pulse train's 2nd-, 3rd-, and 4th-order chirp to compress the pulses. Figure 2b shows a calculation of the pulse intensity profile which would result from complete removal of 2nd- to 4th-order chirp after propagation in the HNLF, as well as fully compressed pulses with flat spectral phase. Both intensity profiles have pulse widths of 130 fs, but the intensity profile compressed by the SLM has a substantial fraction of its energy in satellite pulses at $\pm$2.5 ps with respect to the central pulse. This is due to higher-order chirp. We note that the SLM is convenient but not necessary – numerical simulations indicate that the SPM chirp can be compressed in an appropriate length of SMF by the same mechanism as the chirp imposed by external phase modulation, as discussed above. Results of a simulation of this compression employing the NLSE with Kerr nonlinearity and the full specified dispersion curve for SMF are also shown in figure 2b; after propagation through 1.8 m of SMF, the pulses reach 130 fs duration with satellite pulses at $\pm$2.5 ps, similar to what can be achieved using the SLM as described above. The pulse evolution during propagation in SMF is shown in figure 2c.

In the third stage of the experiment, after pulse compression in the SLM, the pulses are passed through a Mach-Zehnder modulator functioning as an electro-optic gate. The gate is used to selectively transmit pulses and reduce the pulse repetition rate [18, 19]. This facilitates supercontinuum generation by increasing the pulse train's peak-to-average power ratio. While we have shown in Ref. 15 that a coherent octave-spanning supercontinuum may be generated without the gate by amplifying to an average power of 4 W, here we focus on the case where the repetition rate is reduced to 2.5 GHz.

The 2.5 GHz pulse train is amplified to 1.4 W average power and passed through 8 m of hybrid HNLF, yielding the spectrum shown in figure 2d. The hybrid HNLF consists of two sections of fiber with different dispersion characteristics, and we can qualitatively understand the supercontinuum spectrum by considering the two sections independently [20]. The first fiber is 30 cm of highly dispersive HNLF ($D = 6\ \mathrm{ps/nm \cdot km}$), and generates a dispersive wave centered at 1090 nm. Simulated results of propagation in this fiber are shown in teal in figure 2d. The second fiber is 7.7 m of lower dispersion HNLF ($D = 1.5\ \mathrm{ps/nm \cdot km}$), and generates a Raman-self-frequency-shifted soliton centered near 2150 nm. Simulated results of propagation in this fiber are shown in red. Simulations use the LaserFOAM program [21], which employs the generalized NLSE including Raman scattering, self-steepening, and 2nd- to 4th-order dispersion. The two simulations are run independently, and both take as initial conditions 170 fs Gaussian pulses with 350 pJ energy, close to the energy remaining in the pulses exiting the amplifier after accounting for loss between the amplifier and the HNLF.

For the octave-spanning supercontinuum generated in the HNLF to be useful it must be coherent, with resolvable comb modes. The contribution of the noise of the modulation tone f_r to the optical frequency noise of the comb modes scales linearly with mode number N, as measured relative to the seed laser; the contribution to the power spectrum of frequency noise scales as N^2. This presents a challenge for maintaining coherence of the supercontinuum and detecting f_{CEO}. For standard $f-2f$ interferometry, the factor by which the noise of f_r is multiplied to determine its contribution to the f_{CEO} signal is the ratio between the seed frequency and the repetition rate; $N = f_c/f_r = 19340$ for the 10 GHz comb discussed here. This contribution is shown in figure 3a, along with the contribution from the CW seed laser. The noise on f_r is determined by the technical noise from the synthesizer at low Fourier frequencies and approaches a white Johnson-Nyquist phase noise floor of -177 dBm/Hz at high Fourier frequencies. As discussed in Ref. 15, unmitigated multiplication of the white phase noise floor by 19340 leads to frequency fluctuations that are high enough to prevent detection and measurement of f_{CEO}. By passing the pulse train through a Fabry-Perot cavity whose free spectral range is actively stabilized to the comb's mode spacing, the contribution of the Johnson-Nyquist frequency noise at high Fourier frequencies is reduced by the filter cavity's Lorentzian transfer function, leading to resolvable comb modes in the supercontinuum and permitting measurement of f_{CEO}. We use a filter cavity with 7.5 MHz

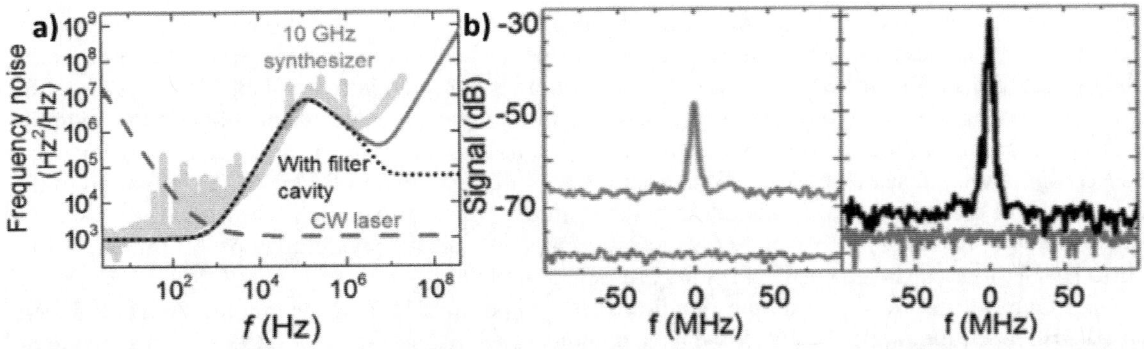

Figure 3 (colour online). (a) Contributions to the spectrum of frequency fluctuations of the carrier-envelope offset frequency detected through $f-2f$ interferometry: model of the CW laser (dashed green), model of the 10 GHz synthesizer multiplied by 19340^2 without filter cavity (solid red, experimental data thick red), and synthesizer multiplied by 19340^2 and the filter cavity transfer function (dotted black). (b) A comparison of detected heterodyne beats between the supercontinuum and a 1319 nm wavelength CW laser without (red, left) and with (black, right) the optical filter cavity. The level of intensity noise on the supercontinuum, measured by removing the 1319 nm CW laser, is shown by the lower gray trace in each plot. Signal-to-noise ratios for the beat are 17 dB without and 40 dB with the filter cavity.

linewidth. The effect of the cavity is shown concretely in figure 3b, where we compare the lineshape of a heterodyne beat between the supercontinuum and a CW laser with 1319 nm wavelength without filter cavity and with the cavity in place. The signal-to-noise ratios for the beat without the cavity and with it are 17 dB and 40 dB, respectively.

Lastly, we propose a simplification and extension of this system to 20 GHz repetition rate, and possibly beyond. The initial bandwidth of an EOM comb scales with the comb's mode spacing and with the phase-modulation index β_{m}, which determines the number of modes generated. By modestly increasing both parameters, the spectral broadening for supercontinuum generation can be accomplished in a single piece of fiber because sufficiently short (several hundred femtoseconds) optical pulses are supported by the initial comb spectrum. The system we propose is shown in figure 4a, where four phase modulators are used with an intermediate booster amplifier to achieve a phase-modulation index of $\beta_{\mathrm{m}} = 45\pi/2$. The calculated bandwidth for this comb is 25 nm, and the simulated spectrum is shown in blue

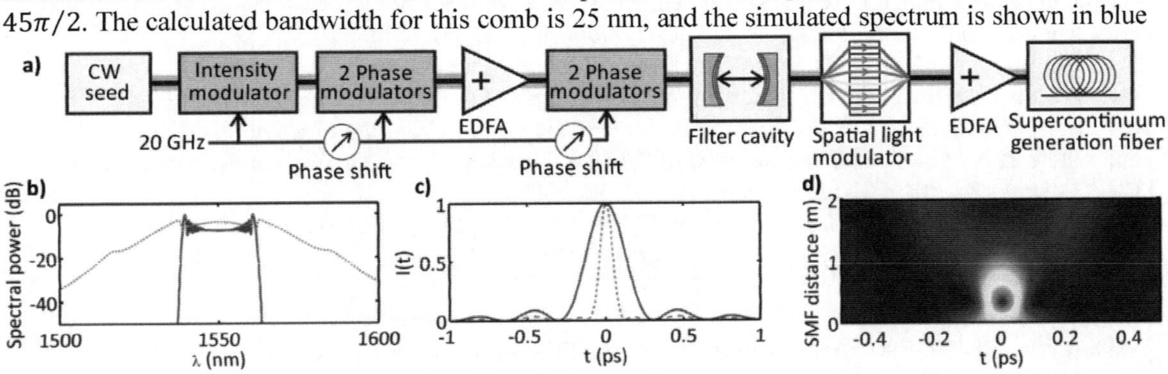

Figure 4 (colour online). (a) Proposed experimental setup for generation of ~100 fs pulses to seed supercontinuum generation at a 20 GHz repetition rate. (b) Calculations of the EOM comb spectrum at 20 GHz with $\beta_m = 45\pi/2$ (solid blue) and the SPM-broadened spectrum after propagation in 20 cm of low-dispersion HNLF at 8 W average power (dotted red). (c) Time-domain intensity profiles corresponding to the spectra in (b). The trace for the SPM-broadened spectrum includes pulse compression in an appropriate length of SMF, with a resulting pulse width of 92 fs. (d) Simulated compression of the pulses in SMF after SPM chirp-generation in 20 cm of HNLF.

in figure 4b. After passing the pulse train through the filter cavity for suppression of frequency fluctuations, we can exploit our precise modelling of the EOM comb to use the SLM to compress the pulses to the transform-limited duration of 275 fs. Using a commercially available 10 W EDFA, the pulses can be amplified to 500 pJ energy and launched into a piece the low-dispersion HNLF, with the resulting spectrum plotted in figure 4b. After propagation through an appropriate length of SMF for pulse compression, pulses with 92 fs FWHM duration result, as shown in figure 4c. A simulation of the compression of the pulses in SMF including both nonlinearity and dispersion is shown in figure 4d. We expect the compressed, ~350 pJ pulses to generate coherent octave-spanning supercontinuum in the hybrid HNLF, in accordance with our results presented above.

Our system provides a simple and direct route to octave-spanning supercontinuum at 2.5 GHz and 10 GHz repetition rates. In Ref. 15, we have demonstrated that spectra generated with this technique are coherent and suitable for optical frequency metrology. The comb's wide mode spacing and the tunability of the comb's repetition rate and offset frequency, via tuning of the comb's seed laser, make the system ideal for optical arbitrary waveform generation, astronomical spectrograph calibration, and fast acquisition of precise spectroscopic data. The extension of this technique to higher repetition rates is limited largely by the challenge of achieving high pulse train average power for nonlinear spectral broadening and by the availability of stable microwave sources for generation of the modulation tone. Finally, we emphasize that this system has the advantage of being well understood and amenable to precise numerical modelling, and that the EOM comb system can be constructed from commercially available parts using mature technology from the telecom industry.

We thank Travis Briles and Tara Drake for helpful comments on the manuscript and M. Hirano for providing the HNLF. This work is supported by the DARPA QuASAR program, AFOSR, NASA, and NIST. DC acknowledges support from the NSF GRFP under Grant No. DGE 1144083.

References

[1] Udem T, Holzwarth R, and Hansch T 2002 *Nature* **416** 233
[2] Diddams S 2010 *Journal of the Optical Society of America B* **27** B51
[3] Steinmetz T *et al.* 2008 *Science* **321** 1335
[4] Ycas G *et al.* 2012 *Optics Express* **20** 6631
[5] Stowe M *et al.* 2008 *Advances in Atomic, Molecular, and Optical Physics* **55** 1
[6] Jiang Z, Huang C-B, Leaird D and Weiner A 2007 *Nature Photonics* **1** 463
[7] Ellis A and Gunning F 2005 *Photonics Technology Letters* **17** 504
[8] Delfyett P *et al.* 2006 *Journal of Lightwave Technology* **24** 2701
[9] Bartels A, Heinecke D and Diddams S 2009 *Science* **326** 681
[10] Murata H *et al.* 2000 *IEEE Journal of Selected Topics in Quantum Electronics* **6** 1325
[11] Metcalf A *et al.* 2013 *IEEE Journal of Selected Topics in Quantum Electronics* **19** 3500306
[12] Ishizawa A *et al.* 2010 *Electronics Letters* **36** 1343
[13] Wu R, Supradeepa V, Long C, Leaird D and Weiner A 2010 *Optics Letters* **35** 3234
[14] Wu R, Torres-Company V, Leaird D and Weiner A 2013 *Optics Express* **21** 6045
[15] Beha K *et al. Preprint* arXiv:1507.06344
[16] Myslivets E, Kuo B, Alic N and Radic S 2012 *Optics Express* **20** 3331
[17] Agrawal G 2007 *Nonlinear Fiber Optics* (Burlington: Academic Press)
[18] Cole D, Papp S and Diddams S 2013 *Preprint* arXiv:1310.4134
[19] Mandridis D *et al.* 2010 *Applied Optics* **49** 2850
[20] Dudley J, Genty G and Coen S 2006 *Reviews of Modern Physics* **78** 1135
[21] Amorim A *et al.* 2009 *Optics Letters* **34** 3851

Journal of Physics: Conference Series **723** (2016) 012036 doi:10.1088/1742-6596/723/1/012036

Advanced two-way satellite frequency transfer by carrier-phase and carrier-frequency measurements

Miho Fujieda, Tadahiro Gotoh and Jun Amagai

National Institute of Information and Communications Technology, 4-2-1 Nukui-kitamachi, Koganei, Tokyo, 184-8795 Japan

E-mail: miho@nict.go.jp

Abstract. Carrier-phase measurement is one of the ways to improve the measurement resolution of two-way satellite frequency transfer. We introduce two possible methods for carrier-phase measurement: direct carrier-phase detection identified by Two-Way Carrier-Phase (TWCP) and the use of carrier-frequency information identified by Two-Way Carrier Frequency (TWCF). We performed the former using an arbitrary waveform generator and an analog-to-digital sampler and the latter using a conventional modem. The TWCF measurement using the modem had a resolution of 10^{-13} and the result agreed with that obtained by GPS carrier-phase frequency transfer in a 1500 km baseline. The measurement accuracy may have been limited by the poor frequency resolution of the modem; however, the TWCF measurement was able to improve the stability of conventional two-way satellite frequency transfer. Additionally, we show that the TWCP measurement system has the potential to achieve a frequency stability of 10^{-17}.

1. Introduction

Recently developed optical clocks require a highly stable frequency transfer technique [1]. Optical-fiber frequency transfer enables optical-clock comparison without any degradation of the clock accuracy of 10^{-17} or better [2]. However, this method is physically dependent on existing optical-fiber networks. The procurement of optical fiber link is sometimes difficult from a viewpoint of the rental cost of optical fibers. Additionally, the use of submarine cables is necessary for intercontinental frequency transfer, whose availability remains unclear. As a result, we occasionally have to depend on satellite-based frequency transfer techniques such as GPS frequency transfer and two-way satellite time and frequency transfer (TWSTFT) [3, 4]. In TWSTFT, paired earth stations transmit and receive modulated signals by code sequences via a geostationary satellite and determine the propagation delay of the signal transmitted from the counterpart station from the code phase. The time difference between the paired stations is calculated from the difference between the two delays. The measurement is typically carried out using a commercially available instrument for signals modulation and demodulation called a modem. The stability, which is at the low 10^{-15} level in one day, is insufficient in comparison with that of optical clocks [5]. Since the measurement resolution for a code phase is inversely proportional to the chip rate, a higher chip rate gives a higher measurement resolution. At the same time, the signal bandwidth becomes broader, which means that a higher satellite-link cost is necessary. Thus, it is not easy to obtain better measurement resolution by increasing the chip rate. In a recently reported frequency transfer by GPS precise point positioning (PPP) with integer ambiguity resolution, an accuracy in fractional frequency uncertainty of 1×10^{-16} was obtained in a few days [3]. We also implemented a carrier-phase measurement to improve the stability of TWSTFT and established a carrier-phase two-way satellite frequency transfer technique (TWCP) [6]. So far, we have used an analog-to-digital (A/D) sampler [7] to detect the carrier phase and have succeeded in improving the measurement resolution to the 10^{-13} level. Additionally we performed TWCP measurement with a very long baseline of 9000 km and confirmed that it was operational, and the result agreed well with that obtained by GPS carrier-phase (GPS CP) frequency transfer [8]. To achieve the general versatility of advanced two-way satellite frequency transfer measurement, we perform measurement using TWCF data recorded by a modem (here, the A/D sampler is not used). This may enable the improvement of the measurement resolution since all the participating stations in TWSTFT own the modem. In TWCF, the calculation of frequency difference between paired stations requires the Doppler coefficient in addition to carrier-frequency data of four signals [9]. The Doppler coefficient can be given from the ranging measurement using the code-phase since the code-phase measurement is simultaneously carried out by the modem. The thus obtained results and a comparison between TWCP and TWCF are introduced in this paper.

2. Measurement setup

2.1. TWCP

Figure 1(a) shows a schematic of the earth station used for TWCP measurement. We typically use a 1.8 m parabolic antenna. A signal modulated by a code sequence is generated at a frequency of 70 MHz by an arbitrary waveform generator (AWG). The measurement scale is the carrier phase. However, a chip rate of about 128 kHz is used to assist the signal tracking. The signal bandwidth has previously been limited to 200 kHz when the waveform file is generated. This bandwidth is helpful for reducing the satellite-link cost and is about 10 times narrower than that used in conventional TWSTFT. The phase of the reception signal is sampled by an A/D sampler and the carrier phase is detected. The U/C, D/C, AWG and A/D sampler are phase-locked to the external reference at 10 MHz. Additionally, the head of the code sequence generated by the AWG is synchronized to an external 1 pulse-per-second (pps) signal. The sampling by the A/D sampler is also carried out synchronously to the 1 pps signal.

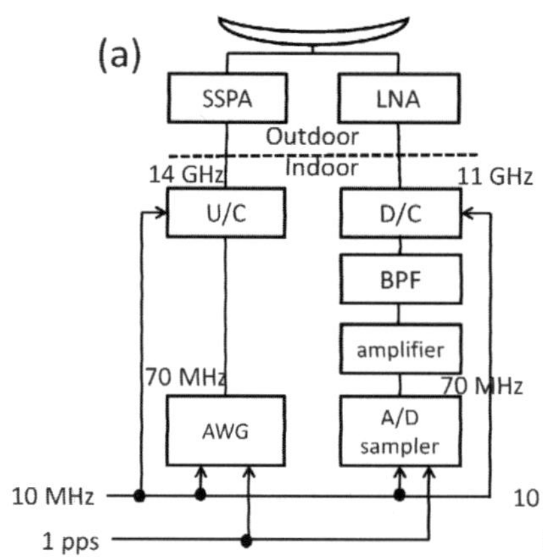

Figure 1. Configuration of earth stations for TWCP (a) and TWCF (b). SSPA: solid-state power amplifier, LNA: low-noise amplifier, U/C: frequency upconverter, D/C: frequency downconverter, AWG: arbitrary waveform generator, A/D: analogue to digital, BPF: bandpass filter.

2.2. TWCF

Figure 1(b) shows a schematic of the earth station for TWCF measurement, which is identical to that for conventional TWSTFT measurement. The modem generates a modulated signal by a code sequence. The chip rate can be chosen from different options. In the TAI time-transfer network, 1 MHz or 2.5 MHz is typically adopted. The BPF is installed in the transmission and reception paths to limit the signal bandwidth.

3. Measurement results

3.1. Common clock measurement

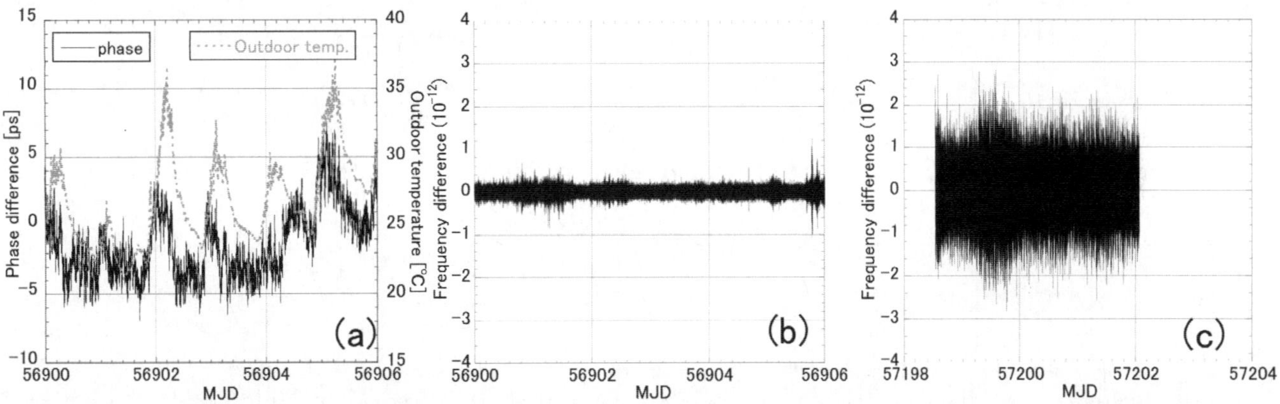

Figure 2. Results of common clock measurement. (a), (b): Phase and frequency differences obtained by TWCP, (c): frequency difference obtained by TWCF.

We evaluated the systematic instability of TWCP and TWCF by common-clock measurements which are free from the stability of the reference clock, ionosphere delay and the effects of the satellite orbit. Figure 2 depicts the results of the common clock measurements by TWCP and TWCF via a satellite. Two independent earth stations at National Institute of Information and Communications Technology (NICT) and the same uplink and downlink frequencies were used for both measurements, where 1 data point per second was available. For the TWCF measurement, a chip rate of 2.5 MHz was adopted. Figures 2(a) and (b) show the time and frequency differences obtained by TWCP, respectively. Figure 2(c) depicts the frequency difference obtained by TWCF. It can be clearly seen in Figure 2(a) that the daily phase variations were synchronized with those of the outdoor temperature. The mean frequency differences in Figures 2(b) and (c) were -3.5×10^{-18} and -6.0×10^{-16}, respectively. Figure 3 shows the modified Allan deviation of the common clock measurements. The TWCF result appears to be slightly noisier than the TWCP result, however, measurement resolution at the 10^{-13} level was obtained, which reached the 10^{-16} level after an averaging time of a few hours. The resolution of the carrier frequency was limited to the mHz level by the modem, which determined the TWCF resolution at the 10^{-13} level. However, the TWCF result obtained is better than those previously reported [9, 10]. The green inverted triangles depict the TWCP result of internal loop measurement configured using the AWG, U/C, D/C and A/D sampler, that is, without the satellite. From the other measurement result carried out using the AWG and A/D sampler, we found that the phase noises due to the U/C and D/C determine the slope of the common-clock instability. The long-term instability of TWCP is clearly limited by the phase noise. However, the result indicates that TWCP has the potential to reach a frequency stability of 10^{-17}.

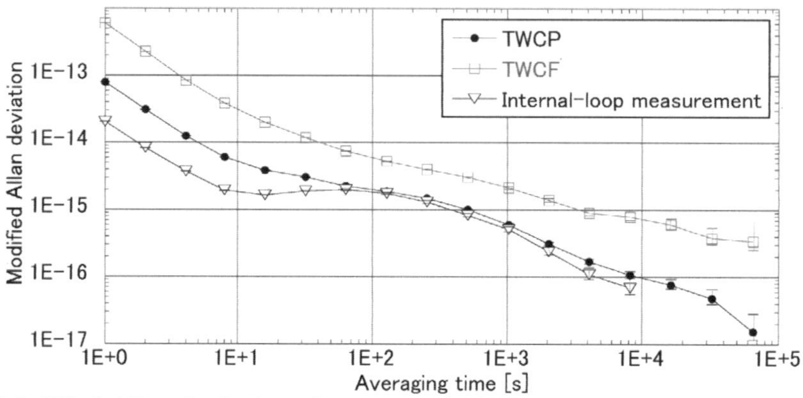

Figure 3. Modified Allan deviation of common clock measurements. Internal-loop measurement refers to the common clock measurement by TWCP without the satellite, that is, configured using the AWG, U/C, D/C and A/D sampler.

3.2. TWCF measurement with zero baseline between UTC(NICT) and a H-maser

To evaluate the measurement accuracy, we measured the frequency difference between UTC(NICT) and a hydrogen maser (H-maser) by TWCF and using a phase comparator. Two earth stations were connected to UTC(NICT) and the H-maser signals. The resultant frequency differences almost agreed as shown in Figure 4(a). The frequency stability expressed in terms of modified Allan deviations is depicted in Figure 4(b). Owing to the high stability of the H-maser, we obtained stability at the 10^{-16} level. The mean frequency differences obtained by TWCF and using the phase comparator were -698.0×10^{-15} and -696.1×10^{-15}, respectively, when the outliers were removed. The difference was -1.9×10^{-15}. On the other hand, we performed the similar measurement by TWCP to evaluate its accuracy. We confirm that the frequency difference between the results obtained by TWCP and using the phase comparator is 1×10^{-16} [11].

3.3. TWCF measurement with 1500 km baseline between UTC(NICT) and a H-maser

As the next step, we performed TWCF measurement between the NICT Koganei headquarters and our Okinawa site. Okinawa is in the southwest of Japan and about 1500 km away from Koganei. There is an earth station with a 1.8 m dish and a H-maser at the Okinawa site. We measured the frequency difference with respect to UTC(NICT) by TWCF and the GPS carrier phase (GPS CP) [12] and compared the results. Unfortunately the measurement period was limited to two days because a typhoon hit Okinawa and during which the SSPA in the station broke. Figure 5(a) shows the frequency difference between UTC(NICT) and the H-maser obtained by TWCF and the GPS CP. The ionosphere delay was not compensated in the TWCF result. The data rate of GPS CP was every 120 s. There was a frequency jump at MJD 57212 because we manually changed the frequency of the H-maser at the Okinawa site. Both results show good agreement before and after the frequency change. The mean values are summarized in Table 1 and agree within 3×10^{-15}. The frequency stability, before the frequency change, is depicted in Figure 5(b). The short-term stability was limited by the poor stability of the H-maser. The bump at approximately 1000 s is attributed to the phase variations in

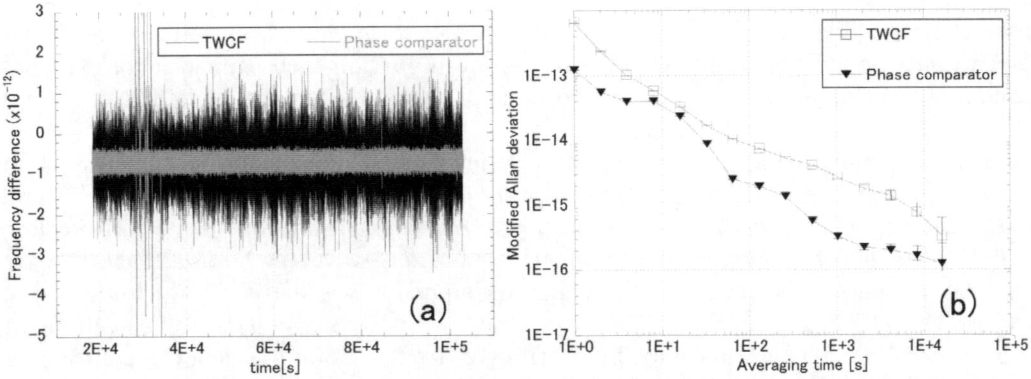

Figure 4. Results of the frequency-difference measurement with zero baseline by TWCF and using the phase comparator. Frequency difference of UTC(NICT)-H-maser (a) and modified Allan deviations (b).

the U/C and D/C, which are caused by the variation of the indoor temperature of 2 °C at the Okinawa site. As shown by this result, we confirm that the TWCF measurement was consistent with the result obtained by GPS CP and that the stability was higher.

Table 1. Means of frequency difference before and after the frequency change of the H-maser.

Frequency difference (10^{-15})	GPS CP	TWCF	Δ
Before	456.6	457.3	-0.7
After	-177.8	-180.8	3.0

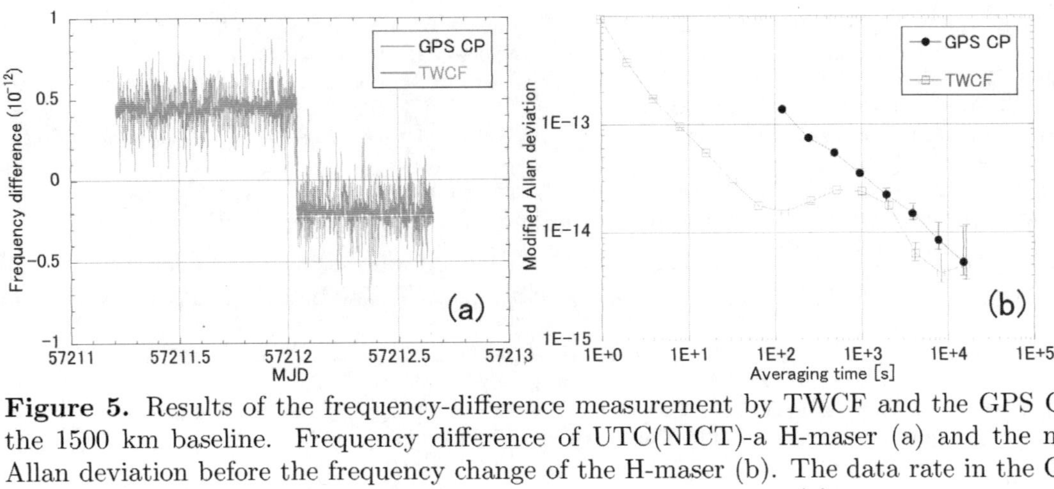

Figure 5. Results of the frequency-difference measurement by TWCF and the GPS CP with the 1500 km baseline. Frequency difference of UTC(NICT)-a H-maser (a) and the modified Allan deviation before the frequency change of the H-maser (b). The data rate in the GPS CP is every 120 s. The 1 s data of TWCF are averaged over 100 s in (a).

3.4. Discussion

We demonstrated that TWCF enables higher frequency-transfer stability than conventional TWSTFT. However, the measurement accuracy is also an important factor in clock comparison. We compared our measurement results with others obtained by a phase comparator and a GPS CP. As a result, it is considered likely that the TWCF accuracy in fractional frequency uncertainty will remain at the low 10^{-15} level. To obtain the results shown in this paper, instruments other than the modem were commonly used in both TWCF and TWCP measurements. Therefore, the modem appears to have a limitation. For confirmation, we measured the frequency difference between two continuous-wave (cw) signals at 70 MHz generated by the modems. The resultant frequency difference was 1×10^{-12}, which can cause a frequency error of 10^{-5} Hz in the uplink frequency, which corresponds to measurement inaccuracy at the 10^{-15} level. This may be attributed to the poor frequency resolution of the modem. We will investigate this by carrying out further measurements.

4. Conclusions

Conventional TWSTFT has been carried out by code phase measurement. Carrier-phase measurement is one of the ways to improve the measurement resolution without increasing the satellite-link cost. We introduced two possible methods for carrier-phase measurement: direct carrier-phase detection (TWCP) and the use of carrier-frequency information (TWCF). The former was enabled using an AWG and an A/D sampler and the latter was performed by employing a modem used in conventional TWSTFT. We confirmed that both TWCP and TWCF gave us higher stability than that obtained by conventional TWSTFT. However, the measurement accuracy of TWCF using the modem appears to remain at the 10^{-15} level owing to the lack of the frequency resolution, and improvement of the hardware may be necessary to reach a frequency stability of 10^{-16}. On the other hand, we showed the high potential to achieve a frequency stability of 10^{-17} by TWCP. To eliminate as much of the error source as possible such as that in common-clock measurement, careful attention should be paid to the environment of the outdoor and indoor instruments.

References

[1] Ushijima I, Takamoto M, Das M, Ohkubo T and Katori H 2015 *Nat. Photonics* **9** 185
[2] Droste S, Ozimek F, Udem Th, Predehl K, Hansch T W, Schnatz H, Grosche G and Holzwarth D 2013 *Phys. Rev. Lett.* **111** 110801
[3] Petit G, Kanj A, Loyer S, Delporte J, Mercier F and Perosanz F 2015 *Metrologia* **52** 301
[4] Kirchner D 1999 *The Review of Radio Science* (Oxford University Press) chapter 2 pp 27-44
[5] Bauch A, Achkar J, Bize S, Calonico D, Dach R, Hlavac R, Lorini L, Parker T, Petit G, Piester D, Szymaniec K and Uhrich P 2006 *Metrologia* **43** 109
[6] Fujieda M, Gotoh T, Nakagawa F, Tabuchi R, Aida M and Amagai J 2012 *IEEE Trans. Ultrason. Ferroelectr. Freq. Control* **59** 12 2625
[7] Kondo T, Koyama Y, Ichikawa R, Sekido M, Kawai E and Kimura M 2008 *J. Geodetic Soc. Jpn.* **54** 233
[8] Fujieda M, Piester D, Gotoh T, Becker J, Aida M and Bauch A 2014 *Metrologia* **51** 253
[9] Fonville B, Matsakis D, Pawlitzki A and Schaefer W 2004 *Proc. 36th Annual Precise Time and Time Interval (PTTI) Meeting* (Washington, DC) p 149
[10] Kanj A, Achkar J and Rovera D 2012 *Proc. European Frequency and Time Forum* (Gothenburg, Sweden) p 477
[11] Fujieda M, Ido T, Hachisu H, Gotoh T, Takiguchi H, Hayasaka K, Toyoda K, Yonegaki K, Tanaka U and Urabe S 2015 submitted to *IEEE Ultrason. Ferroelectr. Freq. Control*
[12] Gotoh T, Fujieda M and Amagai J 2007 *Proc. IEEE international Frequency Control Joint Symposium with the 21st European Frequency and Time Forum* (Geneva, Switzerland) p 1188

Journal of Physics: Conference Series **723** (2016) 012037 doi:10.1088/1742-6596/723/1/012037

A direct comparison between two independently calibrated time transfer techniques: T2L2 and GPS Common-Views

G D Rovera[1] , M Abgrall[1], C. Courde[2], P Exertier[2], P Fridelance[3], Ph Guillemot[4], M Laas-Bourez[2], N Martin[2], E Samain[2], R Sherwood[5], J-M Torre[2] and P Uhrich[1]

[1] LNE-SYRTE, Observatoire de Paris, PSL Research University, CNRS, Sorbonne Universités, UPMC Univ. Paris 06, 61 avenue de l'Observatoire, 75014 Paris, France

[2]Geoazur, CNRS, Observatoire de la Côte d'Azur, UNS, 06560 Sophia-Antipolis

[3]Phusipus Intégration, Saint Vallier, France

[4]CNES French Space Agency, Paris, France

[5]NERC Space Geodesy Facility Herstmonceux, United Kingdom

E-mail: daniele.rovera@obspm.fr

Abstract. We present a direct comparison between two satellite time transfer techniques on independently calibrated links: Time Transfer by Laser Link (T2L2) and Common-Views (CV) of satellites from the Global Positioning System (GPS) constellation. The GPS CV and T2L2 links between three European laboratories where independently calibrated against the same reference point of the local timescales. For all the links the mean values of the differences between GPS CV and T2L2 are equal or below 240 ps, with standard deviations below 500 ps, mostly due to GPS CV. Almost all deviations from 0 ns are within the combined uncertainty estimates. Despite the weak number of common points obtained, due to the fact that T2L2 is weather dependent, these results are providing an unprecedented sub-ns consistency between two independently calibrated microwave and optical satellite time transfer techniques.

1. Introduction

The most common time transfer technique used for remote time scale comparisons is based on Common-Views (CV) of satellites from the Global Positioning System (GPS) constellation, one of the Global Navigation Satellite Systems (GNSS) currently in operation. Time Transfer by Laser Link (T2L2) is an optical two-way technique developed by the laboratory Géoazur of Observatoire de la Côte d'Azur (OCA), Valbonne, France, with the support of the French Space Agency CNES.

During Autumn 2013, a mobile laser ranging station called FTLRS, developed by OCA, was implemented in Observatoire de Paris (OP), Paris, France, for tests on time and frequency transfer in parallel with GPS CV. To this aim, FTLRS was connected to the active H-maser used as local oscillator for the LNE-SYRTE Primary Frequency Standards, LNE-SYRTE being the OP laboratory designated by the French National Metrology Institute (NMI) Laboratoire National de Métrologie et d'Essais (LNE) for Time and Frequency Metrology activities. OP was then included in the regular set of T2L2 sessions, and it was possible to build a direct comparison between GPS CV and T2L2. For that purpose, a double campaign for the characterization of the link offsets was organized by LNE-SYRTE for the GPS CV, and by OCA for T2L2. An additional third station was included: NERC Space Geodetic Facility (SGF) in Herstmonceux, United Kingdom. In the three stations both time transfer techniques where using a reference time scale materialized by a 1 PPS pulse coherent with a 10 MHz signal. The time scale at OP and SGF was a free running H-maser, while in OCA the H-maser signal was steered by a micro phase stepper, in order to keep the local time scale close to Coordinated Universal Time (UTC). For the scope of this experiment all the measurements have been made by taking as common reference marker of the time scale the crossing of the 1 V level of the 1 PPS signal on a well defined connector of a 1 PPS signal distribution amplifier.

2. T2L2 link

The T2L2 principle is derived from classical satellite laser telemetry techniques, with a dedicated on-board instrumentation able to time tag the laser pulses reaching the satellite [1]. Measurement are timing triplets based on the time of transmission of a laser pulse from the ground, the time of arrival of the pulse on-board the satellite, and the time of arrival on the ground of the photons from that given pulse reflected by the satellite. The time transfer between two stations is obtained from the combination of such triplets, after computation of a fit for each satellite path producing one normal point of the time scale difference. An improvement of at least one order of magnitude in time transfer uncertainty over the existing microwave techniques is expected, the stability of this optical technique provides the capability to compare today's most accurate frequency standards.

T2L2, based on a ground network of laser ranging stations and on a dedicated space segment, is currently in its operational phase since the launch in 2008 of the dedicated payload hosted by the JASON-2 satellite [2, 3, 4].The full accuracy of a T2L2 link can only be achieved when JASON-2 is visible at the same time from both the ground station of the link. Some attempt for non common view time transfer has been made. But in this case the accuracy is strongly limited by the stability of the on-board quartz oscillator and does not present a noticeable improvement against traditional time transfer techniques.

The years 2012-2013 were especially devoted to T2L2 accuracy estimation. A first ground to ground validation was achieved by a short baseline experiment [5]. A deep analysis of the instrumental bias has also been performed [6], and a ground to ground time transfer accuracy budget was established [7]. For a typical T2L2 time transfer link the extended uncertainty ($k = 2$) is estimated at 138 ps. Although many laser ranging station participate to T2L2 experiment we only consider here the links between OCA, SGF and OP, because the GPS links have been calibrated for these three stations only. One of the goals of the T2L2 experiment, which is weather dependent, is indeed the one-off calibration of other time transfer techniques.

3. GPS CV link

In each station, a GPS receiver collects measurements of the differences between the local time scale and a satellite clock, which can be referred to GPS Time, the common satellite system time scale. Measurements can be based on P-code data broadcast on both GPS microwave carriers, allowing for the ionosphere-free combination based on the software developed at the Royal Observatory of Belgium (ROB) and made available by the Bureau International des Poids et Mesures (BIPM) that we call TAIP3 here [8].

The data processing consists in averaging outliers out [9], in order to build a time difference for each standard epoch according to the Common Generic GNSS Time Transfer Standard (CGGTTS) format [10]. Time transfer between remote time scales is obtained by computing the CV differences between similar CGGTTS data recorded at the same epochs. Another geodetic technique applied to time transfer between remote time scales, the Precise Point Positioning

(PPP), as was developed by National Resources Canada (NRCan) [11] among other groups, provides a better short term stability because measurements are based on the phase of the carrier frequencies. But the carrier does not contain any identifiable time marker, and therefore the time transfer uncertainty remains limited by the detection capability of the time tag associated with the code. In addition, the short term stability is not the issue in the comparison with another time transfer technique. This is why we disregard PPP in the following, any GPS time transfer being limited by the same uncertainty on the determination of the hardware delays.

4. Direct comparison between both techniques

During the comparison period, all GPS stations have operated almost continuously. By using the offsets between remote time scale reference points obtained from the GPS calibration campaign, the time differences between the time scales have been computed by GPS CV for the three links OCA-OP, OCA-SGF and OP-SGF. The classical GPS CV are 16 min sampled. But we also computed a filtered GPS time transfer by a moving average over 13 consecutive CV samples. This is providing 16 min sampled 3.5 h averaged points exhibiting a lower noise without smoothing out artificially physical effects included in the data. On the other hand, the T2L2 normal points are providing the time differences between two remote time scales at epochs depending on the satellite path and on the available triplets.

For each link, we have plotted both GPS CV time transfer results, together with the T2L2 results. For clarity of the figures, an additional quadratic fit was also removed, in order to take into account the H-maser differential drifts: the parameters of the fit are provided inside each figure. GPS CV are producing 16 min sampled time scale differences which are not necessarily synchronous with any T2L2 normal point epochs. For the computation of the differences between both techniques, in order to detect any potential deviation, two adjacent GPS CV points have been interpolated to produce one GPS CV result synchronous with the corresponding T2L2 normal point epoch. The combined uncertainties were computed by building a simple quadratic sum between the GPS CV link uncertainties given in [12], and the ground to ground T2L2 uncertainty given in [7]. We obtain 1.1 ns for OCA-OP and OCA-SGF, and 0.8 ns for OP-SGF.

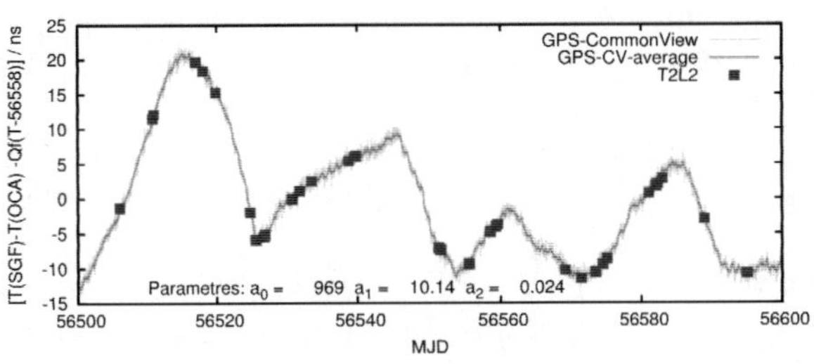

Figure 1. Time differences between SGF and OCA by GPS CV and T2L2 with a quadratic fit removed (colour online).

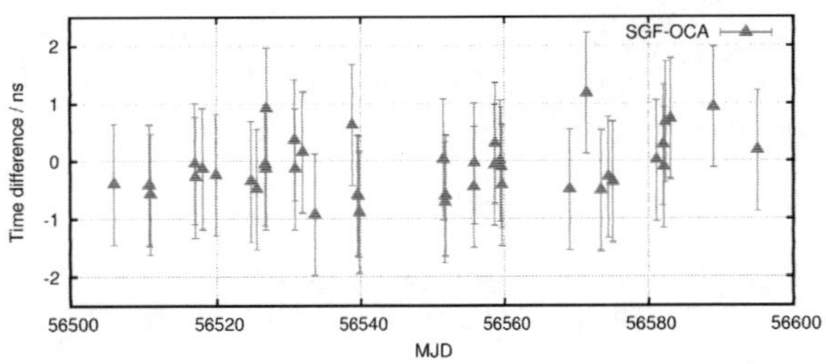

Figure 2. Difference between both techniques on the link SGF-OCA. The error bars represent the combined uncertainty (colour online).

Figure 1 shows such time scale differences issued from both techniques on the link SGF-OCA: the most noisy continuous green line shows the GPS CV 16 min sampled time transfer, the smoothed continuous red line shows the additionally 3.5 h averaged GPS CV time transfer, and the blue squares are the T2L2 time differences. The deviations between both techniques for the links SGF-OCA are reported in Figure 2.

Similarly, the time differences between the time scales and the differences between both techniques are reported in Figure 3 and 4 for the link SGF-OP and in Figure 5 and 6 for

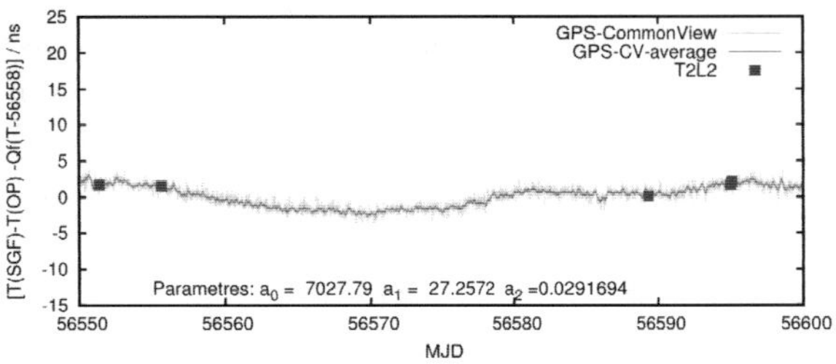

Figure 3. Time differences between SGF and OP by GPS CV and T2L2 with a quadratic fit removed (colour online).

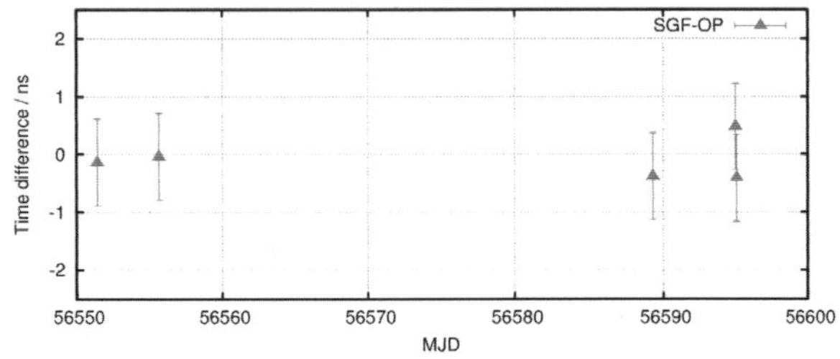

Figure 4. Difference between both techniques for the link SGF-OP. The error bars represent the combined uncertainty (colour online).

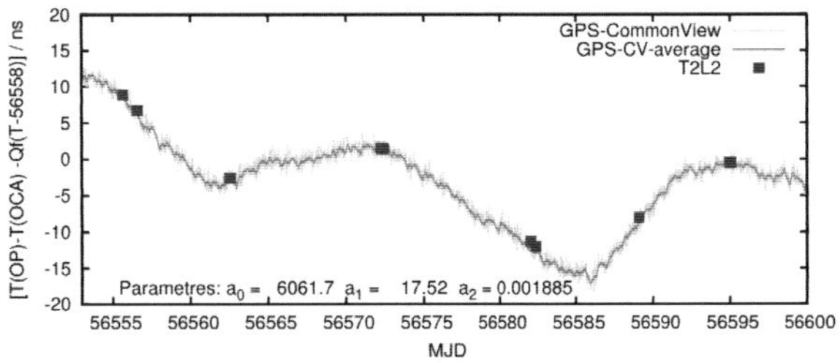

Figure 5. Time differences between OCA and OP by GPS CV and T2L2 with a quadratic fit removed (colour online).

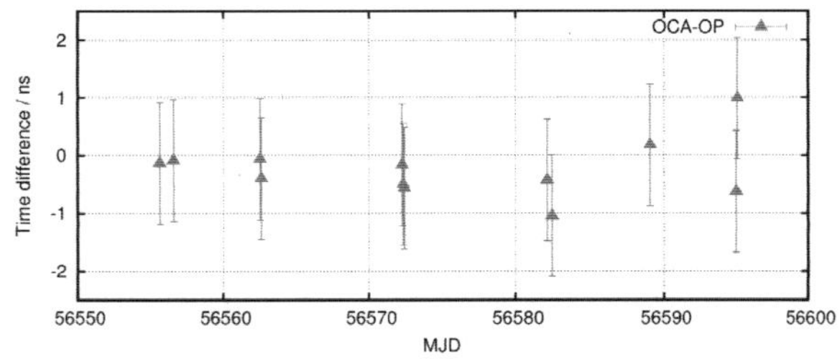

Figure 6. Difference between both techniques for the link OCA-OP. The error bars represent the combined uncertainty (colour online).

the link o OCA-OP. Note that in the plots including OCA, the shape of the curves is due to the manual steering of the H-maser in order to keep the local time scale close to UTC. One can see that there is only a weak number of T2L2 points on the plots. This is mostly due to the fact that T2L2 is weather dependent, but also to some technical issues which limited the measurements for some links. Clearly, the OCA-SGF link is exhibiting the largest number of common points compared to the two other links. But we also see in Figure 1, 3 and 5 that the discrepancy between both independently calibrated techniques remains remarkably small over the whole period of measurements for all the links.

When properly calibrated, the two techniques are measuring the same time deviations between remote time scale reference points, hence the difference is expected to be close to 0 ns. What we see in Figure 2, 4 and 6 is more accurately the remarkable consistency between both techniques.

The mean values of the three data sets are reported in Table 1, together with the standard deviations and the combined uncertainties. We consider here the standard deviations obtained from the 16 min sampled GPS CV and the ones obtained after the additional smoothing on 13 consecutive samples. We see that the average differences between GPS CV and T2L2 are equal or below 240 ps. Despite the weak number of common points obtained, this is clearly showing a largely sub-ns consistency between the two independently calibrated techniques: all deviations of the mean values from 0 ns are within the combined uncertainties. In addition, we do not detect any significant trend nor periodic term, even when considering the OCA-SGF link exhibiting the highest number of data. When considering this link only, where the number of available points seems significant enough to build a consistent mean value of the differences together with the related standard deviation, we can see that the moving average on GPS data reduces the noise

Table 1. Mean values of the time differences between GPS CV and T2L2.

Link	Number of points	Mean value (ns)	Standard Deviation (ns)		Combined Uncertainty (ns)
			Raw GPS CV	Filtered GPS CV	
OCA-OP	12	0.24	0.48	0.25	1.1
OCA-SGF	42	0.09	0.49	0.37	1.1
OP-SGF	5	0.10	0.32	0.32	0.8

significantly. But, as expected, the GPS CV are setting not only the noise level but also the uncertainty for such data.

5. Conclusion

For the two techniques GPS CV and T2L2, two independent relative calibration campaigns were achieved in Autumn 2013 between the three European stations OCA, OP and SGF. By using the corrective offsets obtained for each link between remote time scale reference points, common to both microwave and optical satellite techniques, a direct comparison provided an unprecedented consistency at a sub-ns level. The mean values of the time differences between GPS CV and T2L2 are equal or below 240 ps, with a standard deviation below 500 ps, mostly due to GPS CV. The resulting average differences are clearly negligible with respect to the combined uncertainties between both techniques, which are between 0.8 and 1.1 ns depending on the link considered. Although the estimated uncertainty of T2L2 is much lower than the estimated uncertainty of GPS CV, the good agreement between the two techniques confirms that the independent link calibrations have been carried out properly and that the uncertainty budgets have been established in a rigorous way, even when conservative. This result provides a full validation of the GPS CV link uncertainty budget computations, and a confirmation of the T2L2 performances for ground to ground time transfer . It is also a demonstration of the T2L2 capability for the sub-ns calibration of other time transfer techniques. The decommissioning of the space segment of T2L2, which was considered in 2015, has been postponed, and the operation extended at least until the end of 2016, allowing for further measurements. It should also be noted that a similar instrument called European Laser Timing (ELT) [13] will be hosted on-board the International Space Station (ISS) inside the Atomic Clock Ensemble in Space (ACES) module [14] . The launch of the ACES module is currently planned for early 2017. As was proven by this experiment, optical satellite techniques like T2L2 should be kept as possible candidates for the validation of long distance time transfer through optical fiber links, provided that the required portable stations be will be made available in the reference sites and that the required flight instruments would be implemented on-board additional space segments in the future. Unfortunately today there are no laboratories that are equipped with a T2L2 station and with a fiber time transfer link, therefore to benefit of the full accuracy of a T2L2 link for the validation of a fiber link [15] the deployment of two mobile T2L2 stations is required.

References

[1] Fridelance P, Samain E and Veillet C 1997 *Experimental Astronomy* **7** 191–207 ISSN 0922-6435 URL `http://dx.doi.org/10.1023/A3A1007982512087`
[2] Samain E, Weick J, Vrancken P, Para F, Albanese D, Paris J, Torre J M, Zhao C, Guillemot P and Petibon I 2008 *International Journal of Modern Physics D* **17** 1043–1054

(*Preprint* http://www.worldscientific.com/doi/pdf/10.1142/S0218271808012681) URL http://www.worldscientific.com/doi/abs/10.1142/S0218271808012681

[3] Samain E, Exertier P, Guillemot Ph, Laurent Ph, Pierron F, Rovera G D, Torre J M, Abgrall M, Achkar J, Albanese D, Courde C, Djeroud K, Bourez M, Leon S, Mariey H, Martinot-Lagarde G, Oneto J L, Paris J, Pierron M and Viot H 2011 Time transfer by laser link T2L2: Current status and future experiments (invited) *Frequency Control and the European Frequency and Time Forum (FCS), 2011 Joint Conference of the IEEE International* pp 1 –6 ISSN 1075-6787

[4] Exertier P, Samain E, Martin N, Courde C, Laas-Bourez M, Foussard C and Guillemot Ph 2014 *Advances in Space Research* **54** 2371 – 2385 ISSN 0273-1177 URL http://www.sciencedirect.com/science/article/pii/S0273117714005274

[5] Laas-Bourez M, Courde C, Samain E, Exertier P, Guillemot Ph, Torre J M, Martin N and Foussard C 2015 *Ultrasonics, Ferroelectrics, and Frequency Control, IEEE Transactions on* **62** 255–265 ISSN 0885-3010

[6] Samain E, Vrancken P, Guillemot P, Fridelance P and Exertier P 2014 *Metrologia* **51** 503 URL http://stacks.iop.org/0026-1394/51/i=5/a=503

[7] Samain E, Exertier P, Courde C, Fridelance P, Guillemot Ph, Laas-Bourez M and Torre J M 2015 *Metrologia* **52** 423 URL http://stacks.iop.org/0026-1394/52/i=2/a=423

[8] Defraigne P and Petit G 2003 *Metrologia* **40** 184 URL http://stacks.iop.org/0026-1394/40/i=4/a=307

[9] Rovera G D, Chupin B, Abgrall M and Uhrich P 2013 A simple computation technique for improving the short term stability and the robustness of GPS TAIP3 common-views *European Frequency and Time Forum International Frequency Control Symposium (EFTF/IFC), 2013 Joint* pp 827–830

[10] Defraigne P and Petit G 2015 *Metrologia* **52** G1 URL http://stacks.iop.org/0026-1394/52/i=6/a=G1

[11] Kouba J and Héroux P 2001 *GPS Solutions* **5** 12–28 ISSN 1080-5370 URL http://dx.doi.org/10.1007/PL00012883

[12] Rovera G, Abgrall M, Courde C, Exertier P, Fridelance P, Guillemot Ph, Laas-Bourez M, Martin N, Samain E, Sherwood R, Torre J M and Uhrich P 2014 A direct comparison between two independently calibrated time transfer techniques: T2l2 and gps common-views *Precision Electromagnetic Measurements (CPEM 2014), 2014 Conference on* pp 666–667 ISSN 0589-1485

[13] Schreiber K, Prochazka I, Lauber P, Hugentobler U, Schafer W, Cacciapuoti L and Nasca R 2010 *Ultrasonics, Ferroelectrics, and Frequency Control, IEEE Transactions on* **57** 728–737 ISSN 0885-3010

[14] Salomon C, Cacciapuoti and Dimarq N 2007 *International Journal of Modern Physics D* **16** 2511–2523 (*Preprint* http://www.worldscientific.com/doi/pdf/10.1142/S0218271807011681) URL http://www.worldscientific.com/doi/abs/10.1142/S0218271807011681

[15] Lopez O, Kanj A, Pottie P E, Rovera G D, Achkar J, Chardonnet C, Amy-Klein A and Santarelli G 2013 *Applied Physics B-lasers and Optics* **110** 3–6

Perspectives of Time and Frequency Transfer via Satellite

W. Schäfer, T. Feldmann

TimeTech GmbH
D-70563 Stuttgart, Germany

Wolfgang.Schaefer@timetech.de

Abstract. This paper reviews key characteristics and requirements for Two-Way Satellite Time and Frequency Transfer (TWSTFT) links. The outline and perspective of two concepts are described: 1. The classical bend-pipe transponder, without special equipment on board, with the potential of 10^{-17}/day frequency transfer uncertainty; 2. The ACES-MWL (Atomic Clock Ensemble in Space Microwave Link) like active on-board equipment, with the potential down to few 10^{-19}/day, using both high-rate modulation and carrier-phase. An outlook to advanced links is given. The aim of these proposed links is to provide a unique platform for the demanding requirements in the field of time and frequency metrology and to support improved orbit determination and relativistic geodesy for the next decades.

1. Introduction

Two-Way Satellite Time and Frequency Transfer TWSTFT is in regular use since the early 1980s, mostly using a modem, which has originally been developed at the Technical University (TU) in Berlin [1]. During its introduction, its stability and accuracy was roughly 1 to 2 orders of magnitude better than existing clocks, whereas today it lacks performance by few orders compared to modern clocks. This is due to the dramatic improvement of microwave frequency standards and optical clocks over the last decades.

Today, TWSTFT is mainly used to link metrological institutes in US, Asia and Europe in support of the generation of Coordinated Universal Time UTC [2]. It reaches a 10^{-15} uncertainty level [7] with modest modulation rates of 1 to 2.5 MChip/s. Ground station networks use the CDMA (Code Division Multiple Access) properties of Pseudorandom Noise (PN) Codes to exchange signals via a single regional geo-stationary transponder. Each station determines the time-of-arrival of signals from the remote sites. The difference between simultaneous readings of a link provides the wanted clock offset [3]. The main advantage of TWSTFT is still seen in the capability to cover intercontinental distances and being a true alternative to links based on Global Navigation Satellite Systems (GNSS). Although ground optical fibers have a promising perspective both for frequency and time transfer [9], they are still limited to continental distances and only very few optical fiber links between selected institutes are available.

Hardware with improved performance has been developed for the ACES (Atomic Clock Ensemble in Space) mission [10], i.e. the ACES MicroWave Link (MWL) [11]. ACES will embark a cold atom clock (Pharao) and an active hydrogen maser on board the International Space Station (ISS), to be launched in 2017. The on-board clocks are compared to ground using the MWL instrument at performance levels in the 10^{-17} region of frequency uncertainty. The performance improvement has been achieved by carrier phase measurements, with phase continuity from pass to pass. Results of laboratory tests show, that the engineering model meets the ACES mission performance requirements.

This renders the ACES-MWL architecture a candidate for even higher performance missions in the near future. Preliminary link simulations for the proposed STE-QUEST mission [12] suggest, that an uncertainty level of 10^{-18} is within reach for ground-to-space links, while ground-to-ground links are less critical and laboratory measurements show that few 10^{-19} are possible.

There are exciting new microwave technologies [15], which suggest a further growth potential using very high modulation rates in the GChip/s region and Extremely High Frequencies (EHF, 30 GHz to 300 GHz), but still providing tolerance to attenuation caused by fog and clouds [16].

An outlook is given for optical free-space communication links [17], currently explored mainly in space-to-space configuration, but which might become relevant for space-to-ground applications as well. The latter is limited to favorable weather conditions, however.

2. Historical Background

2.1. Basic Principles of Remote Time and Frequency Transfer

TWSTFT [3] is based on the principles of coherent communication, which was - at the time of development - well known from pioneering work at JPL in the frame of deep space links [4]. Two simultaneous emissions in bi-directional configuration allow suppression of most link-induced errors. Systems developed so far are MITREX [1] (1977), SATRE [5] (1990), PRARE [6] (1983, for geodetic applications) and the ACES-MWL [11] (2003). In brackets, year of start of development.

The overall performance of such systems is primarily determined by the noise characteristics of the delay-lock tracking loops (DLL), which are used to re-construct the phase of the modulation of the received signal by correlation with a locally generated replica of the received PN signal. In the short-term, i.e. for integration times up to approximately 100 s, thermally-induced white phase noise dominates, averaging well with time. In the medium-term to 1000s, flicker phase noise is limiting, which is best overcome with high modulation rates. In the long-term above 1000s, environmental effects, in particular temperature, impact the performance, which require a well-engineered design of all elements involved.

2.2. Key concepts of current hardware implementations

Some key concepts have been developed already in the 1980s at TU Berlin, including

- Use of PN coded signals for ambiguity resolution in favor of tone-based ranging systems;
- Binary Phase Shift Keying (BPSK) modulation with fully suppressed carrier for best efficiency;
- Use of truncated maximum-length PN codes to fit into a decade system for 1pps generation;
- Use of non-coherent DLL to become independent from potential carrier instabilities, including those from Doppler and from superposed data modulation;
- Use of a PN phase shift modulation to add 1pps markers without need for carrier recovery;

These key elements are still present in actual designs, while most performance improvements have been achieved since MITREX by increasing the modulation bandwidth, by using modern wide-band electronics with high intrinsic delay stability and by coherent carrier recovery and demodulation. The latter allows for precise Doppler measurements and data transmission.

2.3. Standard Two-Way Satellite Time and Frequency Transfer (TWSTFT) and Ranging

TWSTFT is based on simultaneous bi-directional exchange of signals through geostationary communication satellites, using Ku- and X-band frequencies. There is no special equipment on board. Most links are via commercial satellites, for which transponder bandwidth has to be paid for. For cost reasons, low bandwidth signals at 1 MChip/s modulation are used in the metrological networks for links between US and Europe and within Europe, the Asia to Europe links operate at a more favorable 2.5 MChip/s rate. These links contribute to the generation of UTC by the BIPM with regularly scheduled operations [8].

At chip rates of roughly 250 MChip/s, TWSTFT has the potential of 10^{-17} / day frequency transfer uncertainty and a systematic time transfer uncertainty of well below 1 ns. However, due to cost reasons, such links would be limited to dedicated experiments.

A good example for higher precision is ranging operation at 20 MChip/s, which exhibits 2 cm ranging noise via geo-stationary satellites at C/No of 43 dBHz, even when the transponder is loaded with regular TV-signals and the ranging signal is 30 dB below main traffic in CDMA mode. For ranging, the round-trip signal delay is measured, where the link delay does not cancel as in the bi-directional case (see Fig 1.)

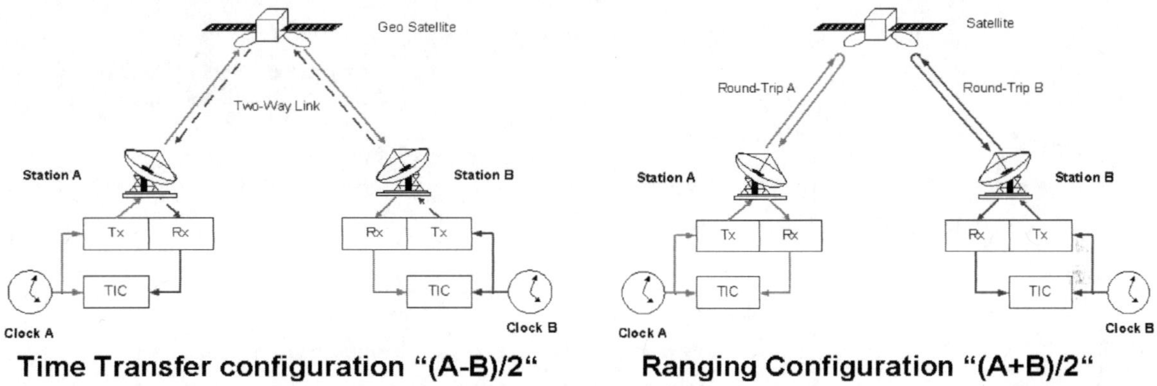

Figure 1. Bi-directional time and frequency transfer (left) and satellite ranging (right)

It is interesting to note, that ranging and TWSTFT can be performed simultaneously and with identical equipment. This allows the combination of ranging for geodesy and time transfer for metrology.

3. Active Systems on Satellites

In the case of commercial transponders, the unknown, free-running local oscillator required for signal translation does not allow for unambiguous use of the signal carrier. A 2-way carrier phase (2-WCP) configuration was designed [13], mainly to explore the significantly higher precision available from the carrier, but 2-WCP lacks strict coherency between modulation and carrier. Hence this method is limited to frequency transfer, as it does not allow unambiguous recovery of the carrier phase after signal loss. To add full carrier phase capability with phase continuity after signal loss, active on-board systems have been developed, which use a fixed phase relationship between modulation and carrier.

3.1. PRARE on-board ERS-2 and ACES MWL

The Precise Range and Range Rate Equipment (PRARE) [6] was actually a geodetic instrument based on time-interval metrology. It was intended to combine the best available concepts at that time, including dual band delay measurements and carrier phase data. It was the first ranging system, where the measurement was initiated at the satellite. It operated 12 years successfully on board of ESA's environmental satellite ERS-2 in support of the main radar altimeter instrument, providing high precision orbital data. Round-trip range measurements were in X-band to up to four ground stations simultaneously with CDMA and a PN-code rate of 10 MChip/s. An additional S-band downlink was used to determine the first order ionosphere delay. Ranging noise using modulation was 0.9 cm (corresponding to 54 ps round-trip delay noise) and the velocity noise using carrier Doppler was 0.015 mm/s respectively at 15 s integration intervals. (data courtesy GFZ, Potsdam).

PRARE formed the baseline for the ACES Microwave Link (MWL), which shall be remarkably better than usual GNSS and TWSTFT techniques. Its performance is more than 10 times improved with respect to PRARE and its layout on ground is optimised for bi-directional operation rather than for round trip ranging. Ground stations feature a dedicated design, with antenna radome and well

Journal of Physics: Conference Series **723** (2016) 012038 doi:10.1088/1742-6596/723/1/012038

controlled temperature for the front-end and electronics. The MWL will be installed as part of ACES on the ISS. It uses mainly the bi-directional Ku-band signals, with an auxiliary S-band downlink. Modulation rates are 100 MChip/s in Ku-band and 2.5 MChip/s in S-band (Figure 2, left).

The fully coherent architecture is optimized for unambiguous carrier phase measurements, where the modulation is used mainly for time transfer, synchronisation and carrier cycle identification. Ionospheric delay is accounted for by differential delay in Ku vs S-band and removal of ionospheric variations is by differential carrier phase. Ground tests have demonstrated stability of 2 x 10^{-16} for the modulation and few parts in 10^{-17} for the carrier at 10000 s integration time (Figure 2, right).

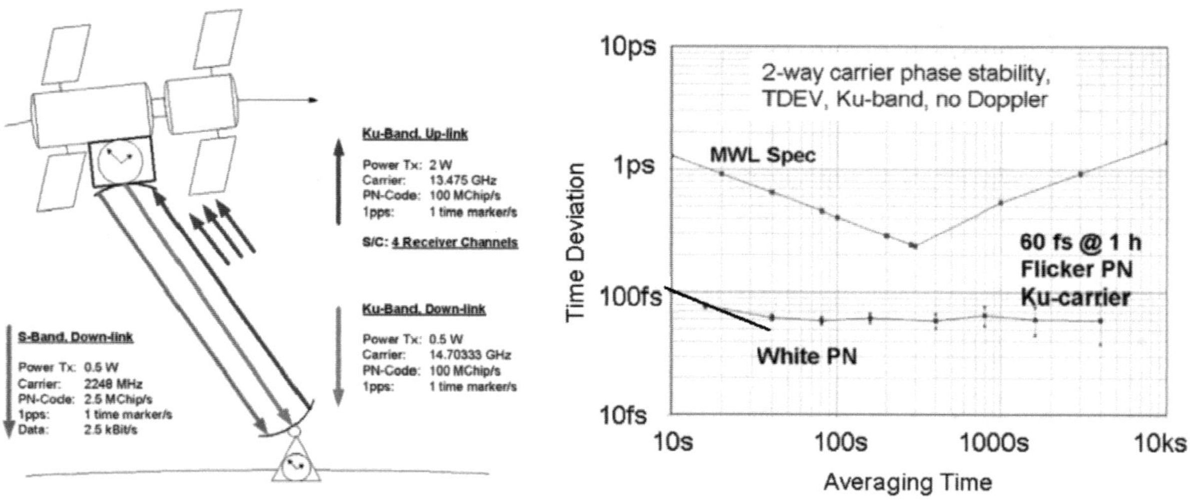

Figure 2. ACES signal characteristic (left) and 2-way laboratory test results compared to specifications.

3.2. *MWL design for HEO mission*

A new Microwave Link is currently researched for satellite missions on Highly Elliptical Orbits (HEO) with apogee of about 50000 km and perigee of about 900 km, such as the proposed STE-QUEST [12] mission. Intended frequency stability is 10^{-18} at 100000 s averaging time. A systematic frequency uncertainty 5 x 10^{-19} at 1 week is required for ground-to-ground clock comparisons involving two ground stations. Tentative operation is in Ka-band with modulation at 250 MChip/s and X-band as auxiliary downlink. Key concepts from the ACES-MWL will be re-used, and new components for Ka-band are under development. End-to-end simulations show feasibility to meet these requirements. The major challenges here are potential systematic frequency errors due to relative acceleration between space and ground.

4. A Possible Path to the Future: Laser Communication Terminals (LCT)

Laser communication technology can be regarded as the logical evolution of microwave links. Being bi-directional by design [17] to fulfill strict pointing requirements for the optical beams, they feature very high bit-rate communication and thus positive signal-to-noise ratio. This shows their very promising capabilities, while retaining the bi-directional architecture, at significantly increased signaling rate and an improved link budget. Typical C/No is around 100 dBHz, with BPSK at Gbit/s rates, which results in 4 to 5 orders of magnitude improvement compared to ACES-MWL modulation performance. Even if the optical carrier itself would not be used for metrology, measurements on the modulation alone are very attractive, providing inherent ambiguity removal. Operation under very high Doppler and Doppler-rate has been successfully demonstrated e.g. between two Low Earth Orbiting (LEO) satellites TerraSAR-X and NFIRE, with 5.625 Gbit/s data rates [18].

Journal of Physics: Conference Series **723** (2016) 012038 doi:10.1088/1742-6596/723/1/012038

It is expected, that optical communication link technology will further evolve in the near future, at the same rate as the demand for link capacity is ever increasing. The next step has been achieved for bi-directional links between GEO and LEO satellites at 1.8 Gbit/s data rates [19] and further reductions in complexity and cost are likely for this novel concept.

Current LCT hardware does not yet implement precise time-of-arrival measurements, but the perspective is clearly visible, once such additional capabilities would be incorporated.

Apart from the coherent and continuous LCT designs, pulsed lasers like the ELT (European Laser Timing) [20] as part of ACES payload on the ISS are promising even under light cloudy weather conditions. Their stability performance is not yet considered comparable to continuous links. The limitation is mainly by the available measurement rate. However, pulsed laser links with quasi-instantaneous operation are important means for the calibration of continuous links, both for ranging using the established Satellite Ranging Method (SLR) or for time transfer, using ELT-like configurations. With clocks being synchronized by TWSTFT, one-way laser ranging and time-transfer becomes feasible, which is not limited in distance by on-board laser reflectors.

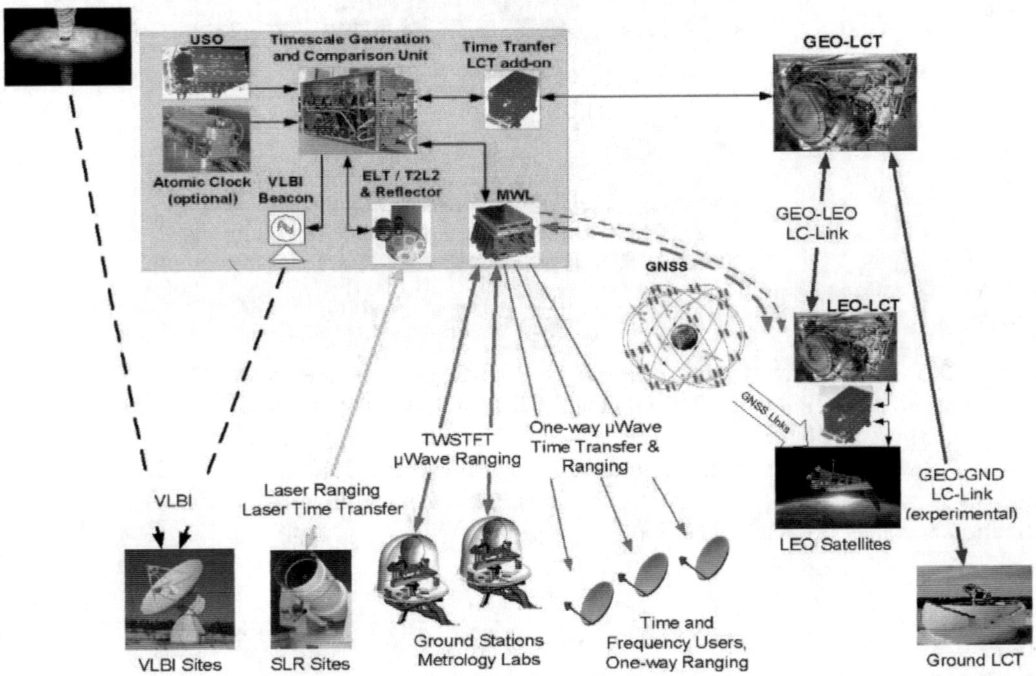

Figure 3. Candidate GETRIS architecture.

5. Geodesy and Time Reference in Space: The GETRIS Concept

The GETRIS concept (Figure 3) has been proposed to fulfill demanding time transfer needs as well as geodetic purposes. It is a geostationary system with sensor co-location on board and on ground. This allows for optimum combination and comparison of different technologies, mainly microwave and optical, both pulsed and continuous. VLBI is used to link terrestrial and celestial reference frames. Optical systems will be used for dedicated experiments and for the delay calibration of microwave links, which serve as everyday workhorse under all-weather conditions. GETRIS aims to detect and solve for systematic errors, which might exist in current geodetic (DORIS [13], GNSS, SLR) and metrological systems. Three spacecrafts would be needed for earth coverage. The planned European Data Relay Satellite System (EDRSS) would be an ideal candidate to host such payload. The application of relativistic geodesy would be feasible regarding observation time and ground coverage.

5.1. Alternative Spacecraft Locations

A recent study performed for ESA showed the feasibility to place a clock at a Lagrange point between Earth and Sun, with link to Earth. Equally well, a package at Lagrange Point L1 between Earth and Moon could provide worldwide coverage with a single spacecraft only. Lagrange orbits are currently discussed as data relay for the exploration of the dark side of the Moon or, more speculative, to assemble missions to Mars.

In contrast to reflective systems, like Lunar Laser Ranging, an active system on board allows for significantly smaller Earth stations, without scarifying performance, and even one-way operation.

6. Advanced Microwave Links

Links in the EHF region (i.e. 30 to 300 GHz) are deemed to be an excellent alternative to optical space-ground links, still with high tolerance to rain and clouds. The technology for such links is available, developed for high-speed feeder links, terrestrial and space-to-ground [15], [16].

Figure 4. High gain horn antenna at 240 GHz [15]

Figure 4 shows the set-up of a propagation experiment between Aalen television tower (Germany) and a 15 km remote location using 240 GHz and with data rates of several GBit/s. The experiment has shown that the link will also work in extreme weather conditions like rain and heavy fog. Although the atmospheric attenuation increases dramatically with increased frequency [22], several "radio windows" are available, where attenuation is relatively low. With GBit/s data rates, carrier cycle identification becomes feasible also in the EHF region.

		Ground to GEO			Ground to Lagrange Point (Moon)		
Frequency	GHz	50	70	200	50	70	200
Modulation	MChip/s	250	500	2000	500	500	2000
Transmitter power	W	3	3	3	6	6	6
Satellite antenna diameter	cm	2.6	1.8	0.6	18	13	4
Ground antenna diameter	m	1.2	1.2	3.0	0.73	0.63	0.73
Receiver C/N_0	dBHz	50	50	48	49	47	49
Code jitter @ 1s	**ps**	**18**	**9**	**3**	**10**	**12**	**2.6**
Carrier jitter @ 1s	**fs**	**14**	**10**	**5**	**17**	**14**	**4**

Table 1. Link budget for EHF links to GEO and to an Earth-Moon Lagrange point

Table 1 shows example link budgets for EHF links between Geostationary Orbit (GEO) and ground at 50, 70 and 200 GHz. The atmospheric attenuation is estimated considering ITU recommendations

[22]. The satellite antenna provides full Earth coverage. The noise (jitter) on the carrier is at fs level, thus the links have the potential to reach 10^{-19} frequency uncertainty within 1 day, ground-to-ground.

Table 1 contains link budgets for the Earth-Moon Lagrange point L1 as well. These links reach similar performance levels with twice the RF power.

The links shown in Table 1 are particularly suited for time transfer in support of the generation of UTC. Key is the available bandwidth in the EHF range, which allows for chip rates in excess of 100 MChip/s. The DLL tracking loops used in current projects like PRARE (10 MChip/s) and ACES-MWL (100 MChip/s) can be adapted to the chip rates shown and allow for an unambiguous transmission of time signals, with long-term stability, i.e. weeks and months, at the ps level. Calibrated links between UTC laboratories at the 10 ps level become feasible using a transportable TWSTFT station [21].

7. Conclusion

This contribution tries to conclude on the most promising way forward for time and frequency transfer via satellite, mainly in support of a worldwide network to link clocks for the generation of UTC over inter-continental distances and to match current clock performance with room for growth. Operational links still operate at less than optimal conditions for cost reasons. The advanced link designed for the ACES mission introduces new concepts, in particular full carrier coherency and thermally stabilized ground equipment. It requires an active element on-board, in contrast to traditional bend-pipe commercial transponders.

The only way towards better performance is the use of higher frequencies, both for modulation and for the carrier. The need for ever-increasing communication bandwidth goes into the same direction, with promising results both in microwave and optical link technology.

While optical links appear to be the logical choice for best bandwidth and highest carrier frequency, the full coherency between modulation and carrier is not as obvious as in the microwave region, but more severe, weather dependency limits their use to few favorable locations. On the other hand, EHF technology is likely to overcome weather problem at similar data rates, and significant advances readily appear [15]. Drawing on the experience with ACES-MWL, carrier cycle identification even at EHF frequencies is feasible.

Hence, it is suggested to go the microwave way for the next generation of an operational system. Continuous laser technologies are certainly attractive for specialized missions and space-to-space links. Pulsed lasers are still the foundation for calibration purposes, both for timing and ranging.

Regarding orbits, LEO and HEO show significant disadvantages for an operational system, due to limited visibility and due to extreme Doppler shifts. It is thus recommended to use benign orbits, like GEO or, even more promising, at Lagrange point L1 between Earth and Moon. Link budgets show modest instrument requirements. Concepts with synergies between geodetic and metrological applications are proposed, which would make such systems attractive for a number of communities.

References

[1] Hartl P et al 1983 High Accuracy Global Time Transfer Via Geosynchronous Telecommunication Satellites with Mitrex, *Z. Flugwiss. Weltraumforsch.* **7** Heft 5

[2] Piester D et al 2011 Remote atomic clock synchronization via satellites and optical fibers, *Adv. Radio Sci.* **9** pp 1-7

[3] Kirchner D 1999 Two-Way Satellite Time and Frequency Transfer (TWSTFT): Principle, Implementation, and Current Performance, *Rev. of Radio Sci. 1996-1999* (Oxford University Press) pp. 27-44

[4] Gill W 1966 A comparison of Binary Delay-Lock Tracking-Loop Implementations, *IEEE Trans AES-***2,** *Iss 4,* pp 415-424

[5] Online: http://www.timetech.de/index.php?page=products_overview&pgraph=6&subpage=description_twstft_1 (last accessed 25 May 2016)

[6] Falck C et al 2013 Betrieb des PRARE Bodensegmentes für ERS-2, Abschlussbericht 2003, *Scientific Technical Report* **04/20** *GFZ German Research Centre for Geosciences*
[7] Bauch A et al 2006 Comparison between frequency standards in Europe and the USA at the 10^{-15} uncertainty level *Metrologia* **43** pp 109-120
[8] Bauch A et al 2011 *Rapport BIPM-2011/01* (Bureau International des Poids et Mesures, Sévres)
[9] Lopez A et al 2013 Simultaneous remote transfer of accurate timing and optical frequency over a public fiber network *Appl. Phys. B* **110** pp 3-6
[10] Salomon C et al 2001 Cold atoms in space and atomic clocks: ACES *Comptes Rendus de l'Académie des Sciences - Series IV* – Physics **2** Iss 9 pp. 1313-1330
[11] Delva P et al 2012 Time and frequency transfer with a MicroWave Link in the ACES/PHARAO mission *Proc. EFTF 2012* pp 28-35
[12] Schiller S, Tuckey P and Rasel E 2014 The STE-QUEST Mission: A space test of the Equivalence Principle in the quantum domain *ESA Cosmic Vision M3 Selection, Paris*
[13] Online: http://ids-doris.org/ (last accessed 26 Mar 2016)
[14] Fonville B et al 2004 Development of Carrier-Phase-Based Two-Way Satellite Time and Frequency Transfer (TWSTFT) *Proc PTTI* **36** pp 149-164
[15] Boes F et al 2014 Ultra-broadband MMIC-based wireless link at 240 GHz enabled by 64 GS/s DAC *Proc.* **39th** *IEEE IRMMW-THz* Tucson AZ pp 1-2
[16] Lewark U et al 2013 Link budget analysis for future E-band gigabit satellite communication links (71-76 and 81-84 GHz) *CEAS Space Journal* **4** Iss 1-4 pp 41-46
[17] Lange R 2003 online: *http://www.dlr.de/rd/Portaldata/28/Resources/dokumente/RK/lange-laser_communication_terminals.pdf* (last accessed 25 Mar 2016)
[18] Gregory M et al 2010 TESAT Laser Communication Terminal Performance Results on 5.6 GBit Coherent Inter-Satellite and Satellite to Ground Links *Int. Conf. Space Optics, Rhodes, Greece 4-8 October 2010*
[19] Zech H et al 2015 LCT for EDRS: LEO to GEO optical communications at 1,8 Gbps between Alphasat and Sentinel 1a, *Proc. SPIE* **9647**, *Unmanned/Unattended Sensors and Sensor Networks XI; and Advanced Free-Space Optical Communication Techniques and Applications*, 96470J (October 29, 2015)
[20] Schreiber U et al 2009 The European laser timing experiment (ELT) on-board ACES, *Proc. IFCS EFTF 2009* pp 594-599
[21] Feldmann T et al, 2013 TWSTFT Calibration Involving Four Sites Using a Mobile Station on a Trailer *Proc. Joint UFFC, EFTF and PFM Symposium 2013* pp 485-491
[22] Recommendation ITU-R P.676-3 Attenuation by atmospheric gases

Impact of turbulent phase noise on frequency transfer with asymmetric two-way ground-satellite coherent optical links

Clélia Robert[1], Jean-Marc Conan[1], Peter Wolf[2]

[1]Onera - The French Aerospace Lab, F-92322 Châtillon, France
[2]SYRTE, Observatoire de Paris, PSL Research University, CNRS, Sorbonne Universités, UPMC Univ. Paris 06, LNE, 61 avenue de l'Observatoire, 75014 Paris, France

E-mail: Peter.Wolf@obspm.fr

Abstract: Bidirectional ground-satellite laser links suffer from turbulence-induced scintillation and phase distortion. We study how turbulence impacts on coherent detection capacity and on the associated phase noise that restricts clock transfer precision. We evaluate the capacity to obtain a two-way cancellation of atmospheric effects despite the asymmetry between up and down link that limits the link reciprocity. For ground-satellite links, the asymmetry is induced by point-ahead angle and possibly the use, for the ground terminal, of different transceiver diameters, in reception and emission. The quantitative analysis is obtained thanks to refined end-to-end simulations under realistic turbulence and wind conditions as well as satellite cinematic. These temporally resolved simulations allow characterizing the coherent detection in terms of time series of heterodyne efficiency for different system parameters. We show that Tip/Tilt correction on ground is mandatory at reception for the down link and as a pre-compensation of the up link. Good correlation between up and down phase noise is obtained even with asymmetric apertures of the ground transceiver and in spite of pointing ahead angle. The reduction to less than 1 rad^2 of the two-way differential phase noise is very promising for clock comparisons.

1. Introduction

Optical bi-directional links between ground stations and spacecraft in Earth or solar system orbits are of interest for a number of applications ranging from telecommunications to navigation, geodesy, time/frequency metrology and fundamental physics [1-4]. Such links promise high data rates and low phase noise in long distance clock comparison or ranging measurements. In particular, low noise, long distance clock comparisons in Earth orbit and in the solar system are of strong interest for experimental gravitation, relativistic geodesy, and time/frequency metrology [2-5]. The dominating limiting effect for such optical links is likely to be atmospheric turbulence, and in this work we study its impact on two-way clock comparisons by numerical simulations.

Modern long distance clock comparisons take advantage of two-way compensation schemes that allow mitigation of any source of phase noise that is reciprocal on the two channels, as demonstrated in optical fiber links [6, 7] and free space optical links over short horizontal distances [8–10]. Simply put, the phase difference between the two distant clocks is obtained from the half difference of the up and down link measurements, leading to a complete cancellation of any phase noise that is identical on the two links. In our case, it is then essential to characterize the reciprocity of phase fluctuations between the up and down link, and to quantify any residual phase noise resulting from non-reciprocity. As discussed in section 2, one can show that the principle of reciprocity applies to propagation through turbulence [11–13]. Implications in terms of reciprocity of the optical link have been extensively studied [9,14–21]. In practice, there are several reasons that limit the link reciprocity in ground-satellite laser links:

- Large satellite velocities imply a point ahead angle for the ground station, which means that the up and down links will cross different turbulent volumes.
- Technical considerations (available laser power on ground and on board, pointing requirements, stray light,...) may favor different aperture sizes for reception and emission on ground.
- Finite light velocity implies that the turbulent layers are crossed at different times by the up and down signal, implying a change in the turbulent layers due to wind.

Of these asymmetries the first two are largely dominant, since typical propagation times across the turbulent atmosphere are of order 100 μs making the third effect very small. Hence for our study, we aim at providing quantitative estimates of the residual phase noise after two-way compensation due to pointing ahead and transceiver asymmetry. Refined end-to-end simulations in the temporal mode - where the phase screens are moved deterministically following wind profiles and satellite cinematic - are in our opinion the only way to study realistic scenarios, for which the layered turbulent structure (profiles of C_n^2 and L_0) plays a central role.

2. Principle of reciprocity for modelling of two-way ground-satellite optical links

Before discussing the link reciprocity, let us come back to the general principle of reciprocity in propagation through turbulence, a well-known physical property [11–13]. In short, wave propagation being a linear process, there is a unique functional that allows to describe by an integral equation the forward propagation of the complex field from a plane Π_1 (terminal T1) to a plane Π_2 (terminal T2), and the backward propagation from Π_2 to Π_1:

$$\mathcal{E}_{2S} = \int h(\mathbf{r}_1, \mathbf{r}_2)\mathcal{E}_{1E}(\mathbf{r}_1)d\mathbf{r}_1 \quad (1)$$
$$\mathcal{E}_{1S} = \int h(\mathbf{r}_1, \mathbf{r}_2)\mathcal{E}_{2E}(\mathbf{r}_2)d\mathbf{r}_2 \quad (2)$$

where, for the forward propagation, $\mathcal{E}_{1E}$ is the emitted complex field in plane Π_1 and $\mathcal{E}_{2S}$ the received complex field in plane Π_2, $\mathcal{E}_{2E}$ and $\mathcal{E}_{1S}$ being the emitted and received fields for the backward propagation. $h(\mathbf{r}_1, \mathbf{r}_2)$ is the functional that characterizes the propagation medium, here the propagation through a given turbulence state at a given time. The implicit assumption is that turbulence does not evolve during the time of propagation through the turbulent volume.

In heterodyne detection one is interested in a complex quantity $\psi(t)$ called "complex coupling": the overlap integral on the pupil aperture P of the received signal electromagnetic field ($\mathcal{E}_S$) and the local oscillator field ($\mathcal{E}_L$) that reads $\psi(t) = \int_P \mathcal{E}_L^\star \mathcal{E}_S d\mathbf{r}$. Using (1) and (2) those couplings can be written as functions of the fields emitted at the opposite terminals

$$\psi_1 = \int \mathcal{E}_{1L}^\star(\mathbf{r}_1)\left(\int \mathcal{E}_{2E}(\mathbf{r}_2)h(\mathbf{r}_1, \mathbf{r}_2)d\mathbf{r}_2\right)d\mathbf{r}_1 \quad (3)$$
$$\psi_2 = \int \mathcal{E}_{1E}(\mathbf{r}_1)\left(\int \mathcal{E}_{2L}^\star(\mathbf{r}_2)h(\mathbf{r}_1, \mathbf{r}_2)d\mathbf{r}_2\right)d\mathbf{r}_1 \quad (4)$$

where integral orders have been exchanged to obtain a reciprocal expression of ψ_2, expressed in the plane Π_1. The complex couplings at both sides of the optical link can therefore be expressed as overlap integrals in the T1 aperture plane. They both involve complex fields resulting of the propagation from T2 to T1. These propagated fields, the terms between parenthesis, correspond on one hand to the natural propagation of $\mathcal{E}_{2E}$, the field emitted by T2, and on the other hand to a fictive back propagation through turbulence of $\mathcal{E}_{2L}^\star$, the T2 local oscillator complex conjugate.

It is interesting to note that in the particular case where $\mathcal{E}_{1E} = \mathcal{E}_{1L}^\star$ and $\mathcal{E}_{2E} = \mathcal{E}_{2L}^\star$, symmetry on each terminal between the emission and the local oscillator, then ψ_1 is equal to ψ_2. Under these conditions the link reciprocity is perfect, meaning that both sides of the link have strictly identical heterodyne efficiency and phase noise.

Again, these developments are simply a transposition to heterodyne detection of the work of J. Shapiro (see [17] and references therein), where we have local oscillator modes in place of fiber modes, and where our expressions deal with complex coupling instead of its square modulus, since we care about phase noise in our application.

Let us now restrict our discussion to the case of ground-satellite links. Such a link has several specificities:

1. very long distance of propagation (thousands to tens of thousands of kilometers), hence clearly corresponding to a far field case,
2. turbulent volume is concentrated in the first few tens of kilometers near ground; note that the time of propagation through turbulence remains very short (around 100 μs), even if the total propagation time to the satellite is large,
3. high satellite velocity leads to a point ahead angle between the downward and upward beams.

Items 1 and 2 imply that the turbulence coherence area in the satellite plane is much larger than the T2 (satellite) pupil size. Therefore, in the previous expressions, on can neglect the dependence in $\mathbf{r}_2$ of $h(\mathbf{r}_1, \mathbf{r}_2)$, and approximate it by $h(\mathbf{r}_1, 0)$ [23]. The complex couplings then read:

$$\psi_1 = \int \mathcal{E}_{1L}^{\star}(\mathbf{r}_1) h(\mathbf{r}_1, 0) d\mathbf{r}_1 \left(\int \mathcal{E}_{2E}(\mathbf{r}_2) d\mathbf{r}_2\right) \tag{5}$$

$$\psi_2 = \int \mathcal{E}_{1E}(\mathbf{r}_1) h(\mathbf{r}_1, 0) d\mathbf{r}_1 \left(\int \mathcal{E}_{2L}^{\star}(\mathbf{r}_2)\, d\mathbf{r}_2\right) \tag{6}$$

In both expressions the terms between parenthesis are constant factors independent of turbulence. This implies that the analysis of turbulent effects does not depend on T2 modes. Looking at (2), one can see that $h(\mathbf{r}_1, 0)$ corresponds to the downward propagation of a plane wave through turbulence. The fact that one can model the downlink with a downward plane wave propagation, neglecting the Gaussian nature of the beam coming from the satellite, is actually a usual approximation in ground-space optical links, both for incoherent and coherent detection. The reciprocal expression of ψ_2 shows that, based on the same approximation, the uplink coupling can also be calculated from this downward plane wave. Actually, this is true only when neglecting the point ahead effect mentioned in item 3. Accounting for point ahead is however easy and only implies that $h(\mathbf{r}_1, 0)$ is not exactly the same term in Eq. (5) and (6): it corresponds to the result of the propagation, through the same turbulent volume, of two downward plane waves with two propagation directions separated by the point ahead angle. Say on-axis for downlink in Eq. (5) and off-axis at point ahead for uplink in Eq. (6). There is therefore no need to simulate the upward propagation of a Gaussian beam.

3. Model and methods

PILOT is our numerical tool based on Monte-Carlo split-step propagation techniques: simulation of optical propagation through discrete turbulent random phase screens and Fresnel propagation. This code is routinely used for endoatmospheric applications [23], astronomy [24], and ground-space telecommunications modelling [25].

Figure 1 displays a sketch describing the optical propagation through turbulence for ground-space optical links. The turbulence volume can be described by discrete layers (perpendicular to the line of sight) introducing a phase perturbation on the optical beam. In a given layer, the turbulent phase statistics are well known (Von Karman spectrum), with turbulence strength characterized locally by the so-called "refractive index structure parameter" denoted C_n^2 and the turbulence outer-scale L_0. The integrated strength along the line of sight of these wavefront distortions is quantified by the Fried parameter r_0, always expressed in plane wave in this article. Even if turbulence induces locally phase only perturbations, diffraction effects during the beam propagation progressively transform phase effects in amplitude perturbations (so-called scintillation). The scintillation pattern produced by a layer at distance L is granular with a characteristic size $\sqrt{\lambda L}$, where λ is the wavelength. Phase and scintillation effects of course limit the performance of optical time/frequency comparisons.

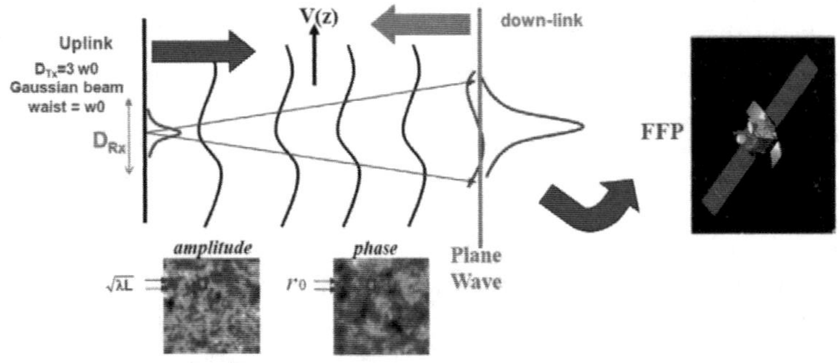

Figure 1: Turbulence effects on ground-space optical propagation: up link (Gaussian beam) and down link (plane wave). Photograph of the satellite from http://commons.wikimedia.org/wiki/File:UFO_satellite_2.jpg. Color online.

For down links, considering the beam divergence of the laser coming from the satellite and the small size of the ground receiver aperture, one can assume plane wave propagation. The up link channel is slightly more complex since it requires accounting for Gaussian beam propagation through the turbulent atmosphere. In this case, turbulence still induces phase and scintillation perturbations, as well as beam wander effects. However, we make use of the reciprocity principle to compute the up link performance from plane wave downward propagation coupled to a Gaussian reception mode. We thus compute two downward plane wave propagations along two axis separated by the point ahead angle.

We run the wave-optic propagation code PILOT at the working wavelength of λ = 1064 nm. We use a vertical C_n^2 profile covering the [0,20 km] altitudes modeled by a Hufnagel-Valley profile above the boundary layer (at about 1.5 km) linked to the Monin-Obukov similitude laws below. We use 10 phase screens distributed non-uniformly. Five of them are put in the first km, inside the boundary layer, to sample the significant turbulence. The other five are placed at 4, 6, 10, 14 and 18 km. We consider an oblique line of sight with 20° elevation. We study a moderate turbulence condition with r_0 = 13 cm (in plane wave and along this line of sight), corresponding to observation from an astronomical site at high altitude, e.g. the Calern observatory used for lunar and satellite laser ranging (1300 m altitude). We set an outer scale L_0 = 8 m which is a realistic value [26] and we found furthermore a good convergence of the 2 first moments of the heterodyne efficiency m after having varied L_0 on a [1-25] m range. Influence of turbulent parameters r_0 and L_0 is beyond the scope of this article and will be studied in a forthcoming publication [27].

Using our propagation code, we get time series of the electromagnetic fields for the signal. The associated statistical quantities $|\mathcal{E}_S(\mathbf{r},t)|$ (modulus) and $\arg[\mathcal{E}_S(\mathbf{r},t)]$ (phase) are used to derive link metrics such as the phase noise of the signal at the intermediate frequency, and heterodyne efficiency. They can be deduced from respectively the argument and the squared modulus of the complex quantity $\psi(t) = \int_P \mathcal{E}_L^* \mathcal{E}_S d\mathbf{r}$ the overlap integral on the pupil aperture P of the electromagnetic field propagated and the local oscillator. So, we can derive temporal series of phase noise and heterodyne efficiency through Eq. (7) and Eq. (8), and last but not least, with and without Tip/Tilt correction.

$$\varphi(t) = \arg[\psi(t)] \tag{7}$$

$$m(t) = \frac{|\psi(t)|^2}{|\psi_0(t)|^2} \tag{8}$$

where $\psi_0(t)$ is the overlap integral in the absence of turbulence.

For the down link, the overlap integral computes the coupling between the on-axis downward plane wave and the ground local oscillator mode. While for the up link, and by application of the reciprocity principle (see section 2. above), we actually compute an overlap integral between the ground emission mode and a downward plane wave. Taking an on-axis plane wave corresponds to neglecting the point ahead effect, otherwise one has to use an off-axis downward plane wave at the point ahead angle. We verified consistency of this reciprocal up link modelling by comparing the results with those obtained with a real up link propagation. From these considerations it is obvious that, in the absence of point ahead, up and down link detection is identical if the ground emitter mode is equal to the ground local oscillator mode.

In practice to avoid phase jumps of 2π we will proceed by successive increments:

$$\varphi(t + dt) - \varphi(t) = \arg[\psi(t + dt)\psi^*(t)]. \tag{9}$$

The cumulative sum on all instants of the phase increment in Eq. (9) provides the phase noise temporal trajectory. Simulations can include full Tip-Tilt correction at the ground receiver aperture for the down link. The same Tip-Tilt is used as a pre-compensation of the up link at the ground emitter.

4. Performance evaluation for time/frequency transfer

We consider a satellite in highly elliptic Earth orbit (inspired by the STE-QUEST proposal [4]), at perigee (z = 3200 km) since this corresponds to the worst case in terms of pointing ahead angle (*paa* = 63 µrad), and to the largest velocity of the satellite (v = 9 km/s). To evaluate the effect of the pointing ahead, we compare in the next subsection an on-axis two-way link (with point ahead angle artificially set to zero) to a two-way link (with pointing ahead angle *paa* = 63 µrad). We model a temporal set of 8000 turbulent electromagnetic fields using the propagation code PILOT. The duration of the series is 8 s with 1 ms sampling steps.

4.1. Heterodyne efficiency vs system parameters

We study here the heterodyne efficiency for various system parameters under moderate turbulence condition. We consider two values of the ground telescope diameter $D_{g\text{-rec}}$ = [0.15, 0.4] m (receiver). The down link is coupled to a Gaussian LO of waist w_0, using the usual truncation rule of Gaussian beams $w_0 = D/3$. For the up link, the two Gaussian beams emitted from ground have waists respectively w_0 = 0.05 m and w_0 = 0.133 m, corresponding to two values of the emitter diameter $D_{g\text{-emit}}$ = 0.15 m and $D_{g\text{-emit}}$ = 0.4 m. As recalled in Sect. 2, due to reciprocity the up and down link have by construction identical performance, in particular they have the same heterodyne efficiency, if the ground emitter mode is equal to the ground local oscillator mode (diameter and waist) and in the absence of point ahead.

Table 1: Synthesis of the mean heterodyne efficiency $\langle m\rangle$ and its fluctuation $[\sigma_m^2/\langle m\rangle^2]$ for the up and down link, the two Gaussian modes, with and without Tip/Tilt correction. Performance of the up link is given with/without pointing ahead.

Gaussian modes LO for down link or waist of uplink	**Down link & Up link no pointing ahead**		**Up link pointing ahead**
w_0 **= 0.050 m**	Turbulent	0.60 [0.14]	0.60 [0.14]
	Tip/Tilt control	0.79 [0.03]	0.78 [0.03]
w_0 **= 0.133 m**	Turbulent	0.18 [0.84]	0.18 [0.84]
	Tip/Tilt control	0.53 [0.07]	0.51 [0.07]

Journal of Physics: Conference Series **723** (2016) 012039 doi:10.1088/1742-6596/723/1/012039

The performance of the heterodyne efficiency in various cases is summarized in Table 1. Full turbulence performance presents lower mean value and larger variance than in the case with Tip/Tilt control (in this work we assume ideal tip/tilt control, the effect of imperfections being the subject of future work). Without Tip/Tilt control again, we observe relatively large variances of the heterodyne efficiency for up and down links using 0.4 m aperture (w_0 = 0.133 m). For such aperture, Tip/Tilt correction is mandatory to maintain reasonable signal to noise ratio (S/N), even at astronomical sites with moderate turbulence. Applying down link Tip/Tilt to up link is efficient, and essential for reciprocity. Note that the efficiency of Tip/Tilt pre-compensation is slightly minored at the percent level in presence of pointing ahead, but remains nonetheless very efficient. The slight reduction of performance is caused by tip-tilt anisoplanatism that induces a difference between on-axis and off-axis tip-tilt at point ahead. The fact that this effect is minor may seem surprising since point-ahead angle is larger than the isoplantic angle in our case. One has however to remember that the isoplanatic angle is not a relevant parameter to describe the modal normalized angular correlations of Zernike modes. Tip-tilt is known for instance to have a large angular correlation (see for instance Fig. 3 in [24]): in our case the normalized correlation is better than 95% at point ahead angle. *Figure 2* shows, as an example, the heterodyne efficiencies for the 40 cm diameter receiver downlink, with and without tip/tilt correction.

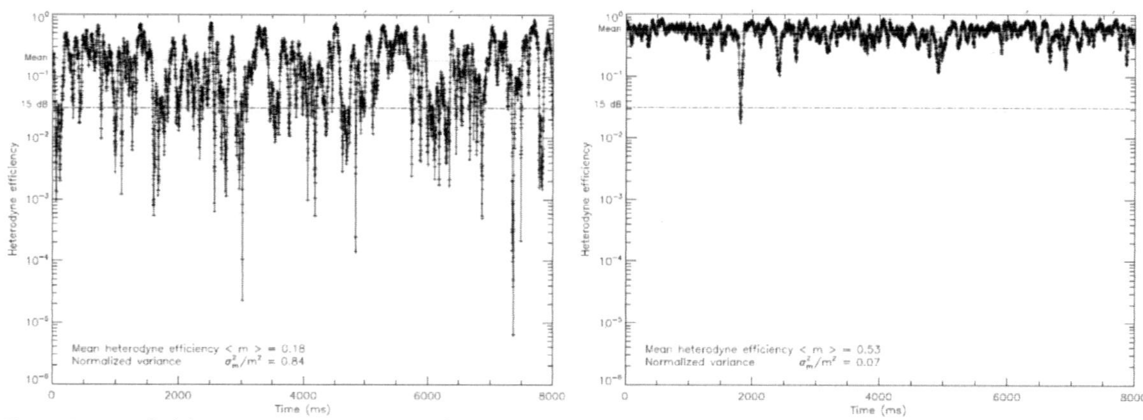

Figure 2: Down link (D_{g-rec} = 0.4 m, w_0 = 0.133 m) time series of heterodyne efficiency, without (left) and with (right) tip/tilt control. Horizontal line in red shows the detection level. Green horizontal line shows the mean heterodyne efficiency $\langle m \rangle$. Due to reciprocity (see sect. 2) uplink performance (with point ahead) is similar with only a slight degradation due to a minor reduction of tip/tilt correction efficiency (see table 1). Color online.

We indicate on *Figure 2* the loss of 15 dB in S/N which we consider as a "conservative" extinction threshold. Obviously extinctions of the signal happen more often without Tip/Tilt control and can be very deep, leading to a S/N loss of up to 10^6. With tip/tilt control the situation is much improved with only very moderate loss of S/N.

4.2. *Two-way phase noise compensation*

High performance clock comparison techniques use two-way signals to compensate for any phase noise that is reciprocal in the two directions. Consider two clocks A and B whose phase difference is to be determined. A typical two-way link then consists of a signal going from A to B whose phase at emission is measured on clock A and whose phase at reception is measured on clock B using e.g. a heterodyne detection. Provided the phase shift accumulated during propagation $\delta\phi_{A\to B}$ is known the phase difference of the clocks $\Delta\Phi_{AB}$ can be deduced from the measurements. Similarly, the return signal allows deducing the quantity $\Delta\Phi_{BA} = -\Delta\Phi_{AB}$ from a knowledge of $\delta\phi_{B\to A}$. Forming the half difference of the two links gives the clock phase difference affected by the link noise via the quantity $(\delta\phi_{A\to B} - \delta\phi_{B\to A})/2$, i.e. any noise on the links that is reciprocal cancels in the two-way clock comparison. As mentioned earlier, reciprocity of turbulent phase noise is perfect provided that the links are symmetric.

But this is never the case in practice, because of the necessary point ahead angle which can be much larger than the isoplanatic angle (here ≈25 μrad), and technical considerations that may require different apertures for emission and reception on the ground.

Figure 3 shows individually the effect of the two asymmetries. In the presence of a point ahead angle of 63 μrad, the phase noise of the up and down links as well as their difference is plotted for the case where the apertures are symmetric ($D_{g\text{–}emit} = D_{g\text{–}rec} = 0.4$ m) with tip/tilt correction on the down link and the same tip/tilt correction applied as pre-compensation on the up link. Also plotted is the case of asymmetric apertures on the ground ($D_{g\text{–}emit} = 0.15$ m, $D_{g\text{–}rec} = 0.4$ m) and no point ahead angle. The latter is not a realistic situation as point ahead is unavoidable, but it shows the different residual noise behaviour coming from the aperture asymmetry. We see that the up and down links are highly correlated (ie. good reciprocity) in both cases. The residual noise due to the asymmetry has variances of less than 1 rad^2 in both cases, significantly less than on the individual up or down links. The structure of the residual phase noise is different in the two cases, but with similar magnitude.

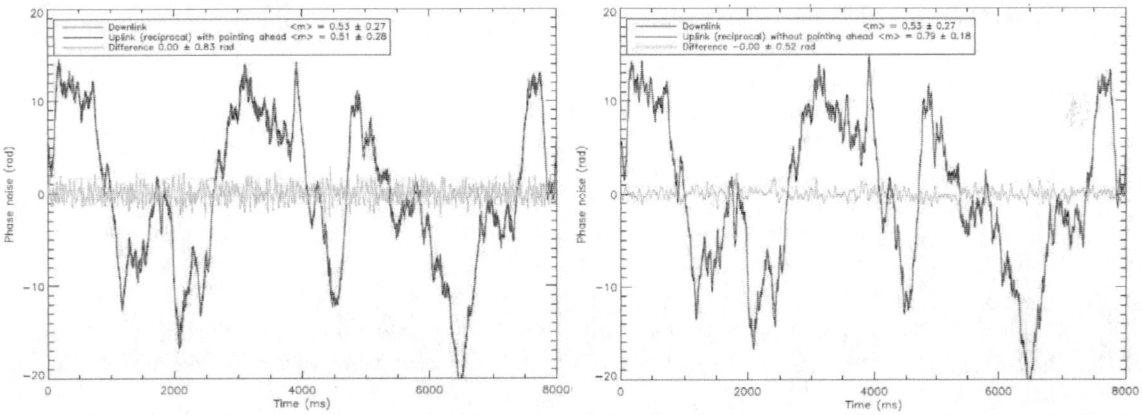

Figure 3: Two-way compensation on phase noise (phase noise/rad vs. time/ms) in the case of symmetric apertures with point ahead angle (left) and asymmetric apertures without point ahead angle (right), with tip/tilt correction for both. Color online.

Finally, *Figure 4* shows the modified Allan deviation of the down link, and the two-way combination, (up – down)/2, for the case of symmetric apertures on the ground (0.4 m) and a point ahead angle of 63 μrad. We see a very steep decrease of the frequency noise with averaging time, reaching 1.4x10^{-17} in fractional frequency after only 1 s averaging time. For asymmetric apertures the long term performance is somewhat degraded (7x10^{-17} at 1 s averaging) [27], nonetheless these results are very promising for optical free space time/frequency comparisons.

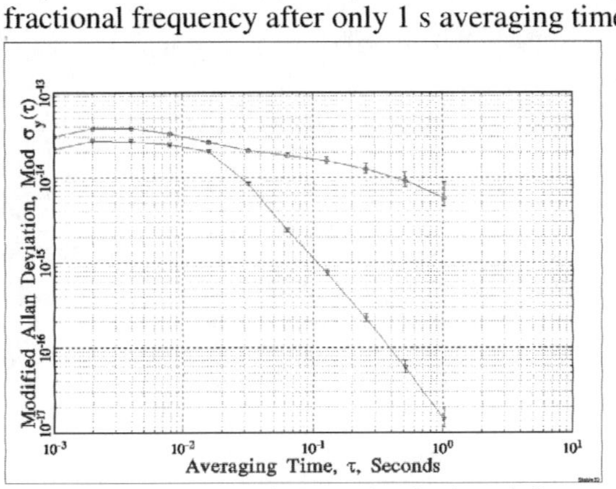

Figure 4: Modified Allan deviation for the down link, (top) and the two way combination (bottom) for symmetric apertures and a point ahead angle of 63 μrad. Color online.

5. Conclusion

In summary, we show that the down link must be tip/tilt corrected to improve the heterodyne efficiency, thus avoiding large fluctuations in signal to noise ratio with frequent extinctions. This is usually the case in all optical links as tip/tilt correction is used for active tracking. We also show that the same tip/tilt correction can be efficiently applied as pre-compensation on the up link. Then good correlation between up and down phase noise is obtained even with asymmetric apertures of the ground transceiver and in spite of the point ahead angle. For a moderate turbulence site, we quantify the reduction of the two-way differential phase noise at less than 1 rad^2. These simulation results are very promising for the feasibility of ground-satellite coherent optical links for time/frequency transfer application with outstanding performance.

Acknowledgements: This work was supported by the French Space Agency CNES (Centre Nationnal d'Etudes Spatiales). The authors thank also Marie-Thérèse Velluet, Nicolas Védrenne and Vincent Michau for the fruitful exchanges they have had with them.

References

[1] Seel, S., *et al.* 2011 in *Aerospace Conference 2011* IEEE, 1–7.
[2] Christophe, B., *et al.* 2012, Experimental Astronomy, **34**, 203-242.
[3] P. Wolf *et al.* 2009, *Experimental Astronomy*, **23**, 651-687.
[4] see http://sci.esa.int/ste-quest/
[5] Reynaud, S., *et al.* 2008, *Phys. Rev*. D, **77**, 122003.
[6] Lopez, O., *et al.* 2012, *Opt. Express* **20**, 23518–23526.
[7] Predehl, K., *et al.* 2012, Science, **336**, 441–444.
[8] Djerroud, K., *et al.* 2010, *Opt. Lett.*, **35**, 1479–1481.
[9] Giorgetta, F. R., *et al.* 2013 *Nature Photonics*, **7**, 434–438.
[10] Sinclair, L. C., *et al.* 2014, *Phys. Rev.* A, **89**, 023805.
[11] Shapiro, J. H., 1971, *JOSA*, **61**, 492–495.
[12] Lutomirski, R. and Yura, H., 1971, *Applied Optics*, **10**, 1652–1658.
[13] D. Fried and H. Yura. 1972, *J. Opt. Soc. Am.*, **62**, 600-602.
[14] Shapiro, J. H., 1974, *Applied Optics*, **13**, 2614–2619.
[15] Shapiro, J. H. 1975, *JOSA*, **65**, 65–68.
[16] Shapiro, J. 1978, in *Laser Beam Propagation in the Atmosphere*, 171–222, Springer.
[17] Shapiro, J. H. and Puryear, A. L. 2012, *J. of Opt. Comm. and Networking*, 4, 947–954.
[18] Puryear, A. L., *et al.* 2013, *J. of Opt. Comm. and Networking*, 5, 888-900.
[19] Parenti, R. R., *et al.* 2012, *Optics Express*, **20**, 21635–21644.
[20] Giggenbach, D., *et al.* 2012 *Appl. Opt.*, **51**, 3145–3152.
[21] Minet, J., *et al.* 2013, *Journal of Optics*, **15**, 022401.
[23] Velluet, M.-T., *et al.* 2005, *SPIE proc.*, **5981**, 188–196.
[24] Robert, C., *et al.* 2006, *JOSA*, **A23**, 613–624.
[25] Védrenne, N., *et al.* 2012, *proc. Int. Conf. Sp. Opt. Sys. (ICSOS)*, Sept. 9–12, Ajaccio, France.
[26] Ziad, A., *et al.* 2004, *Appl. Opt.*, **43**, 2316–2324.
[27] Robert, C., Conan, J.-M. and Wolf, P. 2016, Phys. Rev. A, **93**, 033860.

Methods of time series preparation based on UTC and UTCr scales for predicting the [UTC-UTC(PL)]

L Sobolewski[1],**W Miczulski**[1]

[1] Institute of Metrology, Electronics and Computer Science, University of Zielona Gora, prof. Z. Szafrana 2, 65-246 Zielona Gora, Poland

E-mail: l.sobolewski@imei.uz.zgora.pl

Abstract. The article presents the results of methods of preparation of two time series on the quality of predicting the [UTC-UTC(PL)] for the Polish Timescale UTC(PL) using GMDH neural networks. The first time series (TS1) was based on the [UTC-UTC(PL)] deviations designated by the BIPM. In the second time series (TS2) the deviations designated by the BIPM on the basis of the UTC and UTC Rapid scales were applied. The obtained results indicate that the use of time series for predicting the [UTC-UTC(PL)], based on deviations determined by the UTC and UTC Rapid scales, allowed to obtain more accurate predictions.

1. Introduction

The organization responsible for determining the quality of national timescales UTC(k) is the International Bureau of Weights and Measures (BIPM, Bureau International des Poids et Mesures). Every month for each UTC(k) the $xb(t)$ deviations are determined by the BIPM according to relation

$$xb(t) = UTC(t) - UTC_k(t), \quad (1)$$

defining the differences of the national timescale relative to UTC. These deviations are calculated as averages per day on days MJD (Modified Julian Date) ends with the digits 4 and 9. The process of calculating the UTC scale, based on specific algorithms [1], is very complex and time-consuming. The measurement data for the previous month are sent to the BIPM from the National Metrology Institutes NMIs. The assembly, proper preparation of measurement data and calculation of the UTC scale by the BIPM, cause a delay in publishing by the BIPM "Circular T" bulletin containing calculated deviations of UTC(k) in relation to UTC for a given month. This bulletin is published between the 8th and 12th day of the following month. Considering the delay in publication of the "Circular T" bulletin, ensuring the highest possible compatibility of UTC(k) in relation to the UTC is only possible by predicting the [UTC-UTC(k)]. The designated value of the prediction would be the basis for correcting the UTC(k) scale. Part of the NMIs laboratories are predicting the [UTC-UTC(k)]. In the literature for predicting the [UTC-UTC(k)] only methods that are based on statistical models are presented, for example [2, 3, 4, 5].

Polish Timescale UTC(PL) is the local implementation of Coordinated Universal Time (UTC). Central Office of Measures (GUM, Główny Urząd Miar) is responsible for maintaining the UTC(PL). The physical realization of the UTC(PL) is based on one of the selected caesium atomic clocks available in the GUM. The system implementing the UTC(PL) is equipped with a control device (Phase-Microstepper Austron 2055), enabling the introduction of appropriate adjustments to the time scale. For predicting the [UTC-UTC(PL)], to 13.12.2012, a laborious procedure based on analytical linear regression was used in the GUM.

The Institute of Metrology, Electronics and Computer Science of the University of Zielona Gora, in cooperation with the GUM, began a research on the application of neural networks for predicting the [UTC-UTC(PL)] in 2006. This choice was due to the properties of neural networks [6]. For predicting the GMDH (Group Method of Data Handling), GRNN (Generalized Regression Neural Network), MLP (Multilayer Perceptron) and RBF (Radial Basis Function) neural networks were used. The input data for these neural networks were prepared in the form of time series as described in section 2.1. The results of the studies, presented in [7, 8 and 9], shows that the best results were achieved for GMDH neural network using time series analysis method.

Delay in publication of the $xb(t)$ deviations by the BIPM adversely affect on the result of the prediction, and consequently on maintaining the best convergence of the UTC(k) with the UTC. Therefore, in order to expedite the transfer of information about the differences of the UTC(k) in relation to the UTC, BIPM developed *A Rapid UTC* project. On the basis of UTC Rapid (UTCr) scale, every Wednesday on the BIPMs ftp server the $xbr(t)$ deviations are published, determined according to the relationship

$$xbr(t) = UTCr(t) - UTC_k(t) \tag{2}$$

for the previous week, for each UTC(k). On 1 July 2013 this project became official.

In this paper, on the example of UTC(PL) scale and using GMDH neural networks, the results of research on the methods of preparation of two time series on the quality of predicting the [UTC-UTC(PL)] are presented. For the preparation of these time series the $xb(t)$ and $xbr(t)$ deviations were used.

2. Preparation of the input data for neural network

Predicting the [UTC-UTC(PL)] based on neural networks requires a process of training, the quality of which depends on the number of training data and the method of its preparation [10, 11]. Aiming to achieve the most accurate predictions forces also in the prediction process to maintain the smallest possible time horizon between the date of the determination of the prediction, and the date of the last known value of the time series. Thus, there is also a necessity to supplement the time series with new data published by BIPM, which means that the repetition of the training process of the neural network is needed.

2.1. Input data prepared on the basis of deviations determined according to the UTC scale

The basis of preparing the input data for neural network based on the UTC scale was the relation

$$x_1(t) = xa(t) + xb(t) = UTC(t) - clock_{PL}(t), \tag{3}$$

where:

$xa(t)$ – measurements of the phase time between 1 *pps* signal from the UTC(PL) and a single caesium clock realizing the UTC(PL) ($clock_{PL}$), determined on each day according to relation

$$xa(t) = UTC_{PL}(t) - clock_{PL}(t), \tag{4}$$

$xb(t)$ – values of deviations determined by the BIPM for the UTC(PL) in relation to the UTC according to relation (1). For the set of $xb(t)$ deviations, an interpolation function PCHIP (Piecewise Cubic Hermite Interpolating Polynomial), available in MATLAB was used. A mathematical model was determined that allows to extend a set of [UTC-UTC(PL)] deviations by calculating its values for each day of analysed data set.

The data set $x_1(t)$, called time series TS1, characterize on each day the time instability of $clock_{PL}$ in relation to the UTC. Time series TS1 was created based on a historical data of $xa(t)$ and $xb(t)$ from MJD 55899 to MJD 56694. Determining the value of [UTC-UTC(PL)] prediction (marked further as $xb_p(t)$) was carried out from MJD 56514 to MJD 56694. For the first $xb_p(t)$ prediction designated by the GMDH neural network on day (t_{pred}) MJD 56514 (August 2013) the training data set for time series TS1 was available from day t_0 (figure 1) MJD 55899 (December 2011) to day $t_{n(i)}$ (figure 1) MJD 56504 (July 2013). This means that time series TS1 in the training process of the GMDH neural network contains 605 elements. The required minimum number of elements of times series TS1 ensuring the achievement of predictions with best possible accuracies, determined on the basis of studies [11], is 570. For the predictions designated in 6 consecutive months the number of training data were increased each time by approximately 30 elements, which required re-run of training of the neural network.

2.2. Input data prepared on the basis of deviations determined according to the UTCr scale

The basis for preparation of second time series (TS2) were the $xb(t)$ and $xbr(t)$ deviations designated based on the UTC and UTCr scales. This time series consisted of two subsets of elements prepared according to the rules defined on figure 1. The first subset contains a group of data (from 1 to i) having a values of time series TS1, designated based on relation (3), from day t_0 to day $t_{n(i)}$, for which the last value of $x_1(t_{n(i)})$ before the publication day ($t_{pub(i)}$) was known. The second subset was a complement of time series TS2 by a group of data having values determined based on relation

$$x_2(t) = xa(t) + xbr(t) = UTCr(t) - clock_{PL}(t)\,, \tag{5}$$

between days $t_{n(i)}$ and t_{nr}, published by the BIPM on day t_{pubr}. The publication day of $xbr(t)$ deviations could also be the day (t_{predr}) of designating the $xb_p(t)$ prediction. The values of $xa(t)$ subset, as in the case of data prepared on the basis of the UTC scale, are the measurements of the phase time between the UTC(PL) and $clock_{PL}$.

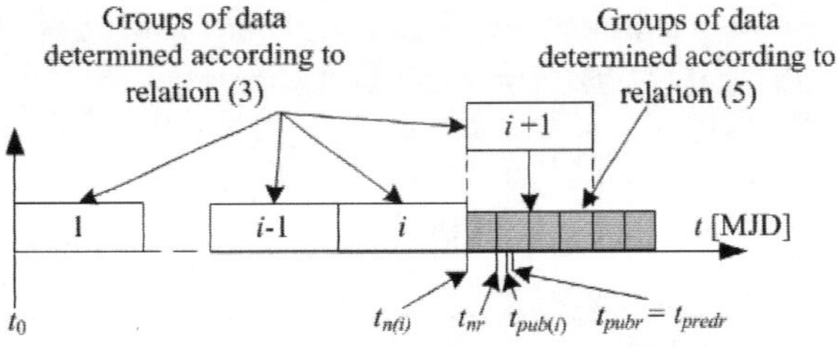

Figure 1. Creation of time series TS2.

By adding to the time series TS2 the successive groups of data calculated based on the relation (5), it is possible to predict the next values of $xb_p(t)$ with an interval of 7 days. At the time of publication of a new "Circular T" bulletin containing values of $xb(t)$ deviations, the new group of data i+1 (figure 1) according to relation (3), which for the respective days replaces the existing data set based on relation (5) is determined.

The first subset of time series TS2 was created based on measurements of $xa(t)$ and $xb(t)$ as for time series TS1, and the second subset for $xbr(t)$ from MJD 55964 to MJD 56694. Predicting the $xb_p(t)$ was carried out form MJD 56499 to MJD 56694. The first value of $xb_p(t)$ prediction was designated on MJD 56499, so the training data set contains 595 elements. For predictions designated in 29 consecutive weeks the number of training data were increased each time by 7 elements.

The selected period of predicting [UTC-UTC(PL)] for time series TS1 and TS2 includes the time interval from MJD 56486 (07.12.2013) to MJD 56546 (09.11.2013), for which the corrections of the UTC(PL) scale using Phase-Microstepper were not carried out.

3. Research results

The studies were carried out on the basis of prepared time series TS1 and TS2. GMDH neural network used for research belongs to a group of self-organizing networks. In the training process of this network the selection of a degree of the polynomial transfer function of the neuron and the automatic selection of its structure (the number of hidden layers and the number of neurons in these layers) was chosen. In most cases of training the neural network the selected activation function of the neuron was a linear function. Then the obtained structures of the GMDH neural network consisted of two or one hidden layer having two neurons.

For each month (for TS1) and each week (for TS2) the predictions of $xb_p(t)$ on the MJD days ending with digits 4 and 9 were designated. For time series TS1 in a given month a several predictions (from 4 to 7) with prediction horizon of 10, 15, 20, 25, 30, 35 and 40 days could be designated. However for time series TS2 in a given week one or two predictions could be designated. Figure 2 presents a sample set of designated $xb_p(t)$ predictions for time series TS2. For comparison also the values of $xb(t)$ and $xbr(t)$ deviations at the same days were presented. A similar course of $xb_p(t)$ prediction values and $xb(t)$ deviations were obtained for time series TS1.

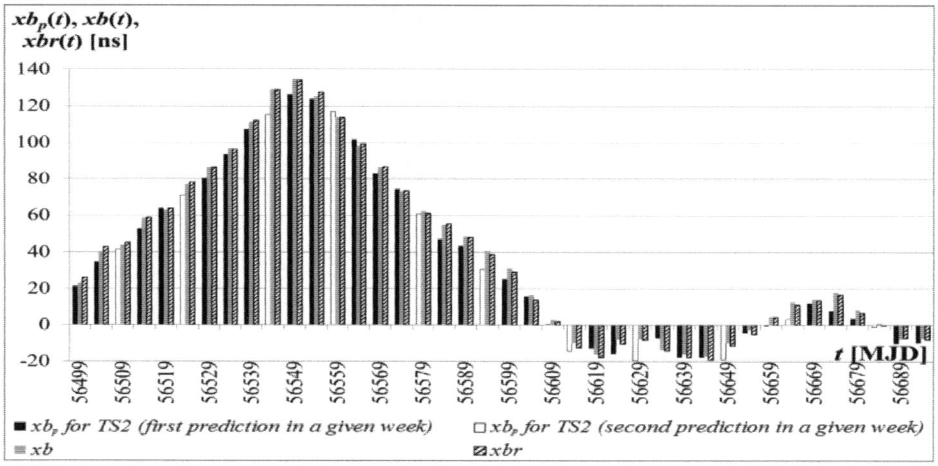

Figure 2. Values of $xb_p(t)$ predictions and the values of $xb(t)$ and $xbr(t)$ deviations designated by the BIPM for time series TS2.

Figure 3 presents the calculated values of r residuals, determined according to relation:

$$r = xb(t) - xb_p(t)\,, \tag{6}$$

for time series TS1 and TS2. The residuals (r) determine the differences between the predicted values of $xb_p(t)$ and the $xb(t)$ published by the BIPM for the same day of prediction. In the vast number of cases the designated first predictions in a given month (TS1) or week (TS2) have smaller values of residuals than the consecutive predictions in the same period of time. Practically this means, that only first designated predictions could be used for correcting the UTC(PL) scale. This fact decided that the further analysis of research results was related only to these predictions.

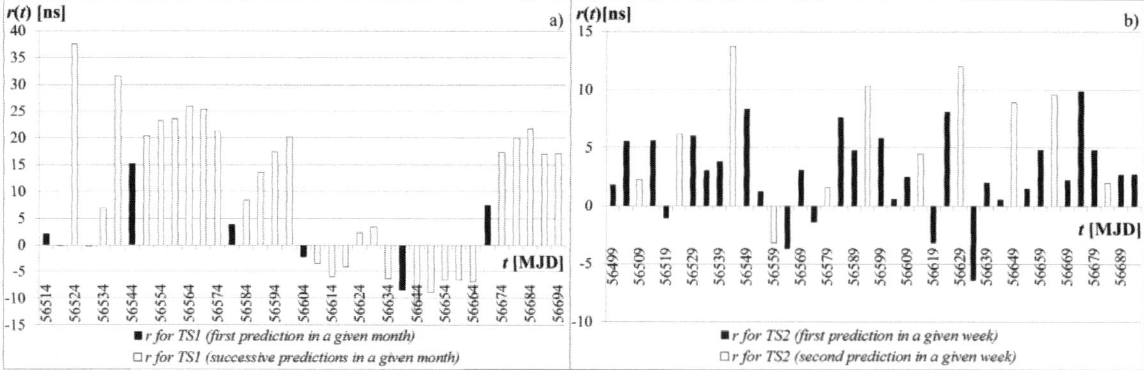

Figure 3. Values of $r(t)$ residuals for time series TS1 a) and for time series TS2 b).

The obtained values of the *r* residuals were used to evaluate the quality of predictions, which was performed on the basis of selected measures of quality of predictions [12]: Mean Error (*ME*), Mean Absolute Error (*MAE*), Mean Squared Error (*MSE*), Root Mean Squared Error (*RMSE*) and linear correlation coefficient between the $xb(t)$ and $xb_p(t)$. Also a three components of a *MSE* error (MSE_1, MSE_2 and MSE_3) were designated, which determine the quality of designated predictions. The MSE_1 component represents the prediction load, MSE_2 is associated with insufficient flexibility of the prediction (so the inaccuracy of estimation of the variation of the predicted variable). The MSE_3 component reports of an error related to the insufficient compliance of a change in prediction direction with a change in direction of predicted value. The results of calculations are presented in Table 1.

Table 1. Values of measures of quality of prediction for examined time series.

Measures of quality of prediction	***Time series***	
	TS1	***TS2***
r_{max} [ns]	15.2	9.9
r_{min} [ns]	-8.0	-6.4
ME [ns]	3.0	-2.9
MAE [ns]	6.6	4.0
MSE [ns^2]	64	22
MSE_1 [ns^2]	8.9	8.4
MSE_2 [ns^2]	37	0.2
MSE_3 [ns^2]	19	14
RMSE [ns]	8.0	4.7
correlation coefficient	0.995	0.997

The presented values of $xb_p(t)$ predictions (figure 2) and values of residuals (figure 3) and calculated values of measures of quality of prediction (Table 1) lead to the following conclusions:

- The comparison of the values of residuals and all of the measures of quality of prediction for analysed time series show that the best results of predicting the $xb_p(t)$ were obtained for time series TS2. This is due to the short prediction horizon, which was achieved by introducing into this time series the values of $xbr(t)$ deviations determined by BIPM.
- The calculated values of correlation coefficient for both time series close to unity, and the values of $xb(t)$ and $xb_p(t)$ (figure 2) show a very good convergence between these values. It also means that absence of correction of the UTC(PL) scale or its use does not affect the predicting process of the [UTC-UTC(PL)] by GMDH neural network.
- Very good convergence also occurs between the values of $xbr(t)$ and $xb(t)$ (figure 2) determined by the BIPM. For analysed period of time the differences between these values were in the range of -3.2 ns to +2.7 ns. This fact decided on trying to use the values of $xbr(t)$ deviations for constructing time series TS2.
- The obtained values of residuals for time series TS2 are much smaller than the residuals obtained for time series TS1. The highest residual value for time series TS2 is equal to 9.9 ns, and for time series TS1 15.2 ns. Larger values of residuals for time series TS1 would be expected due to the longer prediction horizon for first prediction in a given month (form 10 to 20 days). It was dictated by using for construction of this time series only the values of $xb(t)$ deviations. Hence, for this time series the larger values of individual measures of quality of prediction, mainly *MSE* error and its component MSE_2 were obtained.
- Predictions of the $xb_p(t)$ designated based on time series TS2 allow to obtain the values of residuals not exceeding the limit of ±10 ns. This means that predicting the $xb_p(t)$ based on data

prepared using this method can contribute to qualify the UTC(PL) to the first group of timescales.

- Comparison of values of *ME* and *MAE* errors and value of MSE_1 component between time series TS1 and TS2 indicates that designated predictions are underestimated (high number of positive values of *r* residuals).
- Time series TS2 allows the highest flexibility in determining the prediction (the lowest values of MSE_2) in relation to time series TS1. This means that variation of predicted values is good compared to the variation of observed values.
- Time series TS2 also lead to smaller values of MSE_3 error, which indicates the higher degree of compliance of a change in prediction direction with a change in direction of predicted value.

4. Conclusions

Our research on methods of preparation of time series for predicting the $xb_p(t)$ has utilized the fact that the introduction of the UTCr scale by the BIPM allows the determination of prediction for the UTC(PL) with a several-day prediction horizon, compared to the horizons of predictions from 10 to 20 days in the case of UTC scale. This enables also more frequent prediction of $xb_p(t)$ (once a week) compared to one prediction a month in the case of UTC scale. This decreased to a large extent a divergence between the $xb_p(t)$ predictions designated using GMDH neural network and $xb(t)$ deviations published by BIPM. The results at a level not exceeding the limit of ±10 ns and the lack of impact of the correction procedure of the UTC(PL) scale on the results of predicting [UTC-UTC(PL)] by GMDH neural network encourages the authors to develop a procedure for predicting the $xb_p(t)$ for the UTC(PL) based on this neural network.

The observed very good convergence between $xbr(t)$ and $xb(t)$ values for UTC(PL) has also been confirmed for other national time scales realized based on single caesium atomic clocks, e.g. UTC(AUS) in Australia, UTC(BEV) in Austria, UTC(CNMP) in Panama, UTC(LT) in Lithuania and UTC(NRC) in Canada. This allows using GMDH neural network for predicting the [UTC-UTC(k)] based on time series TS2 for other timescales than UTC(PL).

References

[1] Arias E F, Panfilo G and Petit G 2011 Timescales at the BIPM *Metrologia* **48** pp 145–153

[2] Bernier L G 2003 Use of the Allan Deviation and Linear Prediction for the Determination of the Uncertainty on Time Calibrations Against Predicted Timescales *IEEE Transactions on Instrumentation and Measurement* **52** pp 483–486

[3] Czubla A, Konopka J and Nawrocki J 2006 Realization of atomic SI second definition in context UTC(PL) and TA(PL) *Metrology and Measurement Systems* **2** pp 149–159

[4] Davis J A, Shemar S L and Whibberley P B 2011 A Kalman filter UTC(k) prediction and steering algorithm *Proc. Joint IEEE FCS EFTF* (*San Francisco, USA, 2-5 May 2011*) pp 779–784

[5] Panfilo G and Tavella P 2008 Atomic clock prediction based on stochastic differential equations *Metrologia* **45** pp 108–116

[6] Norgard M, Ravn O, Poulsen N and Hansen L 2000 *Networks for Modelling and Control of Dynamic Systems* (London: Springer)

[7] Luzar M, Sobolewski Ł, Miczulski W and Korbicz J 2013 Prediction of corrections for the Polish Timescale UTC(PL) using artificial neural networks *Bulletin of the Polish Academy of Sciences: Technical Sciences* **61** pp 589–594

[8] Sobolewski Ł 2013 Predicting the Polish timescale UTC PL based on the corrections designated by the UTC and UTCr scale *Proc. Joint UFFC EFTF and PFM* (*Prague, Czech Republic, 21-25 July 2013*) pp 658–661

[9] Sobolewski Ł 2014 Predicting the corrections for the Polish Timescale UTC(PL) using GMDH and GRNN neural networks *Proc. European Frequency & Time Forum* (*Neuchatel, Switzerland, 23-26 June 2014*) pp 475–478

[10] Masters T 1993 *Practical neural networks recipes in C++* (San Diego, CA, USA: Academic Press)
[11] Miczulski W and Sobolewski Ł 2012 Influence of the GMDH neural network data preparation method on UTC(PL) correction prediction results *Metrology and Measurement Systems* **19** pp 123–132
[12] Pindyck R S and Rubinfeld D C 1998 *Econometric Models and Economic Forecasts* (Boston: McGraw-Hill)

Journal of Physics: Conference Series **723** (2016) 012041 doi:10.1088/1742-6596/723/1/012041

Rapid evaluation of time scale using an optical clock

T Ido, H Hachisu, F Nakagawa, and Y Hanado

Space-Time Standards Laboratory, National Institute of Information and Communications Technology (NICT), Koganei, Tokyo, 184-8795, Japan

E-mail: **ido@nict.go.jp**

Abstract. Feasibility of steering a time scale using an optical clock is investigated. Since the high stability of optical frequency standards enables rapid evaluation of the scale interval, the requirement for the continuous operation is mitigated. Numerical simulations with the input of real calibration data by a ^{87}Sr lattice clock indicated that the calibrations once in two weeks maintain the time scale within 5 ns level using a currently available hydrogen maser at NICT. "Optical" steering of a time scale by the intermittent calibrations frees an optical frequency standard from being dedicated to the steering, enabling other applications using the same apparatus.

1. Introduction

The recent progress of optical clocks is so rapid that some atomic clocks based on optical clock transitions have clearly surpassed microwave clocks in terms of the accuracy and stability. Following the advancement of optical clocks, the community of time and frequency standards has begun seeking unprecedented applications such as sensors for geodetic science. While these are still a long way from practical use, maintaining a time scale using an optical clock might be a reasonable choice. Time scales have been so far maintained by microwave clocks, such as commercially available Cs beam clocks or hydrogen masers (H-masers). Cs fountain frequency standards are referenced to realize the ultimate accuracy of the International Atomic Time (TAI). The fountains have become mature technology these days, and some laboratories almost always operate fountains and adjust their time scale in real time according to the error that fountains detect [1,2]. For laboratories without Cs fountains, on the other hand, it is not easy to begin the development of a Cs fountain from scratch only for the purpose of time scale maintenance. Thus, in the near future, an optical clock might be required to play a role in maintaining a time scale. From this viewpoint, we studied the feasibility of the "optical" steering of time scale. Assuming that the performance of currently available H-maser at NICT remains as demonstrated in the past, our numerical simulation employing real data gathered from a Sr lattice clock has revealed that the operation of optical clocks twice in a month will keep the time scale within the ±5 ns level.

2. UTC: reference time scale that local time scales synchronize

Universal Coordinated Time (UTC) is a time scale with the same scale interval as the TAI. TAI is based on the average of more than 400 commercial atomic clocks operated in various metrological or astronomical institutes worldwide. The interval obtained by the average of the 400 clocks is adjusted to the SI second by referring to the calibrations provided mainly by caesium fountain standards operated in national metrological institutes. The averaging and adjustment are virtually performed by a numerical process. There is no real signal of UTC. On the other hand, participating institutes normally maintain their own real time scale as close as possible to UTC. This local realization of the UTC is called UTC(k), where k denotes the name of the institute. Bureau International des Poids et Mesures (BIPM) is in charge of the numerical process of calculating the TAI. The "tick" of the TAI is once in five days since this numerical process is performed for the moments at 0:00 UTC on every five days. BIPM notifies participating institutes of the time difference of UTC and their local realization, UTC(k). The notification is performed in a monthly published report, the so-called Circular T. The institutes steer the scale interval according to this notification. Note that BIPM also publishes rapid UTC (UTCr), where the time differences are reported daily. However, the shorter average of a single day suffers from rather large link uncertainty.

The difficulties of this steering are partly due to the latency of the Circular T. The time-keeping institutes are unable to obtain the time difference UTC − UTC(k) in real time. In other words, for the duration of one month, the scale interval of UTC(k) is basically determined by their local frequency standards. Hence, locally available atomic frequency standards are important for the time scale. Caesium fountain frequency standards were first developed in the 1990s, and now the operation of some standards in advanced laboratories is so stable that an up-time ratio of more than 90% has recently been maintained [1]. Thus, some institutes can maintain accurate scale interval of the time scale in real time [1, 2]. This improvement was clearly witnessed in the record of UTC − UTC(k) in the past ten years, which is shown as figure 1. The reductions of the fluctuation in UTC − UTC(PTB) and UTC − UTC(OP) are found at around MJD 55500 and MJD 56200, respectively, where the two institutes initiated the steering of UTC(k) by referring to real-time evaluation using a reliably operated Cs fountain.

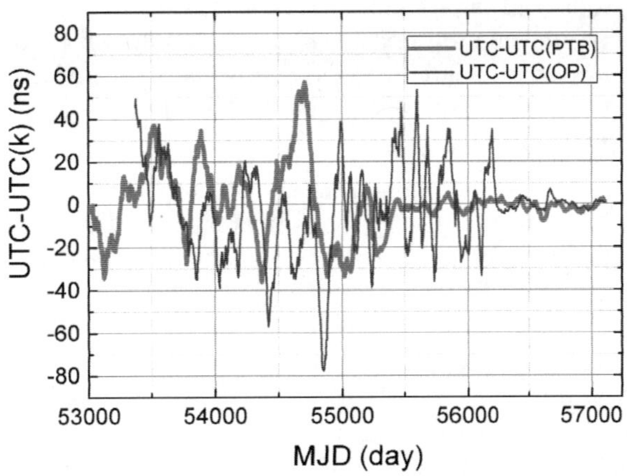

Figure 1. (colour online) Time difference of UTC and its various local realization UTC(k).

3. Real-time estimation of UTC(k)-UTC using a Sr lattice clock

The detail of the ^{87}Sr lattice clock at NICT is described in [3]. The configuration of the lattice is a vertically oriented one-dimensional lattice. Thanks to the vacuum chamber made of >20 cm-thick aluminium alloy, rather homogeneous distribution of the temperature was observed over the chamber. The total systematic uncertainty as an optical standard amounts to 8.6×10^{-17}, which is sufficient to steer time scale. On the other hand, the source oscillator of UTC(NICT) is one of commercial H-masers contained in the NICT maser array. The H-maser frequency is steered with reference to the weighted average of eighteen commercial caesium clocks. The long term stability of this averaged Cs frequency is low 10^{-15} level around 30 days. This stability is not sufficient to maintain the |UTC − UTC(NICT)| below 10 ns during one month.

Absolute frequency measurements of an optical clock transition are normally performed when one builds an optical frequency standard. Particularly, in the case of ^{87}Sr lattice clocks, more than ten frequency measurements were already reported. On the basis of these data, we are able to calculate the weighted mean frequency with an uncertainty of less than 1×10^{-15}. In addition, optical clocks have better short-term stability than microwave standards. Thus, a lattice clock has the ability of the real-time evaluation of the scale interval of UTC(k). We investigated this potential capability using the data of five consecutive days of measurement (MJD 57079-57083), which we obtained for the latest absolute frequency measurement [3].

The uncertainty of the absolute frequency of ^{87}Sr lattice clock transition has recently been reduced in some laboratories, where the state-of-the-art Cs fountain standards serve as a reference of the SI second. Two measurements [4, 5] have an uncertainty of $< 5\times10^{-16}$, and the fractional difference is less than 1×10^{-16}. Thus, it is reasonable to assume the weighted average of the two measurements as the absolute frequency of the Sr lattice clock transition. Because of the lack of a state-of-the-art Cs fountain, our frequency measurement first leads the frequency of the ^{87}Sr lattice clock transition with reference to UTC(NICT). Conversely, this measurement allows us to evaluate the frequency of UTC(NICT) on the basis of the Sr lattice clock. We measured frequency for 10000-20000 s per day for five consecutive days. The instability of the 10000-20000 s measurement deviates from high 10^{-16} to low 10^{-15}, depending on which H-maser is used in the maser array. Assuming that the offset of the UTC(NICT) frequency from that of UTC is constant until the following day's update, it is possible to estimate the gain or loss of UTC(NICT) during the five days.

The result is summarized in figure 2. The data, from which the scale interval of UTC(NICT) is deduced, was originally taken from the latest absolute frequency measurements [3]. By integrating the fractional offset frequency of UTC(NICT), the gain of UTC(NICT) since 0:00 UTC on MJD57079 is estimated as drawn in red, where the slope to the first point is assumed to be same as that between first and second points. Large dots indicate the update of the scale interval by the operation of the optical clock, and red lines are the resultant estimation, indicating the real time estimation of UTC(NICT) − UTC. UTC(NICT) has an instability of 1×10^{-15} at one day. There are five estimations in figure 2, which lead to the error of $5^{1/2} \times (10^{-15} \times 86400$ s$) = 0.19$ ns at the end of fifth day. Besides that, the evaluation of the UTC(NICT) frequency has an uncertainty due to the uncertainty of the ^{87}Sr clock transition frequency (5×10^{-16}) as well as the type-A uncertainty of HM4-UTC(NICT) link which was measured to be 2×10^{-16} in 10000 s average. In total, we evaluate the uncertainty of the time difference on MJD 57084 to be 0.29 ns.

The validity of the estimation was confirmed by the comparison against two values. One is the UTC(NICT)-UTC(USNO) derived from the GPS P3 all-in-view methods. The data is expressed as small dots in figure 2. One dot is the mean of 15 min. The overall trends fairly agree with the prediction. The other data to be compared is the time difference reported in Circular T. The UTC-UTC(NICT) at 0:00 on MJD 57079 and on 57084 is found in Circular T No. 327. From these two values, we can calculate the change in UTC − UTC(NICT) during these five days, which is shown as blue empty circles in figure 2. The change is 1.9 ns for the five days, whereas the estimation by the Sr lattice clock is 1.7 ns, showing reasonable agreement. Note that the scale unit of the UTC is not identical to the SI second. The mean difference in scale unit of TAI from the SI second for MJD57079-57109 is 2.7×10^{-16}. Considering the link uncertainty, the uncertainty of UTC-UTC(NICT) is 9.8×10^{-16} for five-day average [6]. This frequency uncertainty corresponds to $(9.8\times10^{-16}\times5\times86400) = 0.42$ ns of the uncertainty of the variation of the five-day time difference, which is consistent with the 0.2 ns difference.

Figure 2 demonstrates that 10000 s of lattice clock operation each day is sufficient to maintain the time scale within a 1 ns time difference. However, it is not yet practical to dedicate one system of an optical clock to a time scale. The quest for high accuracy is under way in most laboratories, which is not compatible with daily routine operation of the time scale maintenance. This limitation may be resolved by a stable transfer oscillator such as a H-maser. An H-maser with highly predictable behaviour can be responsible for the down time of optical clocks. To pursue this strategy of intermittent operations of an optical clock, we performed some numerical simulations on the basis of the data of real optical frequency measurements and those of H-masers. The relevant data related to H-masers are available as records of the Japan Standard Time (JST) system.

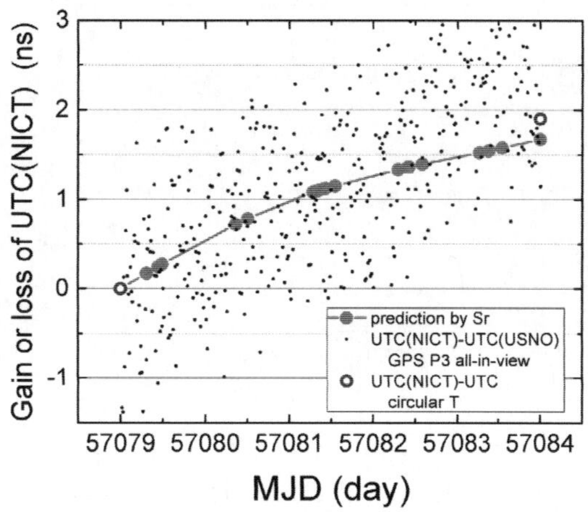

Figure 2. (colour online) Real-time estimation of the gain or loss of a time scale in reference to an optical clock.

4. Capability of keeping a time scale using an optical clock and a highly predictable source oscillator

In our recent campaign of absolute frequency measurements [3], measurements based on an H-maser (HM4) were performed in four campaigns which are temporarily separated by 20–30 days. Conversely, this measurement allows us to obtain the scale interval of the HM4 based on the Sr lattice clock. A sample of the evaluation of HM4 frequency by the ^{87}Sr lattice clock is shown in figure 3. Four campaigns were performed on MJD 57060–57064, 57079–57084, 57108, and 57124–57129. On these days, we obtained the calibration of the HM4 frequency on the assumption that the ^{87}Sr lattice clock generates an optical frequency of 429 228 004 229 873.11 Hz, which is the weighted average of the results shown in [4, 5]. The calibrations were expressed as four groups of green dots. All points are derived from the calibration of 10^3-10^4 s. The type-A uncertainty of each point is $< 1\times10^{-15}$. Whole points may have an offset of $<5\times10^{-16}$ due to the error in assuming the ^{87}Sr clock frequency. Overall points indicate a fairly stable linear frequency drift of HM4. This was also confirmed by the instability against UTC, which is shown as the inset of figure 3. Thus, we can estimate the frequency of HM4 on any instant by interpolating past data by a linear regression. The red thick line indicates such a possible estimation. Using the data taken in the nearest 45 days, the linear drift of the H-maser is estimated, and we interpolate it to estimate the HM4 frequency on that date. Since the data of the first 4 days (MJD 57060-67064) is insufficient to estimate the overall drift rate of the HM4, the red thick line is probably steeper than the real drift rate for the first 30 days. However, the data on MJD 57079, which is a point after 15 days of down time, allows the system to notice that the estimation of the drift rate was higher. Later than that, the estimation of the HM4 frequency becomes fairly good. The blue line is another estimation of HM4 frequency provided by the JST system. JST is a product composed by steering an H-maser frequency in reference to an ensemble time scale of 18 commercial Cs clocks. Since the ^{87}Sr lattice clock has an accuracy of $<1\times10^{-16}$, the discrepancy of green dots and the blue line is predominantly attributed to the Cs ensemble. The maximum discrepancy often amounts to 5×10^{-15}, which easily causes the time difference of more than 10 ns in 20 days.

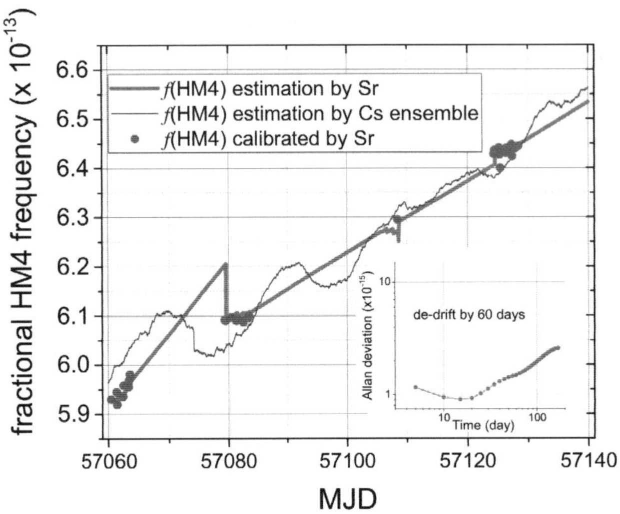

Figure 3. (colour online) Estimation of a H-maser (HM4) frequency by intermittent operations of an optical clock. The inset shows the instability of the HM4 against UTC, where the drift is removed by the fitting of 60 days beforehand.

The data shown in figure 3 enables the simulation of steering time scale using an optical clock. We assume in the simulation that the HM4 signal is supplied to a phase micro-stepper (PMS), where the frequency error estimated as the red thick line in figure 3 is compensated; thus, the calibrated output of the PMS is steered by an intermittent operation of an optical clock. A simulation by this method is feasible in our environment because the relative time difference of HM4 vs UTC(NICT) is monitored and recorded in the JST system on every second, thus enabling the calculation of the time error of the optically steered time scale against UTC. Note that UTC-UTC(NICT) is only available every five days. We can compare the optical-based time scale to UTC at 0:00 UTC on every five days.

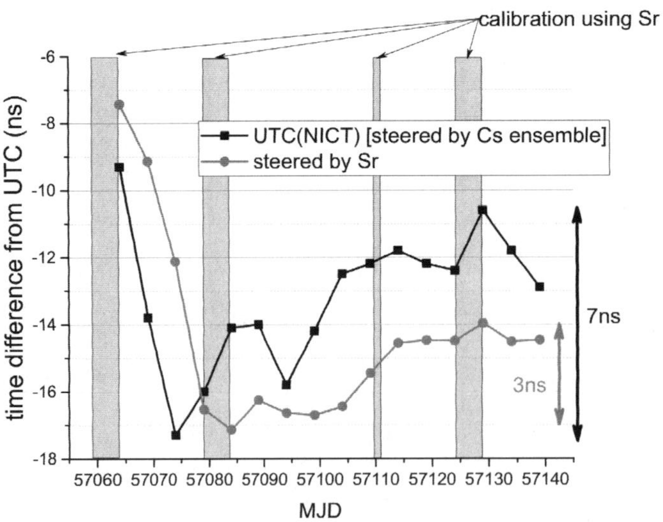

Figure 4. (colour online) Simulation of a steering of a time scale using an optical clock (circle). Filled squares are the results reported in Circular T. Note that manual adjustment was made on MJD 57074 for UTC(NICT).

The result of the comparison against UTC is shown in figure 4. Thesquares are the real record of UTC(NICT) which are found in Circular T. On the other hand, the circles are the simulation result. The vertical bands represent the duration in which the UTC(NICT) is calibrated by the Sr. We did not make a real physical signal, but the method described above allows us to simulate the possible result of generating a time scale. The estimation of the HM4 overall drift rate with reference to the first four days was affected by a sporadic short-term frequency fluctuation, which causes the increase in the time difference during MJD 57060–57080. After the 2nd campaign that began on MJD 57079, the overall drift of the HM4 frequency is well estimated, realizing a stable time difference. While the current system of UTC(NICT), which steers the H-maser frequency referring to the Cs ensemble, causes a maximum deviation of 7 ns from UTC during the 65 days of MJD 57074-57139, the steering using an optical clock calibration suppresses the deviation to 3 ns. This improvement is found in the instability as well. The 5-day instabilities of squares and circles after MJD 57079 are 2.7×10^{-15} and 1.3×10^{-15}, respectively. .

5. Summary

In summary, we have investigated the feasibility of steering a time scale using an optical clock. We performed three sets of five (or four)-consecutive-day measurements as well as a one-day measurement of ^{87}Sr lattice clock frequency with reference to an identical H-maser over two and a half months. Since the JST system always monitors the time difference between the H-maser oscillator and UTC(NICT), we succeeded in estimating the scale interval of UTC(NICT) in real time with this data. Dealing with the ^{87}Sr clock frequency as a reference, we evaluated the frequency as well as the drift rate of the source oscillator. The estimation agreed well with UTC(NICT) − UTC given by Circular T. Furthermore, the numerical simulation of the steering showed the possibility of steering the time scale using the intermittent operation of an optical clock. We simulated the steering of a time scale, which indicated that the calibration of the H-maser frequency twice a month maintains the time scale in ±5 ns. Since optical clocks are nearly one order of magnitude more stable than microwave frequency standards, the operation of optical clocks in several hours is long enough to evaluate the scale interval of the time scale at the $<10^{-15}$ level. Therefore, intermittent operations of optical clocks can maintain the time scale, allowing an optical frequency standard to be utilized for other applications in extra time. The ability to steer the time scale using an optical clock would allow some laboratories to focus on optical atomic standards. This would be the case particularly for some national metrological institutes, where atomic frequency standards are developed for the first time.

Acknowledgements
The authors thank H. Ito and T. Gotoh for providing us the JST record and GPS P3 all-in-view data, respectively. H. Ishijima, S. Ito, and M. Mizuno provided the necessary technical help for the measurements.

References

[1] Bauch A, Weyers S, Piester D, Staliuniene E, and Yang W, 2012 *Metrologia* **49** 180
[2] Abgrall M, Chupin B, De Sarlo L, Guena J, Laurent Ph, Le Coq Y, Le Targat R, Lodewyck J, Lours M, Rosenbusch P, Rovera D, and Bize S 2015 *Comptes Rendus Physique* **16** 461
[3] Hachisu H and Ido T 2015 *Jpn. J. Appl. Phys.* **54** 112401
[4] Le Targat R *et al* 2013 *Nat. Commun.* **4** 2109
[5] Falke St, Lemke N, Grebing C, Lipphardt B, Weyers S, Gerginov V, Huntemann N, Hagemann Ch, Al-Masoudi A, Häfner S, Vogt S, Sterr U and Lisdat Ch 2014 *New J. Phys.* **16** 073023
[6] Panfilo G, and Parker T 2010 *Metrologia* **47** 552

8th Symposium on Frequency Standards and Metrology 2015 IOP Publishing
Journal of Physics: Conference Series **723** (2016) 012042 doi:10.1088/1742-6596/723/1/012042

Absolute frequency measurement at 10^{-16} level based on the international atomic time

H Hachisu, M Fujieda, M Kumagai, and T Ido

Space-Time Standards Laboratory, National Institute of Information and Communications Technology (NICT), Koganei, Tokyo 184-8795, Japan

E-mail: hachisu@nict.go.jp

Abstract. Referring to International Atomic Time (TAI), we measured the absolute frequency of the ^{87}Sr lattice clock with its uncertainty of 1.1×10^{-15}. Unless an optical clock is continuously operated for the five days of the TAI grid, it is required to evaluate dead time uncertainty in order to use the available five-day average of the local frequency reference. We homogeneously distributed intermittent measurements over the five-day grid of TAI, by which the dead time uncertainty was reduced to low 10^{-16} level. Three campaigns of the five (or four)-day consecutive measurements have resulted in the absolute frequency of the ^{87}Sr clock transition of 429 228 004 229 872.85 (47) Hz, where the systematic uncertainty of the ^{87}Sr optical frequency standard amounts to 8.6×10^{-17}.

1. Introduction

Recent developments of optical frequency standards have led to the achievement of stabilities and accuracies at the 10^{-18} level [1–4]. Thus, it is expected that the SI second is redefined by an optical transition in the future. Towards this goal, it is essential to confirm the reproducibility of the frequency by comparing the frequencies among laboratories. From this point of view, direct frequency comparisons of optical clocks have been demonstrated mostly on-campus in the initial stage. Recently, as the next step, the agreement of distant optical clock frequencies has been intensively studied using an optical fiber [5, 6] or a satellite link [7, 8]. These include our activities; direct comparison of our ^{87}Sr optical lattice clock against a distant clock at the University of Tokyo (UT) using an optical fiber link [5, 9] as well as that against one at Physikalisch-Technische Bundesanstalt (PTB) using a two-way carrier phase (TWCP) satellite frequency transfer [7, 10]. Another requirement for the redefinition is the consistency of the duration "one second" before and after the redefinition. This is realized by rigorous absolute frequency measurements of the new standard frequency. Thus, we performed absolute frequency measurements of an ^{87}Sr lattice clock in reference to International Atomic Time (TAI) [11] in 2012 [12] and 2015 [13]. The recent measurement was performed to improve its uncertainty by carrying out homogeneously distributed intermittent measurements over the five-day grid of TAI [13]. The two measurements contributed to the update of the recommended frequency of an ^{87}Sr clock transition by the Comité International des Poids et Mesures (CIPM) since 2012. These efforts will certainly push forward the optical redefinition of the SI second.

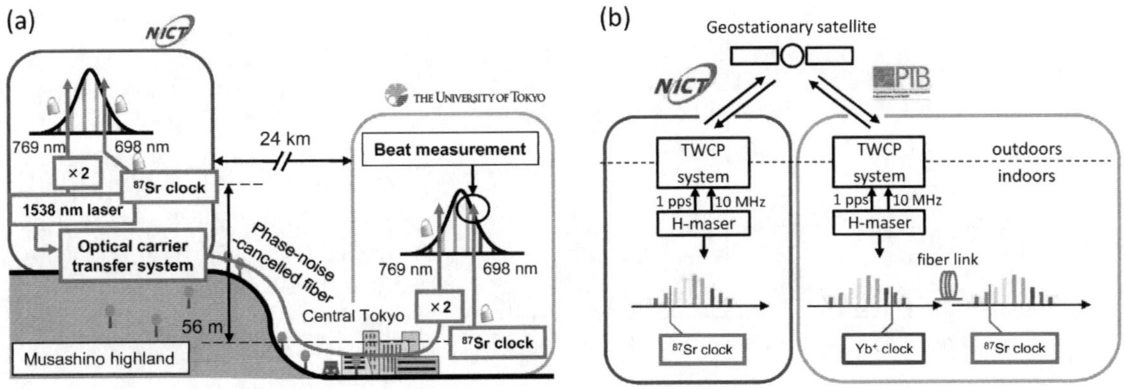

Figure 1. (a) Direct frequency comparison between two distant Sr lattice clocks at NICT and UT using a phase-noise-canceled optical fiber link. The physical distance between two laboratories is 24 km. (b) Intercontinental direct frequency comparison between Sr clocks at NICT and PTB using a satellite link based on two-way carrier phase. The baseline distance of two institutes is 9000 km. The Yb^+ clock acts as a stable transfer oscillator.

2. Direct frequency comparison

2.1. Comparison with UT using an optical fiber link

Using a telecommunication optical fiber link between NICT and UT [9], we performed a direct comparison of two ^{87}Sr clocks at each site [5]. Figure 1(a) shows a schematic diagram of the frequency comparison. The clock frequency at NICT is optically transferred to UT using an optical fiber link through the urban region of Tokyo. At NICT, the Ti:sapphire-based optical frequency comb (Ti:Sa comb) is phase-locked to the clock laser at 698 nm. The telecom laser at 1538 nm is phase-locked to the Ti:Sa comb through its frequency-doubled light at 769 nm and transferred to UT through a 60-km-long optical fiber, to which a phase noise cancellation method is applied. The Ti:Sa comb at UT is phase-locked to the frequency-doubled transferred light. The frequency offset between the two Sr clocks is detected at UT as a beat signal between the clock laser and the nearest component of the Ti:Sa comb.

The observed frequency difference is caused by different systematic shifts of the two clocks, in which the gravitational red shift of 2.62 Hz is the largest. The weighted mean of eleven measurements resulted in the fractional frequency difference between the two distant Sr lattice clocks to be $(1.0 \pm 7.3) \times 10^{-16}$, demonstrating the direct confirmation of the reproducibility of the frequency in distant optical clocks. The uncertainty of the frequency comparison is predominantly limited by the uncertainties of both Sr clocks. Details of the optical fiber link system and the frequency comparison of two distant clocks are described in [9] and [5], respectively.

2.2. Comparison with PTB using a TWCP link

Two ^{87}Sr lattice clocks located at NICT in Japan and at PTB in Germany are directly compared using a satellite-based frequency transfer using the TWCP [7]. The baseline of 9000 km is the longest ever realized in the direct comparison of optical clocks to the best of our knowledge. Figure 1(b) shows a schematic diagram of the comparison, which involves a link system based on the TWCP technique, optical clocks, and frequency combs at each site. The technical details of the TWCP link are described in [10]. The frequency of the lattice clock at each site is counted with reference to a local hydrogen maser (H-maser). A $^{171}Yb^+$ clock based on the $E3$ transition at PTB [14, 15] was employed as a stable transfer oscillator to extend the measurement time

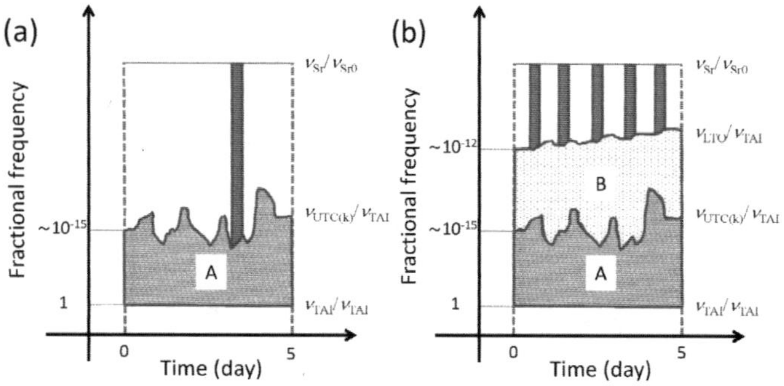

Figure 2. A schematic diagram for an absolute frequency measurement using TAI. (a) The time difference between UTC(k) and UTC at 0:00 (UTC) of day 0 and day 5 is reported by Circular T issued by BIPM, which corresponds to area A. (b) A highly stable local transfer oscillator (LTO) reduces the ambiguity of the LTO frequency. Area B is obtained by a time interval counter.

when the lattice clock at PTB is offline. A total measurement time resulted in 83640 s including the extension by the Yb^+ clock. Both H-masers are linked by the TWCP system. The frequency ratio of the two ^{87}Sr clocks is derived from the ratio of the two local H-masers in real time. We measured the clock frequencies with reference to each local H-maser for several hours per day over four days. Taking into account the uncertainty of the ^{87}Sr lattice clock at NICT and PTB, we concluded that a fractional difference of two distant Sr clocks is $(1.1 \pm 1.6) \times 10^{-15}$. Details of the frequency comparison are described in [7].

3. Absolute frequency measurement based on TAI

In order to maintain the continuity of the scale interval "one second", the absolute frequencies of optical frequency standards need to be rigorously evaluated to the limit of the SI second. It is straightforward to measure the optical frequency by referring to locally available cesium primary frequency standards. The evaluation of the frequency based on TAI, however, is an alternative for laboratories where highly accurate primary frequency standards are not yet available. TAI is accessible even at isolated locations via the Global Navigation Satellite System (GNSS). Thus, H-masers with a satellite-based link to TAI, which we call a TAI link hereafter, have often been employed in optical frequency measurements [12, 16–21].

Every month, the Bureau International des Poids et Mesures (BIPM) reports in Circular T the result of calculated time difference between Coordinated Universal Time (UTC) and the local realization of UTC (UTC(k)) of every fifth day on the TAI grid. The scale interval of UTC is identical to that of TAI. Consequently, only a five-day average of the TAI frequency is available as a reference to which we evaluate optical frequencies. It is ideal to operate optical clocks all through the five-days of the TAI grid. Instead of the continuous operation, however, we utilize an H-maser and pursued a possibility to estimate the difference of the H-maser frequency between the limited time of clock operation and the five-day average. Note that a five-day average of the H-maser frequency can be accurately evaluated from the time difference reported in Circular T.

Figure 2 depicts the scheme of an absolute frequency measurement employing TAI as a frequency reference. Here, the horizontal and vertical axes are time and the fractional frequency relative to TAI, respectively. In addition, ν_{Sr}/ν_{Sr0} is the fractional frequency of Sr relative to a certain reference frequency of Sr based on TAI. This ratio is as stable as TAI. $\nu_{UTC(k)}/\nu_{TAI}$ indicates the temporal frequency variation of UTC(k) with reference to TAI. In the case of UTC(NICT), the frequency is adjusted every 8 h for the 10^{-15} level with reference to Cs ensemble time. Therefore the frequency of UTC(NICT) at the moment of measurement is unknown. On the other hand, the area A turns out to be the time difference between 0:00 (UTC) of day 0 and that of day 5 since that is the integrated fractional frequency difference between UTC(k) and TAI during this period. Thus, it is possible to evaluate only a five-day average of the fractional

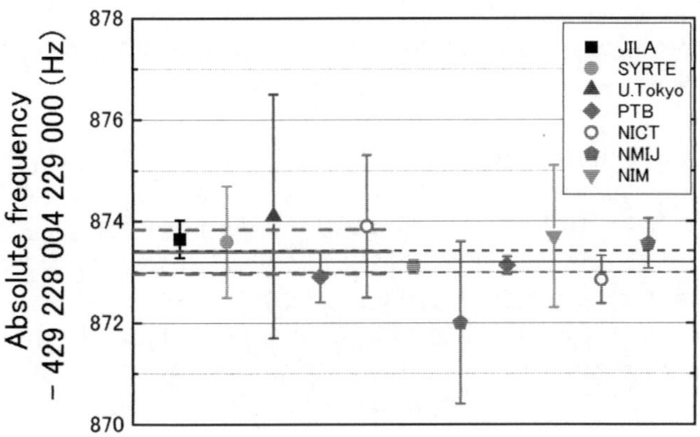

Figure 3. (colour online) Absolute frequencies reported so far. The frequencies can be found in the following publications: JILA [22], SYRTE [23,24], UT [16], PTB [25,26], NICT [12, 13], NMIJ [20, 27], NIM [28]. The second data from right is the result of this work. The CIPM recommendation and its uncertainty are expressed as lines and broken lines, respectively, which were updated in 2012 (thick lines in red) and 2015 (thin lines in blue).

frequency of UTC(NICT) referring to TAI for this period. Each bar in figure 2 represents the operation period of an optical clock which does not cover the whole five days, requiring an additional uncertainty to estimate the frequency of UTC(k) at the instant of measurement. We call this uncertainty the dead-time uncertainty. As shown in figure 2(b), employing a more stable predictable oscillator as an LTO moderates this error. Since the LTO oscillates continuously for five days, the sum of areas A and B in figure 2(b) can be obtained by a time-interval counter. Here, frequency measurement is affected not by the fluctuation of UTC(k) but by that of the LTO. In addition, it is known that carrying out homogeneously distributed measurements over five days reduces the dead time error [13].

We measured the absolute frequency of an ^{87}Sr optical lattice clock in 2012 using a scheme as shown in figure 2(a), which had the uncertainty of 3.3×10^{-15} [12]. In this first measurement, a large dead time uncertainty of 2.6×10^{-15} dominated since we did not take care of the distribution of measurement time over the five-day grid. Furthermore, the inaccuracy of our Sr system has been recently reduced to 8.6×10^{-17}, as briefly described in [13]. Consequently, it is crucial to reduce the TAI link uncertainty in order to evaluate the absolute frequency more accurately. Thus, on the occasion of the recent measurement in 2015 [13], we employed the homogeneously distributed intermittent measurement scheme as shown in figure 2(b).

Three sets of the five-day frequency measurements were performed. The total weighted average of three campaigns was calculated on the basis of the statistical weighting. The uncertainty comprises of the statistical uncertainty, the clock and TAI systematic uncertainties, the link uncertainty of TAI−UTC(NICT), and the dead time uncertainties of H-maser−TAI and TAI−TT (Terrestrial Time) which are estimated by numerical analysis as described in [13]. The total uncertainty is predominantly limited by the uncertainty due to the dead time of TAI−TT frequency difference. The frequency of the TAI that we employed as a reference in this work is the five-day average, whereas the calibration of the TAI with reference to TT is provided in Circular T on basis of one-month average, requiring dead time uncertainty of 1.1×10^{-15} per one measurement of the five-day campaign. Consequently, we have not yet reached the uncertainty at 10^{-16} level. The effort to reduce the link uncertainty performed in this work, however, will pave the way to the measurement at the 10^{-16} level since a few measurement campaigns performed in one month may reduce the TAI−TT dead time uncertainty. This prospect is supported by numerical simulation of the same manner as [13].

Finally, figure 3 summarizes the absolute frequency measurements performed in various institutes worldwide. The result reported here is consistent with other measurements within the uncertainty, indicating the validity of the uncertainty evaluation investigated here.

4. Conclusion

We measured the absolute frequency of the ^{87}Sr clock transition with its uncertainty of 1.1×10^{-15} by reducing the TAI link uncertainty. Three sets of five (or four)-day measurements, where the frequency was measured for 10000 – 24000 s per day, reduced the uncertainty due to the frequency link between TAI and the LTO. In addition, the uncertainties owing to the dead time are evaluated by numerical simulations. These efforts reduced the link uncertainty. Note that the H-maser we employed as the LTO is the most stable in the H-maser array of the Japan Standard Time system. The instability reaches 1×10^{-15} at 1000 – 2000 s and stays a few 10^{-16} for $10^4 - 10^5$ s according to the extended three-corner-hat method using several H-masers. The predictability of the H-maser frequency is another key for the measurement.

The intermittent measurement of an optical frequency standard can be applied to the estimation of the scale interval of TAI, similarly to the role currently played by microwave fountain standards. Cesium or rubidium fountain frequency standards have so far contributed to the calibration of TAI frequency by their continuous operation over 15 days or more. On the other hand, optical clocks together with a stable LTO may not require continuous operation since they reach the same level of instability in a few hours. While a frequency link in principle requires the prolonged signal integration of an LTO, the homogeneously distributed intermittent evaluations and likewise the management of dead time uncertainty may be sufficient for the estimation of TAI.

Acknowledgments

The authors are deeply grateful to T. Takano, M. Takamoto and H. Katori for their collaboration in the direct frequency comparison using an optical fiber link and to St. Falke, N. Huntemann, C. Grebing, B. Lipphardt, Ch. Lisdat and D. Piester for their collaboration in a direct international frequency comparison with the TWCP link. We would also like to thank F. Nakagawa and H. Ito for providing the record of JST. We are also grateful to Y. Hanado and T. Gotoh for the discussion regarding the TAI link. Parts of the Sr system were built by A. Yamaguchi, A. Nogami, S. Nagano, and Y. Li in the early stage of its construction. H. Ishijima, S. Ito, and M. Mizuno provided the necessary technical assistance in the measurements.

References

[1] Chou C W, Hume D B, Koelemeij J C J, Wineland D J and Rosenband T 2010 *Phys. Rev. Lett.* **104** 070802
[2] Hinkley N, Sherman J A, Phillips N B, Schioppo M, Lemke N D, Beloy K, Pizzocaro M, Oates C W and Ludlow A 2013 *Science* **341** 1215
[3] Nicholson *et al.* 2015 *Nat. Commun.* **6** 1215
[4] Ushijima I, Takamoto M, Das M, Ohkubo T and Katori H 2015 *Nat. Photonics* **9** 185
[5] Yamaguchi A *et al.* 2011 *Appl. Phys. Express* **4** 082203
[6] Lisdat C *et al.* 2015 arXiv:1511.07735
[7] Hachisu H *et al.* 2014 *Opt. Lett.* **39** 4072
[8] Leute J *et al.* 2015 arXiv:1507.04754
[9] Fujieda M, Kumagai M, Nagano S, Yamaguchi A, Hachisu H and Ido T 2011 *Opt. Express* **19** 16498
[10] Fujieda M, Piester D, Gotoh T, Becker J, Aida M and Bauch A 2014 *Metrologia* **51** 253
[11] http://www.bipm.org/en/bipm/tai/tai.html
[12] Yamaguchi A, Shiga N, Nagano S, Li Y, Ishijima H, Hachisu H, Kumagai M and Ido T 2012 *Appl. Phys. Express* **5** 022701
[13] Hachisu H and Ido T 2015 *Jpn. J. Appl. Phys.* **54** 112401
[14] Huntemann N, Okhapkin M, Lipphardt B, Weyers S, Tamm C and Peik E 2012 *Phys. Rev. Lett.* **108** 090801
[15] Huntemann N, Lipphardt B, Okhapkin M, Tamm C, Peik E, Taichenachev A V and Yudin V L 2012 *Phys. Rev. Lett.* **109** 213002
[16] Hong F -L *et al.* 2009 *Opt. Lett.* **34** 692
[17] Yasuda M *et al.* 2012 *Appl. Phys. Express* **5** 102401
[18] Madej A A, Dubé P, Zhou Z, Bernard J E and Gertsvolf M 2012 *Phys. Rev. Lett.* **109** 203002
[19] Park C Y *et al.* 2013 *Metrologia* **50** 119

[20] Akamatsu D, Inaba H, Hosaka K, Yasuda M, Onae A, Suzuyama T, Amemiya M and Hong F -L 2014 *Appl. Phys. Express* **7** 012401
[21] Huang Y, Guan H, Liu P, Bian W, Ma L, Liang K, Li T and Gao K 2016 *Phys. Rev. Lett.* **116** 013001
[22] Campbell G K *et al.* 2008 *Metrologia* **45** 539
[23] Baillard X *et al.* 2008 *Eur. Phys. J.* D **48** 11
[24] Le Targat R *et al.* 2013 *Nat. Commun.* **4** 2109
[25] Falke S *et al.* 2011 *Metrologia* **48** 399
[26] Falke S *et al.* 2014 *New J. Phys.* **16** 073023
[27] Tanabe T *et al.* 2015 *J. Phys. Soc. Jpn.* **84** 115002
[28] Lin Y -G *et al.* 2015 *Chin. Phys. Lett.* **32** 090601

Atomic clocks and dark-matter signatures

Andrei Derevianko

Physics Department, University of Nevada, Reno, NV 89557, USA

E-mail: andrei@unr.edu

Abstract. Recent developments in searches for dark-matter candidates with atomic clocks are reviewed. The intended audience is the atomic clock community.

1. Dark matter problem and what we know about dark matter

Multiple astrophysical observations suggest that the ordinary (luminous or baryonic) matter (or simply "us") contributes only $\sim 5\%$ to the total energy density budget of the Universe. Exacting the nature of the two other constituents, dark matter (DM) and dark energy, is a grand challenge to the contemporary state of knowledge. It is believed that dark matter is required for galaxy formations, while dark energy leads to the accelerated expansion of the Universe. The distinction between dark matter and dark energy can be formalized by treating them as cosmological fluids: they have different equations of state, dark matter is being pressureless while dark energy exerting negative pressure. For further details I refer the reader to S. Weinberg textbook on cosmology [1] and also to reviews [2, 3, 4, 5]. Below I primarily focus on dark matter, although some aspects of the discussion, e.g., how an atomic clock may couple to various fields in a detectable way, would also apply to hypothetical dark energy fields.

All the evidence for dark matter (galactic rotation curves, gravitational lensing, peaks in the cosmic microwave background spectra, etc) comes from galactic scale observations. The challenge lies in extrapolating down from the 10 kpc characteristic distances to the laboratory scales. This is a truly vast extrapolation scale and a large number of theoretical models can fit the observations. The prevailing view is that all the theoretical constructs have to conform to the cold dark matter (CDM) model (for the purpose of the following discussion all DM objects move with velocities much smaller than the speed of light). More broadly, Λ-CDM model describes the large-scale structure formation of the Universe [6].

Our galaxy, the Milky Way, is embedded into a DM halo and rotates through the halo. Astrophysical simulations provide estimates of DM properties in the Solar system (see, e.g., [7]). The DM halo density distribution is usually parameterized by the Navarro-Frenk-White profile that is fit to various astrophysical observations. Based on these fits, the DM energy density in the vicinity of Solar system is estimated to be $\rho_{\rm DM} \approx 0.3\,{\rm GeV/cm^3}$, corresponding to about one hydrogen atoms per three cubic cm. Further, in the DM halo reference frame, the velocity distribution of DM objects is nearly Maxwellian with the dispersion of $v_{\rm vir} \sim 270\,{\rm km/s}$ (referred to as the virial velocity in the literature) and a sharp cut-off at the galactic escape velocity $v_{\rm esc} \approx 650\,{\rm km/s}$. Further, the Milky Way is a spiral galaxy rotating through the DM halo. In particular, the Sun moves through the DM halo at galactic velocities $v_{\rm g} \approx 230\,{\rm km/s}$.

"Dark" universe does not absorb or emit electromagnetic radiation (dark particles are expected to carry no electric charge) and all the evidence is based on the gravitational interaction between the dark and luminous matter. It means that the gravitational interaction between DM objects and between DM and ordinary matter is assumed. But the gravity is weak on laboratory scales, so one needs to introduce additional interactions in order to make DM detectable by our instruments. Atomic clocks ultimately rely on locking electronic counters (time = (period of oscillation) × (number of oscillations)) to frequencies of atomic transitions. If DM affects atomic frequencies, then the DM effects can be potentially measured in terrestrial experiments. In Sec. 2, I review a class of couplings to DM sector ("portals") that translates into either transient or oscillating variations of fundamental constants. Such variations can pull on the transition frequencies and thereby affect atomic clocks.

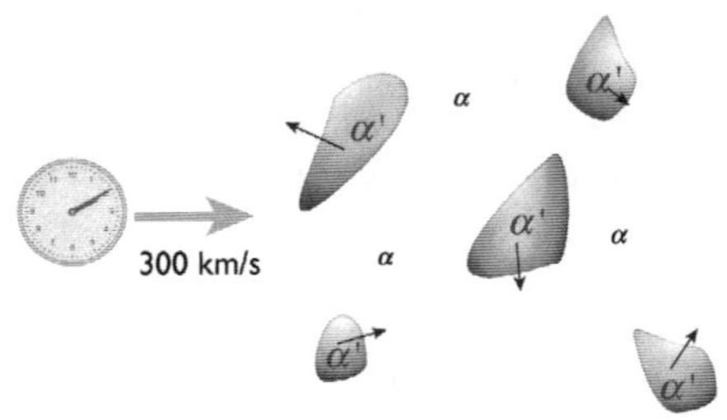

Figure 1. (Colour online) An atomic clock sweeps through the dark matter halo at galactic velocities. Dark matter is assumed to be composed of extended objects (or clumps). If there the difference of fundamental constants (such as the fine-structure constant α in the figure) inside and outside the clumps, the clumps can cause the clock to slow down or speed up [8].

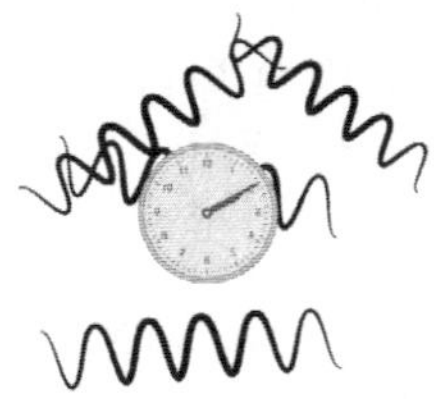

Figure 2. (Colour online) Ultra-light fields can lead to oscillating fundamental constants at the field Compton frequency. By Fourier-transforming a time series of clock frequency measurements, one could search for peaks in the power spectrum and potentially identify DM presence [9].

Considering a wide variety of DM models, it is natural to ask if atomic clocks, being ultra-precise sensors, can be used to verify predictions of certain classes of DM models. There were two recent proposals identifying such possibilities [8, 9]. These two approaches lead to distinct DM signatures and are illustrated in Fig. 1 and Fig. 2.

Fig. 1 depicts collisions of extended DM objects (or DM clumps) with an atomic clock traveling through the DM halo at galactic velocities. It is assumed that DM clumps effectively change the values of fundamental constants (such as the fine structure constant α and/or masses of elementary particles). Such ideas based on correlated measurements using a network of atomic clocks were introduced in Ref. [8] and I review them in Sec. 4. In the DM model of Ref. [8] the clumps are identified with topological defects (TD) formed due to the self-interaction of ultralight fields. DM clumps can be formed in other models as well, so the TD model is not required per se. The question of *microstructure* of DM is an open question [10]. From an observational point of view we can pose a general question: can atomic clocks or other devices detect our motion through the "preferred" DM halo reference frame?

Another possibility (see Fig. 2) is DM composed of ultralight fields, considered by Arvanitaki et al. [9]. Such fields can lead to oscillating fundamental constants at the Compton frequency of the field. By Fourier-transforming a time series of clock frequency measurements, one could hunt for peaks in the power spectrum and potentially identify DM presence.

2. Dark matter portals and new regimes of variation of fundamental constants

A systematic phenomenological approach to DM-SM (Standard Model) sector couplings, is that of the so-called portals [11], when the gauge-invariant operators of the SM fields are coupled to the operators that contain DM fields. We focus on the SM-DM interactions in the form of the linear ($k = 1$, or dilaton) and quadratic ($k = 2$) scalar portals,

$$-\mathcal{L}_k^{\rm int} = \phi^k \left(\frac{m_e c^2 \bar{\psi}_e \psi_e}{\Lambda_{k,e}^k} + \frac{m_p c^2 \bar{\psi}_p \psi_p}{\Lambda_{k,p}^k} - \frac{1}{4\mu_0 \Lambda_{k,\gamma}^k} F_{\mu\nu}^2 + \dots \right) . \quad (1)$$

The terms inside the brackets of Eq. (1) are pieces from the SM sector Lagrangian density. These pieces are weighted with inverses of high-energy scales $\Lambda_{k,X}$ which parametrize unknown coupling constants. In particular, $m_{e,p}$ and $\psi_{e,p}$ are electron and proton masses and fields ($\bar{\psi} = \psi^\dagger \gamma_0$), and $F_{\mu\nu}$ are the electromagnetic field tensor components. Linear portals were considered in Ref. [9] while quadratic portals — in Ref. [8]. The energy scales are constrained from below from terrestrial experiments and astrophysical bounds. Generally such constraints on the linear portals are far more stringent than those on the quadratic portals [12].

It is instructive to compare the portals (1) with the QED Lagrangian governing atomic physics:

$$-\mathcal{L}^{\rm QED} = -i\hbar c \bar{\psi}_e D \psi_e + m_e c^2 \bar{\psi}_e \psi_e + \frac{1}{4\mu_0} F_{\mu\nu}^2 . \quad (2)$$

It is clear that Eq. (1) when added to $\mathcal{L}^{\rm QED}$ leads to the modulation of fundamental constants by DM fields (notice that in the SI $\alpha = \mu_0 e^2 c/(4\pi\hbar)$):

$$m_{e,p}^{\rm eff} = m_{e,p} \times \left(1 + \frac{\phi(x,t)^k}{\Lambda_{k,e,p}^k}\right), \quad \alpha^{\rm eff} \approx \alpha \times \left(1 + \frac{\phi(x,t)^k}{\Lambda_{k,\gamma}^k}\right) . \quad (3)$$

Masses of quarks and other particles and various couplings (e.g., strong force) are affected in a similar way.

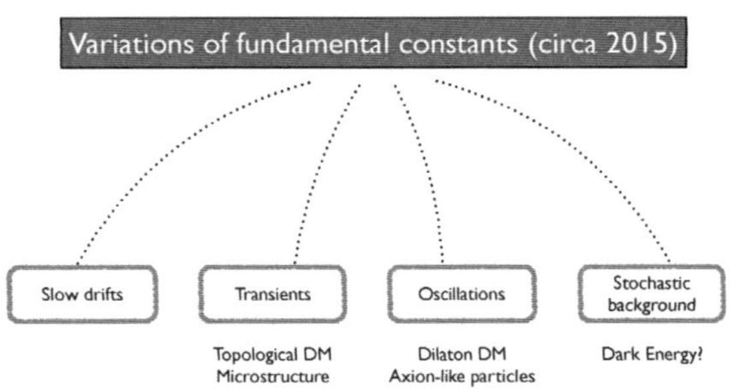

Figure 3. (Colour online) Various scenarios of space-time variations of fundamental constants.

Now I invite the reader to take a closer look at the implications of Eq. (3). For topological defects, outside the clump, by assumption, $\phi \to 0$ and these portals renormalize masses and couplings only when the TD core overlaps with the quantum device. This, via Eq.(3), leads to *transient* variation of fundamental constants as the defect passes through the device. The ultra-light fields oscillate at the Compton frequency, $\phi(\boldsymbol{r}, t) = A\cos(m_\phi c^2 t - k_\phi \cdot \boldsymbol{r} + \cdots)$, where m_ϕ is the mass associated with the field and k_ϕ is the wave-vector of the field. Thereby Eq. (3) leads to *oscillating* fundamental constants.

The subject of space-time variation of fundamental constants has been explored previously by the atomic clock community, however the attention until recently has focused on the searches for slow drifts of fundamental constants [13]. As we see, DM searches with atomic clocks can be translated into searches for new regimes of space-time variation of fundamental constants. These various regimes are summarized in Fig. 3.

Operationally an atomic clock locks onto an atomic transition (frequency f_0). One could parameterize the variation of the clock frequency in terms of sensitivity coefficients K_X [14]

$$\frac{\delta(f_0/U)}{f_0/U} = \frac{\delta V}{V}, \quad V = \alpha^{K_\alpha} \left(\frac{m_q}{\Lambda_{\rm QCD}}\right)^{K_q} \left(\frac{m_e}{m_p}\right)^{K_{me/p}}. \tag{4}$$

Here U is the unit of frequency, m_q is a quark mass, $\Lambda_{\rm QCD}$ is the quantum chromodynamics mass scale, and m_e/m_p is the electron to proton mass ratio. In general, one distinguishes between two broad classes of atomic clocks: microwave and optical clocks. Microwave clocks, such as H, Rb and Cs, operate on hyperfine transitions; these depend both on $\Lambda_{\rm QCD}$ and α. Nuclear-structure-dependent coefficient K_q exhibits non-monotonic behavior [15] and $K_{me/p} = 1$ for hyperfine transitions. K_α grows with nuclear charge due to increasing relativistic effects. Optical clocks are only sensitive to the variation of α. It means that by comparing clocks of various sensitivity and type, one could discern individual terms in the portals (1).

3. Relation to dark-matter models

While the on-going particle physics DM searches focus on particles with masses $\sim 1 - 10^3\,$GeV, here we consider an alternative: ultralight fields. Depending on the initial field configuration at early cosmological times, light fields could lead to DM oscillations about the minimum of their potential, or form stable spatial configurations due self-interaction potentials. The former possibility leads to fields oscillating at Compton frequency (dilaton-type DM [9]) and the latter to the formation of topological defects (TD) such as domain walls, strings and monopoles (TD-type DM [8]). These two models were introduced in Sec. 1.

The maximum values of portal couplings (1) depend on the DM field amplitudes A; these can related to the dark-matter energy density in the Solar system neighborhood in the assumption that such models individually saturate the DM energy density. For example, a gravitationally interacting gas of monopole TDs is a pressureless gas and can account for the observed properties of cold DM [8]. The spatial extent of the monopole d is given by the Compton wavelength $d \sim \hbar/(m_\phi c)$. In this model, monopoles populate DM halo. As discussed in Sec. 1, while the velocity distribution of monopoles is quasi-Maxwellian in the halo reference frame, the Earth is moving through the halo at galactic velocities $v_g \sim 300\,\mathrm{km/s}$. The DM field amplitude inside the monopole can be expressed as $A = \left(\frac{\rho_{\rm DM}(\hbar c)^3}{m_\phi c^2} \frac{\mathcal{T}}{\hbar} \frac{v_g}{c}\right)^{1/2}$, where $\mathcal{T}$ is the average time between consecutive collisions with the Earth (assuming $d \lesssim R_\oplus$, the Earth radius). For dilaton-type DM, $A = \frac{\hbar}{m_\phi c}\sqrt{2\rho_{\rm DM}}$.

4. Dark matter signatures

Inevitable clock noise (especially flicker noise) for a single clock can mimic an encounter with a DM clump. Moreover, even if the DM induced clock glitches are large, an unsuspecting experimentalist is likely to discard the event and attribute it to something perhaps unexplained but mundane (see blog post [16]). In the DM TD searches the solution [8] is to rely on a geographically distributed clock *network* which seeks synchronous propagation of clock "glitches" at galactic velocities through the network. This approach is similar to the proposed magnetometer GNOME network [17] or gravitational wave detection [18]. In the context of clocks, two spatially-separated and initially-synchronized identical clocks are expected to exhibit a distinct de-synchronization and re-synchronization pattern, shown in Fig. 4 for an encounter with a relatively thin (compared to the distance between the clocks) clump. The duration of the characteristic "hump" is given by $\Delta t = l/v$, with l being the distance between the clocks and v being the relative velocity of the encounter. If $v \sim v_g$, $\Delta t \sim 3\,\mathrm{s}$ for a trans-continental network ($l \sim 1,000\,\mathrm{km}$) and $\Delta t \sim 150\,\mathrm{s}$ for clocks onboard navigational satellites ($l \sim 50,000\,\mathrm{km}$).

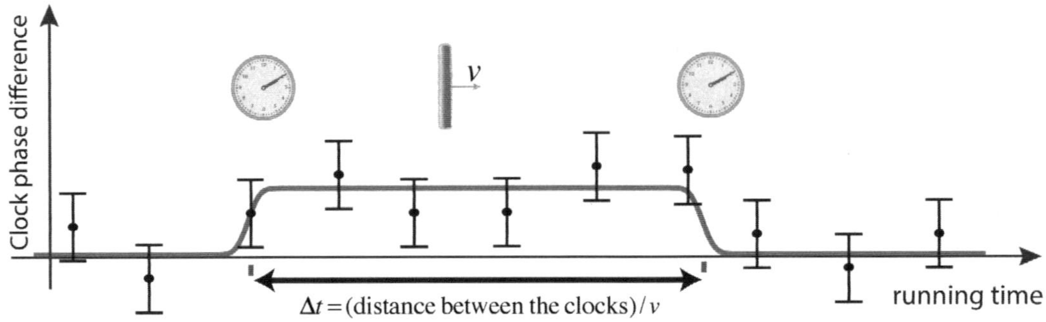

Figure 4. (Colour online) Spatially-separated and initially-synchronized identical clocks are expected to exhibit a distinct de-synchronization and re-synchronization pattern due to an encounter with a dark matter clump.

As indicated in Fig. 4, the measurements are carried out in the presence of clock noise characterized by the Allan variance $\sigma_y(T_m)$, where T_m is the sampling interval. The signal-to-noise ratio [8] reads

$$S/N = \frac{c\hbar\rho_{\mathrm{DM}}\mathcal{T}d^2}{T_m\sigma_y(T_m)\sqrt{2T_m v/l}}\sum_X K_X\Lambda_X^{-2}\,. \tag{5}$$

Here ρ_{DM} is the DM energy density in the vicinity of the Solar system and $\mathcal{T}$ is the characteristic time between subsequent DM encounters. The above S/N ratio has been recently re-evaluated [19] taking into account time-transfer link noise. Notice that the derived S/N is for *two* identical clocks. The statistical confidence in the event can be improved dramatically with many nodes in the network and if one could detect a synchronous propagation of the sought signal through the network.

To fully disentangle the energy scales Λ_α, Λ_q, and $\Lambda_{me/p}$ one would require at each node at least two microwave clocks and one optical clock or three microwave clocks of different types. As we track transient variation of fundamental constants, it is sufficient to have identical clocks on different nodes. This differs from the search for a slow-drift-in-time of fundamental constants where a typical experiment (for example, [13]) uses two co-located clocks with different sensitivity coefficients.

The noise scales as $T_m^{3/2}\sigma_y(T_m)$. Typically $\sigma_y(T_m)$ scales down as $1/\sqrt{T_m}$ with increasing T_m. Then $S/N \propto 1/T_m$ and it is beneficial to work with shorter measurement intervals. The minimum time between consecutive measurements is determined by several factors: in lattice clocks [20], this would be an atomic ensemble preparation time (about 1 second), in microwave fountain clocks [21] it the time of flight across interrogation chamber (also about 1 second). The lattice clocks [20] may be best suited for TDM search due to their stability and accuracy. The statistical advantage of lattice clocks comes from a large number of atoms being interrogated simultaneously. As the detection schemes would benefit from improved short-term stability, we believe that the search would greatly benefit from advances in Heisenberg-limited time-keeping with entangled atoms.

What are the requirements for the network? Ideally, it should be large and dense. Large in order to maximize the collision cross-section and rate with the defects and dense in order for the smallest DM clumps to be "caught in the net". For example, the distance between two nearby

GPS satellites in the same orbital plane is $\sim 3 \times 10^4$ km, meaning that this is the smallest size the GPS constellation network can be sensitive to.

If the DM events are not observed, it could mean that either the DM model is incorrect, or that the clocks are not sensitive enough to measure the effects. In the latter case, setting $S/N = 1$ establishes the limits on the energy scales (or coupling constants) entering the portals (1). An example of such an analysis can be found in [8] for quadratic portals; it is clear that the existing constraints on Λ_X can be dramatically improved even with the existing GPS network of relatively low-accuracy atomic clocks. While we (the GPS.DM observatory [22]) have initiated efforts on mining a decade-long GPS data set for DM "clump" signatures, the DM search sensitivity can be improved further with a trans-continental network [23, 24] of high-accuracy laboratory clocks. As to the dilaton-type DM, the first experimental bounds have already appeared [25] and I hope that the nascent dark matter searches with atomic clocks discover invaluable DM signatures in terrestrial experiments. In any case they have a powerful potential of improving existing bounds on exotic interactions.

Acknowledgments
This work was supported in part by the US National Science Foundation.

References

[1] Weinberg S 2008 *Cosmology* (Oxford University Press)
[2] Peebles P J E 2003 *Rev. Mod. Phys.* **75** 559–606
[3] Bertone G, Hooper D and Silk J 2005 *Phys. Rep.* **405** 279–390
[4] Feng J L 2010 *Ann. Rev. Astro. Astrophys.* **48** 495–545
[5] Matarrese S, Colpi M, Gorini V and Moschella U 2011 *Dark Matter and Dark Energy: A Challenge for Modern Cosmology* (Berlin Heidelberg: Springer)
[6] Blumenthal G R, Faber S M, Primack J R and Rees M J 1984 *Nature* **311** 517–525
[7] Nesti F and Salucci P 2013 *Journal of Cosmology and Astroparticle Physics* **2013** 16
[8] Derevianko A and Pospelov M 2014 *Nature Phys.* **10** 933
[9] Arvanitaki A, Huang J and Van Tilburg K 2015 *Phys. Rev. D* **91** 015015
[10] Berezinsky V S, Dokuchaev V I and Eroshenko Y N 2014 *Physics-Uspekhi* **57** 1–36
[11] Essig R, Jaros J J A, Wester W et al. 2013 Working Group Report: New Light Weakly Coupled Particles Preprint 1311.0029
[12] Olive K A and Pospelov M 2008 *Phys. Rev. D* **77** 043524
[13] Rosenband T, Hume D B, Schmidt P O, Chou C W, Brusch A, Lorini L, Oskay W H, Drullinger R E, Fortier T M, Stalnaker J E, Diddams S A, Swann W C, Newbury N R, Itano W M, Wineland D J and Bergquist J C 2008 *Science* **319** 1808–12
[14] Flambaum V V and Dzuba V A 2009 *Can. J. Phys.* **87** 25–33
[15] Flambaum V and Tedesco A 2006 *Phys. Rev. C* **73** 055501
[16] Derevianko A When would an unanticipated new physics event be apparent to an unsuspecting experimentalist? URL `http://wp.me/p2Z9xm-8Z`
[17] Pustelny S, Kimball D F, Pankow C, Ledbetter M P, Wlodarczyk P, Wcislo P, Pospelov M, Smith J R, Read J, Gawlik W and Budker D 2013 *Annalen der Physik* **525** 659–70
[18] Abbott B P and et al 2016 *Phys. Rev. Lett.* **116** 61102
[19] Calonico D, Inguscio M and Levi F 2015 *EPL* **110** 40001
[20] Derevianko A and Katori H 2011 *Rev. Mod. Phys.* **83** 331–348
[21] Wynands R and Weyers S 2005 *Metrologia* **42** S64
[22] GPS.DM observatory URL `http://www.dereviankogroup.com/gps-dm/`
[23] Predehl K, Grosche G, Raupach S M F, Droste S, Terra O, Alnis J, Legero T, Hänsch T W, Udem T, Holzwarth R and Schnatz H 2012 *Science* **336** 441–444
[24] Lisdat C, Grosche G, Quintin N, et al. 2015 Preprint 1511.07735
[25] Van Tilburg K, Leefer N, Bougas L and Budker D 2015 *Phys. Rev. Lett.* **115** 011802

Journal of Physics: Conference Series **723** (2016) 012044 doi:10.1088/1742-6596/723/1/012044

General relativistic effects in quantum interference of "clocks"

M Zych[1], I Pikovski[2,3], F Costa[1] and Č Brukner[4,5]

[1] Centre for Engineered Quantum Systems, School of Mathematics and Physics, The University of Queensland, St Lucia, QLD 4072, Australia
[2] ITAMP, Harvard-Smithsonian Center for Astrophysics, Cambridge, MA 02138, USA
[2] Department of Physics, Harvard University, Cambridge, MA 02138, USA
[4] Institute for Quantum Optics and Quantum Information, Austrian Academy of Sciences, Boltzmanngasse 3, A-1090 Vienna, Austria.
[5] Faculty of Physics, University of Vienna, Boltzmanngasse 5, A-1090 Vienna, Austria

E-mail: m.zych@uq.edu.au

Abstract. Quantum mechanics and general relativity have been each successfully tested in numerous experiments. However, the regime where both theories are jointly required to explain physical phenomena remains untested by laboratory experiments, and is also not fully understood by theory. This contribution reviews recent ideas for a new type of experiments: quantum interference of "clocks", which aim to test novel quantum effects that arise from time dilation. "Clock" interference experiments could be realised with atoms or photons in near future laboratory experiments.

1. Introduction

Quantum mechanics and general relativity have successfully passed numerous experimental verifications. Yet, all experiments that detected gravitational effects on quantum systems are still fully consistent with the non-relativistic, Newtonian theory [1–3]. Similarly, in the present-day tests of general relativity the degrees of freedom displaying general relativistic effects can consistently be described by classical physics [4–9]. Therefore, there is a fundamental interest in finding feasible experiments that could probe the regime where the relevant degrees of freedom requite both quantum and general relativistic description.

This contribution reviews recent proposals for testing the overlap between quantum theory and general relativity with composite quantum particles, which have dynamical internal degrees of freedom (DOF). Crucially, such systems can be seen as ideal "clocks". Their study has revealed novel quantum effects from the relativistic time dilation arising in interference experiments with such quantum "clocks" [10–16]. According to quantum theory any system can propagate along multiple paths in superposition, while general relativity predicts that the time elapsed for a system depends on its path, which is simply the effect of time dilation. Therefore, in the regime where both theories apply, time dilation will generally entangle the internal states of a "clock" to the path. In interference experiments with "clocks" this will cause loss and revivals (for periodic "clocks") of the interference fringes [10]. Such experiments could be realised with atoms or molecules, with their electronic or vibrational energy levels as "clocks", neutrons or electrons – with "clock" implemented in spin precession, etc. Analogous effects arise also for photons [11] due to the Shapiro effect [17]. Moreover, the interferometric visibility is affected stronger for larger systems. For macroscopic systems the loss of visibility can become very rapid and, for all practical purposes, irreversible – time dilation can thus result in decoherence [12].

The prime motivation to study the effects of proper time on quantum systems is the fact that they are particularly promising for laboratory experiments: Already current techniques allow for interference experiments with molecules comprising more than 800 atoms [18]. More generally, understanding how general relativity can affect quantum experiments will simply be crucial for many future tests aimed at preparing and detecting quantum states of macroscopic systems [19–21]. Future large-scale quantum technologies involving e.g. quantum communication through satellites [22,23] will also have to account for general relativistic effects – as much as the present-day classical technologies, such as the Global Positioning System, have to incorporate such effects for their correct functioning. The impact of general relativity on future quantum communication and information technologies is already today an active research field [24]. Finally, tests of general relativity are performed with ever increasing precision [9] and quantum effects will unavoidably become relevant in future tests at some scale.

2. Composite particles as "clocks"

For a relativistic particle the square of its four momentum p^μ, $\mu = 0,..,3$ is an invariant quantity, which describes the energy of the particle in its rest frame [25]: $-H_{rest}c^2 = \sum p^\mu g_{\mu\nu} p^\nu$, where $g_{\mu\nu}$ is the metric tensor, with signature $(-,+,+,+)$, and c is the speed of light. For a composite system, H_{rest} comprises not only the rest mass mc^2 but also any binding or kinetic energies of the internal degrees of freedom – the system's entire internal Hamiltonian H_{int} – and thus $H_{rest} = mc^2 + H_{int}$. The energy in an arbitrary reference frame is given by the p_0 component and for a static symmetric metric reads $H \equiv cp_0 = \sqrt{-g_{00}(c^2 p_j p^j + H_{rest}^2)}$.

Legendre transform of H yields the Lagrangian $L = L_{rest}\sqrt{-g_{\mu\nu}\dot{x}^\mu \dot{x}^\nu}/c$, where L_{rest} is the rest frame Lagrangian of the composite system (and a Legendre transform of H_{rest}), $x^\mu(t)$ are the coordinates of the system's world line, and the over-dots denote derivative with respect to the coordinate time t. Furthermore, the proper time element along $x^\mu(t)$ reads $d\tau := \sqrt{-g_{\mu\nu}dx^\mu dx^\nu}/c$ and $\sqrt{-g_{\mu\nu}\dot{x}^\mu \dot{x}^\nu}/c \equiv \dot{\tau}$ is simply the "speed" of proper time with respect to the coordinate time t. This yields $L \equiv L_{rest}\dot{\tau}$. The action of a relativistic composite system along a world line γ is therefore $S = \int_\gamma L_{rest}\dot{\tau}dt \equiv \int_\gamma L_{rest}d\tau$, whereas the proper time elapsing along γ is given by $\int_\gamma \dot{\tau}dt \equiv \int_\gamma d\tau$. Thus, internal degrees of freedom of a composite point-like system evolve according to proper time along the system's world line – "measuring" the elapsing proper time. In this sense relativistic point-like particles with internal dynamics can be seen as ideal clocks.

Journal of Physics: Conference Series **723** (2016) 012044 doi:10.1088/1742-6596/723/1/012044

Any physical system thus far used as a clock in tests of time dilation was of course not fundamentally point-like. For the above framework to apply it is only required that the additional effects on the system due to its finite size are negligible. This is the case if the linear dimension of the system is sufficiently small compared to the length scale where the space-time curvature is non-negligible. In general, the higher the precision of the clock, the smaller the four-volume it can occupy before it becomes necessary to include its finite size, see also ref. [26]. The above conditions also allow to define a generally covariant notion of the centre of mass for an extended system [27].

To the lowest order in the post-Newtonian expansion (for a static symmetric metric) [28] the Hamiltonian for a composite system reads [12]

$$H = H_{\text{ext}} + H_{int}\left(1 + \frac{\Phi(x)}{c^2} - \frac{p^2}{2m^2c^2}\right), \tag{1}$$

where $H_{\text{ext}} = mc^2 + p^2/2m + m\Phi(x) + f(p,\Phi)/c^2$ describes the external DOFs and $f(p,\Phi)/c^2$ incorporates the relevant relativistic corrections and any relevant external forces (e.g. required for trapping the "clock"). In the above approximation

$$\dot{\tau} \approx 1 + \frac{\Phi(x)}{c^2} - \frac{v^2}{2c^2} \tag{2}$$

and the interaction terms $-H_{int}\frac{p^2}{2m^2c^2}$ and $H_{int}\frac{\Phi}{c^2}$ in eq. (1) can immediately be recognised as special relativistic (velocity dependent) and gravitational time dilation of internal dynamics, respectively.

The classical theory of composite particles can be canonically quantised to yield a Hamiltonian description of a quantum "clock" in curved space-time (or, equivalently, a path integral quantisation of the clock action can be carried out). Note, that in quantum theory both external and internal DOFs are described by operators. Since the particle in general propagates along multiple world lines in superposition, the interactions describing time dilation in quantum theory lead to entanglement between internal and external degrees of freedom of the "clock". This yields the novel phenomena in interference experiments with composite systems [10–12] which we review in this contribution.

The two terms describing time dilation in eq. (1) can also be justified by considering the mass-energy equivalence (for the mass of the particle and its internal energy) [29,30]. It entails that changing the particle's internal energy by E changes its mass m by E/c^2, i.e. $m \to m + E/c^2$. The principle applies to any internal energy eigenstate and in quantum theory, due to its linear structure, the principle also applies to any superposition of internal energy eigenstates, thus in quantum theory $m \to m\mathbb{1} + H_{int}/c^2$, where $\mathbb{1}$ is the identity operator on the state space of the internal degrees of freedom. Incorporating the above into the non-relativistic quantum Hamiltonian $mc^2 + \frac{p^2}{2m} + m\phi$ yields the interaction terms present in (1), see ref. [13] for further details and for the new possible tests of the Einstein Equivalence Principle in quantum theory.

3. Interference of "clocks"

So far time dilation was tested by comparing time measured by clocks that followed different, effectively classical, world lines [6–9]. Such tests can be seen as realisations of the famous "twin paradox", a thought experiment where twin siblings embark on separate voyages from the same event in space-time and when their world lines meet again, they discover that they have aged differently. The age of each twin depends on the proper time elapsed along his/her path. Here we discuss a quantum version of the time dilation experiments: a single "clock" (a particle with internal dynamics, e.g. an atom in a superposition of two internal energy levels) which follows in superposition two world lines along which different proper time elapses [10]. This can be seen as a quantum version of the twin paradox, where a "quantum only child" embarks on two different voyages in superposition and when the world lines are recombined, the traveller has aged in superposition by different amounts of time – becoming older-and-younger than himself, in a superposition.

For the realisation of the "quantum twin paradox" consider a Mach-Zehnder interferometer where the quantum "clock" follows in superposition two fixed paths[1] γ_1 and γ_2, see figure 1. The state of the particle inside the setup can be described by

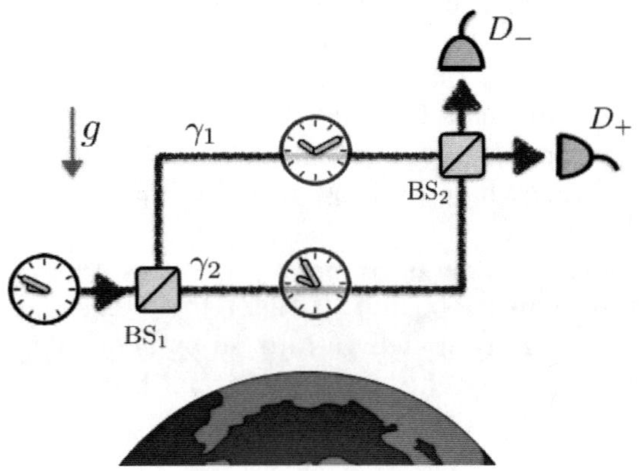

Figure 1: Mach-Zehnder interferometer for testing the quantum version of the twin paradox with interfering "clocks". The setup consists of two beam splitters ($BS_{1,2}$), two detectors $D_\pm$ and is stationary in a laboratory frame on earth, in a gravitational field g. The interferometric paths γ_1 and γ_2 are arranged such that different proper time elapses along each of them. The "clock" can be any small particle with time-evolving internal degrees of freedom. Such an interference experiment will not only display a phase shift, but also (periodic – for a periodic "clock") modulations of the interferometric visibility: The interferometric visibility will be diminished to the extent to which the information about the path becomes available from the "clock" due to time dilation.

a superposition $\frac{1}{\sqrt{2}}(i|\Psi_1\rangle + |\Psi_2\rangle)$ where the states $|\Psi_1\rangle$, $|\Psi_2\rangle$ are associated with the paths γ_1, γ_2, respectively. With the Hamiltonian (1) we can write these states explicitly

$$|\Psi_i\rangle = e^{-\frac{i}{\hbar}\int_{\gamma_i} dt \left[H_{ext}+H_{int}\left(1-\frac{p^2}{2m^2c^2}+\frac{\Phi(x)}{c^2}\right)\right]}|x^{in}\rangle|\tau^{in}\rangle. \quad (3)$$

where $|x^{in}\rangle$ and $|\tau^{in}\rangle$ denote the initial states of the external and the internal DOFs, respectively. Using eq. (2) the internal state evolving along γ_i can be written as

$$|\tau_i\rangle = e^{-\frac{i}{\hbar}\int_{\gamma_i} dt\, H_{int}\dot{\tau}}|\tau^{in}\rangle, \quad (4)$$

where we used $\frac{v^2}{2c^2} \equiv \frac{p^2}{2mc^2}$ (for a particle on a fixed path, in the small velocity regime). This can be further simplified into $|\tau_i\rangle = e^{-\frac{i}{\hbar}\int_{\gamma_i} d\tau\, H_{int}}|\tau^{in}\rangle$ where we explicitly see that

[1] We neglect the proper time effects due to the finite spread of the "clock's" wave function along each path as compared to the effects stemming from the proper time difference between the two paths. This is justified for the experiment realised with massive systems like atoms or molecules.

Journal of Physics: Conference Series **723** (2016) 012044 doi:10.1088/1742-6596/723/1/012044

the internal state of the particle following path γ_i evolves according to the proper time along that path. In the interferometric experiment the paths are finally recombined, which yields the following probabilities for registering the particle in the detectors $D_\pm$

$$P_\pm = \frac{1}{2} \pm \frac{1}{2}|\langle\tau_1|\tau_2\rangle| \cos(\Delta\phi), \tag{5}$$

where $\Delta\phi$ is the total relative phase, which can also be explicitly evaluated from eq. (3), see also sec. 6. In eq. (5), the quantity $\mathcal{V} = |\langle\tau_1|\tau_2\rangle|$ describes the visibility of this interference pattern and explicitly reads

$$\mathcal{V} = |\langle e^{-\frac{i}{\hbar}H_{int}\Delta\tau}\rangle|, \tag{6}$$

where $\Delta\tau := \int_{\gamma_2} d\tau - \int_{\gamma_1} d\tau$ is the proper time difference between γ_1 and γ_2, and the expectation value is taken with respect to the initial internal state of the "clock", here $|\tau^{in}\rangle$.

The visibility $\mathcal{V}$ remains maximal if there is no time dilation between the interfering paths, in particular it is always maximal in the non-relativistic limit (when eq. (1) reduces to $H_{int} + p^2/2m + m\Phi(x)$). On the other hand, $\mathcal{V}$ vanishes in the classical theory – even in curved space-time – no interference takes place for classical particles. Modulations in the visibility described by eq. (6), are a quantum effect that arises only due to the relativistic time dilation and can thus test the regime where both quantum and general relativistic phenomena play a role.

Equation (6) quantifies to which extent an internal state evolving along γ_1 becomes distinguishable from that evolving along γ_2 due to time dilation. If the "clock" has a characteristic evolution time $t_\perp$ after which its initial state evolves into an orthogonal one, also called the orthogonalisation time, the visibility vanishes if $\Delta\tau = t_\perp$. Both $\mathcal{V}$ and $t_\perp$ can be expressed in terms of internal energy moments $\langle H^n_{int}\rangle$:

$$\mathcal{V} = \left|\sum_{n=0}^{\infty}\left(\frac{-i\Delta\tau}{\hbar}\right)^n \langle H^n_{int}\rangle\right| \tag{7}$$

and $\pi\hbar/t_\perp \leq (2\langle H^n_{int}\rangle)^{\frac{1}{n}}$, for $n > 0$ [31], which is valid also for mixed states. Keeping only up to second moments in the right-hand side of eq. (7) gives:

$$\mathcal{V} \approx \sqrt{1 - (\Delta\tau\Delta H_{int}/\hbar)^2}, \tag{8}$$

where $\Delta H_{int} = \sqrt{\langle H^2_{int}\rangle - \langle H_{int}\rangle^2}$ is the internal energy variance. In the same approximation $t_\perp \geq \frac{\pi\hbar}{2\Delta H_{int}}$ [32–35].

The which-way information revealed by the "clock" can be quantified by the distinguishability of the paths, here: $\mathcal{D} = \sqrt{1 - |\langle\tau_1|\tau_2\rangle|^2}$, see [10] for details. This yields the well-known quantitative expression of quantum complementarity: $\mathcal{V}^2 + \mathcal{D}^2 = 1$, see [36–38].

Journal of Physics: Conference Series **723** (2016) 012044 doi:10.1088/1742-6596/723/1/012044

Equation (6) has been derived here for pure states but it directly generalises to mixed internal states[2] [12] – it only depends on the internal energy distribution, not on the purity of the state. For pure internal states the internal energy distribution gives the rate of internal state change (given by $t_\perp$) and the visibility modulations can be interpreted as due to which-way information that becomes available from the "clock". Mixed states, on the other hand, in general cannot be directly interpreted as "clocks" – e.g. thermal states are stationary – and visibility modulations have two equivalent interpretations: as time dilation of internal dynamics or redshift of internal energy values, see ref [12] for further discussion. For mixed states in general one has loss of visibility without the gain in which-way information, which might become classically "scrambled", and $\mathcal{V}^2 + \mathcal{D}^2 \leq 1$.

Finally, we stress that in the relativistic scenario considered here, the visibility can remain maximal only if the internal state of the interfering particle is an exact eigenstate of its internal Hamiltonian. Such a state only evolves by a global phase and can be seen as a "switched-off clock".

4. Interferometric visibility – examples

Here we discuss interferometric visibility in three experimentally promising implementations of the "clock", and where additionally eq. (6) admits a closed analytic expression.

Periodic "clocks" Consider a "clock" periodically evolving between just two mutually orthogonal states with a period $2t_\perp$. We expect $\mathcal{V} = 0$ for $\Delta\tau = (2k+1)\,t_\perp$, $k = 0, 1, 2, \ldots$, when the "clock" stores maximal which-way information, and $\mathcal{V} = 1$ for $\Delta\tau = 2k\,t_\perp$ – when no which-way information is available. In general for a periodic "clock" eq. (6) reads [10]

$$\mathcal{V} = \left|\cos\left(\frac{\pi\Delta\tau}{2t_\perp}\right)\right|. \tag{9}$$

Such a periodic "clock" can be realised as a superposition of two energy eigenstates $\frac{1}{\sqrt{2}}(|E_1\rangle + |E_2\rangle)$ for which $t_\perp = \pi\hbar/|E_1 - E_2|$. Note that $\nu = 2\pi/2t_\perp \equiv$ is simply the angular frequency of this "clock". A physical implementation of such a periodic "clock" is possible e.g. in internal energy levels of an atom, as is routinely employed in atomic clocks.

Continuous-variables: Photons as "clocks" A "clock" can also be implemented in the position of a photon. For a fixed observer, light passing closer to a massive object moves slower than light passing further from the mass – this effect is known as Shapiro delay [17] (and has been tested in the classical regime by measuring round-trip times of radar [17, 39] and radio waves [40] in the Solar System). If a single photon is used in the setup in figure1, the amplitude following path γ_2 is predicted to reach the final beam splitter (BS_2) later than then the amplitude which followed γ_1, even for paths of equal local lengths[3]. For a Gaussian distribution of frequencies in the photon wave-packet $f(\nu) = (\frac{a^2}{\pi})^{1/4}\exp(\frac{-a^2}{2}(\nu-\nu_0)^2)$ the visibility reads [11] $\mathcal{V} = e^{-\left(\frac{\Delta\tau}{2a}\right)^2}$. Defining

[2] The most general initial state to which eq. (6) applies is $\rho(0) = \rho_{cm}(0) \otimes \rho_{int}(0)$ with $\rho_{cm}(0) = |\psi_{cm}(0)\rangle\langle\psi_{cm}(0)|$, where $|\psi_{cm}(0)\rangle$ is a superposition of two path modes just after BS_1.

[3] The path lengths can be calibrated by maximising interference contrast when the interferometer is placed such that both paths are at the same gravitational potential.

as distinguishable the photon wave packets with overlap $1/e$ or smaller, the precision of a Gaussian "clock" is $t_\perp = 2a$. Such a photon-clock is non-periodic and therefore the visibility is also a non-periodic function of the clock's precision and of the time dilation between the paths.

Thermal states under time dilation – mixed "clocks"

Here we consider a particle in a thermal internal state: $\rho_{in} = e^{-\beta H_{int}}/\mathcal{Z}(\beta)$, where $\mathcal{Z}(\beta) = \mathrm{Tr}\{e^{-\beta H_{int}}\}$ is the partition function, $\beta = 1/k_B T$, k_B is the Boltzmann constant, and T denotes temperature, and eq. (6) reads $\mathcal{V} = \left|\frac{\mathcal{Z}(\beta + i\Delta\tau/\hbar)}{\mathcal{Z}(\beta)}\right|$ (note that $\beta + i\Delta\tau/\hbar$ is effectively a "complex temperature"). As an example we consider that internal DOFs comprise N independent harmonic modes (such as normal modes of a molecule, a nanosphere, or a bulk macroscopic system) [12]. In this case $H_{int} = \sum_{i=1}^{N}(n_i + \frac{1}{2})\hbar\omega_i$, where n_i are the number operators, and ω_i – the mode frequencies and the visibility reads

$$\mathcal{V} = \Pi_{i=1}^{N}\left|\frac{1 - e^{-\beta\hbar\omega_i}}{1 - e^{-(\beta + \frac{i\Delta\tau}{\hbar})\hbar\omega_i}}\right|, \tag{10}$$

In a large temperature limit, $1/\beta >> \hbar\omega_i$, the energy variance of a harmonic mode is β^{-1} and in the approximation (8) the visibility reads $\mathcal{V} \approx \left(1 - (\Delta\tau/\beta\hbar)^2\right)^{N/2}$, which for large N becomes

$$\mathcal{V}(t) \approx e^{-\left(\sqrt{\frac{N}{2}}\frac{k_B T \Delta\tau}{\hbar}\right)^2}. \tag{11}$$

As much as for pure internal states the visibility can be expressed in terms of their orthogonalisation time, for thermal states it can be expressed in terms of heat capacity, in the Einstein's model given by $(\Delta H_{int})^2/(k_B T^2) \approx N k_B$, see ref. [12] for further details; and ref. [41] for the analysis using the Debye model of solids.

5. Gravitational and inertial time dilation effects

Time dilation between the paths of an interferometer will in general depend both on inertial and gravitational effects, $\Delta\tau = \oint dt \left(1 - \frac{p^2}{2m^2c^2} + \frac{\Phi(x)}{c^2}\right)$, where $\oint \equiv \int_{\gamma_1} - \int_{\gamma_2}$, in full analogy to classical tests [6, 7]. Here we discuss two different geometries of the paths that allow isolating either the effects of gravitational time dilation or the special relativistic (inertial) time dilation.

Gravitational time dilation – lifting the "clock" Consider the paths as sketched in figure 1, where acceleration and deceleration in the vertical direction, as well as horizontal velocity (which might also be zero when the system is trapped) are the same for both γ_1 and γ_2. This assures that inertial effects give the same contribution to the proper time elapsing along both trajectories, and the proper time difference stems from the gravitational potential alone. In the approximation of homogeneous gravitational field with acceleration g we thus obtain

$$\Delta\tau = \frac{ght}{c^2}, \tag{12}$$

where h is the height difference between the paths and t is the laboratory time which the "clock" spends at the fixed heights. Figure 2 shows interference pattern predicted for

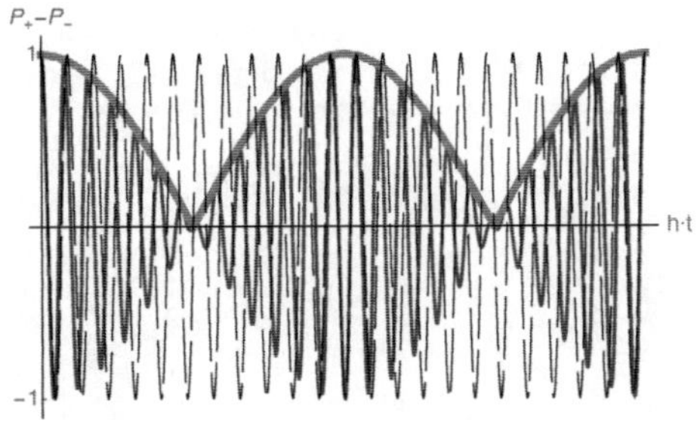

Figure 2: "Clock" interference under gravitational time dilation, plotted as a function of the interferometer area ht [m·s], where h is the height difference between the paths and t is the time for which the "clock" is kept in superposition. In a non-relativistic theory only a relative phase is acquired but the visibility remains maximal (dashed, black line), as well as in a relativistic theory for a "switched-off clock" (particle in an internal energy eigenstate). In a relativistic case with a "switched-on" "clock" (particle in a superposition of internal energy levels) the visibility periodically drops and revives (thick red line), and a correction to the phase shift arises (blue line). Both quantum theory and the notion of gravitational time dilation are necessary to explain the visibility modulations in this experiment.

the experiment realised with a periodic "clock" under gravitational time dilation. In an earth based experiment the only parameter defining $\Delta\tau$ in eq. (12) is the space-time area ht enclosed by the paths of the interferometer and eq. (9) reads $\mathcal{V} = \cos(\frac{\nu g h t}{2c^2})$. For a "clock" with frequency $\nu = 10^{15}$rad/s (which can be realised e.g. in Strontium) the space-time area of the interferometer required to observe full loss and revival of the interference contrast is ~ 10 m·s, where we used $g \approx 10\text{m s}^{-2}$ and $c \approx 3\cdot10^8\text{m s}^{-1}$. The simple cosine law describing loss and revivals of the interference contrast is a key desirable feature of an experiment with a two-level "clock", as the revival of the visibility would allow for an unambiguous distinction between time dilation effects and unavoidable experimental imperfections. While such an experiment is still very challenging, it appears within reach of next generation matter-wave interferometers [21]. An analog version of this test has been recently realised in interference of Rubidium, where the "clock" was implemented via spin precession in an external magnetic field and where time dilation was simulated by preparing magnetic fields of different strengths along the two paths [42].

The visibility in such a gravitational test, but with the interfering particle in a mixed internal state, is plotted in figure 3. The model employed here describes the internal state as N harmonic modes with equal frequency at a temperature $T = 300$K. The visibility thus given by eq. (10) with $\omega_i \equiv \nu$, for $i = 1, .., N$ and $\Delta\tau$ as in eq. (12). Revivals of the visibility occur for such a mixed state as well and depend on the lowest frequency for the considered example (right panel of figure 3). Moreover, the more modes the system comprises, the faster the visibility becomes negligible with increasing the superposition size h or time t (left panel of figure 3); and the shorter is the duration of the revival peaks (right panel of figure 3). For a system with Avogadro number of particles in a superposition of $h = 1$ mm, the visibility is estimated to drop below 1% already after $\sim 2\,\mu$s. Moreover, the frequency of the lowest phonon mode decreases with the size of the system and therefore the revival time becomes increasingly long: For the model above with $\nu \sim 500$rad/s the revival time becomes 10^{17}s, the estimated age of the Universe. See ref. [12] for details on how time dilation can effectively result in decoherence. In the general case, for a system containing modes of arbitrary frequencies (as opposed to all modes at the same frequency) the revival time will be given by the fundamental

Journal of Physics: Conference Series **723** (2016) 012044 doi:10.1088/1742-6596/723/1/012044

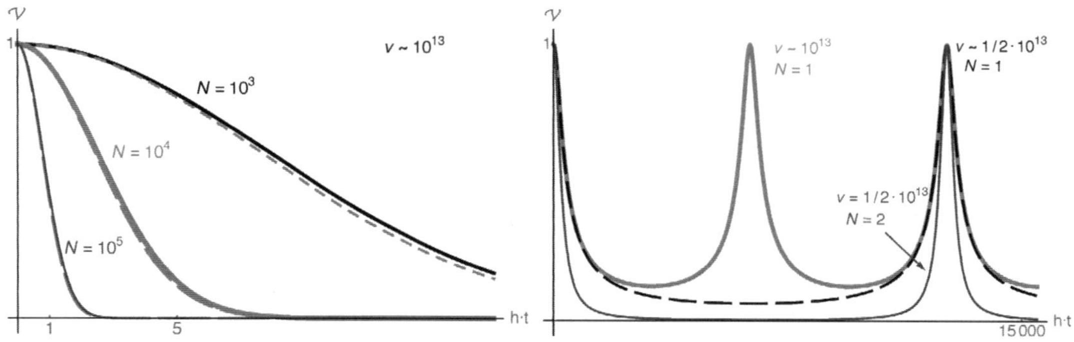

Figure 3: Visibility $\mathcal{V}$ in an interference experiment with a particle in a thermal internal state at $T = 300\text{K}$ under gravitational time dilation. $\mathcal{V}$ is plotted as a function of the space-time area $h \cdot t$ [ms], with h the height difference between the paths and t – time for which the amplitudes are kept at different heights. Left: the visibility loss is faster for systems with higher number N of thermalised modes. Continuous lines represent the analytic result, eq. (10) for $\nu \sim 10^{13}$rad/s. Dashed lines represent the approximation obtained for high T and large N, eq. (11). Right: revivals of the visibility occur for any system of finite size. The revival time is a decreasing function of the lowest frequency of the system: the thick red line shows the visibility for a single harmonic mode at frequency 10^{13}rad/s, dashed black line – for half that value and the thin blue line illustrates how the pattern looks for the latter frequency but two modes. Note that the revivals are shorter the more modes the system comprises.

frequency (the greatest common divisor of all the system frequencies), which in general will be much longer than the period corresponding to the lowest frequency.

Special-relativistic time dilation – rotating the "clock" As an illustrating example, we consider a possible implementation of "clock" interference in the presence of special-relativistic time dilation, i.e. due to the relative velocity between two trajectories. We analyse a scenario in which the relative velocity is due to circular motion of one of the two amplitudes.

Consider a setup where an interferometer is mounted on a rotating platform, figure 4. The "clock" is first sent through a beam splitter which creates a coherent superposition of two amplitudes. One amplitude is sent to a trap near the centre of rotation, while the other is sent to a trap at a distance R from the centre. The two amplitudes are kept in the respective traps for some time t (as measured in the laboratory frame) and then sent back to the beam splitter, where they are recombined to produce interference fringes.

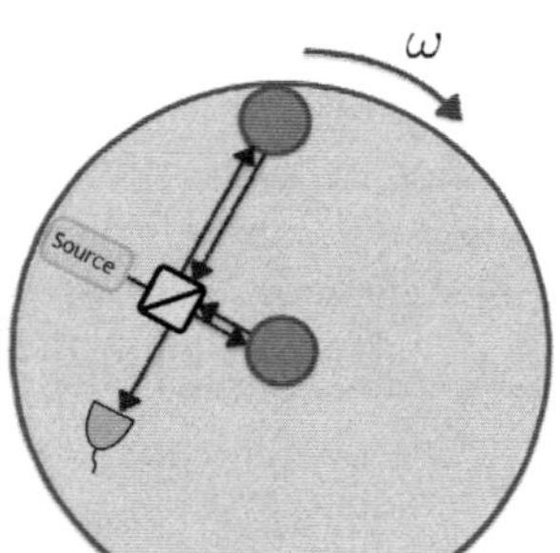

Figure 4: Setup for testing special-relativistic time dilation effects on interference of "clocks". A particle in a superposition of internal energy levels (a "clock") is sent to an interferometer mounted on a rotating platform. Each amplitude is stored for a laboratory time t in a trap, one near the centre and one near the rim of the platform. The two amplitudes are then recombined. The rotation velocity of the platform induces time dilation between the two traps, which affects the interferometric visibility.

If the platform rotates at an angular velocity ω, the outer trap moves with linear velocity $v = \omega R$. Therefore, the outer amplitude accumulates a proper time $\tau \approx t(1 - \frac{\omega^2 R^2}{2c^2})$, while the amplitude in the centre accumulates proper time t (laboratory time), giving the proper time difference

$$\Delta\tau \approx t\frac{\omega^2 R^2}{2c^2}. \tag{13}$$

(For simplicity, we assumed that the time needed to transfer the amplitudes to the traps and back is negligible with respect to t.) For an equal superposition of two internal energy levels, with transition frequency $\nu = |E_1 - E_2|/\hbar$, the visibility is modulated as

$$\mathcal{V} = \left|\cos\left(\frac{\nu\omega^2 R^2}{2c^2}t\right)\right|. \tag{14}$$

To estimate the magnitude of the effect, consider a platform rotating with angular velocity $\omega \sim 100$rad/s, a clock with optical transition frequency $\nu \sim 10^{15}$ rad/s, and a platform with radius $R \sim 1$ m. Complete loss of visibility then occurs if the amplitudes are recombined after $t \sim 0.1$ s.

In comparison, the time dilation produced by a difference $\Delta\Phi$ in gravitational potential is $\Delta\tau \approx \Delta\Phi t/c^2$, which reproduces (13) for the effective centripetal potential $\Delta\Phi = \frac{\omega^2 R^2}{2}$. Performing two clock-interference experiments: one in the gravitational field and the second on a rotating platform, would constitute a new test of the equivalence principle – in a hitherto untested regime where both external and internal DOFs of test system require relativistic as well as quantum description.

6. Discussion

Here we address key questions discussed during the 8th Symposium on Frequency Standards and Metrology, (Potsdam, October, 2015).

Why should one measure visibility and not the phase? For the visibility modulation it is crucial that internal energy contributes to the total mass as an operator. The relative phase, eq. (5), can be written as[4] $\Delta\phi_{GR} = (mc^2 + \langle H_{int}\rangle)\Delta\tau/\hbar$. Thus, first of all, a theory where only the mean internal energy contributes to the total mass ($m \to (m + \langle H_{int}\rangle/c^2)\mathbb{1}$) suffices to explain the phase shift – but does not predict the visibility loss. Secondly, in the non-relativistic, Newtonian, limit the phase shift does not vanish but is given by $\Delta\phi_N \approx m\Delta\Phi t/\hbar$ (for the gravitational experiment), which is fully explained by a Newtonian gravitational potential in Euclidean space-time – with absolute time – as confirmed in numerous experiments [1,43,44]. Moreover, even beyond the non-relativistic limit, the phase shift $\Delta\phi_{GR}$ can be explained by a phase $\Delta\phi'_N$ stemming from a modified gravitational potential, i.e. where Newtonian gravity is modified but one still considers an Euclidian space-time. More generally, a simple model is given in ref. [11] that reproduces results of the existing tests of time dilation, such as [8, 9], as well as

[4] The expression is valid for any state with internal energy distribution symmetric around the mean $\langle H_{int}\rangle$ and up to additional phases stemming from the non-gravitational forces required to control the paths, e.g. to trap the "clock".

predicts correct relativistic phase shifts for quantum interference of "clocks", but which does not predict any loss of visibility. Observation of the modulations in the visibility described by eq. (6) would simultaneously disprove non-relativistic, Newtonian gravity, as well as a classical description of the world lines of the "clocks". This has not been achieved thus far and cannot be achieved by the measurements of the phase shift alone.

In atomic interferometers there is no time dilation between the paths because atoms are in free fall. In the specific, Kasevich-Chu, atom interferometers [45] proper time is indeed the same along the paths, to lowest order in relativistic corrections. This is however a specific feature of a particular setup and needs not to be true for interferometers with different geometry. For example, time dilation will not vanish in interferometers with asymmetric paths, or when using Bloch lattices [46]. Importantly, the issue of vanishing proper time difference is unrelated to whether the paths are in free fall or not. Time dilation can be non-vanishing even between two free falling world lines, and interferometric visibility for this scenario has been studied in [16].

In full analogy to spin echo it is always possible to "reverse" the effect of time dilation and "bring the coherence back". In order to have high interference contrast the paths must be arranged such that the proper time difference between them is small compared to the characteristic time scale of the system (which in general depends on its internal Hamiltonian and state). For a two-level atom such a "gravitational echo" could be a feasible way to assure high visibility. However, when the internal energy spread grows, e.g due to large number of internal modes, even small fluctuations in the interferometer area lead to a substantial loss of visibility, figure 3 (right). For increasingly composite systems, the precision with which the paths need to be controlled thus grows, but this precision will be limited due to the Heisenberg uncertainty. For sufficiently large systems time dilation induced loss of visibility can become for all practical purposes irreversible, even though it stems from fundamentally reversible unitary dynamics.

For a discussion of time dilation in quantum theory and additional topics, we refer the reader to ref. [47].

7. Conclusion

Quantum effects have been demonstrated with complex systems comprising hunderds of atoms [18, 48–51]. The regime where general relativity affects internal dynamics of such systems might soon allow testing the interplay between quantum mechanics and general relativity. Composite quantum systems also offer intriguing new insights into the notion of time in quantum theory: They can describe a clock that runs different proper times in superposition and show that clocks generally must become entangled due to their unavoidable gravitational interactions [52]. Finally, considering systems with quantised internal energy calls for a suitable quantum formulation of the Einstein Equivalence Principle (EEP) and its new quantum tests [13]. Composite quantum particles subject to relativistic effects are particularly suited for exploring the joint foundations of quantum theory and general relativity.

Acknowledgments

M.Z. would like to thank the organisers of the 8th Symposium on Frequency Standards and Metrology 2015 for hospitality and participants of the conference for discussions,

This work was supported by the NSF through a grant to ITAMP, the Austrian Science Fund (FWF) through the Special Research Program Foundations and Applications of Quantum Science (FoQuS) and Individual Project (No. 2462), the Australian Research Council Centre of Excellence for Engineered Quantum Systems through grant number CE110001013 and by the Templeton World Charity Foundation, grant TWCF 0064/AB38.

References

[1] Colella R, Overhauser A and Werner S 1975 *Phys. Rev. Lett.* **34**(23) 1472–1474
[2] Nesvizhevsky V V, Börner H G, Petukhov A K, Abele H, Baeßler S, Rueß F J, Stöferle T, Westphal A, Gagarski A M, Petrov G A *et al.* 2002 *Nature* **415** 297–299
[3] Jenke T, Geltenbort P, Lemmel H and Abele H 2011 *Nat. Phys.* **7** 468–472
[4] Will C M 2014 *Living Rev. Relativity* **17** 4
[5] Pound R and Rebka G 1960 *Phys. Rev. Lett.* **4**(7) 337–341
[6] Hafele J C and Keating R E 1972 *Science* **177** 166–168
[7] Hafele J C and Keating R E 1972 *Science* **177** 168–170
[8] Chou C W, Hume D, Rosenband T and Wineland D 2010 *Science* **329** 1630–1633
[9] Nicholson T L, Campbell S L, Hutson R B, Marti G E, Bloom B, McNally R, Zhang W, Barrett M, Safronova M, Strouse G, Tew W and Ye J 2015 *Nat. Commun.* **6** 6896
[10] Zych M, Costa F, Pikovski I and Brukner C 2011 *Nat. Commun.* **2** 505
[11] Zych M, Costa F, Pikovski I, Ralph T C and Brukner C 2012 *Class. Quantum Grav.* **29** 224010
[12] Pikovski I, Zych M, Costa F and Brukner Č 2015 *Nat. Phys.* **11** 668–672 ISSN 1745-2473
[13] Zych M and Brukner Č 2015 Quantum formulation of the einstein equivalence principle (*Preprint* arXiv:1502.00971)
[14] Brodutch A, Gilchrist A, Guff T, Smith A R H and Terno D R 2015 *Phys. Rev. D* **91**(6) 064041
[15] Gooding C and Unruh W G 2014 *Phys. Rev. D* **90**(4) 044071
[16] Gooding C and Unruh W G 2015 *Found. Phys.* 1–13 ISSN 0015-9018
[17] Shapiro I I 1964 *Phys. Rev. Lett.* **13**(26) 789–791
[18] Eibenberger S, Gerlich S, Arndt M, Mayor M and Tüxen J 2013 *Phys. Chem. Chem. Phys.* **15** 14696–14700
[19] Schmöle J, Dragosits M, Hepach H and Aspelmeyer M 2016 *ArXiv e-prints* (*Preprint* 1602.07539)
[20] Schnabel R 2015 *Phys. Rev. A* **92**(1) 012126
[21] Kovachy T, Asenbaum P, Overstreet C, Donnelly C A, Dickerson S M, Sugarbaker A, Hogan J M and Kasevich M A 2015 *Nature* **528** 530
[22] Scheidl T and Ursin R 2012 *Proc. International Conference on Space Optical Systems and Applications (ICSOS), Ajaccio, Corsica, France, 2012* pp 2–4
[23] Scheidl T, Wille E and Ursin R 2013 *New J. Phys.* **15** 043008
[24] Mann R B and Ralph T C 2012 *Class. Quantum Grav.* **29**
[25] Weinberg S 1972 *Gravitation and cosmology: Principle and applications of general theory of relativity* (John Wiley and Sons, Inc., New York)
[26] Zych M 2015 *Quantum Systems under Gravitational Time Dilation* Ph.D. thesis University of Vienna, Austria
[27] Dixon W 1964 *Il Nuovo Cimento* **34** 317–339
[28] Poisson E and Will C 2014 *Gravity: Newtonian, Post-Newtonian, Relativistic* (Cambridge University Press) ISBN 9781107032866
[29] Einstein A 1905 *Annalen der Physik* **323** 639–641
[30] Einstein A 1911 *Annalen der Physik* **35** 898–908
[31] Zieliński B and Zych M 2006 *Phys. Rev. A* **74**(3) 034301
[32] Mandelstam L and Tamm I 1991 *Selected Papers* (Springer Berlin Heidelberg) pp 115–123
[33] Fleming G 1973 *Il Nuovo Cimento A* **16** 232–240 ISSN 0369-3546
[34] Margolus N and Levitin L B 1998 *Physica D: Nonlinear Phenomena* **120** 188–195 ISSN 0167-2789 proceedings of the Fourth Workshop on Physics and Consumption

[35] Kosiński P and Zych M 2006 *Phys. Rev. A* **73**(2) 024303
[36] Wootters W K and Zurek W H 1979 *Phys. Rev. D* **19**(2) 473–484
[37] Greenberger D M and Yasin A 1988 *Phys. Rev. A* **128** 391–394 ISSN 0375-9601
[38] Englert B G 1996 *Phys. Rev. Lett.* **77**(11) 2154–2157
[39] Shapiro I I, Ash M E, Ingalls R P, Smith W B, Campbell D B, Dyce R B, Jurgens R F and Pettengill G H 1971 *Phys. Rev. Lett.* **26**(18) 1132–1135
[40] Bertotti B, Iess L and Tortora P 2003 *Nature* **425** 374–376
[41] Carlesso M and Bassi A 2016 Decoherence due to gravitational time dilation: analysis of competing decoherence effects (*Preprint* arXiv:1602.01979)
[42] Margalit Y, Zhou Z, Machluf S, Rohrlich D, Japha Y and Folman R 2015 *Science* **349** 1205
[43] Zawisky M, Baron M, Loidl R and Rauch H 2002 *Nuclear Instruments and Methods in Physics Research Section A: Accelerators, Spectrometers, Detectors and Associated Equipment* **481** 406–413 ISSN 0168-9002
[44] Peters A, Chung K Y and Chu S 1999 *Nature* **400** 849–852
[45] Tino G M and Kasevich M A 2014 *Proceedings of the International School of Physics - Enrico Fermi* (IOS)
[46] Tino G M 2014 *Proceedings of the International School of Physics - Enrico Fermi* ed Tino G M and Kasevich M A pp 457–493
[47] Pikovski I, Zych M, Costa F and Brukner Č 2015 Time dilation in quantum systems and decoherence: Questions and answers (*Preprint* arXiv:1508.03296)
[48] Arndt M, Nairz O, Vos-Andreae J, Keller C, Van der Zouw G and Zeilinger A 1999 *Nature* **401** 680–682
[49] Hackermüller L, Hornberger K, Brezger B, Zeilinger A and Arndt M 2004 *Nature* **427** 711–714
[50] Gerlich S, Eibenberger S, Tomandl M, Nimmrichter S, Hornberger K, Fagan P J, Tüxen J, Mayor M and Arndt M 2011 *Nat. Commun.* **2** 263
[51] Haslinger P, Dörre N, Geyer P, Rodewald J, Nimmrichter S and Arndt M 2013 *Nat. Phys.* **9** 144–148
[52] Castro-Ruiz E, Giacomini F and Brukner Č 2015 Measuring time with physical clocks (*Preprint* arXiv:1507.01955)

Journal of Physics: Conference Series **723** (2016) 012045 doi:10.1088/1742-6596/723/1/012045

Direct measurement of the $\gamma_{\mathrm{He}}/\gamma_{\mathrm{Xe}}$ ratio at ultralow magnetic field

Isaac Fan, Silvia Knappe-Grüneberg, Jens Voigt, Wolfgang Kilian, Martin Burghoff, Detlef Stollfuss, Allard Schnabel, Gerd Wübbeler, Olha Bodner, Clemens Elster, Frank Seifert, Lutz Trahms

Physikalisch-Technische Bundesanstalt (PTB), Abbestrasse 2-12, 10587 Berlin, Germany

E-mail: isaac.fan@ptb.de

Abstract. A co-located ^{3}He and ^{129}Xe nuclear spin free precession measurement at sub-μT magnetic field was carried out in a magnetically shielded environment. The uncorrected quotient of the gyromagnetic ratios between neutral ^{3}He and ^{129}Xe atoms is determined to be 2.754 082 81(07), accounting for only statistical error. Our measurement shows that this ratio has a stability of $1.4\times10^{-5}/\sqrt{\tau}$, demonstrating the ability to reach the current precision limit of the quotient in a 10000 s of averaging time τ. This precision is enough for the next-generation EDM search in neutral ^{129}Xe atoms based on a similar comagnetometer scheme.

1. Introduction

The interaction of the nuclear electric dipole moment ($\vec{d}$) with an external electric field ($\vec{E}$) creates a measurable energy $\vec{d}\cdot\vec{E}$ for an atom or molecule despite the Schiff's screening effect [1]. In the nuclear spin free precession setting, this $\vec{d}\cdot\vec{E}$ term corresponds to an additional shift to the Larmor frequency. We are currently involved in a collaboration in the search of non-vanishing electric dipole moment (EDM) in neutral ^{129}Xe atoms [2]. The experiment will be based on the free spin precession in the gaseous phase at ultralow magnetic field ($\approx \mu$T) in a heavily shielded environment such as the ones located in Berlin [3] and Munich [4].

The expression for determining the atomic electric dipole moment of ^{129}Xe is given as follows:

$$d_{\mathrm{Xe}} = \frac{h}{4E}\left[(f_{\mathrm{Xe}}^{\Uparrow\Uparrow} - f_{\mathrm{Xe}}^{\Uparrow\Downarrow}) - \gamma_{\mathrm{Xe}}\Delta B_0(t)\right], \tag{1}$$

where the arrows $\Uparrow\Uparrow$ and $\Uparrow\Downarrow$ indicate the relative orientations of the externally applied electric and magnetic fields, $\Delta B_0(t)$ is the temporal drift of the magnetic field during measurements, and E is the applied electric field that is stable in time. The signature of an EDM would be a change in the ^{129}Xe precession frequencies (f_{Xe}) that is correlated to the inversion of the relative orientation of the electric and magnetic fields. However, a similar signature (i.e. a fake EDM signal) can also result if the magnetic field term drifts in time.

To cancel out the common magnetic systematic and other noise sources, simultaneous precessions of two atomic species located in the same sample space are used (e.g. see the previous EDM search in ^{129}Xe which dealt with frequencies in the microwave regime [5].) Based on the Zeeman effect ($f_{\mathrm{Larmor}} = \gamma \cdot B_0$), the issue of magnetic field instability is alleviated by

Table 1. Gyromagnetic ratios for ^{3}He and ^{129}Xe derived from measurements in the literature. All values are stated without shielding corrections.

	Latest	Source
γ_p	42.576 3866(10) MHz/T	Ref [9]
γ_{Xe}	11.776 7392(10) MHz/T	Combine Ref [7] and [9]
γ_{He}	32.434 100 83(90) MHz/T	Combine Ref [8] and [9]
γ_{He}/γ_{Xe}	2.754 081 35(22)	Ref. [6]
γ_{He}/γ_{Xe}	2.754 081 60(31)	Ref. [7],[8] via γ_p
γ_{He}/γ_{Xe}	2.754 081 60(18)	Ref. [7],[8] via Eqn. 5

frequency comparison of the comagnetometer species, ^{129}Xe and ^{3}He, according to the following expression:

$$d_{\mathrm{Xe}} = \frac{h}{4E}\left[(f_{\mathrm{Xe}}^{\Uparrow\Uparrow} - f_{\mathrm{Xe}}^{\Uparrow\Downarrow}) - \frac{\gamma_{\mathrm{Xe}}}{\gamma_{\mathrm{He}}}(f_{\mathrm{He}}^{\Uparrow\Uparrow} - f_{\mathrm{He}}^{\Uparrow\Downarrow})\right]. \quad (2)$$

It is evident from Eqn. 2 that a crucial parameter is the quotient of the gyromagnetic ratios $\frac{\gamma_{Xe}}{\gamma_{He}}$, or equivalently $\frac{\gamma_{He}}{\gamma_{Xe}}$. In this report, we present data and analysis of a *direct* measurement of the $\frac{\gamma_{He}}{\gamma_{Xe}}$ ratio at 0.4 μT and discuss the impact of its precision on the EDM search.

2. Current status of $\frac{\gamma_{He}}{\gamma_{Xe}}$

The most precise value available for the ^{129}Xe nuclear magnetic moment was derived from the ratio of Larmor frequencies of ^{129}Xe and ^{1}H at 1.5 T [7]. The ^{1}H reference signal was obtained by replacing the 0.98 bar Xenon sample with a demineralized H_2O sample of supposedly the same spherical shape. The frequency ratio was determined to be:

$$\frac{f_{^{129}\mathrm{Xe}}(\mathrm{gas, sphere}, 22.5°\mathrm{C})}{f_{^1\mathrm{H}}(\mathrm{H_2O, sphere}, 22.5°\mathrm{C})} = 0.276\ 602\ 600(17). \quad (3)$$

For ^{3}He and ^{1}H, an even more precise value of the Larmour frequency ratio was derived at 0.1 T [8]. The internal field difference caused by the cell shape differences was eliminated by replenishing the 4 mbar of ^{3}He with H_2O in the same cell consecutively. Both ^{3}He and H_2O signals were compared to a common water NMR reference standard. This ratio was determined to be:

$$\frac{f_{^3\mathrm{He}}(\mathrm{gas, sphere}, 25°\mathrm{C})}{f_{^1\mathrm{H}}(\mathrm{H_2O, sphere}, 25°\mathrm{C})} = 0.761\ 786\ 1313(33). \quad (4)$$

Equation 4 leads to the most precise value for the helion magnetic moment and is accepted in the 2010 evaluation of Fundamental Physical Constants [9]. Dividing it by Eqn. 3 yields the quotient of interest for the EDM search in ^{129}Xe:

$$\frac{f_{^3\mathrm{He}}/f_{^1\mathrm{H}}}{f_{^{129}\mathrm{Xe}}/f_{^1\mathrm{H}}} = \frac{\gamma_{\mathrm{He}}}{\gamma_{\mathrm{Xe}}} = 2.754\ 081\ 60(18). \quad (5)$$

It should be noted that the uncertainty of 6.5×10^{-8} is dominated by the uncertainty of the gyromagnetic ratio of ^{129}Xe. The value for the quotient of gyromagnetic ratios can also be deduced from the absolute numbers of the respective gyromagnetic ratios provided that the shielded proton gyromagnetic ratio is known.

Values of gyromagnetic ratios from various measurements are summarized in Table 1. The uncertainty for the quotient, however, will be slightly bigger for the value derived from the proton than the value shown in Eqn. 5 due to the contribution of the uncertainty of the shielded proton gyromagnetic ratio. In Ref. [6], a high-field NMR measurement at 11.7586 T was done separately for ^{3}He and ^{129}Xe with the same set-up to derive a $\frac{\gamma_{He}}{\gamma_{Xe}}$ value of 2.754 081 35(22), which is in agreement with Eqn. 5 with a comparable precision.

3. Experiment

The measurement was carried out in the magnetically shielded room in Berlin (BMSR-2) [3]. The nuclear spins of ^{3}He were polarized by means of metastability exchange optical pumping, whereas the nuclear spins of ^{129}Xe were polarized by means of spin exchange optical pumping. ^{3}He and ^{129}Xe atoms were mixed in a single cylindrical cell (diameter=58 mm, length=60 mm) with the following partial pressure conditions ^{3}He: 3.6 mbar, ^{129}Xe: 5.3 mbar, and N_2: 26.3 mbar. The nitrogen buffer gas pressure was chosen [11] based on the trade-offs between (a) reducing relaxation caused by formation of Van der Waals molecules (need high N_2 concentration) and (b) reducing relaxation caused by magnetic field gradients (need low total gas pressure).

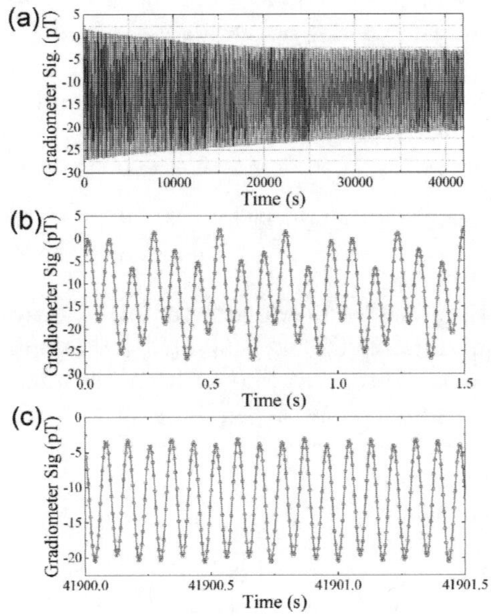

Figure 1. (Colour Online) (a) The software gradiometer signal. The non-zero offset is a derivative of the working point of the SQUID locking circuit. (b) The data and the fit at the beginning of the measurement. (c) The data and the fit near the end of the measurement. The open dots are the measured values. The red lines represent the model-regressed curves.

The spin precession was initiated by a non-adiabatic 90° change of the direction of the guiding field. The guiding field ($B_0 = 350$ nT) was induced by two sets of crossed Helmholtz coils. The signal of the precessing spins was detected with a multichannel vector magnetometer system, containing directly coupled superconducting quantum interferometer devices (dc-SQUID) [10]. More details of the experimental setup are described in Ref. [11] and Ref. [12].

In this measurement, 33 SQUID magnetometer channels were recorded, differing in their positions with respect to the cell. Thus, far-field common-mode backgrounds and mechanical-vibration modes, originating from relative motions of the detectors and the cell, could be suppressed by a software gradiometer. Specifically, signals from two magnetometer channels were combined during the post-processing to form one gradiometer signal. One of the combining channels should be close to the cell to maximize the sensitivity of the spin precession. The other combining channel should be sufficiently far away from the cell such that hardly any precession signal is detected which would otherwise reduce the signal-to-noise ratio (SNR) of the calculated gradiometer signal.

Fig. 1-a shows the superposition of the two decaying sinusoidal signals of ^{3}He and ^{129}Xe. The measurement contains 10,501,760 data points in total. The time interval between subsequent data points is 1/250 s as shown by the open dots in Fig. 1-b and Fig. 1-c. The signal, and

Journal of Physics: Conference Series **723** (2016) 012045 doi:10.1088/1742-6596/723/1/012045

consequently the SNR, is decreasing with time due to the dephasing of the nuclear spins with a transversal relaxation time T_2^*. In this measurement, more than 59 hours and 5 hours of T_2^* times were obtained for ^{3}He and ^{129}Xe, respectively.

4. Frequency Extraction by Regression Model

Precession frequencies ($\omega = 2\pi f$) of ^{3}He and ^{129}Xe need to be extracted to derive $\frac{\gamma_{\rm He}}{\gamma_{\rm Xe}}$. Common methods to extract the frequency estimator from a time-domain precession signal are (a) the zero-crossing counting, (b) the frequency mixing and demodulation, and (c) the time-series regression. Methods (a) and (b) typically involve pre-stage and post-stage filtering processes whereby deducing the impacts on the final uncertainty budget is cumbersome. Therefore, method (c) is the preferred analysis tool. However, the pre-requisite of the time-series regression is that the measurement needs to be partitioned into blocks so that the frequency shift in a block is negligible in comparison to the estimation uncertainty.

A regression model in its most general form can be expressed as $\sum_{j=0}^{d} b_j t^j + \sum_{j=1}^{\lambda}\sum_{k=1}^{\beta}\left[c_{jk}\cos(j\omega_k t) + s_{jk}\sin(j\omega_k t)\right] + \epsilon$, where ϵ is the noise, $[b_j, c_{jk}, s_{jk}]$ are parameters to be fitted, $[d, \lambda]$ are ordering parameters for the regression model, and β is the number of frequency components to be regressed. Setting the ordering parameters d and λ to 1 and allowing the existence of two frequencies, the model to fit is obtained as:

$$\mu(t) = b_0 + b_1 t + \sum_{k=1}^{2}\left[c_k\cos(\omega_k t) + s_k\sin(\omega_k t)\right] + \epsilon. \tag{6}$$

Equation 6 is a linear model [1] in the sense that the response μ is a linear combination of predictors $[t, \cos(\omega t), \sin(\omega t)]$ scaled by corresponding coefficients $[b_1, c_1, s_1]$. For signals within a block, assumptions implicitly made by the model are: (a) the drift in the signal offset is approximated to be linear as represented by the first two terms in Eqn. 6, (b) the amplitude decay is neglected by the fit [2], (c) the noise ϵ is stationary with a covariance that does not depend on the time shift, and (d) the frequency is constant.

Equation 6 can be re-expressed in the matrix form as:

$$\begin{aligned}\mathbf{U} &= \begin{bmatrix} 1 & t & \cos(\omega_1 t) & \sin(\omega_1 t) & \cos(\omega_2 t) & \sin(\omega_2 t)\end{bmatrix} \cdot \begin{bmatrix} b_0 \\ b_1 \\ c_1 \\ s_1 \\ c_2 \\ s_2 \end{bmatrix} + \epsilon \\ &= \mathbf{DP} + \epsilon,\end{aligned} \tag{7}$$

where the design matrix $\mathbf{D}$ is uniquely determined for a given set of frequency values. There are two stages in the frequency estimation process. In the first step, the initial guess of ω's is determined by the peak location in the Fourier spectrum of the whole time series [3]. In the second step, the parameter vector is found by $\hat{\mathbf{P}} = (\mathbf{D}\mathbf{D}^{\rm T})^{-1}\mathbf{D}^{\rm T}\mathbf{U}$ in the linear least-square sense. The variable projection functional to be minimized is the difference between $\mathbf{U}$ and $\mathbf{D}\hat{\mathbf{P}}$. The nonlinear least-square solver used is `lsqnonlin` package in Matlab®. This variable projection optimization technique for nonlinear least-square problems is well documented in Ref. [13].

Typical fits at the beginning and tail parts of the measurement are shown in Fig. 1-b,c. The apparent existence of a frequency beat note pattern in Fig. 1-b is a clear indicator that there are

[1] A comparison can be made to the more familiar expression $A\sin(\omega t + \phi)$, which is nonlinear in this context since there are two variables $[\omega t, \phi]$ inside one functional predictor (the sine function).

[2] The shortest relaxation time of the signal is about 18,950 s, implying a decrease in the signal amplitude of 0.005% at most within one block length assuming an $\exp(-t/T_2^*)$ dependence.

[3] The estimation error in this initial guess is no more than half of the frequency bin size ≈12 μHz.

multiple resonant frequencies. The disappearance of the frequency beat note pattern in Fig. 1-c is caused by the differences in the dephasing rates of the resonant species. A 100-s block length is chosen such that the averaging time is short enough to avoid the detrimental effect on the frequency stability caused by the magnetic field drift. This blocksize results in a total of 420 blocks. The lowest precession frequency of the signal is about 4.16 Hz, corresponding to 416 cycles per block. Each block is treated as an independent dataset for the regression.

The covariances of fitted parameters, hence of the fitting uncertainties, are derived from the first-order partial derivatives of the model (i.e. the Jacobian matrix $\mathbf{J}$) as follows:

$$\mathrm{cov}(\hat{\theta}) = \frac{srs}{dof}(\mathbf{J}^T\mathbf{J})^{-1}, \tag{8}$$

where $\hat{\theta} = (\hat{b_0}, \hat{b_1}, \hat{c_1}, \hat{s_1}, \hat{c_2}, \hat{s_2}, \hat{\omega_1}, \hat{\omega_2})$ are the best-fit parameter values, srs is the sum of the residual squares, and dof is the degree of freedom (the number of data point subtracted by the number of parameters).

5. Results and Discussion

The extracted spin precession frequencies of the two nuclear species are shown in Fig. 2-a,b. Converting the frequency values to the B_0 field yields the expected 0.35 μT.

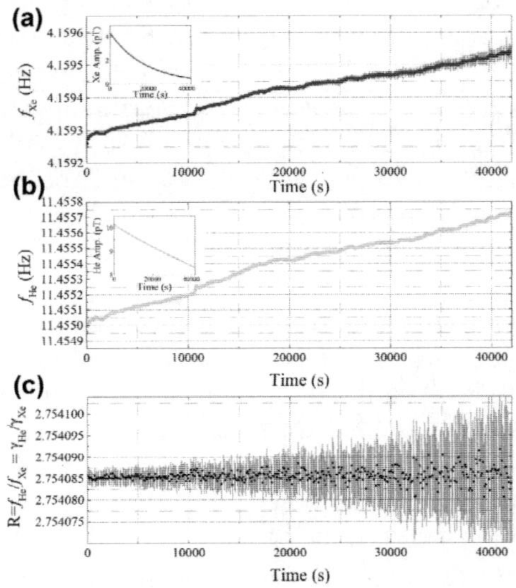

Figure 2. (Colour online) The extracted precession frequencies of (a) ^{129}Xe (blue curve) and (b) ^{3}He (green curve). The insets are the amplitude curves of the respective nuclei. The ratio R of the two frequencies is shown in (c). The blocksize used is 100 s.

Fig. 2-c depicts the quotient of gyromagnetic ratios $R = \frac{\gamma_{\mathrm{He}}}{\gamma_{\mathrm{Xe}}}$. Using the uncertainties of the fitted frequencies, denoting $\sigma_{f_{\mathrm{He}}}$ and $\sigma_{f_{\mathrm{Xe}}}$, the block uncertainty of the ratio of the two frequencies and hence the quotient of the gyromagnetic ratios can be derived using the principles listed in GUM [14] as $\sigma_R = \sqrt{(\sigma_{f_{\mathrm{He}}}/f_{\mathrm{Xe}})^2 + (f_{\mathrm{He}}/f_{\mathrm{Xe}}^2 \cdot \sigma_{f_{\mathrm{Xe}}})^2}$. The uncertainty and the fluctuation of the ratio increases with the measurement time due to the decaying signal amplitude via the T_2^* relaxation mechanism of the nuclear spins.

The overlapping Allan standard deviation of R is shown in Fig. 3. A slope of $1.4 \times 10^{-5}/\sqrt{\tau}$ implies that the common-mode drift in R appears to be successfully removed by the quotient operation. In an averaging time of 10000-s, the stability of R reaches 1.4×10^{-7}.

A weighting factor is assigned to each block according to $\kappa_i = 1/\sigma_{R_i}^2$, which leads to the corresponding weighted mean $\bar{R} = \sum_i R_i\kappa_i / \sum_i \kappa_i$. The standard deviation of the mean is taken as $\delta\bar{R}$. This generic weighted averaging procedure is the standard method when combining

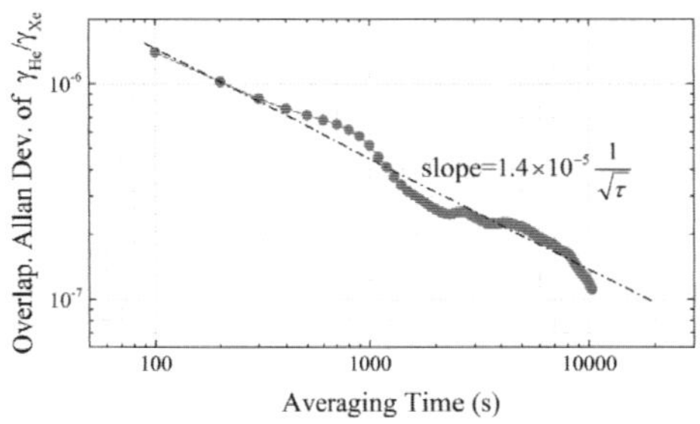

Figure 3. The stability plot of γ_{He}/γ_{Xe} as a function of averaging time. The dashed line is a linear fit to the Allan deviation points from 100 s to 10000 s.

results from different clocks to compute the International Atomic Time [16]. Thus, the $\frac{\gamma_{He}}{\gamma_{Xe}}$ is determined to be 2.754 085 44(07), with a 2.5×10^{-8} relative uncertainty exhibiting less than half the errors as given by Eqn. 5 and the value stated in Ref. [6].

The difference between our value and those from literature arises from the fact that various systematic effects have not been considered. We have to take into account that our measurement was done in a non-inertial frame of reference due to the constantly rotating surface of the Earth. The precession frequencies measured in our laboratory frame were thus modified. The sense of spin rotation can be deduced from the Bloch equations. The rate of the earth rotation fluctuates from day to day. On the day when the measurement was taken, the rotational speed of the earth was 11.605 761 7234(87) μHz [17]. The projection of the earth rotation to the lab frame can be expressed by $sid = \Omega\cdot\cos(\phi)\cdot\cos(\rho)$, where Ω is the angular velocity of the rotating earth, ϕ is the latitude of the setup, and ρ is the angle between B_0 and the North-South axis. The uncertainty of sid can be found by $\sigma_{sid} = \sqrt{(\sigma_\Omega \cos\phi\cos\rho)^2 + (\sigma_\phi \Omega\cos\rho\sin\phi)^2 + (\sigma\rho\Omega\cos\phi\sin\rho)^2}$, where the σ's are the uncertainties in the respective parameters. The latitude of the setup is 52.5164(1)° assuming a perfect spherical model for the Earth. The angle between B_0 and the North-South axis is 28.0(5)°. Therefore, the systematic correction due to the earth rotation is $sid = 6.236(29)$ μHz, with the uncertainty mainly dominated by the accuracy of the B_0 orientation. Considering the sidereal correction, the ratio R is determined to be 2.754 082 81(07).

There are other systematic effects which could arise due to the spin-spin interaction among and between the nuclei. An asymmetry in the spherical shape of the sample cell in the presence of a B_0 field inhomogeneity could cause an additional shift in the precession frequency. Typical known effects include the geometric phase [18], the chemical shift [19], the geometric shift [15], the gravitational shift and the Ramsey-Bloch-Siegert shift [20, 21]. Careful examinations are needed in all of these aspects to derive a corrected R value.

The sensitivity coefficient of ^{129}Xe EDM to $\frac{\gamma_{He}}{\gamma_{Xe}}$ is determined to be 4.14×10^{-46} Jm/V under a 4-kV electric field and a 1-μT magnetic field. Therefore, our current $\frac{\gamma_{He}}{\gamma_{Xe}}$ instability contributes $\leq 10^{-32}$ e·cm to the ^{129}Xe EDM. This precision is well below the 10^{-28} e·cm target value.

6. Conclusion

The direct determination of $\frac{\gamma_{He}}{\gamma_{Xe}}$ has been carried out in a comagnetometer setup under an ultralow ambient magnetic field ($< \mu$T). The resultant precision is higher than the current literature value and will be sufficient for the next generation search of the EDM in ^{129}Xe. Careful and thorough investigations on the systematic effects are underway to understand the remaining deviations from the literature value.

Acknowledgments

The project is funded by DFG under the grant number DFG BU 1037/3-2. The measurement used for this evaluation was performed on Sept. 14th, 2010 in collaboration with the working group of W. Heil from Johannes Gutenberg Universität Mainz. The authors thank the following people for exchanges during the symposium- E. Peik for the fruitful discussion on the temporal variation of fundamental constants using ion clocks, S. Bize for confirming the relation between the g-factor and fundamental parameters, E. Benkler for discussing the interpretations of Allan deviations, and J. Ye for discussing the electron EDM search using molecular ions.

References

[1] Schiff L I 1963 *Phys. Rev.* **132** 2194
[2] The collaboration partners include T.E. Chupp's group at University of Michigan, J. Singh's group at Michigan State University, P. Fierlinger's group at Technische Universität München, and our group at PTB
[3] Bork J, Hahlbohm H D, Klein R, and Schnabel A 2000 *Proc. 12th Int. Conf. on Biomagnetism* (Espoo, Finland)
[4] Altarev I *et al.* 2014 *Rev. Sci. Instr.* **85** 075106
[5] Rosenberry M A and Chupp T E 2001 *Phys. Rev. Lett.* **86** 22
[6] Makulski W 2015 *Magn. Reson. Chem.* **53** 273
[7] Pfeffer M and Lutz O 1994 *J. Magn. Reson.* **108** 106
[8] Flowers J L, Petley B W, and Richards M G 1993 *Metrologia* **30** 75
[9] Mohr P J, Taylor B N, and Newell D B 2012 *J. Phys. Chem. Ref. Data* **41** 043109
[10] Thiel F *et al.* 2005 *Biomed. Technik* **50** 169
[11] Gemmel C *et al.* 2010 *Eur. Phys. J.* D **57** 303
[12] Tullney K *et al.* 2013 *Phys. Rev. Lett.* **111** 100801
[13] Golub G and Pereyra V 2003 *Inverse Problems* **19** R1
[14] Joint Committee for Guides in Metrology, *Evaluation of Measurement Data– Guide to the Expression of Uncertainty in Measurement*, 1st ed. (Joint Committee for Guides in Metrology, Sevres, France, 2008).
[15] Cates G D, Schaefer S R, and Happer W 1988 *Phys. Rev. A* **37** 2877
[16] Tavella P 2008 *Metrologia* **45** S183
[17] Standard Rapid EOP Data since 02. January 1973 (IAU2000) (published by IERS Rapid Service Prediction Centre).
[18] Berry M V 1985 *J. Phys. A* **18** 15
[19] Jameson C J, Jameson A K, and Parker H 1979 *J. Chem. Phys.* **70** 5916
[20] Bloch F and Siegert A 1940 *Phys. Rev.* **57** 522
[21] Ramsey N F 1955 *Phys. Rev.* **100** 1191

A Frequency Metrology approach to Newtonian constant *G* determination using a pair of extremely high *Q* simple pendulums in free decay

A De Marchi
Dept. of Electronics and Telecommunications, Politecnico di Torino, Torino, Italy
E-mail: andrea.demarchi@polito.it

Abstract. It is argued that simple pendulums exhibiting Q values in excess of 10^8 can be realized by using high strength fibres to suspend in vacuum a bob of less than 10^{-3} kg. For a 1 m long pendulum this means damping time constants of several years, long enough to allow experiments in free decay mode, maximizing in this way the expected short term frequency stability. A dual pendulum experiment based on this projection is discussed, which common-modes seismic noise and is expected to yield 10^{-5} uncertainty on Big G. The value of the latter can be obtained from the variation in relative frequency difference between the two pendulums when they are subjected to well controlled variations of the gravitational field. A discussion is given of Type A and Type B uncertainty contributions, and a tentative accuracy budget is projected.

1. Introduction

The measurement of the Newtonian constant G is difficult for the feebleness of the gravitational force both absolute and relative to other forces associated with different phenomena, in the presence of which experiments aimed at measuring G must be carried out. This poses two orders of problems which are typically in competition, as both A and B Type uncertainties must be reduced below the desired level. On one hand, in fact, the need to avoid too low S/N ratios suggests privileging experimental designs yielding as great a signal as possible, even if it implies accepting hard to model features, putting in this way a stress on Type B uncertainty; on the other hand the choice of an approach which can be very well modeled often pays a tribute to S/N ratio, putting a stress on Type A uncertainty. It is no wonder that all efforts to measure G were undertaken with an S/N ratio maximizing approach, in an era when the high resolution now possible in measurements was unthinkable. However, a re-visitation of this trade-off is here suggested, at the light of contemporary amazing Time and Frequency Metrology resolutions. It is claimed that it may offset toward the latter end of the scale the positioning of the best approach choice.

As a matter of fact, while all ten evaluations of G reported in the 19th century, all four made in the first half of the 20th century (only one of which was claimed to have an accuracy better than 0.1%), and all those published after WWII and before 1990 by half a dozen groups, all were obtained with basically the same approach adopted by Henry Cavendish more than two centuries ago [1], substantially different experiments started to crop up thereafter.

Three facts concurred in steering the quest for G into new ways: theoretical doubts raising in the mid-80s on the validity of the Newtonian theory (the fifth force debate), a result with a discrepancy of many standard deviations from the accepted value which was published by a highly respected institution [2], and the discovery by Kuroda [3] of a possible source of error in Cavendish balances, hidden in the an-elasticity of the torsion wire.

As a result, at least a dozen different evaluations have been carried out in as many different groups in the last twenty years, a number of which have adopted a totally different approach from the torsion balance, using as a sensitive device either a beam (Jolly-Poynting) balance, or a free-fall gravimeter, or a simple pendulum, all avoiding in one way or another the torsion wire problem.

In the best experiment with the **beam balance**, carried out in Zurich [4], 1 kg test masses sat on either dish and 4 ton of Mercury were moved up above and down below one of the two. The 1.8×10^{-5} final uncertainty claimed by the group is one of the best to date. A great deal of beautiful error analysis and correction was carried out in this high sensitivity experiment.

The **free-fall gravimeter** is blessed by the absence of suspensions, but no uncertainties better than 10^{-3} have yet been claimed as relative resolution is a problem. In fact, the measured gravitational force is compared directly to the attraction by the whole Earth. Despite the adoption of laser interferometry in one case [5], and atom interferometry in two more [6,7], requirements are so great that measurements remained limited in all cases by Type A uncertainty even after long averages. The positioning of the free-fall approach in the range of possible choices seems to be too biased toward error reduction at all cost, even for modern high resolution measurement technology, because the effect vanishes exactly where one would like to have it most strong, at the center of the active masses.

The **simple pendulum** approach was used in three experiments, two of which in the static mode. In both, two hanging test masses held Fabry-Perot mirrors whose spacing changed as an effect of active masses displacement. The measured quantity was the resonant frequency. In one case the latter was in the microwave range [8] and an accuracy slightly better than 10^{-3} was claimed, in the other it was in the visible [9] and the claimed accuracy was in the low 10^{-5} range. The third attempt, which employed a **dynamic pendulum** approach, is the small scale pilot experiment which was carried out at Politecnico di Torino [10] from 1998 to 2005. This preliminary trial yielded an evaluation of G with a seismic noise limited Type A uncertainty of 3%, nevertheless providing critical indications for this new proposal. Relevant for the accuracy of this and any dynamic experiment is the analysis, carried out within this work, of the frequency change that is induced by variations of the energy stored in the oscillations when the active masses are moved [11]. This approach can be viewed as using a "constrained free-fall", which positions it less far out than the true free-fall toward the same end of the scale. In fact, the signal force is here compared to only a component of Earth's attraction, which also vanishes at rest position. This maximizes the effect exactly where desired, contrary to the free-fall case, as argued in section 3.

In truth, some of the best results ever published to date, with uncertainties in the mid to low 10^{-5} range, have been obtained in this post-Kuroda era with some variation of the **torsion balance**, including both static mode and dynamic mode operation. Two of these were obtained in the static mode, and claimed uncertainty of $1.38\cdot10^{-5}$ [12] and $4.05\cdot10^{-5}$ [13] respectively by avoiding altogether any torsion in the wire with the help of a feedback system that forced the moving apparatus to stay put. Similar accuracy was also claimed by a group at BIPM [14], where a strip was used for a torsion wire, in order to reduce shear strains in deflection, and the residual an-elasticity Kuroda effect was evaluated in dynamic mode operation. In another group, the an-elasticity of the torsion wire was measured separately [15], to evaluate the correction to be made in a dynamic mode measurement.

Unfortunately, in spite of progress made both with the traditional torsion balance approach and with other methods, significant discrepancies persist among different evaluations, which makes the CODATA Group cautious in citing uncertainty. At present, in the most recent CODATA 2014 revision, the relative uncertainty is indicated to be $4.7\cdot10^{-5}$ on a value of $6.67408\cdot10^{-11}$ $m^3kg^{-1}s^{-2}$.

In recognition of the basic metrological principle that the only way to mitigate the risk of embarking unknown systematics in the assessment of a natural object is to obtain compatible results with many totally independent approaches (Karl Popper teaching here!), the present proposal is aimed at reaching the ultimate potential of the dynamic pendulum approach by overcoming with typical Time and Frequency Metrology methods the limitations of the pilot experiment.

A dual pendulum configuration is proposed, aimed at exploiting the fast reduction of statistical uncertainty granted by measuring the variation rate of time delay between the two, which is well known to be equal to their relative frequency difference [16], and at the same time common-moding the relevant 0.2 Hz seismic angular noise due to ocean waves excited Rayleigh waves [10,17] which disturbed period measurement in the pilot experiment. The inferred possibility of reaching pendulum Q values in excess of 10^8 enables free decay operation for months, high frequency stability, and the projection of 10^{-5} uncertainty on G in a few weeks of data taking.

In section 2 the critical point of such high Q values possibility is addressed, as it has direct impact on design, and in section 3 the plan of the experiment is described. In section 4 the projected accuracy budget is presented and shortly discussed, with an indication of expected future improvements.

2. *Q* analysis

To be sure, a well-known truism is that a high Q factor is fundamental for both Type A and Type B uncertainties in an experiment involving an excited resonator, because it generally fosters high short term frequency stability and guarantees for the ability to closely model it. In the torsion balance, for example, a high Q can be taken as a proof of low torsion wire an-elasticity and support high accuracy claims in the analysis of the Kuroda effect, as in the case of the BIPM experiment, which featured a Q of 10^5, the best ever reported in torsion balances for big G determinations.

However, further specific reasons to seek as high a Q as possible in this dual dynamic pendulum configuration include noise filtering action, the need to avoid frequency locking of the two pendulums, and the beauty of long ring-down time constant. In particular, the latter can be as long as several years for a 1m long pendulum with Q in the mid-10^8 range, which would enable a several months long experiment in free decay mode, avoiding both the risk of noise injection with the sustainment mechanism and the frequency drift deriving from the coupling of period and amplitude if the latter is not constant.

Apart from friction on residual air, which however can be reduced below any desired level with adequate vacuum, only two main loss mechanisms were identified in a fibre pendulum realized as pictured in figure 1: the cyclic stretching of the fibre under the action of varying centrifugal force and gravity component along the fibre, and the cyclic bending of the fibre as it wraps and unwraps onto the suspension profile. The latter is assumed circular in figure 1 for ease of implementation reasons, and because a circular profile turns out to produce a minimum at non-zero amplitude in the period versus swing angle curve, which can be exploited to make the pendulum locally isochronous [18].

A first approximation theory can be developed for the contributions of these two mechanisms to pendulum Q by overlooking the cyclic length variations. Expressions for Q_s (stretching) and Q_b (bending) can be rather easily obtained with this simplification, and then combined to predict Q as

$$\frac{1}{Q} = \frac{P_{ds}+P_{db}}{\omega_p E_p} = \frac{1}{Q_s} + \frac{1}{Q_b} , \tag{1}$$

where the index p of resonance angular frequency ω_p and of energy E_p stored in the oscillation indicates that they are relative to the pendulum mode, and dissipated power P_d of the two mechanisms is identified by the second index s or b. Such power can be written for both as a function of pendulum frequency and energy, as in equation (1), but also as a function of eigen-frequency and stored energy of the mode that sustains either loss mechanism, i.e. the stretching and the bending mode. In this latter case the Q value that must be used in the formula is the intrinsic value Q_f of the fibre material, which can in principle be measured by exciting the relevant stretching or bending mode. In practice, this can be easily done on Earth only for the stretching mode because selectively exciting the bending mode to measure frequency and Q would require a gravity free experiment. So it is

$$P_{ds} = \frac{\omega_p E_p}{Q_s} = \frac{\omega_s E_s}{Q_f} , \tag{2}$$

$$P_{db} = \frac{\omega_p E_p}{Q_b} = \frac{\omega_b E_b}{Q_f} . \tag{3}$$

Notice that E_s and E_b represent the energy stored in the stretching and bending modes when they are excited out of frequency at the natural pendulum resonance ω_p and at the actual swing amplitude θ_0.

Now, for the stretching mode, it turns out that the energy E_s stored in it, swapping at double the pendulum frequency between elastic and kinetic energy, is related to pendulum energy by

$$\frac{E_p}{E_s} = \frac{16}{9\theta_0^2 \varepsilon_0} , \tag{4}$$

where ε_0 is the static strain, and the ratio between pendulum and stretching mode frequency is easily seen to be

$$\frac{\omega_p}{\omega_s} = \sqrt{\varepsilon_0} , \tag{5}$$

so that, by combining equations (2), (4) and (5), the following expression is obtained for the contribution to pendulum Q from the stretching mode

$$\frac{Q_s}{Q_f} = \frac{16}{9\theta_0^2\sqrt{\varepsilon_0}} \; . \tag{6}$$

Similarly, for the bending mode, ratios relating energy E_b stored in it to pendulum energy, and its natural frequency to pendulum frequency, can be shown to be

$$\frac{E_p}{E_b} = \left(\frac{\omega_p}{\omega_b}\right)^2 = \frac{LD_b}{D_f^2}\varepsilon_0\theta_0 \; . \tag{7}$$

The frequency ratio can be derived easily by equating the peak potential elastic energy stored in the fibers to the peak kinetic energy calculated in the sine wave approximation. The contribution to pendulum Q from the bending mode can then be calculated by combining equations (3) and (7), and is

$$\frac{Q_b}{Q_f} = \left(\frac{LD_b}{D_f^2}\varepsilon_0\theta_0\right)^{3/2} . \tag{8}$$

The combination of these two contributions leads to the values shown in figure 2 for the total Q of the pendulum as a function of oscillation amplitude θ_0 in various configurations of circular suspension diameter and fibre type. It is here assumed that the mass of the bob be 0.16 g, which was the mass of the 5 mm BK7 ball lens used in the pilot experiment, for which the lowest curve holds.

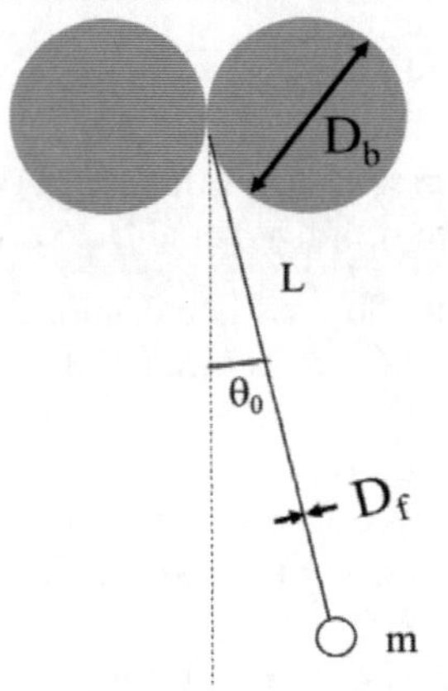

Figure 1. Pendulum configuration and parameters definition. Circular suspension profiles were chosen for ease of precise implementation and coherence between design and theory.

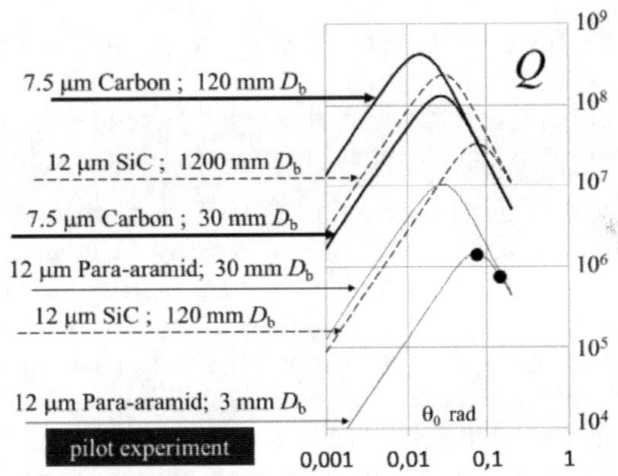

Figure 2. Pendulum Q values calculated for various fibres and suspension diameters. The two dots are measured Q for the pilot experiment. Their agreement with the theory supports its validity, and the confidence that it may be extrapolated to different configurations.

The selection of suitable fibres is mainly guided by equations (6) and (8), summarized in figure 2, which suggest looking through thin high stiffness fibres for Q maximization. However, it must be pointed out here that characterization of mechanical losses in such high strength fibres is typically not to be found in the technical literature. The intrinsic Q of candidate fibre materials was therefore measured in order to produce the results reported in figure 2. This was done by evaluating the Q of a spring-mass resonator in which the spring was a bundle of a reasonably well known number of fibres. Q_f values of 120, 470, and 1000 were found for Para-aramidic, SiC, and Carbon fibres respectively [19].

Clearly, the pendulum configuration should be designed for maximum Q at the desired oscillation amplitude, which in turn must be decided in considering issues like a good match with minimization of period and ease of correlation between measured frequency departure and gravitational effect.

3. Planning the experiment

The experiment which is being implemented includes two 1 m long pendulums oscillating in the same plane about 100 mm apart, and an ensemble of cylindrical Tungsten masses laying either side of such plane, outside the vacuum container, with a small separation in between, where the bob swings inside the container. The bob is a 5 mm diameter BK7 ball lens suspended by two converging fibres which wrap and unwrap on two cylindrical profiles as depicted in figure 1. The arrangement removes the degeneracy with the transversal mode and obliterates undesired Coriolis effects. Active masses are moved periodically back and forth in the two configurations of figure 3, alternately with one pendulum swinging between two pairs of masses and the other in the gap for half the repetition period T_R and vice-versa for the other half. Half the calculated acceleration a_M added by the active masses system, divided by the acceleration due to Earth's gravity a_g, is shown in figure 4 versus pendulum distance from system center. More details are given below, where the near equality between such half ratio and the relative frequency shift is discussed. If perfect equivalence between the two is assumed for the time being, the difference between the two relative frequency differences is twice the sum of a_M/a_g for both pendulums.

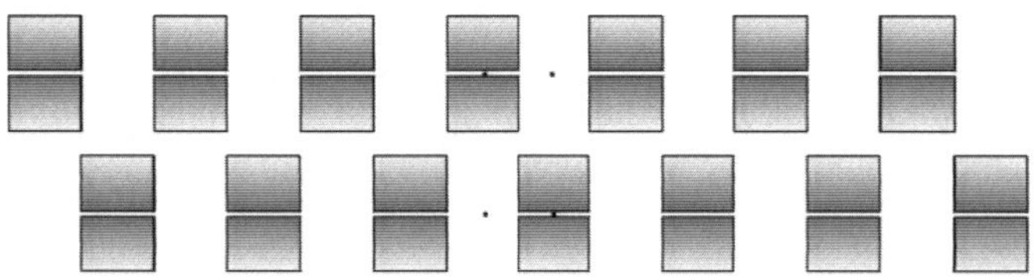

Figure 3. Scheme of the two configurations of active masses. Rest positions of the two bobs are shown by black dots. Subtracting the relative frequency differences in the two configurations doubles the sum of both relative shifts.

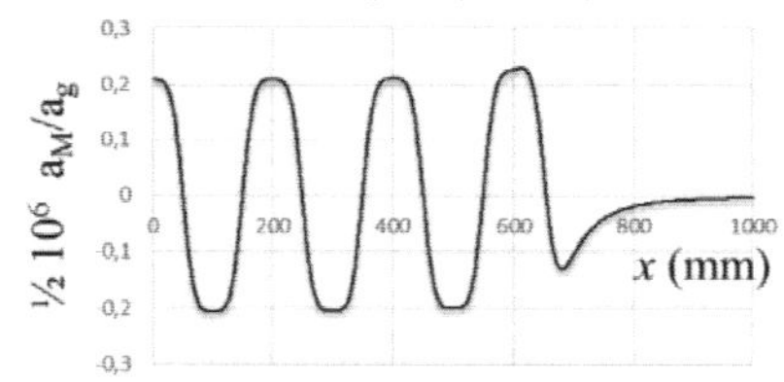

Figure 4. Relative extra acceleration versus distance from center for the Tungsten masses of figure 3. Frequency shifts are roughly equal to ½ a_M/a_g.

In either configuration the time delay rate of change between the two pendulums is measured with two optical high resolution Position Sensitive Detectors (PSD) at the bobs' rest points [10], which in principle yields their relative frequency difference in the most precise measurement of any quantity made available by modern Metrology [16]. This fact, together with the common-moding of angular seismic noise that the dual pendulum solution can provide, makes of this proposed scheme a totally new experiment, and not "just an incremental improvement" on the pilot experiment any more than most *G* determinations have been with respect to the original Cavendish experiment.

A slight mistuning must be imposed between the two pendulums to keep signal injection induced frequency pulling low, but this is really easy given the high projected *Q* values for the pendulums. In fact, even at high coupling (e.g. 10^{-2}) a relative mistuning of 10^{-8} is more than adequate [20,21] to keep pulling below the 10^{-12} level that would start to be relevant for the target 10^{-5} accuracy. It would be hard to avoid such mistuning, but at any rate a rational frequency ratio $(n\text{-}1)/n$ is here planned between the two, in order to get them back in phase every *n* periods, and design the period of such occurrence.

Delay data will be taken at such times in order to guarantee adequate Common Mode Rejection (CMR) of the angular seismic noise that introduced on the single period measurement a white noise level of the order of 3 μs in the pilot experiment. It must be pointed out here for clarity that angle noise in the oscillation plane is not filtered by the pendulum *Q*, as it amounts to a rotation of the laboratory. As ocean waves excited angular noise at 0.2 Hz is sampled at pendulum frequency, such angle noise comes by aliasing to be white delay time noise. Regression on *N* delays will then improve it by $N^{-1/2}$, and division by NT to get the slope will yield a relative frequency resolution $\sigma_y(\tau)$ that improves as $N^{-3/2}$ upon averaging *N* results [16]. If the necessary 2 to $3\cdot10^{-12}$ Type A uncertainty is desired in 10^4 s (about 2.5 hours, or 250 measurements of 40 s), the single shot $\sigma_y(\tau)$ should then be about 10^{-8}, which places at $4\cdot10^{-7}$ the needed single differential delay uncertainty (≈300 ns on single delay). This represents a factor of 10 CMR of angular seismic noise at 0.2 Hz. Such noise should in principle be sampled at the same instant for the two pendulums in order to be perfectly rejected, and the maximum time delay between the two that is acceptable for the desired rejection must be studied.

So, while the *rms* period uncertainty $\delta T_{\rm rms}$ produced in the pilot experiment by the Rayleigh-waves quasi-coherent angle noise of *rms* value $\delta\theta_{\rm rms}$ was

$$\delta T_{rms} = \frac{\sqrt{2}}{\omega_p}\frac{\delta\theta_{rms}}{\theta_0}\ , \tag{9}$$

the *rms* delay uncertainty $\delta(\Delta t)_{\rm rms}$ produced in this experiment by the same angle noise, if the measured delay is Δt, can be easily shown to be

$$\delta(\Delta t)_{rms} = \Delta t\,\frac{\omega_R}{\omega_p}\frac{\delta\theta_{rms}}{\theta_0} \tag{10}$$

when calculated from the time derivative of the quasi-sine-wave angular noise at the frequency $\omega_{\rm R}$ of Rayleigh waves. From the ratio of equations (9) and (10), the maximum delay can be calculated at which a given minimum CMR of the bright angular seismic noise spectral line is obtained in the dual pendulum configuration. For equal amplitude θ_0 in the two experiments, the constraint is

$$\Delta t < \frac{\sqrt{2}}{CMR\ \omega_R}\ . \tag{11}$$

For the actual frequency of Rayleigh waves and the desired CMR, equation (11) indicates 100 ms as the maximum acceptable delay. This is the reason why a rational $(n\text{-}1)/n$ frequency ratio is designed for the two pendulums, so that every n periods they be in phase at the detector location. On the other hand, since every period the slower pendulum's delay increases by about T/n, the minimum value of n that guarantees at least one delay measurement of less than 100 ms at that time is 20, because the pendulum period is about 2 s. The choice of n=21 is proposed here only for discussion purposes. Greater values would offer more useable delay data per coincidence but increase their distance in time.

The relevant time delay slope s is indicated in figure 5 in the delay-versus-time space. Instrumental resolution is clearly not a problem for a desired delay uncertainty of 300 ns, nor is a particularly accurate reference frequency necessary of the time interval counter.

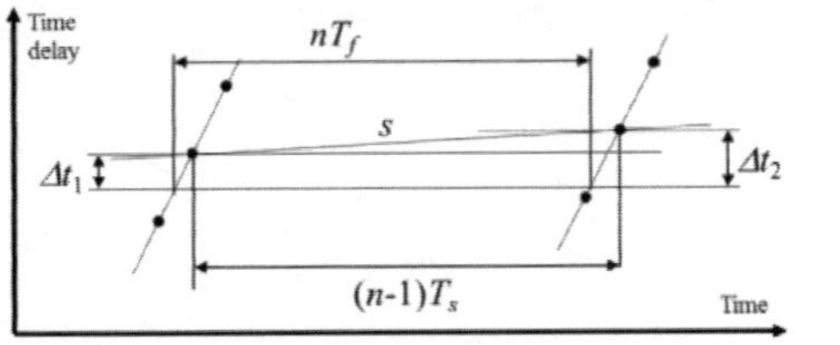

Figure 5. Measured time delay rate (slope s). Dots are ticks from the slow pendulum.

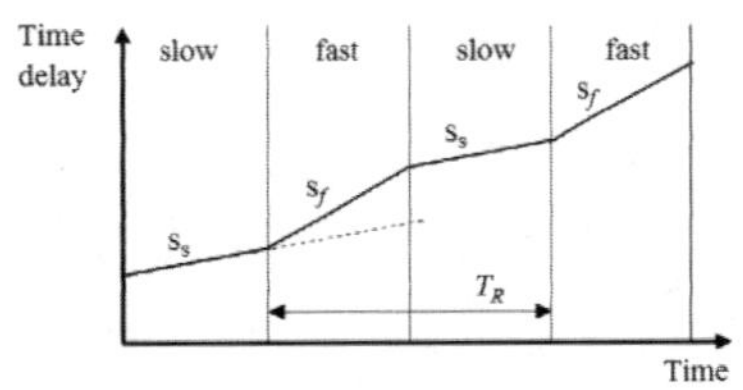

Figure 6. Delay slope changes as active masses are moved from slow to fast pendulum and back.

In figure 6, a picture is given of how time delay slope is expected to change when the center of the active masses system is moved back and forth between left and right pendulums. The relevant information for the extraction of a G value is the difference Δs=s_f-s_s between the slopes measured with active masses centered on the fast pendulum and the slow one. In fact, it can be easily shown that Δs is directly proportional to the average of relative frequency shifts y_{Mf} and y_{Ms} induced by active masses on fast and slow pendulum respectively. With slopes calculated as $(\Delta t_2\text{-}\Delta t_1)/(nT_f)$, the slope difference is

$$\Delta s = \frac{n-1}{n}\frac{T_s}{T_f}2\left(y_{Mf} + y_{Ms}\right), \tag{12}$$

where the ratios T_s/T_f and $(n\text{-}1)/n$ were not simplified because in the actual experimental arrangement they may not exactly compensate as they should.

As for the calculation of the gravitational effect on frequency, the relative extra acceleration given to the bob by the active masses system is the relevant parameter because, inasmuch as the oscillation of the bob's displacement x is a sine wave, by the second law of dynamics its angular frequency is given by $\omega_M^2 = (a_g + a_M)/x$, with $a_g = gx/L$ and ω_M the angular frequency as modified by the active masses.

Both a_g and a_M vanish at both rest positions, but their ratio doesn't, as both are linear in x for small displacements. However, neither is strictly linear, which yields the well-known non-isochronism of the simple pendulum plus, relevant for this experiment, a non-trivial tie between extra gravitational pull and induced frequency shift. So, while it's easy to find frequency and relative shift for small oscillations, as the relative extra acceleration can then be considered constant over the swing, non-trivial calculations are necessary for wider swings. In the small oscillation limit it is $2y_M = a_M/a_g$ at first order, as it can be promptly seen. A quick analysis of the arrangements shown in figure 7, with two equal active masses symmetrically placed on either side of the bob, yields for the effect of the two masses on one pendulum

$$\omega_M = \omega_p(1 + y_M) = \sqrt{\frac{g}{L}}\left(1 + \frac{\rho_M}{\rho_E}\frac{L}{R_E}\left(\frac{R}{a}\right)^3 \Gamma(0)\right) , \qquad (13)$$

where $\Gamma(0)$ is the value taken at x=0 by the geometrical factor function $\Gamma(x)$, which depends on shape and positioning of the active masses system. In equation (13) ρ_M and ρ_E are the densities of active masses and Earth respectively, and R_E is the Earth's radius, L depends on the pendulum that is considered, being L_f for the fast and L_s for the slow one, and R and a are defined in figure 7. For spheres $\Gamma_{sph}(0)=1$ and for cylinders $\Gamma_{cyl}(0)=(a/R)a^2w/\{\tfrac{2}{3}(a^2+w^2)^{3/2}\}$, which comes to the value $3/(4\sqrt{2})$, or roughly ½, if $a \approx w \approx R$.

For wider oscillations the shift can't be figured out so easily as in equation (13) because the relative extra acceleration depends on x, which imposes a non-trivial integration in extracting a G determination from frequency shift data. In order to minimize the impact of this fact on uncertainty, the choice is made here to opt for cylindrical masses because of the considerably higher uniformity of $\Gamma_{cyl}(x)$ as compared with $\Gamma_{sph}(x)$ of the spherical ones. Both geometrical factor functions are responsible for the shape of curves shown in figure 8, and the rationality of the choice appears obvious when considering that Q maximization imposes for the swing amplitude θ_0 values of the order of those highlighted in the figure. The signal size price of choosing cylinders may appear steep in view of the already tiny effect, but the gain in accuracy of G value extraction is so great that it's judged to offset it more than adequately.

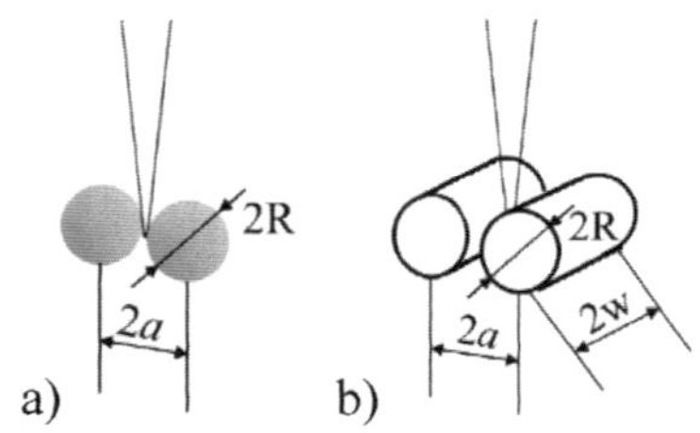

Figure 7. Experimental arrangements for spherical and cylindrical active masses. The bob swings in the centre of the gap between masses.

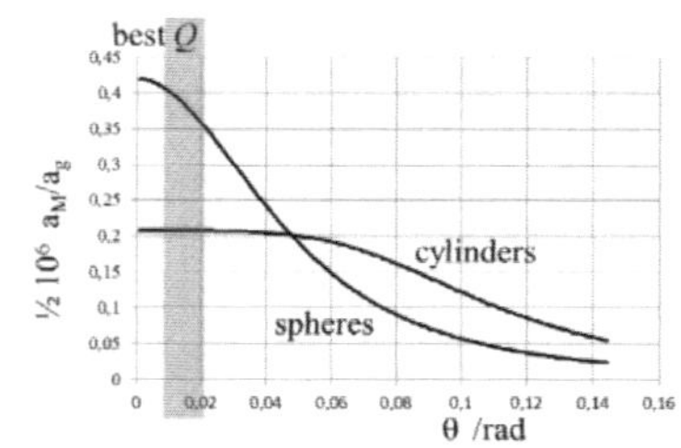

Figure 8. Profile versus θ of the ratio a_M/a_g for a 1 m long pendulum. Tungsten active masses are assumed, with R=50 mm and w/a=1.25.

Full expressions of $\Gamma(x)$ used in figure 8 for both cases, written with $\eta = w/a$ and $\xi = x/a$, are

$$\Gamma_{sph}(x) = (1 + \xi^2)^{-3/2} , \qquad (14)$$

$$\Gamma_{cyl}(x) = \frac{3}{4}\frac{a}{R\xi}\left(\frac{1}{\sqrt{1+(\eta-\xi)^2}} - \frac{1}{\sqrt{1+(\eta+\xi)^2}}\right) , \qquad (15)$$

Equation (15) was used as a basis to calculate the local acceleration profile shown in figure 4 for the case of seven twin cylindrical active masses. In that case, the reported acceleration is calculated as a function of the rest position of the bob, with the aim of illustrating how the frequency shift is symmetric around zero when the system of masses is moved by the separation between pendulums. In fact, it can be easily seen that an infinite periodic structure of active masses would yield a perfectly periodic and symmetric function of position. It turns out (see figures 9 and 10 below) that the system of seven pairs shown in figure 3 is enough to keep well below 10^{-4} the relative acceleration variation around the rest positions of the two pendulums, up to peak swing amplitudes suitable for Q maximization.

4. Accuracy budget

The formula by which the value of the Newtonian constant G can be extracted from the measured value of Δs as illustrated above can be derived by inverting equation (13). For a single pair of cylinders is

$$G = \frac{\pi\,\Delta s}{\rho_M\left(T_f^2+T_s^2\right)}\,\frac{R}{w}\left[\left(\frac{a}{R}\right)^2+\left(\frac{w}{R}\right)^2\right]^{3/2} \tag{16}$$

Periods T_f and T_s in equation (16) are the small oscillation values for fast and slow pendulums, and must be derived from measured periods by taking into account non-isochronism. However, the correction can be made very accurately if the sweet locally isochronous spot of the minimum in the period versus oscillation amplitude curve is chosen as an operation point, because the mathematical model can provide in this case an adequately accurate prediction of the error [18]. To better than 0.1% the relative value of such error turns out to be $-1.8\cdot10^{-5}$ for our 1 m pendulums if $D_b = 20$ mm (see figure 1), and it occurs at $\theta_0 = 0.017$ rad. It appears safe to assume that its contribution to G accuracy may smaller than 10^{-7}.

For the planned seven mass pairs of figure 3 the inversion formula is more complicated because the geometrical factor must take into account all seven contributions. This is not hard for the symmetric positioning, with the bob at the center of the system, but more complex when it turns asymmetric with the bob laying at gap center. Uniformity can be trimmed in both locations by careful design of the cylindrical mass pairs' shape factor $\eta = w/a$. Numerical results for uniformity at the two bob positions are shown for the seven twin masses case in figures 9 and 10. Mass pairs with shape factor $\eta = 1.264$ and length $2w = 100$ mm are assumed, 72 mm in diameter and spaced by 7 mm to host the 5 mm bob and thin vessel walls. Given the uniformity shown for this geometry, it appears reasonable to believe that the integration necessary for connecting G to Δs may be feasible with few percent accuracy, leading to a contribution to accuracy below 10^{-6}. Most of the burden is then left to geometrical accuracy.

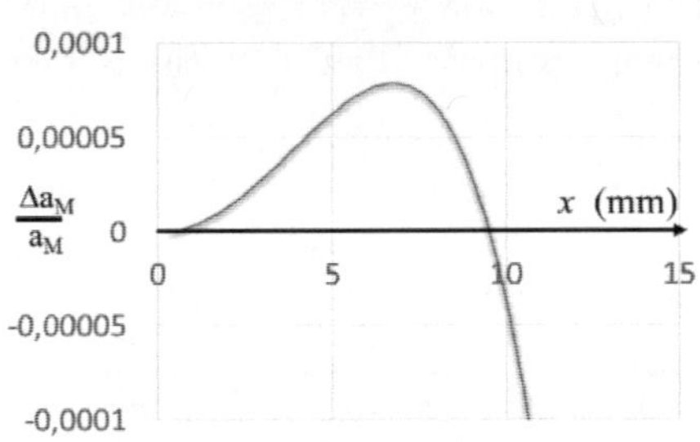

Figure 9. Relative effect at system center from seven twin active masses. Reference for both distance and acceleration is at system center.

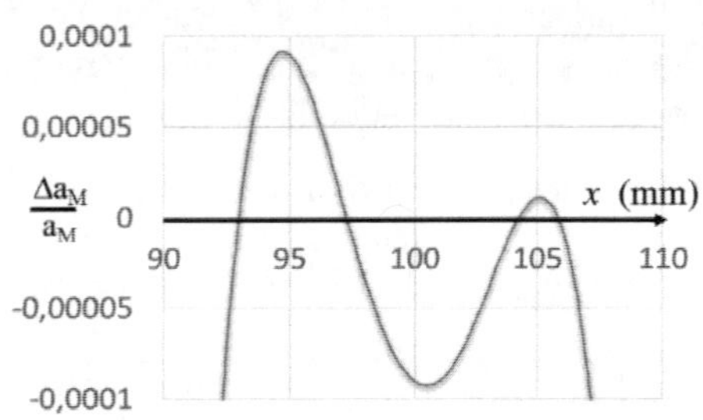

Figure 10. Relative effect at first gap center from seven twin active masses. Distance reference is at system center; acceleration at gap center.

Accuracy on geometry of the active masses system sets the limit for the experiment. In fact, differentiation of the geometrical factor function in equation (16) yields for length and diameter of the active masses a severe constraint of less than 0.5 μm for an accuracy contribution of $6\cdot10^{-6}$ on G. However, given the high accuracy of mass measurements, this condition may be relaxed if R is trimmed to reach a given mass (about 7.9 kg for the described Tungsten masses) after cylinders are cut to length, because the bias imposed on the radius will then be $\Delta R/R = -\Delta w/2w$, where Δw is the unknown length error, and the resulting G bias then indicates the less severe constraint of 1 μm for length uncertainty.

Unfortunately, no such compensation can be done on the spacing between the twin cylinders, and the sensitivity calculation then yields for it a harder requirement of 400 nm uncertainty for $6\cdot10^{-6}$ on G.

Temperature sensitivity of active masses system geometry is also an issue, as their indicated necessary dimensional accuracy is of the order of 10^{-5}, which requires the non-trivial temperature accuracy and stability of 2 K for the Tungsten masses. No particular problem is instead foreseen for the spacing because of its smaller dimension, which allows a temperature tolerance of about 10 K.

More benign is the requirement on positioning of the bob's trajectory with respect to active masses. In fact it turns out that both in the horizontal and vertical direction the extra acceleration features an extreme versus trajectory positioning, as shown in figure 11 and figure 12 respectively: a minimum in the center for the lateral direction, and a maximum a little above masses' gravity centers for the vertical. The vertical displacement of the maximum is due to the vertical pull down that active masses exert if they are lower than the bob, which adds to Earth's gravity and hence to recall force. For cylinders, such maximum is $(a^2+w^2)/3L$ above the masses' gravity centers (more than 2 mm for the planned size), and the relative shift is below $7\cdot10^{-4}$, which imposes an easy 50 μm tolerance to active masses geometry for a 10^{-6} accuracy contribution from uncertainty on its value. Also, trajectory position tolerance for a safe $2\cdot10^{-6}$ uncertainty contribution is ±0.1 mm in both directions for cylinders of the planned size, just a bit wider than what would be needed for the spheres vertically, but more than three times horizontally.

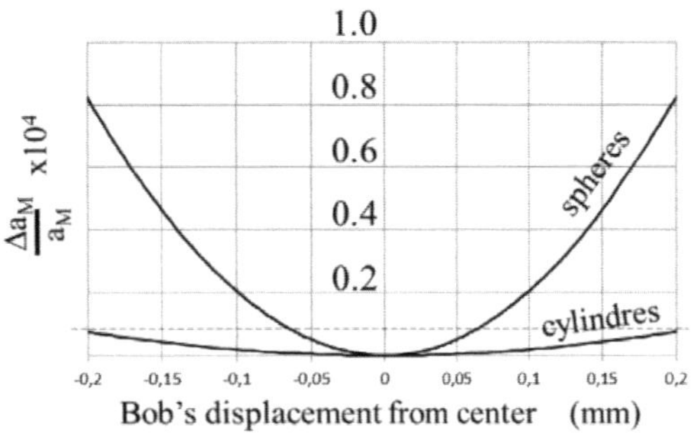

Figure 11. Relative variation of extra pull for lateral displacement of the bob's trajectory from active masses symmetry plane. The dashed line sets bob centering tolerance for 10^{-5} accuracy.

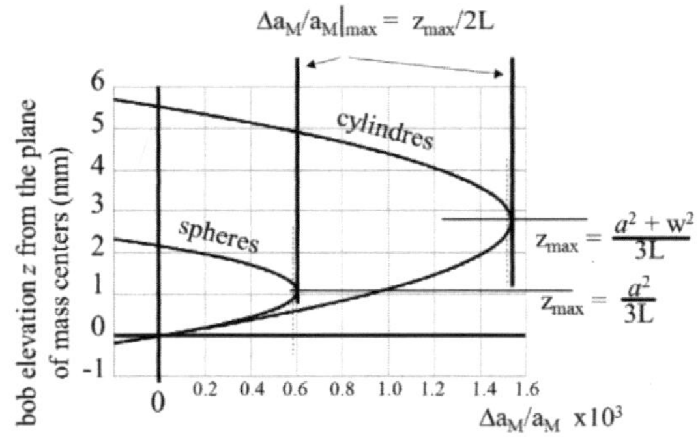

Figure 12. Relative variation of the relevant effect for vertical displacements of the bob's trajectory from the plane of mass centers. Position tolerance for 10^{-5} accuracy is marked by dashed lines.

A summary of the accuracy discussion spelled out here is given in Table 1. It appears that only geometrical uncertainties are expected to be relevant at the level of 10^{-5}, which ushers the possibility of doing even better if resources were to become available to improve geometrical accuracy.

Table 1. Accuracy budget projection based on 1 m long 7.5 μm Carbon fibres, 20 mm diameter suspension profiles, a swing amplitude of 0.017 rad, and a 160 mg bob. Density uniformity is assumed adequate to guarantee positioning uncertainty of the mass center of active masses smaller than 300 nm. For relaxation of dimensional accuracy demand, the diameter of active masses is assumed to be trimmed for accurate mass once the 2*w* length is realized to match the distance between pendulums.

Effect	Relative bias	Uncertainty	conditions
θ dependence of a_M	$<3\cdot10^{-5}$	$<10^{-6}$	Optimization of w/a
Shift at bob's vertical position	$6.7\cdot10^{-4}$	$<10^{-6}$	< 50 μm uncertainty in a , w
bob's vertical position	0	$2\cdot10^{-6}$	0.2 mm full interval tolerance
bob's lateral position	0	$1.7\cdot10^{-6}$	0.2 mm full interval tolerance
Adiabaticity	$<-2.5\cdot10^{-5}$	$2\cdot10^{-6}$	0.017 rad swing amplitude [11]
Non-isochronism	$-1.8\cdot10^{-5}$	$< 10^{-7}$	Operation at minimum of period
Spacing between twin masses	0	$6\cdot10^{-6}$	0.4 μm gap uncertainty
Active masses dimensions	0	$6\cdot10^{-6}$	0.9 μm uncertainty with mass trimming
Active masses density	0	$5\cdot10^{-6}$	
Total Type B		$1\cdot10^{-5}$	$(\Sigma u_i^2)^{1/2}$
Total Type A		$< 3\cdot10^{-7}$	In three months (based on $3\cdot10^{-6}$ in 1 d)

5. Conclusions

A new experiment was presented for the determination of the Newtonian constant, which is based on a Time and Frequency Metrology approach consisting in the measurement of the small frequency difference between two freely oscillating pendulums via their time delay rate of change. A system of dense active masses is moved back and forth between the two, alternately increasing one frequency and reducing the other, and vice-versa. The increase in resolution by averaging is fast in this case because the limiting noise is white delay noise, which yields $\sigma_y(\tau)$ proportional to $\tau^{-3/2}$. This fact is unique among experiments for the determination of *G*, and offsets the poor signal size problem allowing to focus the design on accuracy rather than S/N ratio. A preliminary experiment is in progress to confirm the possibility, indicated by the analysis, of obtaining *Q* values in the 10^8 range for the pendulums in vacuum, and the construction of the dual pendulum experiment is under way. It is expected that an accuracy of 10^{-5} may be obtained for *G* with the proposed approach, and that it may be possible in a metrology laboratory to reduce limiting geometrical uncertainties enough to push it into the 10^{-6} range.

Acknowledgments
The author wishes to acknowledge the encouragement of many colleagues both over the years and recently in this revisited proposal. In particular should be mentioned Robert Drullinger, Stefan Schlamminger and Bill Phillips of NIST, and Valter Giaretto, Rosario Ceravolo and Lamberto Rondoni of the Politecnico di Torino. The author also wishes to acknowledge the support of the US Department of Commerce and NIST through the Precision Measurements Grant Program. Award ID number 70NANB15H348.

[1] Cavendish H 1798 *Philos. Trans. R. Soc. London* **88** 469
[2] Michaelis W, Haars H and Augustin R 1996 *Metrologia* **32** 267
[3] Kuroda K 1995 *Phys. Rev. Lett.* **75** 2796
[4] Schlamminger S, Holzschuh E, Kundig W, Nolting F, Pixley R E, Schurr J and Straumann U 2006 *Phys. Rev. D* **74** 082001
[5] Schwarz J P, Robertson D S, Niebauer T M and Faller J E 1998 *Science* **282** 2230
[6] Fixler J B, Foster G T, McGuirk J M, Kasevich M A 2007 *Science* **315** 74
[7] Lamporesi G, Bertoldi A, Cacciapuoti L, Prevedelli M and Tino G M 2008 *Phys. Rev. Lett.* **100** 050801
[8] Kleinvoss U, Meyer H, Schumacher A and Hartmann S 1999 *Meas. Sci. Technol.* **10** 492
[9] Parks H V and Faller J E 2010 *Phys. Rev. Lett.* **105** 110801
[10] De Marchi A, Ortolano M, Berutto M and Periale F 2002 *Proc. 6th Symposium on Frequency Standards and Metrology (Fife, Scotland, 10-14 September 2001)* p 538
[11] Berutto M, Ortolano M and De Marchi A 2009 *Metrologia* **46** 119
[12] Gundlach J and Merkowitz M 2000 *Phys. Rev. Lett.* **85** 2869
[13] Armstrong T R and Fitzgerald M P 2003 *Phys. Rev. Lett.* **91** 201101
[14] Quinn T J, Speake C C, Richman S J, Davis R S and Picard A 2001 *Phys. Rev. Lett.* **87** 111101
[15] Luo J, Liu Q, Tu L C, Shao C G, Liu L X, Li S Q, Yang Q and Zhang Y T 2009 *Phys.Rev.Lett.* **102**, 240801
[16] Sullivan D, Allan D, Howe D and Walls F L 1990 *Characterization of clocks and oscillators,* NIST Technical note 1337
[17] Berger J, Davis P and Ekstroem G 2004 *J. Geophys. Res.* **109,** B11307
[18] Maffioli L, 2016 *Mathemetical modelization and experimental validation of a simple pendulum for the measurement of the Newtonian constant G*, Tesi di Laurea, Politecnico di Torino
[19] Ceravolo R, DeMarchi A, Pinotti E, Surace C and Zanotti Fragonara L 2016 *Exp.Mech.* in print
[20] Kurokawa K 1973 *Proc. of IEEE* **61**, 1386
[21] Razavi B 2004 *IEEE J. Sol. State Circ.* **39**, 1415

High-Performance Optical Frequency References for Space

Thilo Schuldt[1], Klaus Döringshoff[2], Alexander Milke[1,3], Josep Sanjuan[1] Martin Gohlke[1], Evgeny V. Kovalchuk[2], Norman Gürlebeck[3], Achim Peters[2] and Claus Braxmaier[1,3]

[1] German Aerospace Center (DLR), Institute of Space Systems, Robert-Hooke-Straße 7, 28359 Bremen, Germany
[2] Humboldt-University Berlin, Institute of Physics, Newtonstraße 15, 12489 Berlin, Germany
[3] Center of Applied Space Technology and Microgravity (ZARM), University of Bremen, Am Fallturm, 28359 Bremen, Germany

E-mail: `thilo.schuldt@dlr.de`

Abstract. A variety of future space missions rely on the availability of high-performance optical clocks with applications in fundamental physics, geoscience, Earth observation and navigation and ranging. Examples are the gravitational wave detector eLISA (evolved Laser Interferometer Space Antenna), the Earth gravity mission NGGM (Next Generation Gravity Mission) and missions, dedicated to tests of Special Relativity, e.g. by performing a Kennedy-Thorndike experiment testing the boost dependence of the speed of light. In this context we developed optical frequency references based on Doppler-free spectroscopy of molecular iodine; compactness and mechanical and thermal stability are main design criteria. With a setup on engineering model (EM) level we demonstrated a frequency stability of about $2{\cdot}10^{-14}$ at an integration time of 1 s and below $6{\cdot}10^{-15}$ at integration times between 100 s and 1000 s, determined from a beat-note measurement with a cavity stabilized laser where a linear drift was removed from the data. A cavity-based frequency reference with focus on improved long-term frequency stability is currently under development. A specific sixfold thermal shield design based on analytical methods and numerical calculations is presented.

1. Introduction

High-performance optical frequency references are an essential tool for many applications in metrology, spectroscopy and fundamental physics. Besides on-going improvement with respect to frequency stability, also technology development for future operation in space is currently carried out where crucial design parameters are compactness and rigidity.

Application areas in space include Earth observation, fundamental science as well as navigation and ranging. Thereby, the frequency reference is either needed as laser source for a high-sensitivity inter-spacecraft optical metrology system, as part of a payload enabling tests of fundamental physics or as a high accuracy timebase for global navigation satellite systems (GNSS). Examples are eLISA (evolved Laser Interferometer Space Antenna) [1], a gravitational wave detector using picometer-sensitivity laser interferometric relative distance measurement between spacecraft with a distance of approximately 1 million kilometers; NGGM (Next Generation Gravity Mission) [2], i.e. future missions measuring the Earth's gravitational field requiring a relative distance measurement with nanometer sensitivity between two satellites which are typically 200 km apart; BOOST (BOOst Symmetry Test) [3] / mSTAR (mini SpaceTime Asymmetry Research) [4], a space-based test of Special Relativity carrying out a Kennedy-Thorndike (KT) type experiment, i.e. a test of the boost dependence of the speed of light with an aimed 10 to 100 fold improvement over currently best ground-based determination of the KT coefficient [5] using optical frequency references with frequency instabilities at or below the $1 \cdot 10^{-15}$ level at orbit time.

In the current baseline design, the proposed missions mSTAR and BOOST utilize an optical absolute frequency reference based on molecular iodine and a length reference based on a high-finesse optical cavity, whose resonance frequency is sensitive to the speed of light. The beat-frequency between these two references is measured and analyzed with respect to variations at orbit time of approximately 90 min for the determination of the KT coefficient. While atomic and molecular clocks intrinsically offer high long-term frequency stability, such time scales are challenging for cavity-based frequency references and especially require a specific thermal design. While the frequency references detailed in the following are in particular developed with respect to this application, they are also the basis for many other applications in space as detailed above.

2. Iodine-based frequency references

Optical frequency references based on Doppler-free spectroscopy of molecular iodine near 532 nm are a well-proven technology developed in several laboratories for many years [6, 7, 8], also in compact setups [9, 10, 11, 12]. With setups on laboratory level, frequency stabilities at low 10^{-15} level were demonstrated using modulation transfer spectroscopy (MTS) [8, 13]. For space applications, such systems need to be further developed with respect to compactness and mechanical and thermal stability. In a collaboration of the German Aerospace Center (DLR Institute of Space Systems, Bremen), the Center of Applied Space Technology and Microgravity (ZARM, University of Bremen), the Humboldt-University Berlin and the space company Airbus Defence & Space (Friedrichshafen) two setups on elegant breadboard (EBB) and engineering model (EM) level, respectively, were realized during the last years. These setups are detailed in the following sections.

2.1. Setup on Elegant Breadboard (EBB) level

In a first activity, a spectroscopy setup on elegant breadboard level was realized, based on a laboratory setup at the Humboldt-University Berlin. The light source is a Nd:YAG solid state laser with an output wavelength of 1064 nm, internally frequency doubled to 532 nm (model Prometheus by Coherent, Inc.). The 532 nm output laser beam is split into pump and probe for spectroscopy, both beams pass acousto-optic modulators (AOM) used for intensity stabilization and to generate a frequency shift between pump and probe. Both beams are fiber coupled to the EBB spectroscopy unit. The pump beam is either phase modulated using a fiber-coupled electro-optic modulator (EOM) or frequency modulated using the corresponding AOM.

The EBB spectroscopy unit is assembled on a 550 mm x 250 mm x 50 mm baseplate made of OHARA Clearceram-Z HS with a thermal expansion coefficient (CTE) of $2 \cdot 10^{-8}\,\mathrm{K}^{-1}$. The optics (i.e. mirrors, thin film polarizers, glass plates) are made of fused silica and integrated on the baseplate using adhesive bonding technology with a space-qualified two-component epoxy resulting in a semi-monolithic optical assembly [14], see figure 1. A commercial 30 cm long iodine cell (provided by the Institute of Scientific Instruments of the Academy of Sciences of the Czech Republic, Brno) is used in triple-pass configuration.

Mechanical mounts for fiber outcoupler, waveplates and polarizers are made of Invar for CTE matching between baseplate, mount and optics. Four pairs of AR-coated wedged glass plates (Risley prisms) in pump and probe beam, mounted in precision rotation mounts, enable an alignment of the beam overlap in the gas cell after integration of the optical setup. A commercial pigtailed fiber collimator with an output laser beam diameter of 3 mm is used (provided by OZ

Journal of Physics: Conference Series **723** (2016) 012047 doi:10.1088/1742-6596/723/1/012047

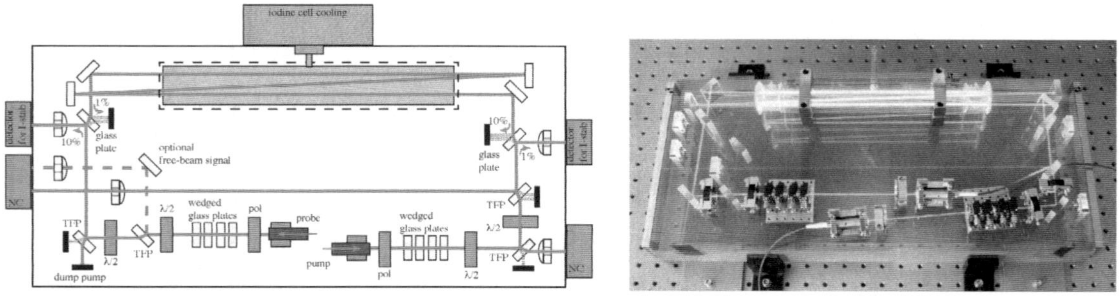

Figure 1. Schematic and photograph of the EBB setup using a commercial 30 cm long iodine cell in triple-pass configuration [15]. The optical components are integrated on a highly stable glass ceramics baseplate using adhesive bonding technology.

Optics). Polarizers are placed directly behind the fiber outcoupling in order to guarantee clean polarization. Additionally, RAM caused by the EOM is detected at a noise-cancelling (NC) detector and removed by feedback to the corresponding AOM in the pump beam. A second noise-cancelling detector is used to detect the MTS signal. Therefor, part of the probe beam is split off before the iodine cell and guided to the NC for balanced detection which allows for shot-noise limited detection. The spectroscopy signal is mixed down with the modulation frequency of about 300 kHz and appropriately filtered. The resulting error signal is fed to a servo control loop actuating the laser frequency via the laser crystal temperature for slow actuation and a PZT mounted to the laser crystal for fast actuation with control bandwidth of about 10 kHz.

The frequency stability of this setup was determined from a beat-note measurement with a ULE cavity stabilized laser. The Allan deviation shows a frequency stability below $1 \cdot 10^{-14}$ at an integration time of 1 s and below $4 \cdot 10^{-15}$ for integration times between 10 s and 1000 s where a linear drift was removed from the beat time record, see figure 2. Residual amplitude modulation (RAM) and residual temperature variations of the iodine cell cooling finger are most probably limiting the frequency stability at longer integration times.

2.2. Setup on Engineering Model (EM) level

In a subsequent activity, the design of the EBB setup was further optimized with respect to compactness and mechanical and thermal stability. A spectroscopy setup on engineering model was developed, based on the experience gained with the EBB setup. The 18 cm x 38 cm x 4 cm baseplate as well as the optics and the iodine gas cell are made of fused silica in order to match the CTE, yielding high dimensional stability under thermal cycling. The optical components are integrated using the same assembly-integration technology (adhesive bonding) as for the EBB setup. A schematic and a photograph of the integrated EM spectroscopy unit are shown in figure 3.

For realizing a compact and ruggedized setup, a special designed compact multi-pass gas cell was realized. The 10 cm x 10 cm x 3 cm fused silica cell is designed for nine-pass of pump and probe beam, corresponding to an interaction pathlength of 90 cm.

Commercial fiber collimators by Schäfter and Kirchhoff (with 3 mm output beam diameter) in combination with Invar mounts are used. A pair of wedged glass plates are placed after each fiber outcoupling, enabling a beam adjustment after integration similar to the EBB setup. The glass plates are mounted to specific mounts made of fused silica. Polarizers and waveplates are glued to mounts made of titanium for CTE matching. As in the EBB setup, part of pump and probe beams are split off before entering the gas cell for power stabilization.

The Allan deviation of the beat note with the ULE cavity indicates a frequency stability of

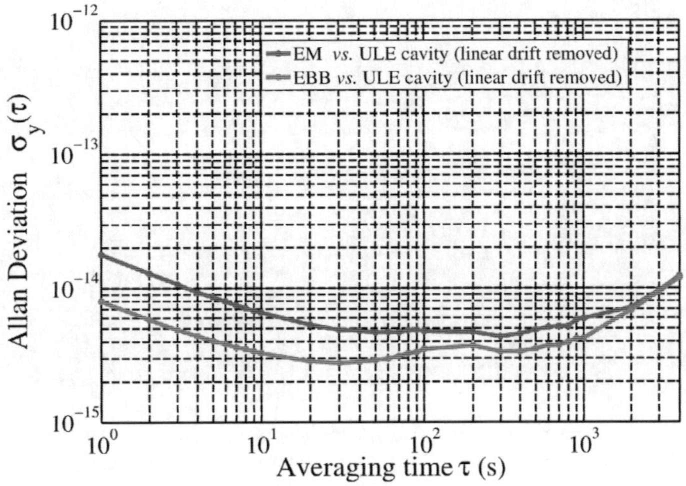

Figure 2. Relative Allan deviation of the beat note frequency between the EBB (green, lower curve) and the EM (blue, upper curve) and a laser locked to a cavity made from ultra-low expansion glass ceramics (ULE). A linear drift of 50 mHz/s, mainly attributed to isothermal creep of the cavity, has been removed from the time record of the beat data. The length of the time record is 76000 s. (colour online)

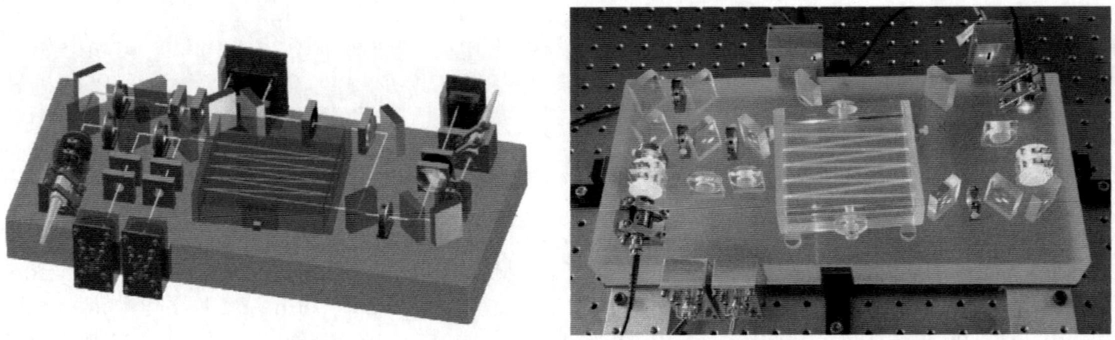

Figure 3. CAD and photograph of the spectroscopy unit on engineering model level using a specifically designed compact gas cell in nine-pass configuration [16].

$5 \cdot 10^{-15}$ for integration times larger than 100 s, similar to the EBB setup, see figure 2. The spectroscopy unit was subjected to thermal cycling from -20°C to +60°C and vibrational loads with sine vibration up to 30 g and random vibration up to 25.1 g_{rms}. The frequency stability was measured before and after the tests where no degradation was observed.

3. Cavity-based frequency reference

Cavity-based frequency references are a very common technology, also already developed in compact and ruggedized setups dedicated to future application in space, see e.g. the JPL cavity development for the Gravity Recovery and Climate Experiment follow-on (GRACE-FO) mission [17] and the ESA project 'High Stability Laser (HSL)' for future Next Generation Gravity Missions (NGGM) [18]. While such systems are usually optimized for highest short-term frequency stability (up to several seconds), improved long-term stability at the low 10^{-15} level at integration times of several 1000 seconds is investigated in a collaboration project of

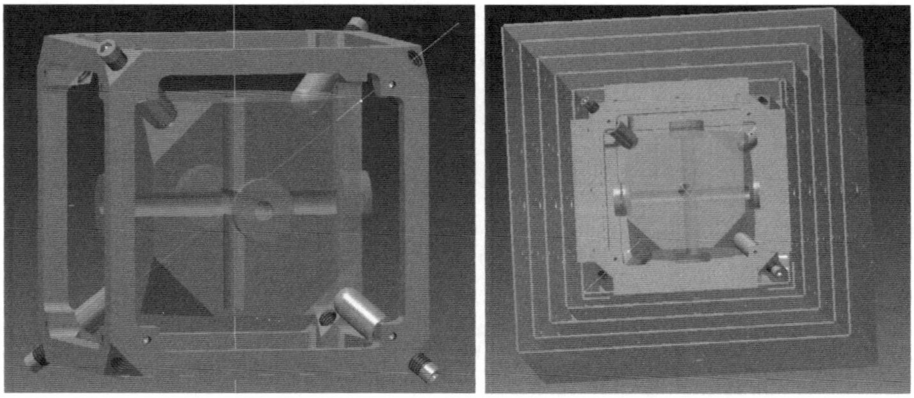

Figure 4. Preliminary cavity mounting and sixfold thermal shielding design. The cubic cavity has a side length of 87 mm and the outermost shield has a length of 242 mm.

the Center of Applied Space Technology and Microgravity (ZARM, University of Bremen), the German Aerospace Center (DLR Institute of Space Systems, Bremen) and the Leibniz-University Hannover to meet the demands of a future KT Mission.

As the frequency stability at longer integration times is mainly limited by residual temperature fluctuations and ultimately by thermal noise, main effort is put on a thermal design of the cavity enclosure. In our experiment, a cubic cavity designed by NPL with a side length of 87 mm [19] is used as it provides highest vibration insensitivity and is already under development for space compatibility within the HSL project. The cavity consists of a spacer made of ultra-low expansion (ULE) glass in combination with fused silica mirror substrates resulting in a theoretical thermal noise limit of $4.5 \cdot 10^{-16}$. The mirrors are equipped with ULE compensation rings in order to obtain a CTE zero crossing temperature of the assembled cavity near room temperature [20].

The cavity mounting is based on experience of NPL in combination with the design of a specific thermal shield. The current baseline foresees a cubic six-stage thermal shield made of 1.5 mm thick highly polished aluminum plates, cf. the CAD shown in figure 4. The temperature stability of the cavity can be further improved by temperature stabilization of the outermost shield at the CTE zero crossing temperature of the cavity. The design of the thermal shield is based on analytical models and numerical simulations, cf. [21, 22]. Different materials such as Teflon, Ultem and Titanium are investigated for the shield supports.

4. Conclusion

We presented our current efforts in realizing compact setups of frequency references based on Doppler-free spectroscopy of molecular iodine and on an optical resonator for future applications in space. Two iodine standards on elegant breadboard and engineering model level were realized, both optimized with respect to dimension, mass and thermal and mechanical stability. A frequency stability below $1 \cdot 10^{-14}$ at an integration time of 1 s and below $4 \cdot 10^{-15}$ for integration times between 10 s and 1000 s was determined in a beat measurement with a ULE cavity setup where the drift attributed to the cavity has been removed. Beside that, we presented first steps on the realization of a cavity setup with enhanced long-term stability. Analytical models were developed for thermal shielding evaluation, resulting in a six-stage thermal shield baseline design. Such optical frequency references are of high interest for a variety of future space missions including the proposed missions mSTAR and BOOST, which require optical frequency references with frequency stabilities at or below the $1 \cdot 10^{-15}$ level at orbit time (90 min). Furthermore, the frequency stability achieved with the iodine references – which in contrast to cavity stabilized systems feature an absolute frequency reference – fulfills the requirements for planned missions like eLISA and NGGM.

Acknowledgments

Financial support by the German Space Agency DLR with funds provided by the Federal Ministry of Economics and Technology (BMWi) under grant numbers 50 QT 1102, 50 QT 1201 and 50 QT 1401 is highly appreciated. The authors thank Ross Williams, Geoffrey Barwood and Patrick Gill from NPL for support and fruitful discussions on the cavity design and Jan Hrabina and Josef Lazar from the Institute of Scientific Instruments of the Academy of Sciences of the Czech Republic (Brno) for helpful discussions on the realization of the multipass gas cell.

References

[1] Amaro-Seoane P, Aoudia S, Babak S, Bintruy P, Berti E, Boh A, Caprini C, Colpi M, Cornish N J, Danzmann K, Dufaux J F, Gair J, Jennrich O, Jetzer P, Klein A, Lang R N, Lobo A, Littenberg T, McWilliams S T, Nelemans G, Petiteau A, Porter E K, Schutz B F, Sesana A, Stebbins R, Sumner T, Vallisneri M, Vitale S, Volonteri M and Ward H 2012 *Classical and Quantum Gravity* **29** 124016

[2] Massotti L, Cara D D, del Amo J, Haagmans R, Jost M, Siemes C and Silvestrin P 2013 *The ESA Earth Observation Programmes Activities for the Preparation of the Next Generation Gravity Mission* (American Institute of Aeronautics and Astronautics)

[3] Milke A, Aguilera D, Gurlebeck N, Schuldt T, Herrmann S, Doringshoff K, Spannagel R, Lammerzahl C, Peters A, Biering B, Dittus H and Braxmaier C 2013 *European Frequency and Time Forum International Frequency Control Symposium (EFTF/IFC), 2013 Joint* pp 912–914

[4] Schuldt T, Saraf S, Stochino A, Döringshoff K, Buchman S, Cutler G D, Lipa J, Tan S, Hanson J, Jaroux B, Braxmaier C, Gürlebeck N, Herrmann S, Lämmerzahl C, Peters A, Alfauwaz A, Alhussien A, Alsuwaidan B, Al Saud T, Dittus H, Johann U, Worden S P and Byer R 2015 *Proceedings of the European Frequency and Time Forum (2015)*

[5] Tobar M E, Wolf P, Bize S, Santarelli G and Flambaum V 2010 *Phys. Rev. D* **81**(2) 022003

[6] Arie A and Byer R 1993 *J. Opt. Soc. Am. B* **10** 1990–1997

[7] Ye J, Ma L S and Hall J L 2001 *Phys. Rev. Letters* **87** 270801

[8] Zang E J, Ciao J P, Li C Y, Deng Y K and Gao C Q 2007 *IEEE Transactions on Instruments and Measurement* **56** 673–676

[9] Hong F L, Ishikawa J, Bi Z Y, Zhang J, Seta K, Onae A, Yoda J and Matsumoto H 2001 *IEEE Trans. Instrum. Measur.* **50** 486–489

[10] Nyholm K, Merimaa M, Ahola T and Lassila A 2003 *IEEE Transactions on Instrumentation and Measurement* **52** 284–287

[11] Schuldt T, Braxmaier C, Müller H, Huber G, Peters A and Johann U 2004 *Proceedings of the 5th International Conference on Space Optics (ICSO 2004)* (ESA Publications) pp 611–617

[12] Acef O, Clairon A, Du Burck F, Turazza O, Djerroud K, Holleville D, Lours M, Auger G, Brillet A and Lemonde P 2010 *Proceedings of the International Conference on Space Optics, ICSO 2012*

[13] Döringshoff K, Reggentin M, Nagel M, Kovalchuk E, Keetman A, Schuldt T, Braxmaier C and Peters A 2012 *Proceedings of the 26th European Frequency and Time Forum (2012)*

[14] Ressel S, Gohlke M, Rauen D, Schuldt T, Kronast W, Mescheder U, Johann U, Weise D and Braxmaier C 2010 *Appl. Opt.* **49** 4296–4303

[15] Schuldt T, Keetman A, Döringshoff K, Reggentin M, Kovalchuk E, Nagel M, Gohlke M, Johann U, Weise D, Peters A and Braxmaier C 2012 *Proceedings of the 26th European Frequency and Time Forum (2012)*

[16] Schuldt T, Döringshoff K, Kovalchuk E, Pahl J, Gohlke M, Weise D, Johann U, Peters A and Braxmaier C 2014 *Proceedings of the 10th International Conference on Space Optics (2014)*

[17] Folkner W M, de Vine G, Klipstein W M, McKenzie K, Spero R, Thompson R, Yu N, Stephens M, Leitch J, Pierce R, Lam T and Shaddock D 2010 *Proceedings of the Earth Science Technology Forum (2010)*

[18] Nicklaus K, Herding M, Wang X, Beller N, Fitzau O, Giesberts M, Herper M, Barwood G P, Williams R A, Gill P, Kögel H, Webster S A and Gohlke M 2014 *Proceedings of the 10th International Conference on Space Optics (2014)*

[19] Webster S and Gill P 2011 *Opt. Lett.* **36** 3572–3574

[20] Legero T, Kessler T and Sterr U 2010 *Journal of the Optical Society of America B* **27** 914

[21] Sanjuan J, Gürlebeck N and Braxmaier C 2015 *Opt. Express* **23** 17892–17908

[22] Dai X, Jiang Y, Hang C, Bi Z and Ma L 2015 *Opt. Express* **23** 5134–5146

Hydrogen molecular ions: new schemes for metrology and fundamental physics tests

Jean-Philippe Karr[1,2], Sayan Patra[3], Jeroen C J Koelemeij[3], Johannes Heinrich[1], Nicolas Sillitoe[1], Albane Douillet[1,2], and Laurent Hilico[1,2]

[1] Laboratoire Kastler Brossel, UPMC-Sorbonne Universités, CNRS, ENS-PSL Research University, Collège de France, 4 place Jussieu, F-75005 Paris, France
[2] Université d'Evry-Val d'Essonne, Boulevard François Mitterrand, F-91000 Evry, France
[3] LaserLaB and Department of Physics & Astronomy, VU University, De Boelelaan 1081, 1081 HV Amsterdam, The Netherlands

E-mail: karr@lkb.upmc.fr

Abstract. High-accuracy spectroscopy of hydrogen molecular ions has important applications for the metrology of fundamental constants and tests of fundamental theories. Up to now, the experimental resolution has not surpassed the part-per-billion range. We discuss two methods by which it could be improved by a huge factor. Firstly, the feasibility of Doppler-free quasi-degenerate two-photon spectroscopy of trapped and sympathetically cooled ensembles of HD^+ ions is discussed, and it is shown that rovibrational transitions may be detected with a good signal-to-noise ratio. Secondly, the performance of a molecular quantum-logic ion clock based on a single Be^+-H_2^+ ion pair is analyzed in detail. Such a clock could allow testing the constancy of the proton-to-electron mass ratio at the 10^{-17}/yr level.

1. Introduction

It has long been suggested that high-resolution spectroscopy of rovibrational transitions in the hydrogen molecular ions H_2^+ and HD^+ can provide an accurate value of the proton-electron mass ratio μ [1, 2]. Recently, a measurement of the (v, L) : $(0, 2) \rightarrow (8, 3)$ vibrational overtone in trapped, cold HD^+ allowed the first determination of μ from molecular vibrations [3] with a relative uncertainty of 2.9×10^{-9} that is mainly limited by Doppler broadening. A promising avenue for further improvement is Doppler-free two-photon spectroscopy [4]; an experiment to probe the $(0, 2) \rightarrow (1, 2)$ transition in H_2^+ is being set up at LKB Paris [5]. We describe an alternative scheme for two-photon spectroscopy of HD^+ with quasi-degenerate counterpropagating photons [6]. The lasers are tuned close to resonance with an intermediate rovibrational level in order to enhance the transition probability, thus avoiding the use of a build-up cavity; the residual first-order Doppler broadening is moreover suppressed by the two-photon Lamb-Dicke effect. The application of this technique to the $(0, 3) \rightarrow (4, 2) \rightarrow (9, 3)$ transitions at 1.44 μm, currently underway at VU University Amsterdam, is discussed in Section 2.
Hydrogen molecular ions were also proposed for the realization of optical molecular clocks with the aim of testing of the time variation of μ [7]. The most stringent laboratory constraints on the $\dot{\mu}/\mu$ variation rate come from frequency comparisons between optical and Cs clocks [8, 9]. The current uncertainty level, slightly above 10^{-16}/yr, is mainly limited by the performance of

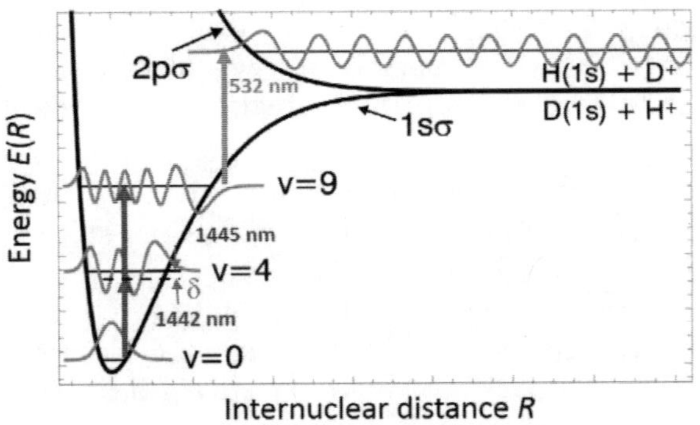

Figure 1. (colour online) Schematic of quasi-degenerate two-photon spectroscopy of the (v, L) : $(0, 3) \rightarrow (9, 3)$ transition in HD^+ (the hyperfine structure of the rovibrational levels is not shown). The transition is detected by resonance-enhanced multiphoton dissociation (REMPD) using an additional laser at 532 nm; the REMPD signal is the fractional loss of HD^+ ions.

Cs standards. A direct frequency ratio measurement between atom- and molecule-based optical clocks would overcome this limitation and enable a nearly pure measurement of $\dot{\mu}/\mu$, independent of variations of the fine structure constant $(\dot{\alpha}/\alpha)$. This motivates the construction of a molecular clock with accuracy and stability reaching the 10^{-17} range. A large number of molecular lines with greatly enhanced sensitivity to μ variations have been discovered over the last decade (see [10] for a review). However, the proposed transitions are often rather exotic in the sense that they involve highly excited states in complex molecules, so that efficient preparation of the initial state is very challenging. The potential of rovibrational transitions from low-lying levels of simple molecules such as H_2^+ and HD^+ [7, 11, 12] or N_2^+ [13, 14], where efficient state preparation by multiphoton ionization has been demonstrated, is thus worth investigating. In Section 3 we discuss the performance of quadrupole transitions in H_2^+. While possible clock transitions in this system were studied in Refs. [11, 12], we bring the discussion further in two main respects. First, important practical considerations for the clock's operation, namely its stability and the requirement of highly efficient internal state preparation of the molecular ions, lead us to select different transitions. Second, we analyze an alternative scheme to cancel the electric quadrupole and quadratic Zeeman shifts, which gave the main contribution to the error budget in [11], and show that it allows improving the clock's accuracy.

2. Two-photon spectroscopy of trapped HD^+ ion ensembles

The spectroscopic scheme for quasi-degenerate two-photon spectroscopy of trapped and sympathetically cooled HD^+ ions is shown in figure 1. Its feasibility was carefully studied in Ref. [6] by solving the optical Bloch equations in the framework of a four-level model. The first three levels are hyperfine sublevels of the $(v, L) = (0, 3)$, $(4, 2)$ and $(9, 3)$ levels respectively, and the fourth one is a virtual level whose population represents the photodissociated fraction, i.e. the REMPD signal. An underlying assumption is that the chosen hyperfine sublevel of the intermediate state $(4, 2)$ is well-isolated, more precisely, that the detuning δ with respect to the one-photon resonance (see figure 1) can be made smaller than the separation with neighboring hyperfine levels. The Zeeman structure is not included, because the linewidths of both 1.4 μm lasers are supposed to be larger than the Zeeman splitting so as to address all Zeeman sublevels and enhance the signal. The Doppler-broadened one-photon lineshape is accurately modeled by injecting realistic ion trajectories from molecular dynamics simulations in the Bloch equations. The following results were established:

(i) Maximal visibility of the Doppler-free signal is achieved for a detuning δ slightly above the Doppler width of the single-photon transitions. Indeed a smaller detuning increases the two-photon transition rate, but δ should remain sufficiently large to avoid competition from the sequential excitation process.

(ii) The two-photon peak is power broadened, with a width of a few hundred Hz at the onset of saturation of the REMPD signal. There is also a lightshift of the order of a few Hz, which can however be canceled if the Rabi frequencies of both single-photon transitions are made equal.

(iii) In order to predict the observed signal in actual experiments where the HD^+ ions are distributed over rotational and hyperfine states within the ground vibrational state $v = 0$, the REMPD rate extracted from the four-level model is included into a set of rate equations describing population redistribution among rotational and hyperfine states by blackbody radiation. The latter enhances the REMPD signal since it continuously recycles ions into the lower state of the transition. It is found that the photodissociated fraction can be made significantly larger than the measurement noise observed in [3].

Quasi-degenerate two-photon spectroscopy of HD^+ thus appears feasible and could reach an accuracy level of about 10^{-14} [6]; the experiment is underway at VU University Amsterdam. We now discuss the choice of suitable hyperfine components. As pointed out in (i) above, it is important to choose the detuning δ away from any transition to intermediate hyperfine levels in $v = 4, L = 2$ by at least the Doppler width. In this way, the first photon of the two-photon transition does not significantly populate any of the intermediate levels, which suppresses the unwanted background signal due to stepwise excitation from $v = 0$ to $v = 9$ via $v = 4$. A good indicator of this suppression factor is the one-photon spectrum of the $(v, L) : (0, 3) \to (4, 2)$ transition at 1442 nm, shown in figure 2. Assuming that the second photon always leads to immediate REMPD - which is a worst case scenario - this spectrum essentially represents the maximum expected background signal due to population in $v = 4$ as a function of the frequency of the first photon. Spectra are shown for two different values of the temperature corresponding to the range of temperatures obtained in previous experiments with the same ion trap [3]. Note that in Ref. [6], ion trajectories are obtained from simulations of sympathetic cooling in an ideal trap, leading to lower temperatures.

Apart from choosing δ such that the maximum expected background signal (figure 2) is low, the strength of the hyperfine component should be as large as possible, and the Zeeman shift as low as possible in order to address all Zeeman components with a narrow-line laser. Figure 2 indicates that several such lines may be probed with sufficient signal-to-noise ratio with intermediate-level detunings of 15-30 MHz. Note that the full two-photon $(v, L) : (0, 3) - (4, 2) - (9, 3)$ Doppler-broadened spectrum may actually be less broadened than the $(v, L) : (0, 3) \to (4, 2)$ transition due to hyperfine state selection during the first step of the excitation. Consequently, even smaller intermediate-level detunings might be feasible. Future work will address the precise shape of the full two-photon spectrum (i.e. both the direct and stepwise contributions) using an extended version of the rate equation approach of Ref [3].

3. Performance of an H_2^+ clock

Before addressing the leading systematic effects affecting an optical clock based on hydrogen molecular ions and the accuracy level at which they can be controlled, other characteristics that influence the choice of suitable rovibrational transitions need to be considered. The first one is the clock's stability, which determines the required integration time to measure the $\dot{\mu}/\mu$ variation rate at the 10^{-17}/yr level. The main source of noise in single-ion clocks is the quantum projection noise, setting a stability limit that is inversely proportional to the clock frequency and to the square root of the interrogation time (see e.g. [15]). This implies, firstly, that the stability limit is lower in HD^+ than in H_2^+ due to the lower lifetimes of the rovibrational states ($\tau \sim 15 - 55$ ms for $v = 1 - 4$ [16], whereas $\tau \sim 10^6$ s in H_2^+ [17]) which limit the interrogation time; this is why we only consider H_2^+ in the following. Secondly, all other characteristics being equal, one should

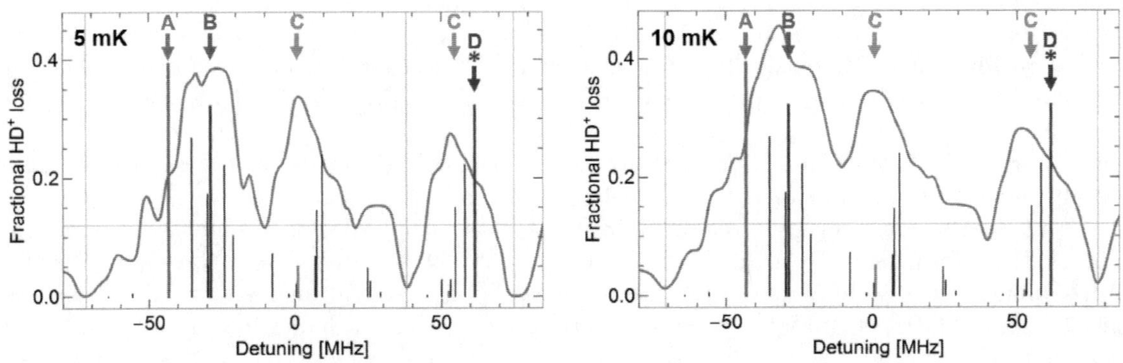

Figure 2. (colour online) Calculated Doppler-broadened single-photon spectra of the (v, L) : $(0,3) \rightarrow (4,2)$ transition at 1442 nm, assuming detection by REMPD. Doppler-broadening, saturation (due to the finite HD^+ sample size) and population redistribution by blackbody radiation lead to a rich spectrum (gray curve). The spectra were calculated for an interrogation time of 10 s, an ambient blackbody radiation temperature of 300 K, and an HD^+ ion temperature of 5 mK (left panel) and 10 mK (right panel) using the method described in Refs. [3, 6]. The underlying hyperfine transitions are indicated by the vertical black sticks. Possible intermediate steps for enhanced two-photon transitions are indicated by the thick vertical sticks and arrows. Transitions with (from left to right) $F = 1$, $S = 2$, and $J = 5$ to $J' = 4$ (A), $F = 1$, $S = 1$, and $J = 4$ to $J' = 3$ (B), and $F = 1$, $S = 2$, and $J = 1$ to $J' = 0, 2$ (C) were identified in [6] as having very small Zeeman shifts at the 10-Hz level (for a 0(2) μT magnetic field, realized with three pairs of external magnetic field coils while using the fluorescence of the Be^+ ions as a magnetic field probe). The line which may be probed with the lowest detuning, i.e. $F = 0$, $S = 1$, and $J = 4$ to $J' = 3$ (D) is labeled with a star. It exhibits a larger Zeeman broadening of 1.5 kHz, but the net Zeeman shift is below 1 Hz. The gray horizontal line indicates the expected achievable two-photon signal (which depends on line strengths and is adjustable through the laser intensity), and vertical gray lines indicate suitable laser detunings for which the Doppler-broadened background will be considerably smaller than the two-photon signal.

look for transitions with the highest possible frequency, i.e. vibrational overtones. Another important feature is the possibility of efficient state preparation in the lower rovibrational and hyperfine state by resonant-enhanced multiphoton ionization (REMPI). H_2^+ ion production in selected rovibrational states with $\sim 90\%$ selectivity has been reported [18], but the population of hyperfine sublevels has not been addressed, which is why it seems preferable to choose levels with the simplest possible hyperfine structure, i.e. $L = 0$ (no structure) or $L = 2, 4...$ (2 sublevels). Finally, one-photon quadrupole transitions have been shown to be more promising than two-photon transitions as the latter have lower frequencies and are affected by larger lightshifts [12].

Some of the main systematic shifts lend themselves to efficient cancellation strategies. It was pointed out that the mean frequency of transition pairs between "stretched states", $(v, L, J = L + 1/2, M_J = \pm(L + 1/2)) \rightarrow (v', L', J' = L' + 1/2, M'_J = \pm(L' + 1/2))$ is free of both linear and quadratic Zeeman shifts [12, 19]; here L is assumed to be an even number, so that the total nuclear spin is zero, and $\mathbf{J} = \mathbf{L} + \mathbf{s_e}$ is the total angular momentum. The quadrupole shift [20] can be nulled by measuring the transition frequencies for three orthogonal directions of the magnetic field [21]. The level of cancellation with this method is limited by the inaccuracy in setting the magnetic field directions, and the quadrupole shift remains the most important limitation of the clock's accuracy unless a cancellation occurs between the lower and upper levels. The latter consideration led to single out the $(v, L) : (0, 4) \to (2, 2)$ transition at $\lambda = 2.61$ μm [12], whose uncertainty budget is studied in more detail hereafter. We also analyze a different method based on the measurement of several symmetric pairs of Zeeman subcomponents, inspired from previous work in atomic clocks [22] where it was shown to yield a much higher degree of cancellation, in addition to nulling the linear Zeeman shift and the tensor part of the Stark shift. This approach is also applicable in H_2^+ since the quadrupole shift has the same form proportional to $J(J + 1) - 3M_J^2$ [20]. However, in contradistinction with the "stretched states" case, the quadratic Zeeman shift is not suppressed and could limit the achievable performance. To solve this issue, we propose averaging instead over *all* the Zeeman subcomponents of the hyperfine structure of the involved rovibrational levels. The quadratic Zeeman shift, which essentially originates from couplings within the hyperfine structure, will then be nulled to a very high degree, while the cancellation of the linear Zeeman shift is preserved. In view of the small hyperfine splitting of even-L states ($\Delta\nu_{\text{hfs}} \sim 100$ MHz [24]) all the transitions can be interrogated with a single laser. To sum it up, we select the following transitions:

- the pair $(v = 0, L = 4, J = 9/2, M_J = \pm 9/2) \to (v = 2, L = 2, J = 5/2, M_J = \pm 5/2)$, for its extremely low quadrupole shift and suppressed Zeeman shift;
- the transitions $(v = 0, L = 0, J = 1/2, M_J = \pm 1/2) \to (v', L' = 2, J', \pm M'_J)$ for canceled quadrupole and Zeeman shifts. We choose $L = 0$ and $L' = 2$ firstly in order to have the simplest possible Zeeman structure, and secondly because the absence of hyperfine structure for $L = 0$ favors efficient state preparation. Five transition pairs are needed for averaging over all states, but this is actually not necessary as a shift-independent frequency can be deduced from the measurement of only three transition pairs (see table 2), exploiting the known dependence of the quadrupole shift on J and M_J.

Our results are summarized in tables 1 and 2. We briefly describe below the methods and hypotheses we have used to estimate the magnitude and residual uncertainties of the leading frequency shifts.

(i) Quadrupole shift. The quadrupole moment of the $(v = 0, L = 0)$ state is zero, and those of the $(v', L = 2)$ states are taken from [20]. As already discussed in [12], those of the $(0, 4)$ and $(2, 2)$ states, which are very close to each other, were recalculated independently in a three-body approach to get a more precise estimate of the quadrupole shift. The latter is evaluated for an electric field gradient $V_{zz} = 0.67 \times 10^8$ V/m^2, similarly to [11, 12]. For the $(0, 4) \to (2, 2)$ transition, the residual uncertainty is equal to 1 percent of the shift, as is typically obtained with the B-averaging method. For the three $L = 0 \to L' = 2$ transitions, assuming that the M_J-averaging method is used we take an uncertainty equal to 10^{-4} times the shift of the most sensitive states, that are the $|M_J| = 5/2$ states. This is a conservative estimate, as an even better level of cancellation was demonstrated in [23].

(ii) Quadratic Zeeman shift. Results of calculations of this shift are reported in [11]. However, we performed independent calculations since the transitions considered here were not included in that reference. Quadratic Zeeman coefficients are obtained by diagonalizing $H = H_{hfs} + H_Z$ where H_{hfs} is the hyperfine structure Hamiltonian [24] and H_Z is the Zeeman Hamiltonian [25] (see also [19]). A first contribution comes from the magnetic field associated with blackbody radiation (BBR); BBR frequencies being much higher than the hyperfine splitting, the resulting shift is very small [26] and may be neglected. The second contribution is from the background magnetic field, for which we choose a small value of 2 μT to reduce the magnitude of the shift. Such a residual field can typically be obtained by building the trap structure with nonmagnetic materials [23]. Similarly to the quadrupole shift, the averaging method we propose for the quadratic Zeeman shift is conservatively assumed to result in cancellation by four orders of magnitude. Note that the averaged shift is not strictly zero due to couplings to other vibrational levels, but such contributions are estimated to be smaller than 10^{-19}.

(iii) Blackbody radiation (Stark) shift. This shift was calculated in [27] for the rovibrational levels of H_2^+. An experimental uncertainty of 5 K on the 300 K BBR temperature is assumed.

(iv) AC Stark shift. The calculation of the lightshift induced by the clock laser is described in [12]; here its expression is slightly different since the tensor part is nulled by the M-averaging procedure:

$$\Delta f_{LS} = -\frac{m_e \lambda^2}{4h} \frac{\Delta\alpha^{if(0)}}{\tau^2 |\Theta_{\varepsilon,n}^{if}|^2}. \tag{1}$$

where m_e is the electron's mass, λ the transition wavelength, $\Delta\alpha^{if(0)} = \alpha_{v',L'}^{(0)} - \alpha_{v,L}^{(0)}$ the dimensionless dynamic *scalar* polarizability difference, τ the duration of the clock's interrogation π-pulse, and $\Theta_{\varepsilon,n}^{if}$ the dimensionless quadrupole transition amplitude, which depends on the propagation direction $\mathbf{n}$ and polarization $\boldsymbol{\varepsilon}$. The latter are adjusted to maximize the amplitude of the weakest Zeeman components. We assume an interrogation time $\tau = 100$ ms [28, 29] and a residual uncertainty equal to 1% of the shift.

(v) Second-order Doppler shift. This is one of the crucial limitations, due to the light mass of H_2^+. For example, in the Al^+/Mg^+ clock the residual uncertainty for this effect is 7.8×10^{-18}; a similar level of control on the trapped ion's residual kinetic energy would lead to a 1×10^{-16} uncertainty in the case of H_2^+. In view of ongoing efforts to improve micromotion minimization techniques further [31], an uncertainty level of 1×10^{-17} may realistically be achieved.

Two other systematic effects were omitted in tables 1 and 2 as they do not contribute significantly to the error budget:

(vi) Scalar Stark shift induced by trap fields. The Stark and second-order Doppler shifts associated with micromotion, being driven by the same electric field, are correlated and should be considered together. According to Eqs. (14) and (24) of [23], the ratio of the Stark and Doppler shifts is given by

$$\frac{\Delta\nu_s}{\Delta\nu_D} = \frac{\Delta\alpha_0}{\hbar\omega_0} \left(\frac{m\Omega c}{e}\right)^2 \tag{2}$$

where $\hbar\omega_0$ is the transition energy , m the H_2^+ ion's mass, Ω the trap frequency and $\Delta\alpha_0$ the static scalar polarizability difference. With the value $\Omega = 2\pi \times 59$ MHz [28, 29] and static polarizabilities from [27], this ratio lies between 0.0014 and 0.002 for the transitions under consideration. The scalar Stark shift thus may be safely neglected.

(vii) Tensor Stark shifts from the clock laser and from trap fields. They are of the same order as the related scalar shifts (as can be seen by comparing scalar and tensor polarizability differences e.g. in table 3 of [12]). Similarly to the quadrupole shift, they are canceled by about four orders of magnitude by the M_J-averaging method [23], and their residual uncertainties are thus negligible with respect to scalar contributions.

Examination of table 1 shows that an accuracy level of a few 10^{-17} represents a realistic goal for an H_2^+ ion clock, using either the $v = 0 \rightarrow 1$ transition at 4.24 μm, or the $v = 0 \rightarrow 2$ transitions at 2.27 and 2.61 μm. The $v = 0 \rightarrow 3$ transition at 1.58 μm would be an ideal choice from the point of view of laser and frequency measurement technology, but as the transitions become weaker with increasing Δv the AC stark shift strongly limits the achievable accuracy. While generalizations of the Ramsey interrogation scheme have been developed to improve light shift cancellation [30], their implementation in a quantum-logic ion clock is challenging. However,

Transition	$(0,4) \to (2,2)$		$(0,0) \to (1,2)$		$(0,0) \to (2,2)$		$(0,0) \to (3,2)$	
Frequency (THz)	114.99		70.64		132.25		190.18	
Source	shift	uncert.	shift	uncert.	shift	uncert.	shift	uncert.
Quadrupole shift	0	0.049	0	1.7	0	1.0	0	0.82
Quadr. Zeeman shift	< 0.01	< 0.01	< 0.01	1.1	< 0.01	0.61	< 0.01	0.46
Blackbody radiation	−12	0.80	−9.3	0.62	−11	0.74	−13	0.88
AC Stark shift	−220	2.2	−5.0	0.050	−200	2.0	−4500	45
Total uncertainty		2.6		2.4		2.7		45

Table 1. Estimated uncertainty budget for quadrupole rovibrational transitions in H_2^+, in parts in 10^{17}. The total uncertainty in the last line also includes a 1.0×10^{-17} contribution from the second-order Doppler shift (see text) and is obtained by adding all contributions quadratically.

M_J	J'	M'_J	$\Delta f_Z^{(1)}$	$\Delta f_Z^{(2)}$	Δf_Q
±1/2	3/2	±1/2	∓33.6 kHz	8.1 Hz	−9.7 Hz
±1/2	5/2	±1/2	∓22.4 kHz	−8.1 Hz	−11.1 Hz
±1/2	5/2	±5/2	∓27.8 Hz	0	13.8 Hz

Table 2. Detailed frequency shifts of the three individual transition pairs to be measured, in the case of the $(v, L) = (0,0) \to (2,2)$ transition. $\Delta f_Z^{(1)}$ and $\Delta f_Z^{(2)}$ are respectively the linear and quadratic Zeeman shifts for a 2 μT B-field. Δf_Q is the quadrupole shift for a field gradient $V_{zz} = 0.67 \times 10^8$ V/m^2. The BBR and scalar AC Stark shift (which are the same for all transitions) respectively amount to -14.7 and -163 mHz. The tensor part of the AC Stark shift (not given here) plays a negligible role since it is of the same order as the scalar part, but is nulled to a much higher degree by the averaging procedure. The shift-independent frequency is $\nu_0 = [3\nu(J' = 5/2, M'_J = \pm 5/2) + \nu(J' = 5/2, M'_J = \pm 1/2) + \nu(J' = 3/2, M'_J = \pm 1/2)]/5$ where $\nu(J', \pm M'_J)$ is itself the averaged frequency of two opposite Zeeman subcomponents. The hyperfine structure splitting between $J' = 3/2$ and $J' = 5/2$ is 92.7 MHz [24].

we note that according to Eq. (1) the lightshift scales like $1/\tau^2$. The $\Delta v = 3$ transition thus becomes practicable if longer interrogation times can be used; for example $\tau = 500$ ms would bring the overall uncertainty down to 2.4×10^{-17}. Finally, it is worth pointing out that, apart from the second-order Doppler shift, residual uncertainties associated with the various effects have been estimated conservatively and may be reduced further, down to the 1.0×10^{-17} mark or even below.

4. Conclusion

The schemes that have been discussed have the potential to improve the current accuracy of molecular ion spectroscopy by several orders of magnitude, with important outcomes in the field of fundamental metrology. Quasi-degenerate two-photon spectroscopy of HD^+ may ultimately reach an accuracy level of 10^{-14} or better; 10^{-12} would actually be sufficient for comprehensive tests of molecular QED theory [32] and an improved determination of the proton-to-electron mass ratio [33]. It is worth noting that this method may be applied to other molecular ions, where similar quasi-coincidences in the rovibrational spectrum are likely to occur. The second possibility, i.e. the realization of an H_2^+ optical clock, is much more challenging. Our accuracy estimates show that it may allow for tests of μ time variation at the 10^{-17}/yr level, improving current limits by one order of magnitude. The main drawback of H_2^+ is its light mass causing a larger second-order Doppler shift. On the plus side, this also leads to higher vibrational transition frequencies, allowing for better clock stability. Finally, a clock based on H_2^+ would represent, like atomic hydrogen masers, a clock with a 'calculable' frequency, in this case providing a direct, 11-digit link from the SI second to the values of fundamental physical constants. Unlike H masers, however, an H_2^+ clock would operate at optical wavelengths and benefit from many of the techniques developed for ion-based optical atomic clocks.

Acknowledgements
The authors are grateful to V.I. Korobov and D. Bakalov for helpful discussions. This work was funded through the Dutch-French bilateral Van Gogh Programme. J.C.J.K. and S.P. thank the Netherlands Organization for Scientific Research (NWO), the Netherlands Technology Foundation (STW), and the Netherlands Foundation for Fundamental Research on Matter (FOM) for support. J.-Ph.K., J.H., N.S., A.D. and L.H. acknowledge funding from Agence Nationale de la Recherche (grant ANR-13-IS04-0002-01) and the European Commission under the Seventh Framework Programme FP7 GA 607491 COMIQ. J.-Ph.K. acknowledges support as a fellow of the Institut Universitaire de France.

References

[1] Wing W H, Ruff G A, Lamb Jr W E and Spezeski J J 1976 *Phys. Rev. Lett.* **36** 1488
[2] Bressel U, Borodin A, Shen J, Hansen M, Ernsting I and Schiller S 2012 *Phys. Rev. Lett.* **108** 183003
[3] Biesheuvel J, Karr J Ph, Hilico L, Eikema K S E, Ubachs W and Koelemeij J C J 2016 *Nature Comm.* **7** 10385
[4] Karr J Ph, Kilic S and Hilico L 2005 *J. Phys. B* **38** 853
[5] Karr J Ph, Bielsa F, Douillet A, Pedregosa Gutierrez J, Korobov VI and Hilico L 2008 *Phys. Rev. A* **77** 063410
[6] Tran V Q, Karr J Ph, Douillet A, Koelemeij J C J and Hilico L 2013 *Phys. Rev. A* **88** 033421
[7] Schiller S and Korobov V 2005 *Phys. Rev. A* **71** 032505
[8] Godun R M, Nisbet-Jones P B R, Jones J M, King S A, Johnson L A M, Margolis H S, Szymaniec K, Lea S N, Bongs K and Gill P 2014 *Phys. Rev. Lett.* **113** 210801
[9] Huntemann N, Lipphardt B, Tamm Chr, Gerginov V, Weyers S and Peik E 2014 *Phys. Rev. Lett.* **113** 210802
[10] Jansen P, Bethlem H L and Ubachs W 2014 *J. Chem. Phys* **140** 010901
[11] Schiller S, Bakalov D and Korobov V I 2014 *Phys. Rev. Lett.* **113** 023004
[12] Karr J Ph 2014 *J. Mol. Spectrosc.* **300** 37
[13] Kajita M, Gopakumar G, Abe M, Hada M and Keller M 2014 *Phys. Rev. A* **89** 032509
[14] Kajita M 2015 *Phys. Rev. A* **92** 043423
[15] Ludlow A D, Boyd M M, Ye J, Peik E and Schmidt P O 2015 *Rev. Mod. Phys.* **87** 637
[16] Olivares Pilón H and Baye D 2013 *Phys. Rev. A* **88** 032502
[17] Olivares Pilón H and Baye D 2012 *J. Phys. B* **45** 065101
[18] O'Halloran M A, Pratt S T, Dehmer P M and Dehmer J L 1987 *J. Chem. Phys.* **87** 3288
[19] Bakalov D, Korobov VI and Schiller S 2011 *J. Phys. B* **44** 025003
[20] Bakalov D and Schiller S 2014 *Appl. Phys. B* **114** 213
[21] Itano W M 2000 *J. Res. Natl. Inst. Stand. Technol.* **105** 829
[22] Dubé P, Madej A A, Bernard J E, Marmet L, Boulanger J S and Cundy S 2005 *Phys. Rev. Lett.* **95** 033001
[23] Dubé P, Madej A A, Zhou Z and Bernard J E 2013 *Phys. Rev. A* **87** 023806
[24] Korobov VI, Hilico L and Karr J Ph 2006 *Phys. Rev. A* **74** 040502(R)
[25] Karr J Ph, Hilico L and Korobov V I 2008 *Phys. Rev. A* **77** 062507
[26] Itano W M, Lewis L L and Wineland D J 1982 *Phys. Rev. A* **25** 1233
[27] Schiller S, Bakalov D, Bekbaev A K and Korobov V I 2014 *Phys. Rev. A* **89** 052521
[28] Rosenband T *et al.* 2007 *Phys. Rev. Lett.* **98** 220801
[29] Chou C W, Hume D B, Koelemeij J C J, Wineland D J and Rosenband T 2010 *Phys. Rev. Lett.* **104** 070802
[30] Huntemann N, Lipphardt B, Okhapkin M, Tamm Chr, Peik E, Taichenachev A V and Yudin V I 2012 *Phys. Rev. Lett.* **109** 213002; Zanon-Willette T, Yudin V I and Taichenachev A V 2015 *Phys. Rev. A* **92** 023416
[31] Keller J, Partner H L, Burgermeister T and Mehlstäubler T E 2015 *J. Appl. Phys.* **118** 104501
[32] Korobov V I, Hilico L and Karr J Ph 2014 *Phys. Rev. Lett.* **112** 103003 and *Phys. Rev. A* **89** 032511
[33] Sturm S, Köhler F, Zatorski J, Wagner A, Harman Z, Werth G, Quint W, Keitel C H and Blaum K 2014 *Nature* **467** 506

Journal of Physics: Conference Series **723** (2016) 012049 doi:10.1088/1742-6596/723/1/012049

Metrology with Atom Interferometry: Inertial Sensors from Laboratory to Field Applications

B Fang, I Dutta, P Gillot, D Savoie, J Lautier, B Cheng, C L Garrido Alzar, R Geiger, S Merlet, F Pereira Dos Santos and A Landragin

LNE-SYRTE, Observatoire de Paris, PSL Research University, CNRS, Sorbonne Universités, UPMC Univ. Paris 06, 61 avenue de l'Observatoire, 75014 Paris, France

E-mail: `bess.fang@obspm.fr`

Abstract. Developments in atom interferometry have led to atomic inertial sensors with extremely high sensitivity. Their performances are for the moment limited by the ground vibrations, the impact of which is exacerbated by the sequential operation, resulting in aliasing and dead time. We discuss several experiments performed at LNE-SYRTE in order to reduce these problems and achieve the intrinsic limit of atomic inertial sensors. These techniques have resulted in transportable and high-performance instruments that participate in gravity measurements, and pave the way to applications in inertial navigation.

1. Introduction

Since the advent of atom interferometers (AI) in 1990s ([1, 2], see also [3] and references therein), tremendous progress has been made to improve the performance as well as the transportability of inertial sensors based on AI. Today, these sensors have already been used in field applications such as geophysics measurements [4, 5], and promise to surpass other state-of-the-art technologies in inertial navigation [6]. Moreover, they are the prime candidates to explore some aspects of fundamental physics such as testing of Einstein Equivalence Principle [7], and detection of gravitational waves [8, 9].

The performance of AI-based inertial sensors is fundamentally limited by the noise in the measurement of the transition probability, i.e. the standard quantum projection noise (QPN) [10], or ideally the Heisenberg limit, between the two interferometric states. Vibration noise on the ground is however several orders of magnitude higher. The sequential operation of typical sensors exacerbates the problem by down aliasing higher-frequency noise, similar to the Dick effect in cold atomic clocks [11]. Dead time, during which the atoms are prepared or detected, results in loss of information of the inertial signal, preventing fast averaging of the vibration noise.

Different strategies have been proposed to overcome these difficulties. When a (nearly) stable quantity such as the gravitational acceleration $\vec{g}$ is measured in a stable environment (e.g. in a laboratory), one may average the noise out with longer measurements or perform active filtering [12, 13]. When transportability is required for measurements at different sites, the bulky active platform must be replaced by passive isolation systems and classical sensors that measure the aliasing effect as proposed in [10] and first demonstrated in [14]. The need to completely

Journal of Physics: Conference Series **723** (2016) 012049 doi:10.1088/1742-6596/723/1/012049

eliminate dead time comes from applications in geophysics and inertial navigation, where the inertial signals would fluctuate in time and the environment is far from stable [15].

In this paper, we discuss several experiments performed in the LNE-SYRTE inertial sensor group to address the issues related to the vibration noise, aliasing and the dead time in order to explore the intrinsic limits of AI-based inertial sensors.

2. Inertial Sensors based on Atom Interferometry

We start with a brief description of common techniques used in cold-atom inertial sensors using alkali atoms and stimulated Raman transitions. These techniques are by no means unique, but are often practiced for their technological maturity. They have been proven reliable in high-performance instruments. However, the methods for reduction of vibration that we will describe in the subsequent sections can be extended to other schemes of atom interferometers.

Before the interferometric measurements start, an atomic source is loaded and laser cooled to reach sub-Doppler temperatures. These atoms are then launched or released into a free fall and enter the interferometric zone. A three-pulse atom interferometer is analogous to an optical Mach-Zehnder interferometer: the matter wave is split, deflected and recombined by counter-propagating Raman beams via two-photon transitions between two hyperfine states [2]. At the output of the interferometer, the population in each state is detected by fluorescence. This gives the transition probability P, which depends on the phase difference $\Delta\Phi$ experienced by the two arms of the interferometer, $P = \frac{1}{2}(1 + C\cos\Delta\Phi)$, where C is the contrast. Such a setup is illustrated in Figure 1 (a).

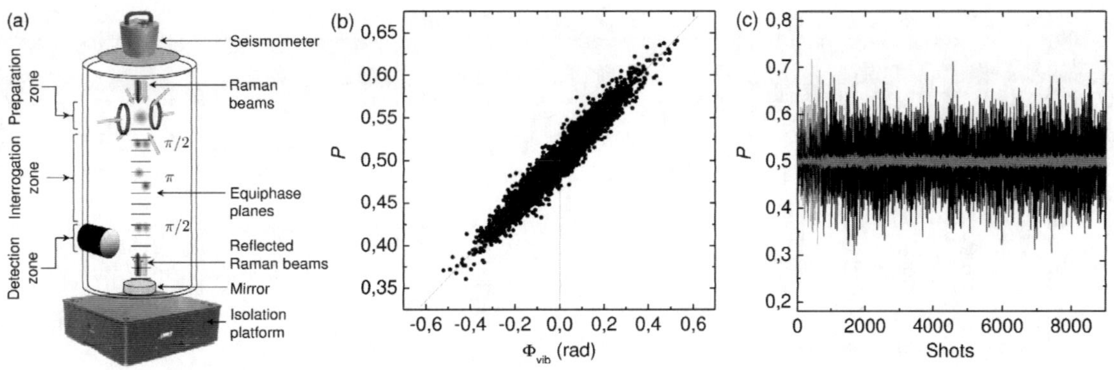

Figure 1. (a) Illustration of an atom gravimeter. It is mounted on a vibration isolation platform for passive isolation. A seismometer is rigidly fixed on the structure in order to record residual vibration noise of the retroreflection mirror. (b) Measured transition probability P vs the calculated vibration phase Φ_{vib} (points). The linear fit (straight line) prompts a simple correction procedure by subtraction. (c) Transition probability P of a series of data points before (black, larger noise) and after (red, reduced noise) correlation (colour online).

The interferometer phase $\Delta\Phi$ is the difference of the atomic phase accumulated along the two paths of the interferometer. It is given by $\Delta\Phi = \phi_1 - 2\phi_2 + \phi_3$, where ϕ_i denotes the phase difference of the Raman lasers at the center of the atomic wave packet during the ith interrogation pulse. Since the Raman beams are often retroreflected, the equiphase planes are defined by the mirror position, making the interferometer sensitive to the displacement in the direction of propagation of the beams [16]. We can thus write the interferometer phase as the sum of the desired signal and the noise (e.g. vibrations), $\Delta\Phi = \Phi_{\text{sig}} + \Phi_{\text{noise}}$.

3. Vibration Isolation and Correlation with Classical Sensors

We demonstrate here that passively isolating the atomic sensor from the environment and performing post correlations with classical sensors allow us to significantly improve the performance of an atom gravimeter and that of an atom gyroscope.

The LNE-SYRTE Cold Atom Gravimeter (CAG) uses free falling rubidium-87 atoms in a three-pulse ($\pi/2$-π-$\pi/2$) configuration. With the Raman beams in the vertical direction, the interferometer is sensitive to the gravitational acceleration $\vec{g}$, i.e. $\Phi_{\rm sig} = -\vec{k}_{\rm eff} \cdot \vec{g}T^2$, where $\vec{k}_{\rm eff}$ denotes the difference between the wave vectors of the two Raman lasers, and T is the time interval between successive Raman pulses.

We operate CAG with a cycling frequency of 3 Hz and a total interrogation time $2T =$ 160 ms [17]. With about 10^6 atoms detected, the ground vibration introduces a noise 10^4 times the standard QPN limit. We passively isolate CAG with an anti-acoustic box and a vibration isolation platform, reducing the vibration noise by about a factor 100. We then acquire the vibration signal of the retroreflection mirror using a seismometer rigidly fixed on the the frame of CAG [see Fig. 1 (a)]. The fluctuating velocity of the mirror during the interrogation results in a shift of the interferometric phase $\Phi_{\rm noise} = \Phi_{\rm vib}$. This can be computed by integrating the acquired velocity with a weighting function that describes the response of the interferometer, known as the time transfer function (see [18] and references therein). Figure 1 (b) shows P versus $\Phi_{\rm vib}$. The linear relationship between these two quantities permits a simple correction procedure by subtraction [14]. The vibration rejection efficiency is about 10 in the present case [see Fig. 1 (c)], and has been measured to be up to 120 without passive isolation.

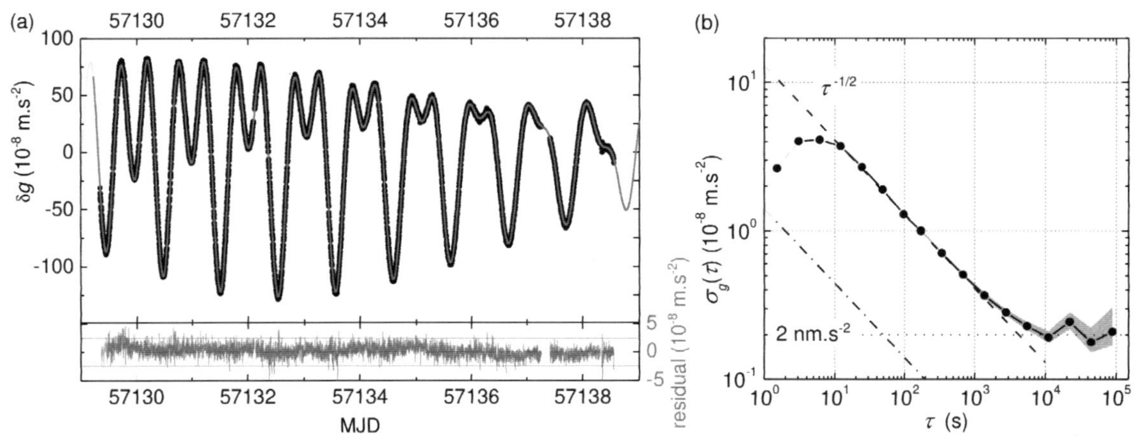

Figure 2. (a) Variation of g around its mean measured value as a function of days. Top: CAG measurement (points) averaged over 170 s and the tide model (line). Bottom: the residual (difference) between the measurement and the model. (b) ADEV of the CAG measurements corrected for the tides and local environmental effects. The shaded region indicates the uncertainty of the measurement, whereas the straight line gives the $\tau^{-1/2}$ scaling characteristic of averaging white frequency noise, where τ is the integration time. Similar trend lines will be given in subsequent figures with no further explanations. The dash-dotted line indicates the calculated QPN limit for typical parameters of 10^6 atoms, 40% contrast, and $T = 80$ ms. (colour online)

We use CAG to record the temporal variation of gravity in our laboratory dedicated to gravity measurements [19]. Figure 2 (a) overlays the measured g and the tide model, showing an excellent agreement throughout the $\pm 10^{-7}$ relative variations around the mean value due to the lunisolar tide. The Allan standard deviation (ADEV) of the measurement corrected for the tides and local environmental effects is shown in Fig. 2 (b), indicating a short-term stability of 13×10^{-8} m.s^{-2} at 1 s and a long-term stability of 2 nm.s^{-2} in 3 hours. This is about 10 times above the calculated QPN limit (dash-dotted line), as a result of the uncorrelated vibration noise. In a vibrationally more stable environment, a best short-term stability of 5.7×10^{-8} m.s^{-2} at 1 s was demonstrated [20]. The accuracy of CAG is 4.3×10^{-8} m.s^{-2} [21], mainly limited by the bias caused by the wave-front aberration of the Raman lasers.

We have also applied the same isolation and correlation techniques to the LNE-SYRTE gyroscope, which uses cesium-133 atoms in a fountain configuration and realizes a four-pulse interferometer. The general idea was first discussed in [22], and the experimental setup was described in [23]. Here, two mechanical accelerometers are used to record the vibration of the experimental structure. A weighted sum of these signals yields the vibration noise $\Phi_{\rm vib}$ at the center of the interferometer. Given the large interrogation time of up to 800 ms and a corresponding macroscopic Sagnac area of 11 cm^2, $\Phi_{\rm vib}$ is often beyond 20 rad peak-to-peak [see Fig. 3 (a)], rendering a linear correction impossible. Instead, we fit 20 data points to a sinusoid in order to obtain the rotation phase $\Phi_{\rm sig} = \Phi_\Omega$. This gives a vibration rejection efficiency of 6, as shown in Fig. 3 (b). We thus achieve a short-term stability of 160 nrad.s^{-1} at 1 s and a long-term stability of 4.4 nrad.s^{-1} after 1300 s.

We have seen that passive isolation and post correlation with classical sensors amount to apply a low pass filter that corrects the atomic sensor from higher-frequency vibration noise. In practice, the efficiency of this method is limited by the intrinsic noise and the nonlinearity of the classical sensor, and by the fact that the recorded vibration noise does not perfectly reproduce the vibration of the retroreflection mirror seen by the atoms. Nevertheless, atomic sensors such as CAG equipped with this system combine high performance and transportability, and can therefore perform precision measurements for metrology at different sites [4]. In particular, CAG has been participating in international comparisons of absolute gravimeters since 2009 [21, 24], and its measurements are in excellent agreement with those from other instruments.

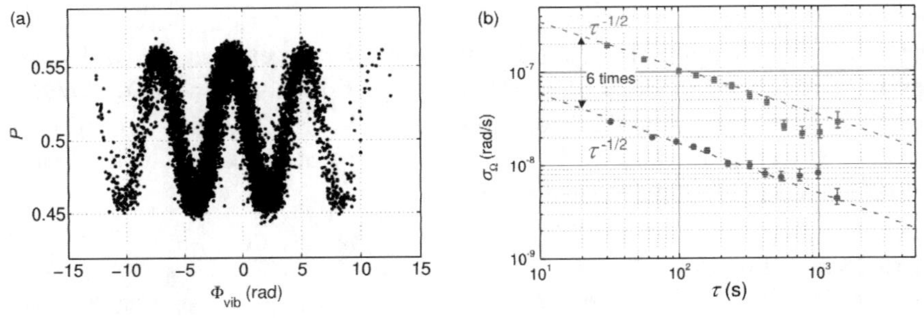

Figure 3. (a) Transition probability P vs the computed vibration phase $\Phi_{\rm vib}$. (b) ADEV of the measured rotation rate before (squares) and after (circles) correlation. (colour online)

4. Vibration Compensation in Real Time and Hybridizing Atomic and Classical Sensors

Environmental stability is rarely guaranteed in field applications. Moreover, inertial navigation requires the sensors to follow the moving frame of the vehicle, hence necessitates the removal of any isolation [25, 6]. Such scenarios, together with the intrinsic high sensitivity of AI-based inertial sensors, result in the scatter of the data points beyond the linear regime, and thus a reduction of sensitivity. We demonstrate a procedure of compensating the vibration phase in real time, and the hybridization of an atomic and a classical accelerometers [26].

We operate a previous prototype of CAG [14, 27] in Paris city. The vibration signal is acquired using a mechanical accelerometer rigidly fixed to the experimental frame. A field programmable gate array (FPGA) based calculator treats the digitized acceleration signal to estimate the vibration phase Φ_{vib}. About 400 μs before the final interrogation pulse, the FPGA commands a phase jump in the direct digital synthesizer (DDS) that controls the phase lock of the Raman lasers. Such a feed forward scheme [see Fig. 4 (a)] allows us to operate the atomic sensor at mid fringe with little scatter. The vibration rejection efficiency can be as high as 60.

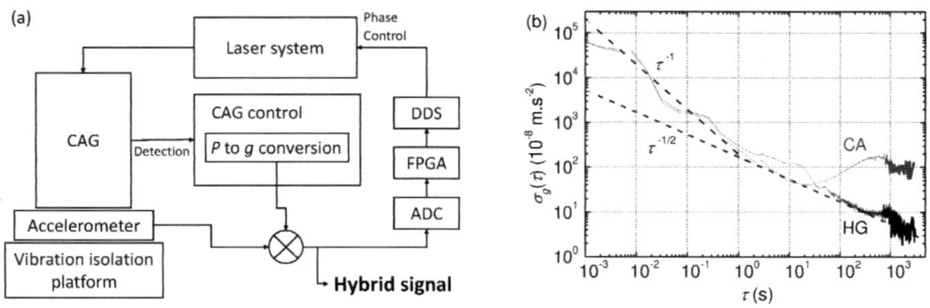

Figure 4. (a) Illustration of the real-time-compensation scheme and the hybridization with classical sensors. (b) ADEV of a classical accelerometer (CA, blue) and a hybrid gravimeter (HG, black). (b) is originally published in [26]. (colour online)

Further more, we perform a hybridization of the accelerometer with CAG. The DC component of conventional accelerometers suffers from drift and temperature dependence that severely degrades the long-term stability of the acceleration measurement. By locking the DC component on the CAG output, we realize a hybrid sensor which combines the large bandwidth (DC to 430 Hz) of the accelerometer and the bias stability and the accuracy of CAG. We show in Fig. 4 (b) the ADEV of the hybrid sensor placed on a vibration isolation platform. The short-term stability of the sensor is determined by the self-noise of the accelerometer and vibrations. The ADEV decreases as τ^{-1}, as high-frequency vibrations are averaged without dead time by the accelerometer. The long-term stability decreases as $\tau^{-1/2}$, indicating that it is set by CAG operating in the sequential mode and limited by the quality of the correlation. The added value of this technique for inertial navigation is assessed in [26].

5. Eliminating Dead Time

We have seen that the long-term stability of a hybrid inertial sensor still suffers from the dead time of the AI operating in sequential mode, which prevents efficient averaging of low-frequency noise (typically due to vibration). We present here a demonstration of a continuous atom interferometer without dead time, which paves the way to the intrinsically limited performance of AI-based inertial sensors.

We operate the fountain gyroscope mentioned in Sec. 3 as an atomic clock [23], using a sequence of two Raman pulses ($\pi/2$-$\pi/2$) so that $\Delta\Phi = \phi_1 - \phi_2$. We add a white noise over a bandwidth of 400 Hz to the local oscillator (LO) which controls the phase lock of the two Raman lasers. The added noise is analogous to the vibration noise in atomic inertial sensors. The cycle time and the Ramsey time are matched to be 480 ms, and each Raman pulse is shared between two atom clouds, one leaving and the other entering the interrogation zone, as shown in Fig. 5 (a). This imprints the same LO noise on consecutive interferometric phases , i.e. $\phi_2^i = \phi_1^{i+1}$, allowing cancellation of the LO noise in the interferometric phase accumulated over multiple shots. As a result, the ADEV of the relative frequency stability of the continuous clock follows

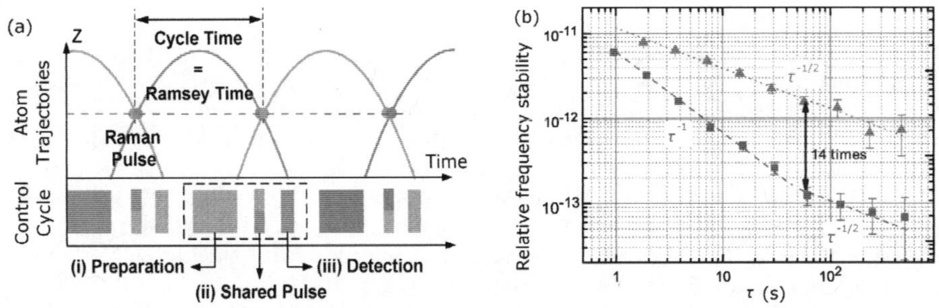

Figure 5. (a) Illustration of the zero-dead-time sequence of the atom gyroscope in the clock mode. (b) ADEV of the atomic clock in normal (triangles) and continuous (squares) mode. This figure is originally published in [23]. (colour online)

the τ^{-1} scaling, as shown in Figure 5 (b) (squares). The same figure shows the ADEV of a sequential clock (triangles), which follows the $\tau^{-1/2}$ scaling. The continuous clock gives a 14-fold gain in the relative frequency stability at 60 s. Thereafter, both are limited by uncorrelated noise (e.g. detection noise), and follow the $\tau^{-1/2}$ scaling. Note that the performance of the continuous clock after 60 s coincides with that of a sequential clock without the additional noise in the LO.

6. Conclusions

We have shown various steps towards exploring the full sensitivity of AI-based inertial sensors, which currently face the problems associated with the vibration noise, aliasing and the dead time between measurements. These problems would only become more severe when the sensors leave the environmentally stable laboratories for field applications. Moreover, the need to measure a fluctuating inertial signal (e.g. in navigation) renders active or passive isolation undesirable. Our efforts to correlate and hybridize atomic and mechanical sensors i) demonstrate AI-based inertial sensors with high sensitivity, ii) improve the dynamics beyond the linear regime, and iii) partially remove the dead time between successive interrogations. They represent a major step forward in reducing the size and the complexity of the sensors, making them suitable for on-board and field applications, and have stimulated the commercialization of atomic inertial sensors [28]. Such instruments now participate in gravity measurements, and have the potential to perform inertial navigations. The ability to fully eliminate the dead time will eventually lead to a new generation of QPN limited high-performance inertial sensors exploiting the full potential of atom interferometry.

Acknowledgments

The authors would like to thank the Institut Francilien pour la Recherche sur les Atomes Froids (IFRAF), the Agence Nationale pour la Recherche (ANR-9-BLAN-0026-01) in the framework of the MiniAtom collaboration, the Action Spécifique Gravitation Références Astronomie et Métrologie (GRAM), the Centre National d'Etudes Spatiales (CNES), the Délégation Générale pour l'Armement (DGA), the LABEX Cluster of Excellence FIRST-TF (ANR-10-LABX-48-01) within the Program "Investissements d'Avenir" operated by ANR, the city of Paris through the Emergence programme (project HSENS-MWGRAV), and the European Metrology Research Programme (EMRP) within the kNOW project for financial support. The EMRP was jointly funded by its participating countries within the European Association of National Metrology Institutes (EURAMET) and the European Union.

References

[1] Riehle F, Kisters Th, Witte A, Helmcke J and Bordé Ch J 1991 *Phys. Rev. Lett.* **67** 177
[2] Kasevich M and Chu S 1991 *Phys. Rev. Lett.* **67** 181
[3] Berman P R 1997 *Atom Interferometry* (San Diego: Academic Press)
[4] Farah T, Guerlin C, Landragin A, Bouyer P, Gaffet S, Pereira Dos Santos F and Merlet S 2014 *Gyroscopy and Navigation* **5** 266
[5] Hauth M, Freier C, Schkolnik V, Peters A, Wziontek H and Schilling M 2013 *Proc. Int. Sch. of Phys. "Enrico Fermi"* vol 188, ed Tino G M and Kasevich M A (Bologna: IOS Press) pp 557-86
[6] Geiger R, Ménoret V, Stern G, Zahzam N, Cheinet P, Battelier B, Villing A, Moron F, Lours M, Bidel Y, Bresson A, Landragin A and Bouyer P 2011 *Nat. Comm.* **2** 474
[7] Altschul B *et al* for the STE-QUEST collaboration 2015 *Adv. Space Res.* **50** 501-5
[8] Dimopoulos S, Graham P W, Hogan J M, Kasevich M A and Rajendran S 2008 *Phys. Rev.* D **78** 122002
[9] Geiger R, Amand L, Bertoldi A, Canuel B, Chaibi W, Danquigny C, Dutta I, Fang B, Gaffet S, Gillot P, Holleville D, Landragin A, Merzougui M, Riou I, Savoie D and Bouyer P 2015 *Proc. 50th Rencontres de Moriond: "100 years after GR"* 21-18 March 2015 La Thuile, Italy
[10] Yver-Leduc F, Cheinet P, Fils J, Clairon A, Dimarcq N, Holleville D, Bouyer P and Landragin A 2003 *J. Opt.* B*: Quantum Semiclas. Opt.* **5** S136-42
[11] Santarelli G *et al* 1998 *IEEE Trans. Ultrason. Ferroelectr. Freq. Control* **45** 887-94
[12] Peters A, Chung K Y and Chu S 2001 *Metrologia* **38** 25-61
[13] Hu Z, Sun B, Duan X, Zhou M, Chen L, Zhan S, Zhang Q and Luo J 2013 *Phys. Rev.* A **88** 043610
[14] Le Gouët J, Mehlstäubler T E, Kim J, Merlet S, Clairon A, Landragin A and Pereira Dos Santos F 2008 *Appl. Phys.* B **92** 133-44
[15] Jekeli C 2005 *Navigation* **52** 1-14
[16] Antoine Ch and Bordé Ch J 2003 *J. Opt.* B*: Quantum Semiclass. Opt.* **5** S199
[17] Louchet-Chauvet A, Farah T, Bodart Q, Clairon A, Landragin A, Merlet S and Pereira Dos Santos F 2011 *New J. Phys.* **13** 065025
[18] Cheinet P, Canuel B, Pereira Dos Santos F, Gauguet A, Leduc F and Landragin A 2008 *IEEE Trans. on Instrum. Meas.* **57** 1141
[19] Merlet S *et al* 2008 *Metrologia* **45** 26574
[20] Gillot P, Francis O, Landragin A, Pereira Dos Santos F and Merlet S 2014 *Metrologia* **51** L15-7
[21] Francis O *et al* 2013 *Metrologia* **50** 257-68
[22] Canuel B, Leduc F, Holleville D, Gauguet A, Fils J, Virdis A, Clairon A, Dimarcq N, Bordé Ch J, Landragin A and Bouyer P 2006 *Phys. Rev. Lett.* **97** 010402
[23] Meunier M, Dutta I, Geiger R, Guerlin C, Garrido Alzar C L and Landragin A 2014 *Phys. Rev.* A **90** 063633
[24] Jiang Z *et al* 2012 *Metrologia* **49** 666-84
[25] Merlet S, Le Gouët J, Bodart Q, Clairon A, Landragin A, Pereira Dos Santos F and Rouchon P 2009 *Metrologia* **46** 87-94
[26] Lautier J, Volodimer L, Hardin T, Merlet S, Lours M, Pereira Dos Santos F and Landragin A 2014 *Appl. Phys. Lett.* **105** 144102
[27] Merlet S, Volodimer L, Lours M and Pereira Dos Santos F 2014 *Appl. Phys.* B **117** 749
[28] Muquans Absolute Quantum Gravimeter http://www.muquans.com/index.php/products/aqg

Journal of Physics: Conference Series **723** (2016) 012050 doi:10.1088/1742-6596/723/1/012050

Mobile quantum gravity sensor with unprecedented stability

C Freier,[1*] M Hauth,[1] V Schkolnik,[1] B Leykauf,[1] M Schilling[2], H Wziontek[3], H-G Scherneck[4], J Müller[2] and A Peters[1]

[1] Humboldt Universität zu Berlin, Institut für Physik, Newtonstr. 15, Berlin, Germany
[2] Inst. für Erdmessung, Leibniz Universität Hannover, Schneiderberg 50, Hannover, Germany
[3] Federal Agency for Cartography and Geodesy, Leipzig, Germany
[4] Onsala Space Observatory, Chalmers University of Technology, Gothenburg, Sweden

E-mail: `christian.freier@physik.hu-berlin.de`, `achim.peters@physik.hu-berlin.de`

Abstract. Changes of surface gravity on Earth are of great interest in geodesy, earth sciences and natural resource exploration. They are indicative of Earth system's mass redistributions and vertical surface motion, and are usually measured with falling corner-cube- and superconducting gravimeters (FCCG and SCG). Here we report on absolute gravity measurements with a mobile quantum gravimeter based on atom interferometry. The measurements were conducted in Germany and Sweden over periods of several days with simultaneous SCG and FCCG comparisons. They show the best-reported performance of mobile atomic gravimeters to date with an accuracy of $39\,\mathrm{nm/s^2}$, long-term stability of $0.5\,\mathrm{nm/s^2}$ and short-term noise of $96\,\mathrm{nm/s^2}/\sqrt{\mathrm{Hz}}$. These measurements highlight the unique properties of atomic sensors. The achieved level of performance in a transportable instrument enables new applications in geodesy and related fields, such as continuous absolute gravity monitoring with a single instrument under rough environmental conditions.

Gravimetry today is based on both absolute and relative measurements[1]. Falling corner-cube absolute gravimeters (FCCG) track the ballistic trajectory of a free-falling test mass using a light interferometer to obtain the local gravitational acceleration. Relative gravimeters are based on a spring-mass system or a superconducting sphere levitating in a magnetic field to measure differential signals with respect to some reference value[1]. While all these instruments have individual merits in advancing the field of gravimetry, they also have characteristic shortcomings. For instance, the best FCCGs[2] have a small total uncertainty of $20\,\mathrm{nm/s^2}$ but require sites with a low level of micro-seismic vibrations to reach their optimal performance [3]. The moving internal test-mass furthermore excites floor vibrations and necessitates massive concrete foundations in order to reduce self-induced vibrations [4]. FCCG were also not built to perform continuous gravity monitoring over extended periods of time due to wear of their moving mechanical parts. Long-term gravity registrations are therefore done with relative gravimeters. Because of their working principle these instruments depend on repeated calibration and drift-rate verification through an absolute gravity reference which is, in the case of SCG, provided by FCCG. Relative and absolute gravity measurements are thus complementary techniques and must usually be used in combination. As shown below, atomic gravimeters overcome many limitations of existing instruments through their ability to generate continuous, absolute gravity measurements over long periods of time with one mobile, stand-alone instrument.

Atom interferometers (AI) using light pulses as beam-splitters and mirrors for atomic matter waves were first realized in 1991[5, 6] and have since become a versatile technology not only for inertial sensing applications[7, 8, 9, 10, 11, 12] but also for precision measurements of the Newtonian gravitational constant[13], the fine-structure constant[14], as well as for proposed and demonstrated tests of the equivalence principle[15, 16] or future gravitational wave detectors[17].

1. The Gravimetric Atom Interferometer

The gravimetric atom interferometer (GAIN) was developed and built at Humboldt Universität zu Berlin to perform mobile high-precision gravity measurements. It is based on interfering ensembles of ^{87}Rb atoms in an atomic fountain configuration and stimulated Raman transitions. Figure 1 shows the working principle of the instrument, a complete description has been published elsewhere[18, 19]. During free fall the atomic matter-waves are split vertically in space, reflected and recombined using a Mach-Zehnder $\frac{\pi}{2}$-π-$\frac{\pi}{2}$ pulse sequence, compare figure 1(b). The output phase of the interferometer depends on the local light phase at the atomic positions during the pulses. This corresponds[18] to a threefold sampling of the ballistic trajectory and thus a g measurement, when neglecting the spatial splitting of the wave-packets during the interferometry sequence. The measurement sequence starts with a Magneto-Optical Trap (MOT) loading

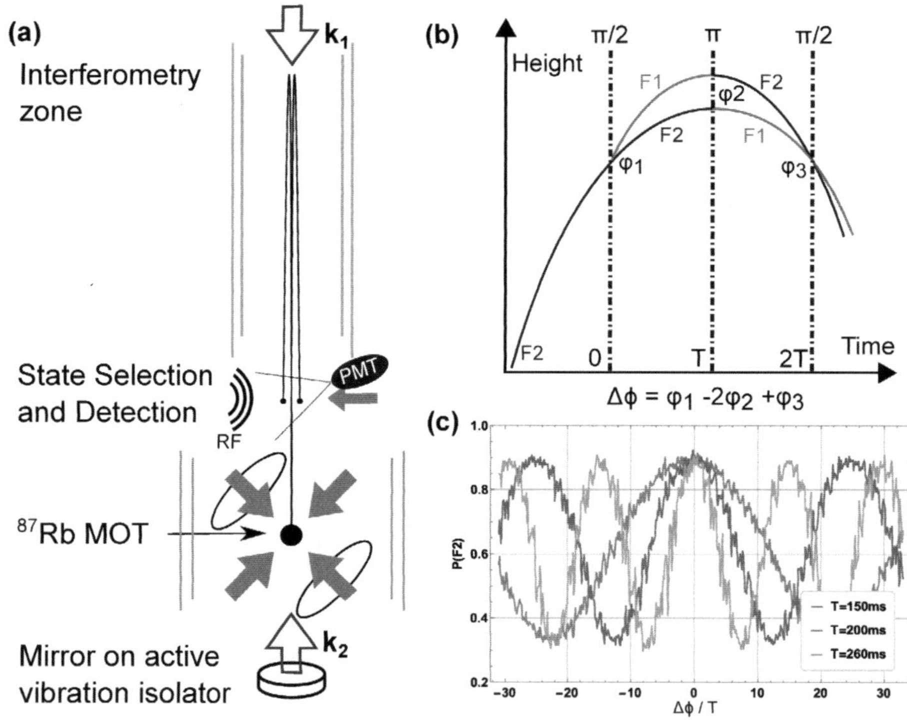

Figure 1: GAIN working principle. **a**, atomic fountain set-up with magneto-optical trap (MOT) and Raman laser beams $\mathbf{k}_1$,$\mathbf{k}_2$ for atom interferometry (AI). **b**, Mach-Zehnder AI sequence under the influence of gravity. **c**, Interferometer fringes encoded in the internal state population (F2) for several values of T. (colour online)

approximately 10^9 laser-cooled ^{87}Rb atoms within 0.6 s. The cloud is then launched upwards using far-detuned moving molasses achieving a temperature of 2 µK and a vertical velocity of 4 m/s. After a previously described process[18] which selects only atoms in the magnetically insensitive sub-state within a narrow vertical velocity class, the cloud reaches the elongated

Journal of Physics: Conference Series **723** (2016) 012050 doi:10.1088/1742-6596/723/1/012050

magnetically shielded interferometry region. The $\frac{\pi}{2}$-π-$\frac{\pi}{2}$ Raman pulse sequence is then conducted with a time $T = 0.26\,\mathrm{s}$ between pulses. Afterwards, the upper state population P_2 is detected using a fluorescence detection system which yields an interference signal $P_2 = \bar{P} + C/2 \cos \Delta\Phi$ with contrast C and offset $\bar{P}$. The next measurement is then started with a cycle time of 1.5 s. The interferometer phase $\Delta\Phi$ is connected to the gravity value g by $\Delta\Phi = (k_{\mathrm{eff}} g - \alpha) T^2 + \Phi_L$, where $\mathrm{k}_{\mathrm{eff}} = |\mathbf{k}_1 - \mathbf{k}_2| \approxeq 2\mathrm{k}$ is the effective wave-number of the Raman transitions, α is a constant frequency chirp rate of the Raman laser which compensates for the time-varying Doppler-shift of the atoms during free fall, and Φ_L is a configurable Raman laser phase which can be used to scan the interferometer fringe. In order to deduce $\Delta\Phi$ at a new measurement site, interference fringes are first scanned by altering the RF chirp rate α for several values of T. Their positions overlap at the desired chirp rate value where $\alpha = k_{\mathrm{eff}} g$ which removes the phase ambiguity due to the fringe pattern, compare above equation and figure 1(c).

The inertial reference is provided by a retro-reflection mirror shown in the bottom of figure 1(a) which therefore has to be decoupled from environmental vibrations. This is realised with an active vibration isolation system[18]. Due to its limited isolation at micro-seismic frequencies of 0.1–0.3 Hz, residual vibrations are contained in its feedback loop error signal. Feeding this signal into a previously demonstrated post-correction algorithm[20] thus further reduces interferometer phase noise by a factor 2 to 4, depending on the micro-seismic background. Special care was taken during the development of GAIN to ensure that the set-up is small and robust enough for transport to other locations with minimal effort. It thus consists of 3 modular units[18] with a respective size of roughly $(1 \times 1 \times 2)\,\mathrm{m}^3$.

2. Mobile gravity comparison campaigns

After completion of the set-up and initial gravity comparisons in Berlin[18], two measurement campaigns were carried out at remote locations. The 1st took place in November 2013 at the Geodetic Observatory in Wettzell, the 2nd in February 2015 at Onsala Space Observatory near Gothenburg, Sweden. Both locations were chosen because the local gravity values are known very accurately due to several measurement campaigns conducted with FCCG during the last years[21]. They furthermore operate SCGs for high-precision monitoring of temporal gravity changes which proved invaluable for characterizing the performance of the GAIN gravimeter. While conducting the 2nd campaign in Onsala, the state-of-the-art FCCG FG5X-220 performed a simultaneous measurement for a direct comparison of the instrumental properties. During both campaigns a small truck was used to transport GAIN to the destination sites. Transport and set-up took approximately 3 days until the first gravity data were produced, a delay which could be shortened in the future to less than a day through optimization of the work-flow and small alterations of GAIN.

3. Gravimeter Performance

A five day segment of the gravity data taken during both campaigns is shown in the upper half of figure 2, showing temporal variations due to earth tides, atmospheric and other environmental effects [1]. In order to characterize the gravimeter's stability, these variations need to be subtracted from the gravity signal. Below a magnitude of $10\,\mathrm{nm/s^2}$ it is not sufficient to use only synthetic models for this purpose because remaining signals caused by incomplete atmospheric correction, tidal modelling and local water storage fluctuations can reach up to a few $10\,\mathrm{nm/s^2}$. We thus use the SCG as a reference to assess instrumental effects in GAIN. As shown in the bottom half of figure 2, both instrument are in excellent agreement. The residuals obtained by subtracting both datasets do not contain visible signals within the error bars of the measurement. The detailed analysis in figure 3 shows that they resemble white noise on time-scales below 10^4 s. Beyond this the relative stability improves further and reaches $0.5\,\mathrm{nm/s^2}$, or 5×10^{-11} g, for a time-scale of 10^5 s or approximately one day. This is the best reported stability

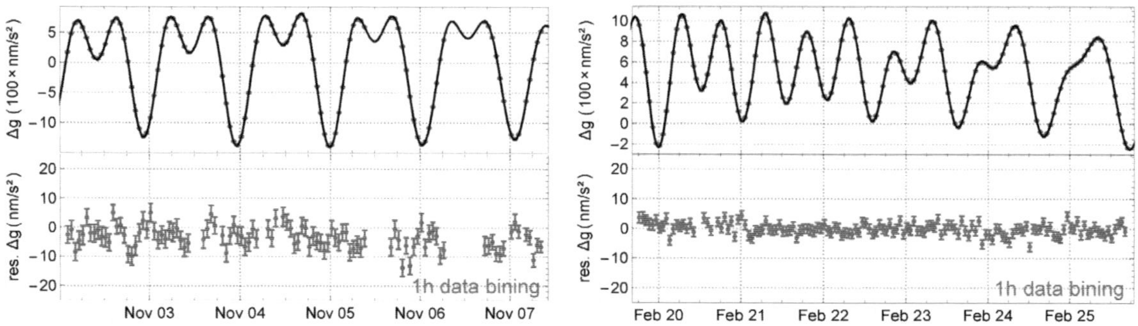

(**a**) Wettzell, Germany (2013) (**b**) Onsala, Sweden (2015)

Figure 2: GAIN gravity data. a, Top: tidal signal measured by GAIN (blue) versus SCG (black) during the 1st campaign. Bottom: difference between GAIN and the SCG signals. **b**, equivalent plot of 2nd campaign. The decreased noise level and improved stability during the 2nd campaign is due to enhancements of GAIN between both campaigns, compare figure 3 **a**.

for an absolute gravimeter to our knowledge and shows the excellent control of any time-variable systematic effects in GAIN as listed in table 1. Comparing SCG and GAIN data on the sub-nm/s^2 level requires SCG scale factor calibration significantly better then 10^{-3} which is beyond standard FCCG results[22]. The scale factor was thus re-obtained for both campaigns using 5 days of GAIN gravity data, yielding values consistent with previous calibrations but improved uncertainties of 4×10^{-4} and 2.6×10^{-4}, respectively. Even under optimized schedules[23], low micro-seismic noise and maximal tidal amplitude, FG5 instruments would need at least three weeks to achieve a similar precision which demonstrates the value of GAIN's improved resolution.

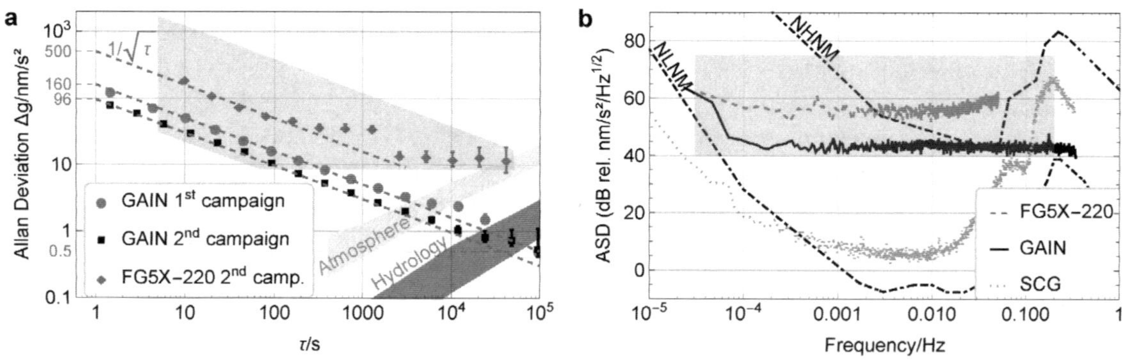

Figure 3: Gravimeter comparison. a, Allan std. deviation of GAIN and FG5X-220 gravity data reduced by SCG signals. The continuous operation of GAIN allows shorter integration time and improved long-term stability compared to the FG5X. The blue-shaded areas highlight common gravity signals due to air-pressure and hydrological effects. **b**, **Amplitude Spectral Density** of GAIN during the 2nd campaign, showing a noise level better than the FG5X but worse than SCG. The FG5X noise level was increased due to elevated micro-seismic vibrations as depicted by high- and low-noise model[24]. The gray areas in the upper half of the images thus denote the expected noise under quiet and noisy conditions[3]. (colour online)

The absolute gravity values were compared to the reference value during the 1st campaign and to the simultaneously measured FG5X-220 result during the 2nd campaign. Both comparisons

Systematic effect	**Offset**	**Error**
	(nm/s^2)	
Raman Wavefronts	28	±22
Coriolis Effect	0	±15
Magnetic Field Effects	0	±10
RF Group delay	0	±10
Self Gravitation	-19	±5
Reference Laser Freq.	12(10)	±5
Synchronous Vibrations	0(-92)	±5(50)
AC Stark Shift (1PLS)	0	±5
Rb background vapour	-5	±3
AC Stark Shift (2PLS)	0	±2
vertical alignment	0(1)	±1
Total	**16(-77)**	**±32(59)**

Table 1: GAIN systematic error budget for the 2nd campaign. The total offset was added to the GAIN raw value to obtain the corrected result. Values for the 1st campaign are denoted in brackets. Synchronous vibrations were eliminated between campaigns by means of a magnetic shield around the MOT region.

First campaign	**Gravity (nm/s^2)**
GAIN raw value	9,808,369,362 ± 01
Systematic offset	-77 ± 59
Height offset	400 ± 10
GAIN corrected	9,808,369,685 ± 60
Reference value	9,808,369,623 ± 18
GAIN-Ref.	**62 ± 62**
Second Campaign	
GAIN raw value	9,817,158,312 ± 01
Systematic offset	16 ± 32
Height offset	727 ± 10
GAIN corrected	9,817,159,055 ± 34
FG5X-220 value	9,817,159,023 ± 20
GAIN-FG5X	**32 ± 39**

Table 2: Absolute gravity comparison of the atomic gravimeter and known gravity reference values during both campaigns, yielding excellent agreement. The vertical gravity gradient was measured independently beforehand and used to account for effective measurement height differences between values.

show excellent agreement within their specified 1σ uncertainties as shown in table 2. The GAIN values are slightly higher in both cases with differences of $(62\pm62)\,\mathrm{nm/s^2}$ and $(32\pm39)\,\mathrm{nm/s^2}$. The achieved level of accuracy is slightly better than another value [25] obtained using atom interferometry and comparable to the level provided by the FG5(X), where individual instruments differ by 20–30 nm/s^2 [1, 25]. The biggest uncertainty is caused by Raman laser wave-front distortions as shown in table 1. They have been reduced by wave-front measurements and a correcting model [26] which was applied to all optical elements in the Raman beam path except the lower vacuum window. In the future this effect could be decreased further by reducing the cloud expansion through advanced cooling techniques or a more complete correction.

4. Discussion

Due to stormy winter weather and the proximity to the Kattegat coastline, micro-seismic vibrations were strongly elevated during parts of the 2nd campaign. This did not affect the GAIN sensor as shown in figure 3(b), but increased the FG5X noise level roughly five-fold as observed before[3, 18, 8]. GAIN's improved tolerance to micro-seismic vibrations is due to its 0.7 Hz repetition rate, as opposed to 0.1 Hz of the FG5X, and more effective isolation system. This shows a general advantage of atomic gravimeters compared to FCCGs which rely on low levels of micro-seismic excitation. Atomic sensors also do not excite vibrations of the floor/pillar as the test-masses involved in the measurement are negligible, which altogether enables low-noise operation not only on dedicated concrete pillars but also on regular flooring. This extends their range to a vastly increased number of measurement sites which, in combination with their continuous, drift-free, absolute measurement, will improve results in applications such as monitoring of tidal, hydrological or geophysical gravity signals with amplitudes in the 1 to 10 nm/s^2 range on long time-scales where spring-based instruments are limited by non-linear drifts. Albeit GAIN can not yet reach the ultimate low short-term noise level provided by SCG, the values demonstrated here are comparable to those of state-of-the-art spring-based relative gravimeters [27] without their long-term drifts. The noise level of mobile atomic gravimeters may in the future be improved through further optimization of vibration isolators[9], hybrid sensor strategies[28] or advanced techniques such as interleaved fountain operation[29]. The comparison between two types of absolute gravimeters based on fundamentally different measurement principles, namely FCCG and atomic gravimeters, is beneficial in its own right and will boost the search for hidden systematic effects and increase the accuracy of today's gravity networks. We further anticipate that the robust and adaptable architecture of cold atom interferometry will enable the development of self-contained, compact and rugged instruments for applications which so far were not accessible to absolute gravimetry. For instance, airborne and shipborne gravity surveys will benefit from the characteristics of atomic gravimeters and gradiometers due to their intrinsic immunity to hysteresis effects which deteriorate the performance of current spring-based instruments during turbulence[30]. Drift-free, continuous operation at places such as underground systems, boreholes or even the sea floor will exceed the operating range of existing spring-based gravimeters and enable new insights in earth sciences and geodesy.

Acknowledgments

This material is based on work funded by the European Commission (FINAQS, Contr. No. 012986-2 NEST), by ESA (SAI, Contr. No. 20578/07/NL/VJ) and by ESF/DFG (EuroQUASAR-IQS, DFG grant PE 904/2-1 and PE 904/4-1).

References

[1] Crossley D, Hinderer J and Riccardi U 2013 *Rep. Prog. Phys.* **76** 046101
[2] Niebauer T M, Sasagawa G S, Faller J E, Hilt R and Klopping F 1995 *Metrologia* **32** 159
[3] Van Camp M 2005 *J. Geophys. Res.* **110** B05406
[4] Charles K and Hipkin R 1995 *Metrologia* **32** 193
[5] Riehle F, Kisters T, Witte A, Helmcke J and Bordé C J 1991 *Phys. Rev. Lett.* **67** 177
[6] Kasevich M and Chu S 1991 *Phys Rev Lett* **67** 181–184
[7] Peters A, Chung K Y and Chu S 2001 *Metrologia* **38** 25
[8] Gillot P, Francis O, Landragin A, Pereira Dos Santos F and Merlet S 2014 *Metrologia* **51** L15
[9] Hu Z K *et al.* 2013 *Phys. Rev. A* **88** 043610
[10] Gauguet A, Canuel B, Lévèque T, Chaibi W and Landragin A 2009 *Phys. Rev. A* **80** 063604
[11] Stockton J K, Takase K and Kasevich M A 2011 *Phys. Rev. Lett.* **107** 133001
[12] Berg P *et al.* 2015 *Phys. Rev. Lett.* **114** 063002
[13] Rosi G, Sorrentino F, Cacciapuoti L, Prevedelli M and Tino G M 2014 *Nature* **510** 518–521
[14] Wicht A, Hensley J M, Sarajlic E and Chu S 2002 *Phys. Scr.* **T102** 82–88
[15] Dimopoulos S, Graham P W, Hogan J M and Kasevich M A 2007 *Phys. Rev. Lett.* **98** 111102
[16] Schlippert D *et al.* 2014 *Phys. Rev. Lett.* **112** 203002
[17] Dimopoulos S, Graham P W, Hogan J M, Kasevich M A and Rajendran S 2009 *Phys. Lett. B* **678** 37–40
[18] Hauth M, Freier C, Schkolnik V, Peters A, Wziontek H and Schilling M 2014 *Proc. of the Int. School of Physics "Enrico Fermi", Course 188, edited by Tino G. M. and Kasevich M. A* (SIF, Bologna and IOS Press, Amsterdam) pp 557–586
[19] Hauth M, Freier C, Schkolnik V, Senger A, Schmidt M and Peters A 2013 *Appl. Phys. B* **113** 49–55
[20] Le Gouët J *et al.* 2008 *Appl. Phys. B* **92** 133–144
[21] Timmen L, Engfeldt A and Scherneck H G 2015 *J. Geod. Sci.* **5** 18–25
[22] Francis O and van Dam T 2002 *Metrologia* **39** 485
[23] Van Camp M, Meurers B, de Viron O and Forbriger T 2015 *J. Geod.* **90** 91–99
[24] Peterson J 1993 Observations and modeling of seismic background noise Open-File Report 93-322 USGS
[25] Francis O *et al.* 2015 *Metrologia* **52** 07009
[26] Schkolnik V, Leykauf B, Hauth M, Freier C and Peters A 2015 *Appl. Phys. B* **120** 311–316
[27] Riccardi U, Rosat S and Hinderer J 2011 *Metrologia* **48** 28–39
[28] Lautier J *et al.* 2014 *Appl. Phys. Lett.* **105** 144102
[29] Meunier M, Dutta I, Geiger R, Guerlin C, Garrido Alzar C L and Landragin A 2014 *Phys. Rev. A* **90** 063633
[30] Forsberg R and Olesen A V 2010 *Sciences of Geodesy - I* ed Xu G (Berlin, Heidelberg: Springer Berlin Heidelberg) pp 83–104 ISBN 978-3-642-11740-4 978-3-642-11741-1

Relativistic geodesy

J Flury

Institut für Erdmessung / Institute of Geodesy, Leibniz Universität Hannover, Schneiderberg 50, 30167 Hannover, Germany

E-mail: flury@ife.uni-hannover.de

Abstract. Quantum metrology enables new applications in geodesy, including relativistic geodesy. The recent progress in optical atomic clocks and in long-distance frequency transfer by optical fiber together pave the way for using measurements of the gravitational frequency redshift for geodesy. The remote comparison of frequencies generated by calibrated clocks will allow for a purely relativistic determination of differences in gravitational potential and height between stations on Earth surface (chronometric leveling). The long-term perspective is to tie potential and height differences to atomic standards in order to overcome the weaknesses and inhomogeneity of height systems determined by classical spirit leveling. Complementarily, gravity measurements with atom interferometric setups, and satellite gravimetry with space borne laser interferometers allow for new sensitivities in the measurement of the Earth's gravity field.

The goal of classical geodesy is the measurement and representation of the geometry of the Earth and of its gravity field. Already 100 years ago the theory of general relativity has shown that geometry and gravity cannot be treated separately, and in recent years relativistic effects have become more and more relevant in geodesy. Relativistic geodesy is an emerging field, and we do not want to attempt a formal definition here. For the time being we consider relativistic geodesy rather generically as the representation and use of the principles of general relativity in geodetic theory and in geodetic applications. This includes the description of curved space-time by four-dimensional coordinate systems, together with the description of trajectories, energy and momentum of test bodies such as satellites, and the description of radiation on geodetic scales and for geodetic purposes. Relativistic geodesy addresses the relativistic description of time and frequencies, and thus deals with the measurement and modeling of the relativistic gravitational frequency redshift in the Earths gravitational field. In particular, it includes chronometric leveling, i.e., the determination of potential differences based on frequency redshift measurements that are obtained from remote comparisons of frequencies generated by ultraprecise atomic clocks. The theory on relativistic geodesy has been studied since several decades [1, 2]. In recent years, optical frequency standards have rapidly improved by orders of magnitude, and breakthroughs have been achieved in long-distance optical frequency transfer. Therefore, chronometric leveling is expected to become very relevant for geodetic applications and geodetic reference frames. It opens the perspective for tying gravitational potential and heights on Earth to an atomic reference, similar to the definition of time through atomic standards. Potsdam as the venue of the 8th FSM conference in 2015 is a very interesting fit for this topic as Helmert House and Einstein Tower both located in close vicinity on Telegraphenberg may be considered as historical symbols for classical and relativistic geodesy.

Quantum sensors such as optical atomic clocks are highly relevant for relativistic geodesy

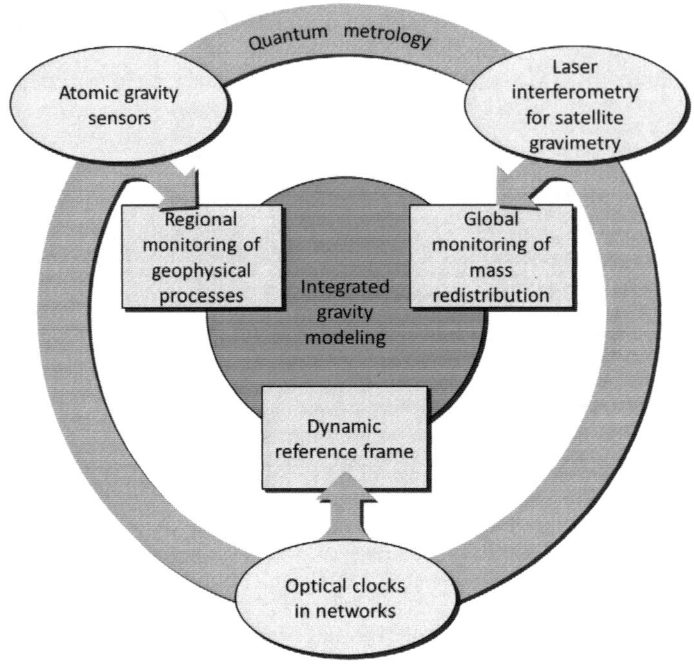

Figure 1. Optical atomic clocks in networks for relativistic geodesy are complementary to other new techniques from the field of quantum metrology and quantum sensors for the measurement and modeling of the Earths gravity field. The complementarity of techniques is in the focus of the Hannover Collaborative Research Center "Relativistic Geodesy and Gravimetry with Quantum Sensors (geo-Q)" (colour online)

because they can achieve extreme sensitivity levels. We consider relativistic geodesy and gravimetry with quantum sensors in a broader context as key elements in the development of new gravitational measurement techniques (Fig. 1). Optical atomic clocks are closely interlinked with atom interferometers with respect to quantum metrology. Atomic gravity sensors based on atom interferometry are being developed for regional monitoring of mass changes with a dedicated sampling of gravity change in space and time. Relativistic geodesy is also strongly linked to the progress in satellite gravimetry where laser interferometry is evolving as exremely precise tool for tracking testmasses in space, in order to determine the large-scale mass variations due to climate, to the Earths water cycle and to processes in the Earths interior.

Relativistic geodesy on Earth is characterized by the properties of the Earth's gravity field. On Earth surface, the gravity anomalies - the spatial variations of gravity acceleration - typically are in the order of 10^{-3} m/s^2, reflecting the internal density structure of the Earth. The geoid - an equipotential surface of the gravity potential that approximates the sea level - deviates from a reference ellipsoid by up to 100 m. Gravity anomalies and geoid undulations are related, the gravity acceleration vector being the gradient of the gravity potential field. Heights on Earth are defined through the gravity potential field, e.g., as the length of the curved plumbline between the Earth surface and the geoid. The determination of heights therefore requires the knowledge of the gravity field. Increasingly, also the temporal change of the gravity field is becoming relevant for geodesy and Earth sciences. Changes in the mass distribution lead to seasonal geoid variations in the order of mm, with long term (secular) changes in the order of mm/y. Temporal

Journal of Physics: Conference Series **723** (2016) 012051 doi:10.1088/1742-6596/723/1/012051

change of gravity acceleration is typically in the 10^{-8} m/s^2 range.

According to General Relativity, a clock at lower altitude - and therefore in a stronger gravity potential - is slowed down with respect to a clock at higher altitude by the relative relativistic frequency shift $\Delta f/f$ that is related to the difference in gravity potential ΔW through

$$\Delta W = \frac{\Delta f}{f} c^2 \tag{1}$$

where c is the velocity of light. The measurement of $\Delta f/f$ using optical atomic clocks connected by optical frequency transfer, enables chronometric leveling (Fig. 2a). A frequency inaccuracy of 10^{-18} corresponds to an error in ΔW of 0.1 m^2/s^2 or a height error of 1 cm. The potential difference ΔW is a quantity that we also obtain from classical geodetic spirit leveling (Fig. 2b). For precision leveling, e.g., in the national first order networks, ΔW between points P along the path of leveling lines is obtained by integrating the product of geometric leveling increments dn measured with leveling rods and gravity g measured with gravity meters:

$$\Delta W_{P_1P_2} = -\int_{P_1}^{P_2} g\, dn \tag{2}$$

Classical precision spirit leveling and chronometric leveling both aim at the observation of potential differences in gravity potential networks. For height networks, a conversion to metric quantities is applied. Orthometric heights H at a point P are derived by dividing the potential difference ΔW_{P_0P} to a datum point P_0 - such as a tide gauge - by the mean gravity $\bar{g}$ along the plumbline in P

$$H_P = -\frac{\Delta W_{P_0P}}{\bar{g}} \tag{3}$$

Alternatively, normal heights are obtained by dividing the potential difference by the mean normal gravity $\bar{\gamma}$ that is determined from a normal gravity field model

$$H_P^N = -\frac{\Delta W_{P_0P}}{\bar{\gamma}} \tag{4}$$

Classical spirit leveling involves a large effort, and national networks are often checked and renewed in intervals of decades only. Over large distances, error propagation in leveling networks is unfavorable. In addition, the choice of tide gauges as datum points is rather arbitrary. Tide gauges do not refer to the geoid but to sea level, which deviates from the geoid by up to 2 meters due to the dynamic sea surface topography related to ocean circulation. Even between neighboring tide gauges, sea level differences may be in the decimeter range. Vertical land movement in some regions further adds to the complexity of providing accurate and stable height networks. As a result, today still a multitude of national and regional height systems exists, which are difficult to combine and which are inhomogeneous in terms of their standards. Homogenization of height systems using space geodesy such as GPS is possible but needs geoid models to derive heights from geometric positions (Fig. 2c). This approach is called gravity field approach to height determination, or GPS-leveling. The determination of the involved geoid model depends on the combination of many different data types such as satellite gravimetry, terrestrial gravimetry and topographic mass distribution. The result is an inherently complex height reference, with a quality that is today typically in the range of few centimeters to few decimeters, although this is expected to further improve. A famous example for height inconsistencies are the height differences between the tide gauges along the Atlantic coast of the US and Canada. Here, inconsistencies between the classical leveling and oceanographic sea level models of up to 60 cm are found. Inconsistencies between the gravity field approach

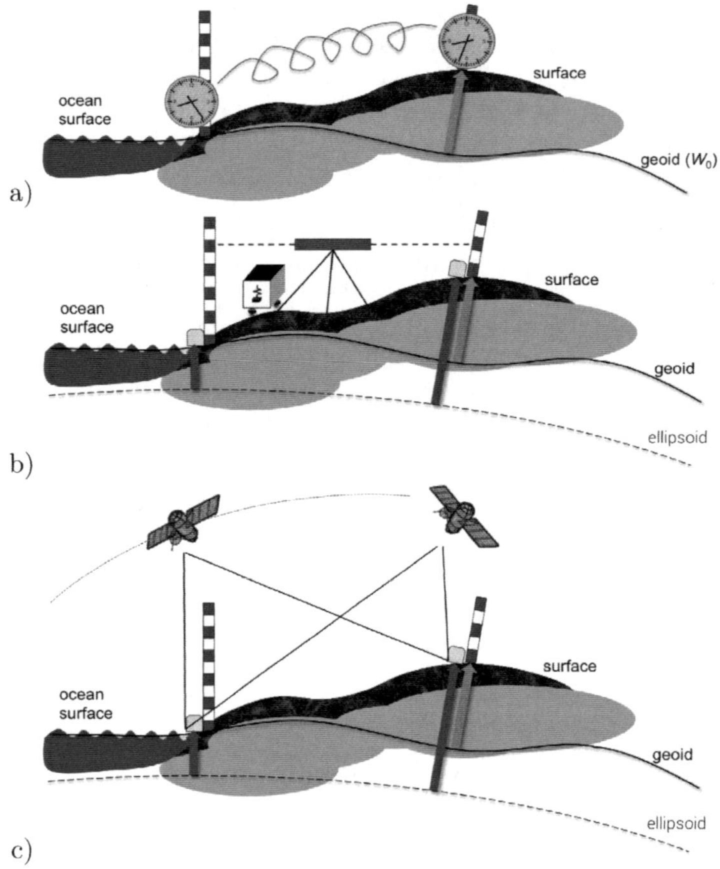

Figure 2. a: chronometric leveling, b: classical spirit leveling, c: leveling with the gravity field approach. Source: P.O. Schmidt (colour online)

and oceanographic models are still up to 30 cm [3]. In Europe, the comparison of leveling networks with the gravity field approach show similar inconsistencies [4]. A consistent world height system at the 1 cm level would be needed for many purposes [5]: for sea level studies, as fundamental geodetic height reference, for engineering, and for geophysical and environmental studies involving land uplift and subsidence. Such a world height system in cm accuracy will, however, be difficult to achieve with classical methods.

Before this background, the current progress in frequency measurement and frequency transfer with optical clocks and phase-stabilized optical fiber is considered as major breakthrough for the determination of heights. Long-distance comparison of the Sr lattice clocks at SYRTE, Paris, and PTB, Braunschweig, has demonstrated a $3 \cdot 10^{-17}$ frequency uncertainty level [6]. The uncertainty of the height difference and the corresponding gravitational redshift is part of this uncertainty. Chronometric leveling with clocks that have been calibrated side-by-side is expected to achieve an accuracy at this level or better, which would correspond to an accuracy of 30 cm or better for long-distance height differences. The next important step will be to use a transportable optical clock for side-by-side calibration with a stationary optical clock and subsequent long-distance frequency comparison through phase-stabilized optical fiber. It is important to note that with calibrated clocks, the quality of chronometric leveling will be determined by the frequency reproducibility instead of the absolute frequency accuracy of the involved clocks. In this respect, the requirements in standard frequency metrology and in relativistic geodesy are

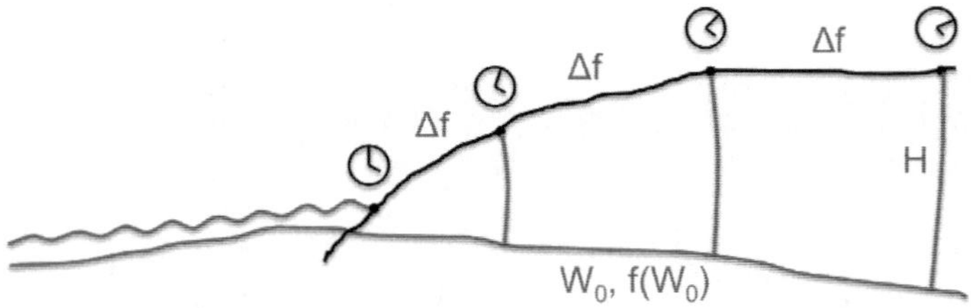

Figure 3. Frequency redshift network providing access to the relativistic geoid definition

substantially different. If proof-of-principle experiments of chronometric leveling are successful, this will allow to tie observations of potential and height differences to atomic standards. Clock networks measuring relativistic frequency differences would make the relativistic geoid definition - the surface where clocks run with the same speed - applicable in geodesy and could provide a well defined height reference that is accessible within continents (Fig. 3). Clock networks are not expected to replace the GPS-leveling approach for height determination. They could evolve as the superior technique for the fundamental networks, whereas GPS-leveling would provide for an economic continuous coverage. The far-reaching perspective may be a global atomic clock network as a homogeneous and stable reference for gravity potential and height systems. This would be strongly complementary to the International Terrestrial Reference Frame that provides based on space geodetic measurements a globally homogeneous geometric reference at the 1 cm level or better [7].

Observations of gravity potential differences in clock networks will be complementary to satellite gravimetry (Fig. 1). The satellite gravimetry missions GRACE and GOCE have provided quantifications for global and regional mass redistribution such as polar ice mass loss and ocean mass increase, as well as precision geoid models for geodesy and modeling of ocean circulation. The GRACE Follow-On satellite mission to be launched in 2017 will further improve the knowledge of the Earths gravity field through satellite-to-satellite tracking with a laser ranging interferometer. The advantage of satellite observations is the global and dense coverage of measurements, which is used to determine complete spherical harmonic gravity models. Chronometric leveling in clock networks will provide completely independent and possibly very precise observations of gravity potential differences in specific points. The fact that chronometric leveling results refer to an atomic standard could also be valuable for the quality assessment of continuous, satellite-based models. The combination of point values with continuous gravity models is, however, not straightforward, and it will be an interesting research question whether and how high quality ground reference for the gravity potential from clock networks can stabilize or improve the satellite models.

Acknowledgements Contributions by Piet O. Schmidt, Christian Lisdat and the geo-Q team are gratefully acknowledged. This work has been funded by the Deutsche Forschungsgemeinschaft in the Collaborative Research Center SFB1128 Relativistic Geodesy and Gravimetry with Quantum Sensors (geo-Q) at Leibniz Universität Hannover.

References

[1] Vermeer M 1983 *Reports of the Finnish Geodetic Institute* **83**(2)
[2] Bjerhammar A 1985 *Bull. Geod.* **59**(3) 207-220
[3] Woodworth PL, Hughes CW, Bingham RJ, Gruber T 2012 *Journal of Geodetic Science* **2**(4) 302-318
[4] Gruber T, Gerlach C, Haagmans R 2012 *Journal of Geodetic Science* **2**(4) 270-280
[5] Ihde J, Sánchez L 2005 *J. Geodyn.* **40**(4) 400-413
[6] Lisdat C, Grosche G, Quintin N, Shi C, Raupach SMF, Grebing C, Nicolodi D, Stefani F, Al-Masoudi A Dörscher S and others 2015 *arXiv preprint* arXiv:1511.07735
[7] Altamimi Z, Collilieux X, Métivier L 2011 *Journal of Geodesy* **85**(8) 457-473

Journal of Physics: Conference Series **723** (2016) 012052 doi:10.1088/1742-6596/723/1/012052

Frequency metrology using highly charged ions

J R Crespo López-Urrutia

Max-Planck-Institut für Kernphysik, Saupfercheckweg 1, 69117 Heidelberg, Germany

crespojr@mpi-hd.mpg.de

Abstract. Due to the scaling laws of relativistic fine structure splitting, many forbidden optical transitions appear within the ground state configurations of highly charged ions (HCI). In some hydrogen-like ions, even the hyperfine splitting of the *1s* ground state gives rise to optical transitions. Given the very low polarizability of HCI, such laser-accessible transitions are extremely impervious to external perturbations and systematics that limit optical clock performance and arise from AC and DC Stark effects, such as black-body radiation and light shifts. Moreover, AC and DC Zeeman splitting are symmetric due to the much larger relativistic spin-orbit coupling and corresponding fine-structure splitting. Appropriate choice of states or magnetic sub-states with suitable total angular momentum and magnetic quantum numbers can lead to a cancellation of residual quadrupolar shifts. All these properties are very advantageous for the proposed use of HCI forbidden lines as optical frequency standards. Extremely magnified relativistic, quantum electrodynamic, and nuclear size contributions to the binding energies of the optically active electrons make HCI ideal toolsfor fundamental research, as in proposed studies of a possible time variation of the fine structure constant. Beyond this, HCI that cannot be photoionized by vacuum-ultraviolet photons could also provide frequency standards for future lasers operating in that range.

1. Introduction

Most of the baryonic matter contained in the universe is virialized within large gravitational potential wells of galaxy clusters that shape the intergalactic medium, and therefore appears in ionization states given by the high temperature of that medium (10^5 K and above) and the atomic number of the corresponding element [1,2]. Within galaxies, the fraction of ionized hot gas is also the largest one; stars basically contain only highly ionized matter. Reionization of the universe is mostly driven by the energy released by stars, quasi-stellar objects, and active galactic nuclei. Therefore, most elements appear naturally as highly charged ions (HCI) with an unbalanced positive charge. In spite of their importance, for both experimental and theoretical reasons our knowledge of HCI is still rather limited.

Crucially, the electronic structures of HCI, exception made of a few isoelectronic sequences, are not accurately known. Only hydrogenlike and lithiumlike ions have been subject to many studies in the x-ray and vacuum ultraviolet (VUV) domains for their theoretical "simplicity"; also the closed-shell isoelectronic sequences, e. g., heliumlike and neonlike, have been investigated due to their strong abundance in hot plasmas. Besides of the difficulties in production and handling of HCI, which have certainly constrained their study, the main interest of the HCI community has rested for the last three decades, on one hand, in the investigation of quantum electrodynamic properties, relativistic electronic structure and on the other hand in the astrophysically abundant ion species. An introductory review on HCI is given in Ref. [3].

As the key spectroscopic tools available for their study were wavelength-dispersive crystal or grating spectrometers and energy-resolving solid-state detectors, the achievable resolution was very limited in comparison with the much more advanced optical and laser spectroscopy of neutrals and singly charged ions. Due to this, the current status of HCI spectroscopy, from the x ray to the optical range, includes only a few transitions for which the wavelengths are known at the ppm accuracy level, and very few with a just sub-ppm-accuracy. An overview of this field is given in Ref. [4].

1.1. Optical transitions in HCI: Experimental status

Arguably, although the forbidden optical transitions of HCI were, unbeknownst to the astronomers of the 19th century observing solar spectra (cf. Ref. [5]), the first forbidden lines seen in spectroscopy, there is only a small set of HCI optical lines in the spectral atlases. Astrophysical observations have yielded a number of HCI coronal lines for a few abundant elements, such as Fe and Ar. Laboratory work has focused on those and a few other examples, such as the titaniumlike [4-6] and boronlike isoelectronic sequences [4, 8-13]. As for the hyperfine transitions in the ground state of hydrogenlike ions [14,15], they have been observed for $^{209}Bi^{82+}$[16], $^{211}Pb^{81+}$ [17] ions using an ion storage ring, and for $^{165}Ho^{66+}$ [18], $^{185,187}Re^{74+}$[19], and $^{203,205}Tl^{80+}$ [20] ions with an electron beam ion trap (EBIT) [21,22]. In both setups, the achieved resolution was in the order of 100 ppm. An overview on the different experiments performed with trapped HCI is given in [4]. HCI are mostly produced and studied in the laboratory with EBITs. In those devices, a monoenergetic electron beam produces, traps and excites HCI in any desired charge state of interest up to bare U^{92+} under well-defined conditions [21,22]. At Max-Planck-Institut für Kernphysik (MPIK) we operate a facility with several EBITs dedicated to fundamental physics research and some astrophysical and plasma applications.

As of today, there is still a very large range of totally unexplored HCI. Many isoelectronic sequences with semi-filled electronic subshells never have attracted any strong interest for the reasons mentioned above. In particular, the plethora of forbidden transitions that appear in their ground states has only recently began to be investigated in the detail needed as a prerequisite for laser excitation studies [23,24]. In view of this data scarcity, in the first place systematic studies of the emission spectra of the isoelectronic sequences of interest under electron-impact excitation conditions are needed. For such studies EBITs are very useful: they provide selectable charge states and easily allow the observation of M1 (magnetic dipole allowed) transitions with decay rates in the range of only 30-500 Hz by using broad-band grating spectrometers and cooled charged-coupled cameras with exposure times of less than one hour. However, while finding new lines in the spectrum of a certain charge state is comparatively easy, assigning them to certain transitions is, in most cases, a far more difficult task. Unfortunately, theoretical calculations are very complex and do not yield the accuracy level which would be desirable for reliable line identification. For these reasons, it becomes necessary to obtain additional information by recording spectra along an isoelectronic sequence. Thereby, smooth changes of the line positions and intensities can be observed and provide clues for the identification process. By carrying out such systematic studies, reliable identifications can be gained. The experimental data obtained with an EBIT on, e. g., M1 and very weak E1 lines at level crossings allows for the determination of the level structure and enables the search for transitions with extremely low decay rates. An example is our recent analysis of the optical spectrum of the Ir^{17+} ion [23], a prime candidate for the search of a time variation of the fine structure constant.

1.2. Production of cold HCI and future applications

Albeit HCI production and study experienced a tremendous bonanza with the introduction of EBITs, a key limitation of those devices (and in general of all sources of HCI) is the high translational temperature of the ions trapped in the deep confining potential generated by the negative space charge of the electron beam itself. Values in the range of MK, or hundreds of eV, with heating resulting from the scattering and ionization interactions with electrons, are typical. Since this was an improvement in comparison with plasma sources, those high values did not much hinder x-ray research hitherto. However, up to now they constitute an unsurmountable problem for high-resolution laser spectroscopy.

Recently, the sympathetic cooling of highly charged ions (HCI) [25,26] in a linear RF trap has been demonstrated in our group by L. Schmöger in collaboration with Physikalisch-Technische Bundesanstalt (PTB), thus, opening this broad class of spectroscopic targets to high precision frequency metrology. For this purpose, a cryogenic RF trap, CryPTEx [27-30] had been developed at MPIK in collaboration with Aarhus University. Upgraded devices are now under construction at MPIK in collaboration with the group of P. O. Schmidt for their future use at PTB in the frequency metrology division. With the rapid development of frequency combs in the VUV region and possibly beyond, further applications of highly charged ions as frequency standards at higher photon energies – the typical realm of HCI spectroscopy – are now under consideration, and experiments are being prepared in our group for exploring those.

2. Laser spectroscopy with HCI

While the first photon excitation experiments on HCI were carried out with E1-dipole allowed transitions at 48 eV energies [31], and subsequently at even higher values, first attempts to excite forbidden transitions with lasers within an EBIT showed that a direct detection of the fluorescence from long-lived, laser-excited metastable states decaying through M1-dipole transitions with lifetimes in the order of several milliseconds was possible. Therefore, for two of the forbidden lines that had already been studied with EBITs under electron beam excitation conditions, laser spectroscopic investigations within the EBIT were also performed by utilizing the so-called magnetic trapping mode [32]. This method consists in turning off the trapping electron beam of the EBIT and using its strong magnetic field and the trapping potentials applied to the drift tubes of such a device to keep the ions confined. This mode of operation is equivalent to that of a Penning trap, and has been used to measure the lifetimes of metastable levels with lifetimes ranging from the microsecond to the millisecond range [4,11,12]. The extremely low residual pressure in the cryogenic vacuum of an EBIT (10^{-13} mbar typically) reduces HCI losses due to charge exchange to negligible levels. Furthermore, ion-ion collisions occur at very far distances in the atomic scale, with the result that quenching of metastable levels through collisions is suppressed. The combination of strong confinement and negligible losses allows for the observation of slow radiative decay channels for seconds and minutes, and delivers very accurate lifetime determinations.

2.1. Laser excitation of forbidden transitions

Examples of HCI that have been studied with laser excitation are the boronlike Ar^{13+}ion [33] and the aluminumlike Fe^{13+} ion (Fe XIV) [34], this latter being the source of the long-unexplained *"green coronal line"* of solar eclipses [5]. For those experiments, a repetitive measurement cycle lasting typically 1 s was used. First, the ions of interest were produced with the electron beam; then, this was turned off and it was waited during a few lifetimes for the decay of the large metastable population populated by the electron beam excitation. A pulsed (8 ns, 100 Hz, 30 mJ, 0.04 cm^{-1} bandwidth) tuneable dye laser then illuminated the ions while a fast shutter closed for ~1 ms the aperture leading to a photomultiplier equipped with a light guide and a filter for the wavelength range of fluorescence. This proved crucial to reduce stray light to acceptable levels. At resonance, and after application of evaporative cooling to the trapped HCI, the laser interaction with a typical ensemble of 10 million trapped ions at a temperature of 20 eV (200 000 K) would generate fluorescence rates of tens per second, well above the dark and stray light count rate. Doppler broadening under these conditions is extremely large, at the level of several GHz. Even with a careful calibration, the resulting experimental wavelength uncertainties are still on the order of hundreds of MHz. Obviously, while needed as a preliminary step, this method is not useful for clock applications. Similar limitations applied elsewhere, and therefore, since more than two decades methods for cooling trapped HCI from the MK level to the mK level have been called for and tried in various settings [35-38].

3. Sympathetic cooling of HCI: A new tool for laser spectroscopy

For direct laser cooling, it is necessary to have fast cycling optical transitions requiring levels of opposite parity and strong overlap near the ground state. In HCI, the next suitable levels are energetically very far from the ground state configuration. Accessing them with optical lasers is, thus, impossible, and cooling has to rely, e. g., on the indirect method of sympathetic cooling. Coulomb interactions between HCI and singly charged ions, which can be directly laser cooled, such as trapped Be^+ remove thermal energy from co-trapped HCI. The choice of Be^+ is motivated by the favourable charge-to-mass ratio of 1/9, sufficiently close to that of typical HCI (~1/20 to ~1/3) to allow for simultaneous trapping in a RF linear trap.

Figure 1. Setup of CryPTEx. The central trap region electrodes are shown in the zoom at bottom left, and a 3D rendering of the trap core on the top right. HCI injection takes place from the top left through the narrow tubes mounted on cryogenic heat shields.

3.1. A cryogenic linear RF trap for sympathetic cooling: CryPTEx

CryPTEx (Cryogenic Paul Trap Experiment) was designed to trap HCI [27,28], and for this purpose the foremost issue is guaranteeing an excellent vacuum. Superior ultra-high vacuum performance of existing cryogenic setups in our group has led us to the choice of such a system. The cryogenic environment shown in Fig. 1 has many advantages, not only for the storage and handling of HCI, but also for the trapping and rovibrational cooling of molecular ions; these advantages were very beneficial for the commissioning experiments carried out with CryPTEx at Aarhus University in collaboration with the Drewsen group [29,30]. The device offers 16 ports with optical access to the central RF trap region with the electrodes operating close to 4 K. Cooling laser beams, a photoionization laser, several excitation lasers, the imaging setup, and injection of well-collimated atomic beams and also of buffer gases are needed to perform these types of experiments.

Beryllium atoms are injected in the form of a strongly collimated atomic beam with a cross section of roughly 1 mm^2. The atomic beam is generated in an oven located in a separated chamber with one intermediate stage of differential pumping. This avoids deposition of beryllium on RF electrode surfaces, and minimizes contamination of the trap chamber. Under normal operation we load an amount of only ~10 mg of pure metal in the oven every year. Most of the evaporated material is collected in a specifically designed, removable housing surrounding the oven, which also includes the first collimating aperture.

A photoionization laser operating at 235 nm crosses the atomic beam at the trap centre and generates the required Be^+ ions. The ion cooling utilizes a 313 nm laser system based on two telecom fibre lasers with subsequent sum-frequency generation and frequency doubling that delivers up to 700 mW at the cooling wavelength; for most experiments only 2 mW are needed. The laser frequency is stabi-

Journal of Physics: Conference Series **723** (2016) 012052 doi:10.1088/1742-6596/723/1/012052

lized with a wavemeter and its intensity with a control system based on a slow regulation with polarizers and a fast regulation with acusto-optic modulators. The laboratory has been temperature stabilized with a home-made control unit which controls the air condition system, reducing the temperature fluctuation to a level below 0.1 K.

For refrigeration, CryPTEx employs a 1 W cryocooler, which brings the temperature of the trap elements and the inner cryogenic shield down to ~4 K, and of the outer cryogenic shield to ~30 K. Turbo-molecular pumps are used in addition to provide a base vacuum in the 10^{-9} mbar range. The laser and ion beams enter the differentially refrigerated parts of the system through narrow tunnels for reduction of both thermal radiation from the room temperature vacuum chamber and ballistic residual-gas inflow. Under cryogenic operation, the pressure at the trap centre was estimated to be $\sim 10^{-15}$ mbar by measuring the lifetime of trapped ions (in excess of 1 day) and the heating and decay rates (at ~mHz level) of rotational excitations in molecular ions. In this way, an "effective" black-body radiation temperature of ~7.6 K was determined. For MgH^+ ions, the rovibrational ground state could be the most populated one. This could also be achieved with a novel method in which rovibrational cooling was accelerated by very rare collisions (a few per second) with a tenuous He buffer gas of only $\sim 10^{10}/cm^3$, that is, four to five orders of magnitude lower density than in typical buffer-gas cooling settings, which did not melt the ion crystal. Rovibrational temperature control through conversion of micromotion in heating rate via this rare collision was also achieved [29,30].

The RF trap operates at 4 MHz and has 24 electrodes of 8 mm diameter, of which 12 are used to confine the crystals. Typical trap frequencies are 400 kHz (radial) and 150 kHz (axial) for the Be^+ ion. The ion crystal is imaged with a lens or a microscope objective onto an intensified CCD. Two photomultipliers with appropriate dichroic mirrors and interference filters are used in parallel to register the Be^+ fluorescence rate and the Ar^{13+} fluorescence. A movable fibre mounted on a manipulator is used as a fiducial mark for the alignment of the lasers and the imaging system.

3.2. External ion source: Electron beam ion trap

For the purpose of HCI production, HYPER-EBIT was used as a source of ions for the present experiment. This device, originally designed for high-intensity operation, became available for the present experiment. In the particular example of Ar^{13+}, an electron beam of ~35 mA at an electron-beam energy of 700 eV is used. Under these conditions, the trapped ions typically reach within the trap a translational temperature of 1 MK. The HCI are dumped from the EBIT with a 1 Hz repetition rate with a total kinetic energy of ~6 keV. The charge states of the ions which are transported to the trap through a ~2.5 m long beamline are selected by controlling their time of flight. The HCI are decelerated prior of entering the CryPTEx chamber by application of high voltages pulses during their passage through specially designed electrodes next to it. After overcoming the bias potential of the RF trap setup, the mass-selected HCI bunches enter the RF trap. An electrode is then biased to close the trap axially. The HCI execute an oscillatory motion along the trap axis in its full length of ~100 mm. During this, they interact several times with a continuously cooled Be^+ ion crystal, thereby loosing kinetic energy until they are finally trapped in crystal sites along the RF node, i. e., the trap axis.

3.3. Diagnostics of HCI cooling process

Images captured in exposures of ~1 s duration by the intensified CCD located below the main chamber show the stopping of HCI and their implantation into the pre-existing Coulomb crystal by the appearance of dark, spherical hollows in the crystal axis that are caused by the Coulomb repulsion of the HCI and the Be+ ions [25,26], as shown in Fig. 2. Ion-ion separations of ~25 μm are typical for Be^+, and about twice as large around an Ar^{13+} ion. We have experimented with different crystal and ion chain configurations, including the one most suitable for application of quantum logic spectroscopy [39] in optical clocks [40], namely one cooling Be^+ ion and one single HCI. For the different setups we have determined the key trap frequencies and carried out different micromotion studies.

The HCI tend to occupy very stable sites in the crystal and show little sensitivity to parameters of the laser cooling which, on the contrary, clearly affect the position and ordering of the surrounding

Journal of Physics: Conference Series **723** (2016) 012052 doi:10.1088/1742-6596/723/1/012052

Be^+ ions. Due to the fact that the RF pseudopotential acts more strongly on the HCI, and to the shielding of electrostatic offset fields, one can presume that their micromotion is even smaller than that of the cooling ions around them. However, this assumption was not used in our preliminary measurements of a conservative upper bound for the HCI temperature of ~200 mK. In summary, the present setup in use is capable of transferring HCI at 1 MK from an electromagnetic trap, HYPER-EBIT to another one, CryPTEx, and refrigerate them down to the level of 200 mK and, presumably, even below that value [25,26].

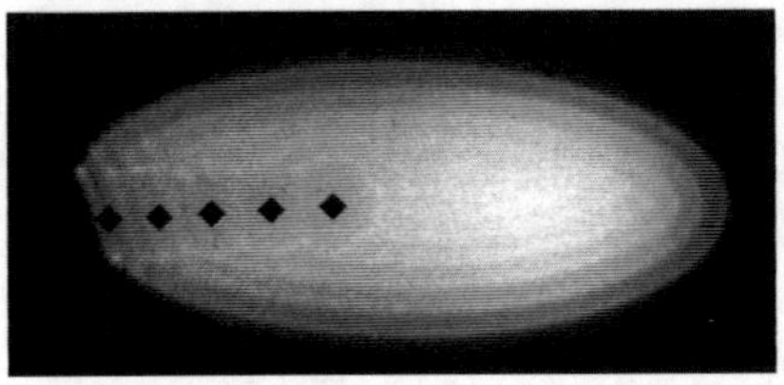

Figure 2. Image of a Coulomb crystal of Be^+ ions with five Ar^{13+} ions (marked with diamonds) implanted in it from the left. The length of the crystal is ~1 mm. A misalignment of the trap axis with respect to the crystal axis due to unbalanced radiation pressure is apparent.

4. Laser spectroscopy of cooled HCI and optical clocks

As a next step, the trapped Ar^{13+} HCI will be excited with a narrow-band CW laser at the resonance wavelength of the M1 ground-configuration transition $1s^2\ 2s^2\ 2p\ ^2P_{3/2} \rightarrow {}^2P_{1/2}$. This transition has been studied in our group before and is the best characterized M1 transition in any HCI [8-10]. However, the wavelengths obtained in earlier studies still have large uncertainties, which will make a long search necessary. The key issue at this moment is the long lifetime of the excited state (9.7 ms [12]), the small number of trapped ions and the total fluorescence detection efficiency of the system (0.4%). Scanning through the spectral range of interest will require long measurement periods. However, once the exact wavelength is found and calibrated by comparison with a GPS-stabilized commercial frequency comb and a Rb vapour cell, we expect that experiments with the Ar^{13+} ion and its isotopic varieties will proceed much faster.

4.1. Studies of the time variation of the fine-structure constant α

Similar experiments with Ir^{17+} will then follow in order to precisely measure all M1 transitions and other forbidden lines for the determination of the sought-after fine-structure constant-sensitive transitions. Furthermore, other HCI species which have been proposed [41-48] will be investigated. Plans for studies of upper bounds for the Lorentz-invariance in an analogous manner to those of Pruttivarasin et al. [49] but with appropriately chosen HCI, as proposed by Safronova, are also aimed at.

4.2. Clock transitions and other future possibilities

Future work aims at implementation of quantum-logic detection of the clock transition for frequency metrology [39]. This method should provide a far more efficient way of interrogation for the very slow transitions of interest that have been proposed in several recent theoretical works [41-48]. If the present combination of methods is successful, we expect that HCI will provide ultra-stable frequency standards and probes for interesting open questions in physics. Beyond that, the study of the intrinsic properties of their electronic structure, which by virtue of scaling laws can be compared to a virtual microscope for relativistic, quantum electrodynamic, and nuclear size effects will certainly also open new possibilities for atomic physics research.

4.3. Next upgrade: CryPTEx-II

Currently, CryPTEx-II, an upgraded version of the current cryogenic Paul trap with improved stray field and vibration suppression is being setup in collaboration between MPIK and PTB. This platform will be used for the implementation of the quantum logic scheme. The setup will include a miniaturized EBIT, which operates at room temperature and utilizes a magnetic circuit with permanent magnets generating a field of 0.85 T. This much-simplified version should suffice to produce the ions, which are currently the aim of this research. A prototype in operation at MPIK has already generated beams of, e. g., Xe^{40+} ions. The PTB device is currently being assembled at MPIK. A deceleration beamline similar to that of CryPTEx is also in the design stage at MPIK and will become part of CryPTEx-II at PTB. Commissioning of the complete setup is expected to take place before the end of the year 2016. A twin setup will also be built at MPIK to support further HCI experiments on its site.

5. Perspectives: Vacuum-ultraviolet range and x-ray domain frequency standards

One of the most important advantages of HCI is their stability at high temperatures and also in the presence of high energy photons. Under those conditions, atoms are obviously not stable, and even their inner-shell transitions become broadened by the coupling to the continuum given by the Auger decay. Not so with HCI, which, depending on their charge state, can in principle survive in photon beams of up to 140 keV, in the extreme case of the U^{91+} ion. While E1 x-ray transitions become very broad due to their short (fs range) lifetimes, there are also various electronic configurations that have metastable states of remarkably long lifetimes.

A good example for a possible VUV clock transition is found in the heliumlike O^{6+} ion, with its lowest excited state $1s2s\ ^3S_1$ decaying to the ground state via a M1 transition at 560 eV (2.2 nm) with a lifetime of ~1 ms [50]. With a transition frequency of ~140 PHz and this lifetime, this atomic resonator possesses a quality factor Q value of approximately 10^{14}. For the M3 decay of the lowest excited state of nickellike $^{129,132}Xe^{26+}$ ions, a transition energy of 1450 eV and a lifetime of ~15 ms were measured with an x-ray microcalorimeter [51]; this yields for $Q \approx 5\times10^{15}$. With even higher-order forbidden transitions, the corresponding Q values can be even higher, and also higher photon energies can be reached.

In principle, sympathetically cooled HCI will be useful as frequency standards at high photon energies for which until now only crystallographic references are available. Those have a several orders of magnitude lower intrinsic accuracy and stability than electronic transitions in HCI. In principle, HCI could support x ray frequency measurements with Hz accuracies, if appropriate x-ray sources based on high-harmonic generation would eventually be introduced. Present work at MPIK includes the development of a high-harmonic frequency comb for the VUV range, which should allow exploring the range of electronic transitions beyond the optical, with a hopefully promising scientific harvest.

Acknowledgments

Work presented in this talk has been carried out at MPIK in collaboration with the team of P. O. Schmidt at PTB, whose current president, J. Ullrich, has always strongly encouraged this research. CryPTEx was developed in collaboration with M. Drewsen (Aarhus University). Authors of individual scientific contributions are acknowledged in the corresponding references. The author expresses, in particular, his strong gratitude to the technical staff at MPIK, which has made with their outstanding skills this research possible, as well as to the always helpful administrative staff of MPIK, and to our electronics workshop. CryPTEx was manufactured in its major part at the MPIK apprentice shop under the guidance of S. Flicker and F. Säubert, who command an excellent team of young technicians. Contributions from our expert central mechanical shop under the direction of T. Spranz and our division technicians are acknowledged. We acknowledge funding from the Max-Planck-Gesellschaft zur Förderung der Wissenschaften e. V. and the Physikalisch-Technische Bundesanstalt.

References

[1] Danforth C W and Shull M J 2008 *Astrophys. J.* **679** 194
[2] Nicastro F et al. 2008 *Science* **319** 55
[3] Gillaspy J D 2001 *J. Phys.* B **34** R93
[4] Crespo López-Urrutia J R and Harman Z 2014 Emission and Laser Spectroscopy of Trapped Highly Charged Ions in Electron Beam Ion Traps, W. Quint and M. Vogel (eds.), *Fundamental Physics in Particle Traps*, Springer Tracts in Modern Physics **256**, (Berlin Heidelberg Springer-Verlag) pp 315-373
[5] Crespo López-Urrutia J R et al. 2002 *Can. J. Phys.* **80** 1687

[6] Morgan C A et al. 1995 *Phys. Rev. Lett.* **74** 1716
[7] Porto J V et al. 2000 *Phys. Rev.* A **61** 545011
[8] Draganić I et al. 2003 *Phys. Rev. Lett.* **91** 183001
[9] Soria Orts R et al. 2006 *Phys. Rev. Lett.* **97** 103002
[10] Soria Orts R et al. 2007 *Phys. Rev.* A **76** 052501
[11] Brenner G. et al 2009 *Astrophys. J.* **703** 68
[12] Lapierre A et al. 2005 *Phys. Rev. Lett.* **95** 183001
[13] Crespo López-Urrutia J R 2008 *Can. J. Phys.* **86** 111
[14] Shabaev V M 1993 *J. Phys.* B **26** 1103
[15] Shabaev V M 1994 *J. Phys.* B **27** 5825
[16] Klaft I et al. 1994 *Phys. Rev. Lett.* **73** 2425
[17] Seelig P et al., 1998 *Phys. Rev. Lett.* **81** 4824
[18] Crespo López-Urrutia J R et al. 1996 *Phys. Rev. Lett.* **77** 826
[19] Crespo López-Urrutia J R et al. 1998 *Phys. Rev.* A **58** 238
[20] Beiersdorfer P et al. 2001 *Phys. Rev.* A **64** 032506
[21] Levine M A et al. 1988 *Phys. Scr.* **T22** 157
[22] Levine M A et al. 1989 *Nucl. Instrum. Methods Phys. Res.* B **43** 431
[23] Windberger A et al. 2015 *Phys. Rev. Lett.* **114** 150801
[24] Bekker H et al. 2015 *AIP Conf. Proc.* **1640** 109
[25] Schmöger L et al. 2015 Science **347** 1233
[26] Schmöger L et al. 2016 *Rev. Sci. Instrum.* **86** 103111
[27] Schwarz M et al. 2012 *Rev. Sci. Instrum.* **83** 083115
[28] Schwarz M et al. 2012 *J. Phys.: Conf. Series* **388** 122005
[29] Versolato O O et al. 2013 *Phys. Rev. Lett.* **111** 053002
[30] Hansen A K et al. 2014 *Nature* **508** 76
[31] Epp S W et al. 2007 *Phys. Rev. Lett.* **98** 183001
[32] Beiersdorfer P et al. 1996 *Rev. Sci. Instrum.* **67** 3818
[33] Mäckel V et al. 2011 *Phys. Rev. Lett.* **107** 143002
[34] Schnorr K et al. 2013 *Astrophys. Journal* **776** 121
[35] Beyer H F et al. 1990 *HITRAP - a facility for experimentation with trapped highly-charged ions at GSI* (Internal report GSI--90-20) (Darmstadt, Germany: GSI)
[36] Poth H and Hasse R W 1991 *19th Int. Symp. on "Cooler Rings and Their Applications"; Tokyo, 1990; T. Katayama and A. Noda, Eds.* (Singapore: World Scientific Singapore) pp 108–123
[37] Quint W et al. 2001 *Hyperfine Interactions* **132** 457
[38] Gruber L et al. 2001 *Phys. Rev. Lett.* **86** 636
[39] Schmidt P O et al. 2005 *Science* **309** 749
[40] Ludlow A, Boyd M, Ye J, Peik E and Schmidt P O 2015 *Rev. Mod. Phys.* **87** 637
[41] Schiller S 2007 *Phys. Rev. Lett.* **98** 180801
[42] Berengut J C et al. 2011, *Phys. Rev. Lett.* **106** 210802
[43] Derevianko A, Dzuba V A and Flambaum V V 2012 *Phys. Rev. Lett.* **109** 180801
[44] Dzuba V A, Derevianko A and Flambaum V V 2012 *Phys. Rev.* A **86** 054502
[45] Yudin V I, Taichenachev A V and Derevianko A 2014 *Phys. Rev. Lett.* **113** 233003
[46] Safronova M S et al. 2014 *Phys. Rev. Lett.* **113** 030801
[47] Safronova M S et al. 2014 *Phys. Rev.* A. **90** 042513
[48] Safronova M S et al. 2014 *Phys. Rev.* A. **90** 052509
[49] Pruttivarasin T et al. 2015 *Nature* **517** 592
[50] Crespo López-Urrutia et al. 1998 *Phys. Rev.* A **58** 238
[51] Träbert E, Beiersdorfer P and Brown G V 2007 *Phys. Rev. Lett.* **98** 263001

Is the time right for a redefinition of the second by optical atomic clocks?

Patrick Gill

National Physical Laboratory (NPL), Teddington, Middlesex TW11 0LW, UK

E-mail: patrick.gill@npl.co.uk

Abstract. Given the dramatic rate of progress in optical atomic clocks over the last decade, this paper presents the current state of play, and considers the possibilities, implications and timescales for a potential redefinition of the SI second in terms of an optical reference transition. In particular, the question of choice of a future standard is addressed, together with the requirements to accurately compare realisations of such standards, both for clocks local to, and remote from each other. Current performances of various optical clock systems are examined and possibilities for moving beyond potential limitations by alternative strategies are outlined.

1. Introduction

The microwave caesium atomic clock has been the basis for realisation of the SI second since 1967. From that time, there have been various clock technology advances, most significantly the advent of laser cooling and its application to cold caesium fountain clocks, the best of which now achieve systematic frequency uncertainties of ~ 1-2 $\times 10^{-16}$ and report regularly to the BIPM. Fountain clocks provide the primary steering corrections for international atomic time (TAI) and universal co-ordinated time (UTC) [1, 2], with subsidiary sub-sets of commercial Cs clocks and hydrogen masers in national laboratories providing local dissemination of time and frequency for high technology and industrial applications. However, since the turn of the century, the pace of optical atomic clock R&D has quickened considerably, with result that several optical clock systems based on different atomic species now achieve lower systematic uncertainties than the best fountain clocks. Already fundamental science is benefitting in a number of cases from the lower uncertainties available from optical clocks, and the question of whether the time is right for a redefinition of the second becomes increasingly relevant. Whilst microwave fountain clocks will continue to underpin traceability for many microwave-based time and frequency applications in navigation, communications and synchronisation, limitations with these systems now arise due to the low microwave frequency and associated extended periods needed to reach 10^{-16} performance. Increasing development of optically-based platforms and architectures, however, can offer alternative and efficient routes and techniques for increased accuracy. Prime examples of this are recent demonstrations of femtosecond comb-based optical-to-microwave down-conversion to provide the microwave local oscillator for probing the Cs fountain [3-5].

2. State-of-the art optical clock performance

Those optical clock species now surpassing the Cs fountain capability include the $^{27}Al^+$, $^{199}Hg^+$, $^{171}Yb^+$ and $^{88}Sr^+$ trapped ion systems and ^{87}Sr, ^{199}Hg and ^{171}Yb atoms in optical lattices, with reported systematic uncertainties in the 10^{-16} to 10^{-18} range. Benchmark frequency instabilities for single cold ions confined within rf traps, and multiple atom systems (~ 10^4 - 10^5 atoms) within optical lattices, are

$3\text{x}10^{-15}\tau^{-1/2}$ and ~$3\text{x}10^{-16}\tau^{-1/2}$ respectively. Recent ^{87}Sr lattice clock results report a relative clock systematic frequency uncertainty of $2.1\text{x}10^{-18}$ [6], and statistical agreement between two cryogenic Sr clocks at $2\text{x}10^{-18}$ [7].

Atom / ion	Clock type	Clock ν THz	Clock λ nm	Lowest published clock systematic uncertainty	Uncertainty of CIPM ν value
^{87}Sr	Lattice	429	698	$2.1\text{ x}10^{-18}$ [6]	$5\text{ x}10^{-16}$
^{171}Yb$^+$	Ion octopole	642	467	$3.2\text{ x}10^{-18}$ [8]	$6\text{ x}10^{-16}$
^{27}Al$^+$	Ion, quantum logic	1121	267	$8.6\text{ x}10^{-18}$ [9]	$1.9\text{ x}10^{-15}$
^{88}Sr$^+$	Ion quadrupole	445	674	$1.2\text{ x}10^{-17}$ [10]	$1.6\text{ x}10^{-15}$
^{199}Hg$^+$	Ion quadrupole	1065	282	$1.9\text{ x}10^{-17}$ [11]	$1.9\text{ x}10^{-15}$
^{40}Ca$^+$	Ion quadrupole	411	729	$3.4\text{ x}10^{-17}$ [12]	$1.2\text{ x}10^{-14}$
^{199}Hg	Lattice	1129	266	$7.2\text{ x}10^{-17}$ [13]	$6\text{ x}10^{-16}$
^{171}Yb$^+$	Ion quadrupole	688	436	$1.1\text{ x}10^{-16}$ [14]	$6\text{ x}10^{-16}$
^{171}Yb	Lattice	518	578	$3.4\text{ x}10^{-16}$ [15]	$2\text{ x}10^{-15}$
^{1}H	Cryogenic beam	1233	243	$4.2\text{ x}10^{-15}$ [16]*	$9\text{ x}10^{-15}$

Table 1: Published systematic fractional frequency uncertainties of various optical clock species, and the fractional frequency uncertainties of the absolute frequency values adopted by the CIPM, following the 2015 Consultative Committee on Time & Frequency (CCTF); [16]*: the published uncertainty is that of the 2466 THz 1S-2S 2-photon transition, whereas the recommended CIPM value corresponds to the 1233 THz single photon frequency.

Table 1 shows the best published uncertainties (column 5) for various atom and ion clock systems. These data have resulted from the derivation of estimated uncertainty budgets for the different species under development at various national measurement and research institutes. This has been achieved by means of measured or theoretically-calculated sensitivities of the clock frequency to environmental perturbations (eg Stark shifts including black-body radiation and light shifts, Zeeman shifts, same-species density shifts and collisions with background gas, and for ions, electric quadrupole shifts) and cold atom/ion residual motional perturbations (eg 2nd order Doppler shifts). Most of these optical candidate-species for a redefinition have already been accepted (with the exception of ^{1}H, ^{199}Hg and ^{40}Ca$^+$) as secondary representations of the second by the CIPM with a stated fractional frequency uncertainty, whereby associated operational clock data (stability, reproducibility and uncertainty) from these systems can now be reported to BIPM on a regular but preliminary basis in order to ascertain longer term performance prior to full contribution to UTC determination. Following the 2015 meeting of the CCTF, the fractional uncertainties shown in column 6 are now adopted by the CIPM. These latter uncertainties are larger than those published in the literature on two counts. First, the optical frequency absolute value has to be related to the current definition and its uncertainty, and, second, the Frequency Standards Working group (WGFS) of the CCTF adopts a more prudent approach when determining universal frequency values from very small data sets such as a measurement from a single laboratory, or a few laboratories, where a normal distribution of results is not available. In such cases, typically an enlargement factor of a few is applied, dependent on circumstances. The basis for such treatment is discussed in a publication in preparation [17].

With the current set of cold atom and ion species that surpass Cs fountain performance, the question of a redefinition of the second in terms of an optical clock species becomes increasingly pertinent. Of course, the caesium microwave standard has proved to be an enduring one over its ~ half-century reign, demonstrating improvements at the rate of a decade per decade during this time, and will undoubtedly continue to play a significant role for microwave applications going forward. However, the averaging

Journal of Physics: Conference Series **723** (2016) 012053 doi:10.1088/1742-6596/723/1/012053

time needed to achieve its current 10^{-16} limiting uncertainty is unwieldy for many measurement applications, and with the perceived difficulties in further significantly reducing this uncertainty, these together clearly identify the higher frequencies of optical clock systems as the future way. The sub-set of species given in Table 1 is of course non-exclusive, and higher frequency options, such as the nuclear transition in $^{229}Th^{+}$, or VUV transitions accessed by high harmonic generation, may also come to the fore in future years. However, considering the current set of optical candidate species, there are some questions that need careful consideration in respect of a redefinition choice, and further ones in respect of when such a redefinition might occur. In the former case, should we decide on the candidate that demonstrates the lowest uncertainty, or alternatively choose the candidate that has collectively been studied most by the standards institutes? Right now, one observes that the candidate that fulfils both these criteria is the ^{87}Sr optical lattice clock, but this may not remain the position going forward, which brings us to redefinition timescales. The progress in uncertainty reduction for many of the candidates is still fast improving, and we need to increase our knowledge of some perturbation sensitivities and limiting uncertainties for improved clock systems and operating algorithms. Further, it would seem unwise to make a redefinition decision until the current fast rate of improvement slows. Finally, it becomes most important to stress-test individual clock performance by real-time comparison of different clocks operated local to and remote from each other. One alternative suggestion to the selection of a "best choice" candidate has been the suggestion of a definition comprising a "universal" value determined from the matrix of individual best values for different species. However, in this scenario, the definition would not be based on a fundamental constant or a "constant of nature" (as in the current definition), but on an aggregation of values for different systems, with little physical significance. Further, this value would change every time individual values were upgraded, and change significantly with the introduction of a new species into the value matrix.

3. Derivation of high accuracy frequency ratios

Given the increasing number of optical clock species with systematic uncertainties lower than that for the Cs fountain clock, absolute determinations of their optical frequency values are at best constrained to that of the Cs fountain uncertainty. A more profitable approach is to directly compare optical frequencies ratios of the optical systems themselves. Here, all frequency comparisons can be quantified in terms of a matrix of frequency ratios, be they optical ratios, optical-microwave ratios (of which optical clocks measured relative to Cs are the most common case) or microwave ratios, and the WGFS of the 2012 CCTF opted to adopt this as a viable way forward to take advantage from the lower optical uncertainties. However, there are some interesting consequences arising. With a full matrix of frequency ratios, absolute frequency values for particular clock species may be derived by various routes eg from a single ratio or combinations of ratios, and this in effect over-constrains the determination of an optimised value for the clock transition frequency. A procedure to deal with this has been proposed recently [18], and is also reported in this Proceedings and should lead to a more statistically comprehensive approach. Nevertheless, at this time, we remain at the intermediate position of combining several Cs-related determinations with one or two all-optical ratios in order to determine values for some clock frequencies, with the Cs uncertainty providing the limitation. An example of this is the absolute value of the $^{27}Al^{+}$ clock transition, which is derived from both a Cs-related measurement of the $^{199}Hg^{+}$ transition and a subsequent $^{27}Al^{+}/\ ^{199}Hg^{+}$ frequency ratio measurement. However, as the number of optical ratio determinations increase, one would expect that significance of the Cs-related values will be reduced within the matrix. It should be said it is still important to include the optical-to-caesium ratios in order to ensure that the matrix values remain consistent with the Cs-related values in order that a discontinuity larger than the Cs uncertainty does not occur at redefinition.

4. Optical clock comparisons

In preparation for a redefinition of the second, and given the range of high performance clocks of different species being researched in various labs internationally, it is highly desirable that individual clock systems are directly compared against each other in real time to provide independent validation, and also to help determine whether there is a "best choice" redefinition candidate. Several NMIs are pursuing a number of clocks with different species and of course femtosecond combs now readily allow such local intercomparisons between different species clocks in the same institution, where the clock uncertainties are not compromised due to the comparison process, with comparison accuracies limited by the uncertainties of the optical clocks themselves eg.[19]. However, point-to-point direct intercomparison of remote clocks in different national laboratories remains a more rigorous goal allowing comparison of clocks of different designs, thereby reducing common mode effects that might arise in the design of clocks built in a single institution. There are a number of different techniques evolving for remote optical clock comparisons, and which are described in more detail in this Proceedings. These include microwave frequency transfer by satellite, optical frequency transfer via ground-to satellite and satellite-to-satellite, portable optical clocks and optical frequency transfer by fibre using dark fibre or dark channels in internet-carrying fibre [20-22]. Microwave frequency transfer by satellite includes traditional 2-way satellite frequency transfer and GPS Precise Point Positioning (PPP) which are limited to comparison fractional accuracies of ~ 10^{-15} per day. A recent 2-way comparison using higher "chip" rates is discussed below, where comparison accuracies in the few x 10^{-16} are expected. With the launch of the ACES mission [23] in 2017, frequency comparison accuracies are expected to achieve 10^{-16} to 10^{-17} assuming lack of optical cycle slip over the 90-minute orbital period, which is increasingly demonstrated by high accuracy clock and comb combinations. On the optical side, free space optical transfer techniques between ground and satellite and satellite to satellite are still in their infancy, but proving experiments point towards accuracies of 10^{-16} per day. There is significant interest in building portable optical clocks for both ground and space operation. Here, there is generally a trade-off between highest clock accuracy and the compactness needed for portability, especially in space. On the ground, optical clock frequency variations due to clock height differences in the cm-range, due to gravitational red shifts, start to compromise clock evaluations. Finally, optical frequency transfer by fibre offers most promise, with frequency stability transfer at the 10^{-18} level in minutes already demonstrated over hundreds of km, dependent on link distance and the need to compensate for fibre phase fluctuations due to temperature and vibration. The main issues with fibre transfer is the point-to-point nature of the process, and the cost of accessing dark fibre or dedicated dark channels in fibre.

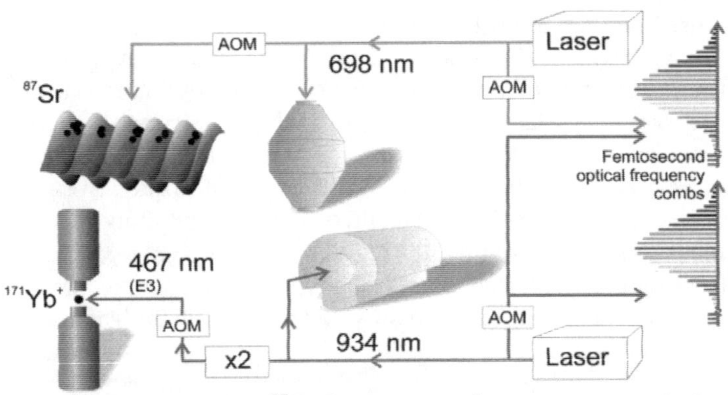

Figure 1: NPL $^{171}Yb^+$ octupole / ^{87}Sr frequency ratio measurement (colour on-line)

Over the past couple of years, international real-time remote optical frequency comparisons have started in earnest. During late 2014, a frequency comparison of the quadrupole transition in single $^{171}Yb^+$ ions clocks at PTB and NPL was performed using GPS PPP, with both clocks related to local hydrogen masers via femtosecond combs [24]. The level of agreement achieved was -1.3 (1.2) x10^{-15}. During summer 2015, a major month-long intercomparison activity between NPL, PTB, SYRTE and INRIM was carried out within the Euramet project on international timescales with optical clocks (ITOC). Several trapped ion and optical lattice clocks at the labs contributed data, with all clocks referenced by comb to H-masers and data relayed via a geostationary SES Astra 3B satellite [25]. During the campaign,

 doi:10.1088/1742-6596/723/1/012053

opportunity was also taken to undertake local comparisons such as the $^{171}Yb^+$ E3 – ^{87}Sr ratio determination at PTB and NPL (figure 1). The campaign itself was a proving ground for the demonstration of optical clock "up-times" which reached in excess of 80% over the month for some clocks, although more work is needed to fully automate clock and comb operation. A large amount of intercomparison data was gathered, with final results expected in spring 2016.

In parallel with the satellite comparisons, international fibre links have been established between PTB and SYRTE, and NPL and LPL / SYRTE. Comparison of the Sr lattice clocks at PTB and SYRTE via the PTB-SYRTE fibre link has recently been reported [26], demonstrating 5×10^{-17} agreement, and comparison of the NPL Sr and $^{171}Yb^+$ clocks with the SYRTE Sr clock is expected shortly, followed by PTB – NPL Sr and $^{171}Yb^+$ clock comparisons via the SYRTE / LPL fibre link. Figure 2 shows the recently-established NPL – SYRTE fibre link.

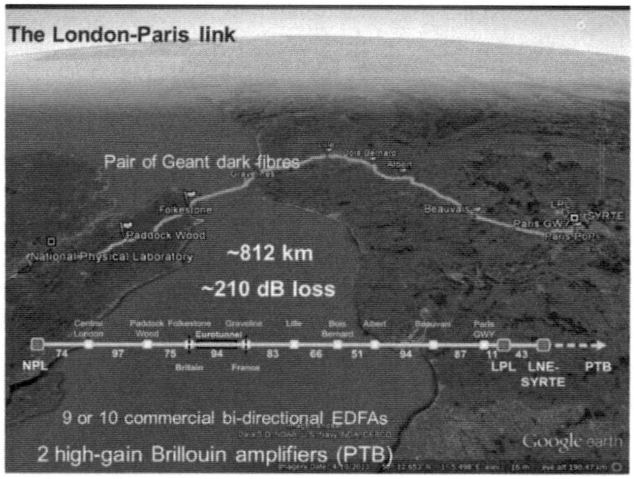

Figure 2: Optical fibre link between NPL and LPL / SYRTE (colour on-line)

5. Next generation local oscillators and clocks

Here I briefly consider some potential developments that may enable better clock stability and reduced uncertainty beyond the current state of the art. On the local oscillator (LO) front, environmental isolation enhancements have led to optical reference cavities with ULE spacers and ultra-high finesse multi-layer dielectric mirror coatings on fused silica substrates achieving thermal noise limited frequency stabilities ~ few x 10^{-16}. Over the past few years, thermal noise has been pushed even lower by two different strategies. These include the use of single crystal silicon cavities operating at 124 K and achieving sub-40 mHz linewidths at 1.5 μm [27], and the use of semiconductor crystalline mirror coatings [28], both of which point to $\leq 10^{-16}$ thermal noise floor. Additionally, we have seen the introduction of the universal synthesiser concept, whereby a master oscillators locked to a very well isolated long ULE cavity (~ 50 cm – 100 cm) is used to anchor a fibre comb, eg [29], from which ultra-stable light from relevant comb modes can be selected to probe clock transitions in different cold atom and ion species. However, cavity, mirror substrate and coating designs look to be approaching limiting thermal noise conditions.

There are other optical local oscillator concepts also under consideration. One of these is spectral hole burning (SHB), for example in the absorption spectrum of a Eu3+ doped Y_2SiO_5 crystal at 4.5 K. A series of spectral holes can be burnt with ~ 50 kHz separation and linewidths as low as 1 kHz with a very compact prototype cryogenic SHB module [30]. These holes are of course limited to particular absorption wavelengths in the crystal, and early research had to contend with the holes broadening and weakening over time, requiring a re-write process. However, a stable writing / probing process has now been established [31], with projected signal-to-noise limited stability reaching below 10^{-16} in 1 s.

Another interesting concept yet to be realised is that of a millihertz-linewidth super-radiant laser based on clock transition photons emitted from lattice-trapped cold atoms within a high-Q optical cavity

[32]. The idea is to operate within the "bad-cavity" limit, where the cavity decay rate κ is much greater than the atom spontaneous decay rate γ, giving rise to an atomic decay linewidth $\Delta\nu$ (atoms) much narrower than the laser cavity decay linewidth $\Delta\nu$ (laser). In this arrangement, energy is effectively stored within the atoms, rather than in the cavity, with result that the clock atoms should become insensitive to cavity length changes by a factor $\Delta\nu$(atoms) / $\Delta\nu$(laser). It is projected that super-radiant laser output powers of picowatts from ~ 10^6 atoms within the cavity should be enough for an atomic phase lock, whilst also avoiding significant cavity pulling effects.

On the cold atom clock front, there are also opportunities for improved clock stability. The first of these is already under study, and targets non-destructive phase measurement of the clock signal by retaining fractions of the cold atom sample through the Ramsey measurement process [33], rather than losing all the atoms during each measurement and having to cool, load and prepare a new sample for the next interrogation cycle. This should allow a reduction in interrogation dead time, thereby reducing the Dick effect. Alternative strategies include an intermediate stage of locking a cavity-stabilised local oscillator to a high signal-to-noise (SNR) cold atom sample in order to further discipline the LO before the high-accuracy probe of the clock atoms in the lattice, or extending this disciplining by a series of nested locks to a number of high SNR samples. One can also consider simultaneous probing of a number (m) of groups of N cold atoms, where one should expect the stability to reduce as a function of $(1 / N^{m/2})$ [34]. Finally, as entanglement techniques become more robust, it can be shown that simultaneous probing of N entangled atoms [35, 36] allows one to surpass the standard quantum limit, and approach the Heisenberg uncertainty limit whereby the instability will scale as (1/N) instead of the usual $(1/N^{1/2})$.

6. Conclusions

I have outlined the significant rate of progress currently occurring in the development of optical atomic clocks, and at this point there is no obvious sign of a slow-down in this rate of progress, with published fractional frequency uncertainty budgets ranging between 10^{-16} and 10^{-18}. However, it is work in progress, with more national measurement laboratories and other research institutes are engaging in the development of such clocks. As clock stability improves for the various clock species, so we expect to see estimation of systematic uncertainties of frequency shifts due to perturbing fields also further reduce. This, together with improving methodologies of real-time local and remote clock comparisons and the adoption and generation of a clock frequency ratio matrix with increased data input, certainly brings us closer to the point where a redefinition of the second looks viable, but this is still at least some years away. One must also recognise that field perturbations will increasingly impinge upon a reducing uncertainty budget. As discussed, the earth's gravitational field at "ground level" will give rise to frequency shifts of order 10^{-18} per cm height difference between clocks, and it's variation both in terms of time and space are beginning to complicate matters. Experiments on Al^+ ion clocks [37] were able to show a gravitational red shift in the laboratory when one clock was raised some 30 cm. As we move to the 10^{-18} domain, are we able to discriminate shifts at this level as due to clock frequency drift, or are they local gravity variations? One solution is the idea of one or more master clocks in space orbit where gravitational shift is reduced. There is space agency interest in this possibility, but it will be ~ a decade before robust space-qualified optical clocks are available. Notwithstanding gravity, environmental control of the clocks to achieve frequency shift uncertainties below 10^{-19} due to the various electric, magnetic and collisional effects will get difficult and need both passive and active control strategies.

So is there a best clock in respect of the choice of atom or ion species? Quite probably the limiting uncertainties for different species will all be within an order of magnitude. Should we go for the best performing clock, or select the species that has seen the most research, or is easier to realise? Does it matter that some national laboratories do not operate a chosen "best" clock species? With local femtosecond comb comparisons and remote optical comparison by fibre, currently we have the high accuracy clock comparison capability that surpasses the uncertainties of the clocks themselves, so the traceability routes between a "primary" clock and other "secondary" clocks would not increase the inaccuracy of the latter.

Finally, the Cs fountain microwave clock will continue to provide traceability at appropriate levels for applications in the microwave domain for near future. However, one can point both to high science applications that can take advantage from lower optical clock uncertainties, and to emerging optical platform architectures that would benefit from a more direct linkage with optical clock capability. This can be seen from the industrial interest in optically-based systems as demonstrated by agencies such as ESA, NASA, DARPA, DSTL and the establishment of national and EU quantum technology initiatives to translate quantum research into quantum-based industries.

Acknowledgments
The author acknowledges discussions with F. Riehle, H.S. Margolis and other members of the CCL-CCTF WGFS, and thanks the NPL Time & Frequency group for their research toward a future redefinition of the second. This NPL activity has been funded by the UK National Measurement System and the European Metrology Research Programme of EURAMET participating countries and the EU.

References
[1] Petit G and Panfilo G 2013 *IEEE Trans. Instrum. Meas.* **62** 1550
[2] Petit G, Arias F and Panfilo G 2015 *Comptes Rendus Physique* **16** 480-488
[3] Lipphardt B et. al. 2009 *IEEE Trans. Instrum. Meas.* **58** 1258-62
[4] J. Millo et al. 2009 *App. Phys. Letters* **94** 141105
[5] Szymaniec K, Lea S N, Gibble K, Park S E, Liu K and Glowacki P 2016 *This Proceedings*
[6] Nicholson T L, Campbell S L, Hutson R B, Marti G E, Bloom B J, McNally R L, Zhang W, Barrett M D, Safronova M S, Strouse G F, Tew W L and Ye J, 2015 *Nature Comms.* **6** 6896
[7] Ushijima I, Takamoto M, Das M, Ohkubo T and Katori H 2015 *Nature Photonics* **9** 185 (2015)
[8] Huntemann N, Sanner C, Lipphardt B, Tamm Chr and Peik E 2016 *Phys Rev. Letters* **116** 063001
[9] Chou C W, Hume D B, Koelemeij J C J, Wineland D J and Rosenband T 2010 *Phys. Rev. Letters* **104** 070802
[10] Dubé P, Madej A A, Tibbo M, and Bernard J E 2014 *Phys. Rev. Letters* **112** 173002 (2014)
[11] Rosenband T et al. 2008 *Science* **319** 1808
[12] Huang Y et al. 2016 *Phys. Rev. Letters* **116** 013001
[13] Yamanaka K, Ohmae N, Ushijima I, Takamoto M and Katori H 2015 *Phys. Rev. Lett.* **114** 230801
[14] Tamm C et al. *Phys. Rev. A* 2014 **89** 023820
[15] Lemke N D et. al 2009 *Phys. Rev. Letters* **103** 063001
[16] Parthey C G et al. 2011 *Phys. Rev. Letters* **107** 230001
[17] Riehle F and P. Gill P 2016 *Manuscript in preparation*
[18] Margolis H S and Gill P 2015 *Metrologia* **52** 628-634
[19] Johnson L A M, Gill P and Margolis H S 2015 *Metrologia* **52** 62–71
[20] Newbury N R, Williams P A and Swann W C 2007 *Optics Letters* **32** 3056
[21] Lopez O et. al. 2010 *Optics Express* **18** 16849
[22] Predehl K et al. 2012 *Science* **336**, 441
[23] Laurent P et al. 2015 *Comptes Rendus Physique* **16** 540-552
[24] Leute J et. al 2016 *IEEE-Trans UFFC* accepted
[25] Gibney E 2015 *Nature News* **522** 16
[26] C. Lisdat et al.2015 *arXiv* 1511.07735
[27] Kessler T, Hagemann C, Grebing C, Legero T, Sterr U, Riehle F, Martin M J, Chen L and J. Ye J 2012 *Nature Photonics* **6** 687
[28] Cole G D, Zhang W, Martin M J, Ye J, and Aspelmeyer M 2013 *Nature Photonics* **7** 644-650
[29] Nakajima Y et al. 2010 *Optics Express* **18** 1667
[30] Thorpe M J, Rippe L, Fortier T M, Kirchner M S and Rosenband T 2011 *Nature Photonics* **5** 688
[31] Cook S, Rosenband T and Leibrandt D R 2015 *Phys. Rev. Letters* **114** 253902
[32] Meiser D, Ye J, Carlson D R and Holland M J 2009 *Phys. Rev. Letters* **102** 163601
[33] Lodewyck J, Westergaard P G and Lemonde P 2009 *Phys. Rev. A* **79** 061401
[34] Borregaard J and Sorensen A S 2013 *Phys. Rev. Letters* **111** 090802
[35] Leibfried D et al. 2004 *Science* **304** 1476
[36] Hosten O et al. 2016 *Nature* **529** 505-8
[37] Chou C W, Hume D B, Rosenband T, Wineland D J 2010 *Science* **329**, 1630

Journal of Physics: Conference Series **723** (2016) 012054 doi:10.1088/1742-6596/723/1/012054

Entanglement with negative Wigner function of three thousand atoms heralded by one photon

Robert McConnell[1], Hao Zhang[1], Jiazhong Hu[1], Senka Ćuk[1,2] and Vladan Vuletić[1]

[1] Department of Physics, MIT-Harvard Center for Ultracold Atoms, and Research Laboratory of Electronics, Massachusetts Institute of Technology, Cambridge, Massachusetts 02139, USA
[2] Institute of Physics, University of Belgrade, Pregrevica 118, 11080 Belgrade, Serbia

E-mail: vuletic@mit.edu

Abstract. Quantum-mechanically correlated (entangled) states of many particles are of interest in quantum information, quantum computing and quantum metrology. Metrologically useful entangled states of large atomic ensembles have been experimentally realized [1, 2, 3, 4, 5, 6, 7, 8, 9, 10], but these states display Gaussian spin distribution functions with a non-negative Wigner function. Non-Gaussian entangled states have been produced in small ensembles of ions [11, 12], and very recently in large atomic ensembles [13, 14, 15]. Here, we generate entanglement in a large atomic ensemble via the interaction with a very weak laser pulse; remarkably, the detection of a single photon prepares several thousand atoms in an entangled state. We reconstruct a negative-valued Wigner function, an important hallmark of nonclassicality, and verify an entanglement depth (minimum number of mutually entangled atoms) of 2910 ± 190 out of 3100 atoms. Attaining such a negative Wigner function and the mutual entanglement of virtually all atoms is unprecedented for an ensemble containing more than a few particles. While the achieved purity of the state is slightly below the threshold for entanglement-induced metrological gain, further technical improvement should allow the generation of states that surpass this threshold, and of more complex Schrödinger cat states for quantum metrology and information processing.

1. Introduction

Many state-of-the-art atomic microwave clocks and interferometers based on atomic ensembles are limited no longer by technical noise, but rather by the fundamental quantum fluctuations of uncorrelated atoms, known as the standard quantum limit (SQL). Entangled states can improve measurement sensitivity beyond this limit using controlled correlations between atoms and represent important resources in quantum metrology. To date, metrologically useful spin-squeezed states [1, 2, 3, 4, 5, 6, 7, 8, 9, 10] have been produced in large ensembles. However, those moderately spin-squeezed states have Gaussian spin distributions and therefore can be viewed as semi-classical. Entanglement enters only to set a reduced amount of spin noise compared to the SQL. Non-Gaussian spin states with a negative Wigner function are however manifestly non-classical, since the Wigner function as a quasiprobability function must remain non-negative in the classical realm. They can be used for realizing quantum sensors which surpass the SQL since the non-Gaussian spin distributions can possess large Fisher information.

Here we generate entanglement in a large atomic ensemble by detecting a single photon that has interacted with the ensemble [16]. An incident vertically polarized photon experiences a weak random polarization rotation associated with the quantum noise of the collective atomic spin. The detection of a horizontally polarized emerging photon then heralds a non-Gaussian entangled state of collective atomic spin (Figure 1) with a negative-valued Wigner function of -0.36 ± 0.08, and an entanglement depth of 90% of our ensemble containing several thousand atoms.

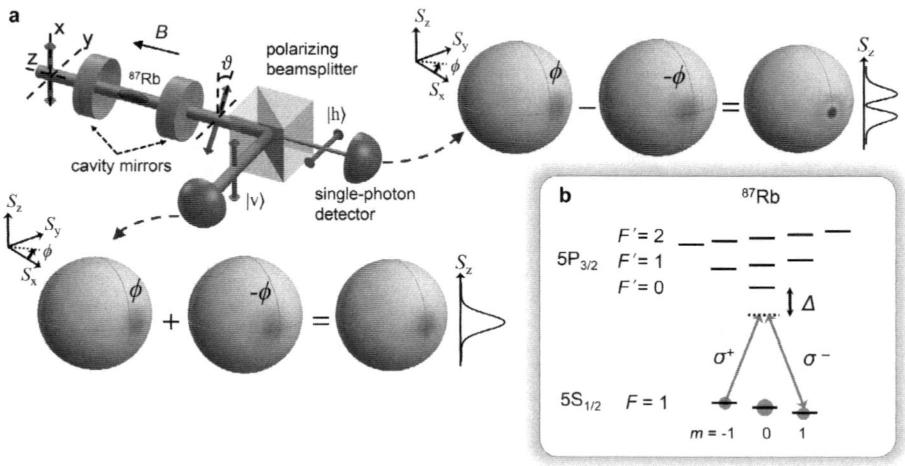

Figure 1. Scheme for heralded entanglement generation in a large atomic ensemble by single-photon detection. (a) Incident vertically polarized light experiences weak polarization rotation due to atomic quantum noise, and the detection of a horizontally polarized transmitted photon heralds an entangled state of collective atomic spin. An optical resonator enhances the polarization rotation and the heralding probability. The atoms are prepared in the product state of maximum S_x, the coherent spin state (CSS), before the heralding detection step. The atomic state is depicted on the Bloch sphere with a radius of $S = S_x$. (b) Atoms in the $5S_{1/2}, F = 1$ hyperfine manifold are coupled to the excited $5P_{3/2}$ manifold via linearly polarized light, decomposed into two circular polarization components $|\sigma^{\pm}\rangle$ that interact with the atomic ground-state *populations*. The outgoing polarization state of the light reflects the quantum fluctuations between the $|5S_{1/2}F = 1, m = \pm 1\rangle$ magnetic sublevels.

2. Photon polarization rotation by quantum spin noise

The pertinent atom-light interaction is enhanced by an optical cavity, into which we load $N_a = 3100 \pm 300$ laser-cooled ^{87}Rb atoms (Figure 1a). The atoms are prepared in the $5S_{1/2}, F = 1$ hyperfine manifold, such that each atom i can be associated with a spin $\vec{f}_i$, and the ensemble with a collective-spin vector $\vec{S} = \sum_i \vec{f}_i$. After polarizing the ensemble ($S_z \approx S$) by optical pumping, the collective spin state is rotated onto the $\hat{x}$ axis by means of a radiofrequency $\pi/2$ pulse, thus the atoms are prepared in the product state of maximum S_x. This (unentangled) initial state that is centered about $S_z = 0$ with a variance $(\Delta S_z)^2 = S/2$ is known as a coherent spin state (CSS). In our experiment, the atoms are non-uniformly coupled to the optical mode used for state preparation and detection, but the relevant concepts can be generalized to this situation [16].

Probe light resonant with a cavity mode, and red-detuned by $\Delta/(2\pi) = -200$ MHz from the ^{87}Rb transition $5^2S_{1/2}, F = 1$ to $5^2P_{3/2}, F' = 0$, is polarization analyzed upon transmission through the cavity. The vertical polarization state of each photon in the incident laser pulse $|\mathrm{v}\rangle = (|\sigma^+\rangle + |\sigma^-\rangle)/\sqrt{2}$ can be decomposed into two circular polarization components $|\sigma^{\pm}\rangle$

that produce opposite differential light shifts between the atomic magnetic sublevels $|m = \pm 1\rangle$. Hence a $|\sigma^{\pm}\rangle$ photon causes a precession of the collective spin vector $\vec{S}$ in the xy plane by a small angle $\pm\phi$ [17], and we denote the corresponding slightly displaced CSS by $|\pm\phi\rangle$. Then the combined state of the atom-light system after the passage of one photon can be written as

$$|\psi\rangle \propto |\sigma^{+}\rangle|+\phi\rangle + |\sigma^{-}\rangle|-\phi\rangle. \tag{1}$$

Conversely, atoms in the states $|m = \pm 1\rangle$ cause different phase shifts on the $\sigma^{\pm}$ photons, resulting in a net rotation of the photon linear polarization if the states $|m = \pm 1\rangle$ are not equally populated. Then the atomic quantum fluctuations between $|m = \pm 1\rangle$ in the CSS randomly rotate the polarization of the input photons $|\mathrm{v}\rangle$, giving rise to a nonzero probability $\propto \phi^2$ for an incident $|\mathrm{v}\rangle$ photon to emerge in the polarization $|\mathrm{h}\rangle = (|\sigma^{+}\rangle - |\sigma^{-}\rangle)/\sqrt{2}$, orthogonal to its input polarization. The detection of such a "heralding" photon projects the atomic state onto $\langle \mathrm{h}|\psi\rangle \propto |\phi\rangle - |-\phi\rangle$, which is not a CSS, but an entangled state of collective spin, namely, the first excited Dicke state [18] $|\psi_1\rangle$ along $\hat{x}$ (Figure 1a). In contrast, if the photon is detected in its original polarization $|\mathrm{v}\rangle$, the atomic state is projected onto $\langle \mathrm{v}|\psi\rangle \propto |\phi\rangle + |-\phi\rangle$, a state slightly spin squeezed [1]. This is because this photon detection outcome is most likely if the atomic spin S_z is near 0, so that the best estimate of the atomic state population distribution is now more tightly concentrated around $S_z = 0$ than it would be for the CSS. When $\phi \ll \sqrt{1/(2S)}$, the atomic state $\langle \mathrm{v}|\psi\rangle$ is essentially identical to the input CSS. The entangled atomic state $|\psi_1\rangle$ is post-selected by the detection of the heralding photon $|\mathrm{h}\rangle$.

In the experiment, the mean photon number in the incident laser pulse $k \sim 210$ is chosen such that the probability for one photon to emerge in heralding polarization $|\mathrm{h}\rangle$ is $p \approx 0.05 \ll 1$. This ensures a very small probability $\propto p^2$ for producing a different entangled state $|\psi_2\rangle$ heralded by two photons [17], a state which, due to our photon detection efficiency of $q = 0.3 < 1$, we would (mostly) mistake for $|\psi_1\rangle$. This admixture of $|\psi_2\rangle$ to the heralded state is suppressed by a factor of $3p(1-q) \approx 0.1$.

3. The effect of the optical dipole trap on the polarization rotation

The atoms are confined on the cavity axis by a far-detuned optical dipole trap at 852 nm with a linear polarization and the trap depth $U/h = 20(3)$ MHz. For our experiments, it is of utmost importance that the probe light polarization rotation ϑ be only caused by the desired effect, the collective atomic spin along the cavity axis, S_z, that exhibits quantum fluctuations around $\langle S_z\rangle = 0$. However, when the trap light polarization is not parallel or perpendicular to the probe, it can also change the probe polarization at 780 nm. Consider the simple case that the atoms are in the $F = 1$, $m = 0$ state. The trap light polarization is at angle $\varphi_{\mathrm{t}}/2$ with respect to the probe (Figure 2a). The probe field $|\sigma^{+}\rangle$, the trap field $|\sigma^{+}\rangle$ and $|\sigma^{-}e^{i\varphi_{\mathrm{t}}}\rangle$ can polarize the atoms through the third-order susceptibility and introduce a phase shift on the probe $|\sigma^{-}\rangle$ field, and vice versa. This is similar to the four-wave mixing process. The probe polarization rotation is proportional to $(U/\Delta)\phi \sin\varphi_{\mathrm{t}}$. This will obscure our desired polarization rotation signal coming from the atomic quantum spin noise in S_z. Only when the linear polarization of the trap light is aligned parallel or perpendicular to the probe light, this effect is diminished.

As a clean diagnostic, we prepare the atoms in the $F = 1$, $m = 0$ state, and measure the polarization rotation ϑ of the probe light as a function of the bias magnetic field B along the cavity axis, with the trap light briefly switched off or present during the measurement. Figure 2b shows that the trap light rotates the probe polarization in addition to the bias field B, resulting in an offset $\mu_B B/h = 3.1(2)$ MHz. Our calculation predicts the offset at $\mu_B B/h = 2.8(5)$ MHz when the trap polarization is at 45° to the probe. By aligning the trap linear polarization parallel to the probe polarization, the trap light no longer contributes to the rotation.

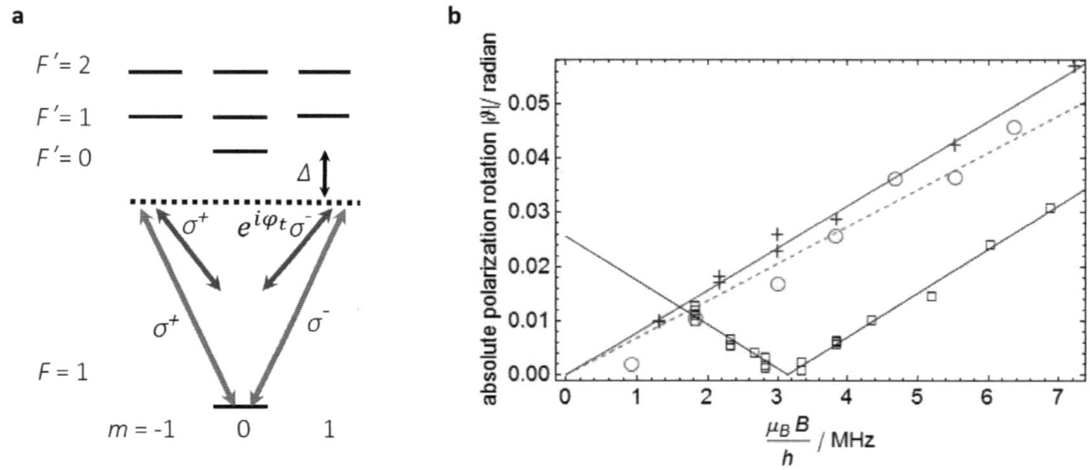

Figure 2. (Colour online) Probe light polarization rotation ϑ with the bias magnetic field B along the cavity axis for 4500 atoms in the $F = 1$, $m = 0$ state. (a) Atoms in the $F = 1$, $m = 0$ state are coupled to the excited states via the linearly polarized probe light (red arrows) and the linearly polarized trap light (blue arrows). The relative angle of the two linear polarizations is $\varphi_t/2$. (b) When the trap light is switched off the expected linear Faraday rotation is observed (red circles). The dashed red line is the prediction. When the trap light is present in the cavity and its linear polarization is not parallel or perpendicular to the probe light, it rotates the probe polarization in addition to the Faraday rotation, resulting in an offset (black squares). The solid black line is the fit. By aligning the trap polarization parallel to the probe, the effect of the trap light is diminished (blue crosses). The solid blue line is the fit.

4. Spin distributions and Wigner function

In order to reconstruct the collective-spin state generated by the heralding event, we rotate the atomic state after the heralding process by an angle $\beta = 0, \frac{\pi}{4}, \frac{\pi}{2}, \frac{3\pi}{4}$ about the $\hat{x}$ axis before measuring S_z. (Thus $\beta = 0$ corresponds to measuring S_z, $\beta = \pi/2$ corresponds to S_y, etc.) The measurement is performed by applying a stronger light pulse in the same polarization-optimized setup used for heralding. As the Faraday rotation angle $\vartheta \ll 1$ is proportional to S_z, and the probability for detecting $|\mathrm{h}\rangle$ photons is proportional to ϑ^2, the measured probability distribution of $|\mathrm{h}\rangle$ photon number, $g(n_\beta)$, reflects the probability distribution of S_β^2. Figure 3a shows that a single heralding photon substantially changes the spin distribution towards larger values of $\langle S_\beta^2 \rangle$. Here we show the measurement for $\beta = 0$ only. For other rotation values β see [16]. We further verify that the heralded state remains (nearly) spin polarized with a contrast of $\mathcal{C} = 0.99^{+0.01}_{-0.02}$, the same as for the CSS within error bars.

From the photon distributions $g(n_\beta)$ we can reconstruct the density matrix ρ_{mn} in the Dicke state basis [18] along $\hat{x}$, where $|n = 0\rangle$ denotes the CSS along $\hat{x}$, $|n = 1\rangle$ the first Dicke state, $|n = 2\rangle$ the second Dicke state, etc. From the density matrix we obtain the Wigner function $W(\theta, \phi)$ on the Bloch sphere [19] (Figure 3e). To accurately determine the Wigner function value on the axis, $W(\theta = \frac{\pi}{2}, \phi = 0) = \sum_n (-1)^n \rho_{nn}$, that depends only on the population terms ρ_{nn}, we average the photon distributions $g(n_\beta)$ over four angles β and thereby reduce the fitting parameters to just $\rho_{nn}, n \leq 4$. This is equivalent to constructing a rotationally symmetric Wigner function from the angle-averaged marginal distribution [20]. We obtain $\rho_{00} = 0.32 \pm 0.03, \rho_{11} = 0.66 \pm 0.04$ with negligible higher-order population terms, giving $W(\frac{\pi}{2}, 0) = -0.36 \pm 0.08$, to be compared to $W(\frac{\pi}{2}, 0) = -1$ for the perfect first Dicke state.

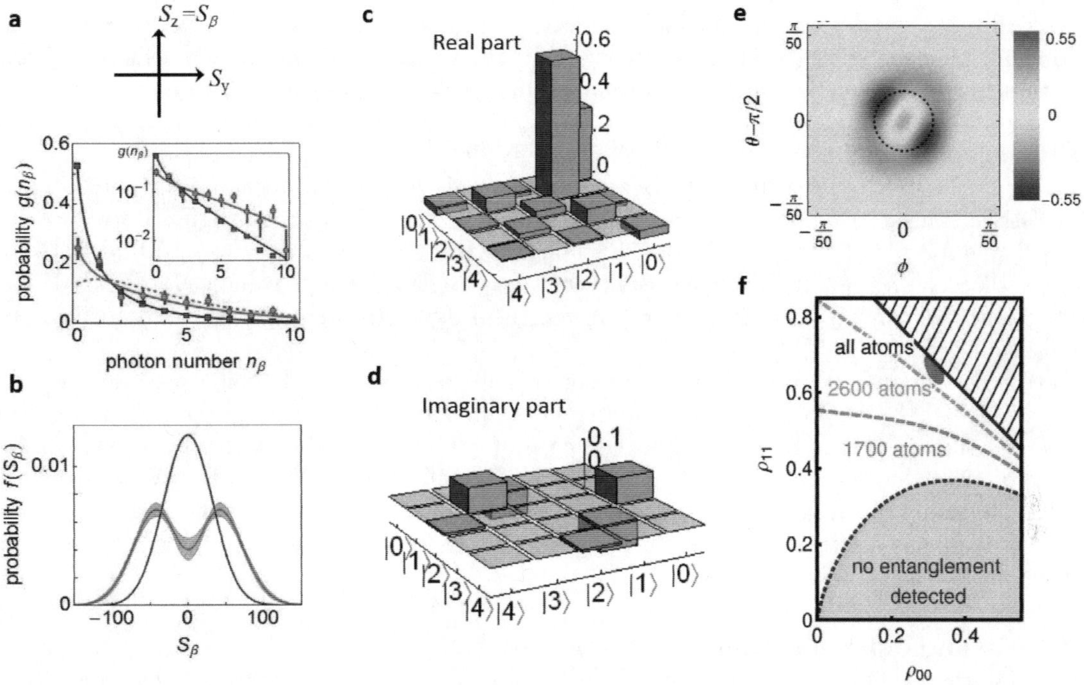

Figure 3. (Colour online) (a) Measured photon distributions $g(n_\beta)$ for no heralding photon detected (blue squares), and for one heralding photon detected (red circles) for rotation angles $\beta = 0$, i.e., $S_\beta = S_z$. For other rotation values β see [16]. Inset: Logarithmic representations of the same data. The solid blue and the dashed red curves are predictions without any free parameters, for the CSS and the perfect first Dicke state, respectively. The solid red line is the fit. (b) Reconstructed collective spin distributions of the heralded state (red) for $\beta = 0$. The spin distributions of the CSS (blue) are for reference. (c),(d) Real and imaginary parts of the reconstructed density matrix elements, in the Dicke state basis along $\hat{x}$, for the heralded state. (e) Reconstructed Wigner function $W(\theta, \phi)$ for the heralded state on the Bloch sphere [19]. θ is the polar angle with respect to $\hat{z}$ and ϕ is the azimuthal angle with respect to $\hat{x}$. The first excited Dicke state and the CSS have $W(\frac{\pi}{2}, 0) = -1$ and $W(\frac{\pi}{2}, 0) = 1$, respectively. To provide a reference scale for the size of the negative region, the black dashed line is the contour at which the CSS has a Wigner function value $1/e$. (f) Entanglement depth criterion [13] for the heralded state, plotted in terms of density matrix elements ρ_{00} and ρ_{11}. The red shaded region represents the 1 s.d. confidence region for the heralded state. Lines represent boundaries for k-particle entanglement in terms of atom number N_a; a state with ρ_{11} greater than such a boundary displays at least k-particle entanglement.

We can also fit the density matrix including the coherence terms simultaneously to $g(n_\beta)$ for all four angles β, without angle-averaging. Since the photon distributions $g(n_\beta)$ measure the distributions of S_β^2, they cannot contain any odd power of S_β. In general, for a given spin distribution $f(S_\beta)$, the distribution function of S_β^2, $G(S_\beta^2) = \frac{1}{2|S_\beta|}\Big(f(S_\beta) + f(-S_\beta)\Big)$. The distribution $f(S_\beta)$ contains even functions of S_β for ρ_{mn} where $m+n$ is even, and odd functions of S_β for ρ_{mn} where $m+n$ is odd, therefore the distribution $G(S_\beta^2)$ and hence the photon number distribution $g(n_\beta)$ contain the contribution of only the even terms of the density matrix ρ_{mn}, and have no information of odd ρ_{mn} terms.

In order to display the Wigner function, we bound the odd ρ_{mn} terms by verifying that the heralding process does not displace the state relative to the CSS. This is accomplished by performing a measurement with an auxiliary probe beam polarized at 45 degrees relative to $|\mathrm{v}\rangle$, such that the difference between the measured $|\mathrm{h}\rangle$ and $|\mathrm{v}\rangle$ photon numbers is proportional to S_z. We find a heralding-light-induced shift $\delta\langle S_z\rangle = -0.2 \pm 1.6$, consistent with zero, and very small compared to the CSS rms width $(\Delta S_z)_{\mathrm{CSS}} \approx 30$. Therefore we set the odd terms to zero, and display the resulting density matrix and corresponding Wigner function in Figure 3c-e. The spin distributions $f(S_z)$ obtained from this density matrix are shown in Figure 3b.

If we calculate $W(\frac{\pi}{2}, 0)$ from the density matrix without angle-averaging, we find $W(\frac{\pi}{2}, 0) = -0.27 \pm 0.08$, within error bars consistent with the angle-averaged value. Note that $W(\frac{\pi}{2}, 0)$ only depends on even ρ_{mn} terms, therefore can be fully determined by the measured photon distributions $g(n_\beta)$.

In order to quantify the minimum number of mutually entangled atoms, we use a criterion derived in Ref. [13] that establishes entanglement depth as a function of the populations ρ_{00} and ρ_{11}. From this criterion, generalized to the case of non-uniform coupling to the measurement light field, we deduce an average entanglement depth of $\mathcal{N}_a = 2910 \pm 190$ out of $N_a = 3100$ atoms (Figure. 3f) using the angle-averaged density matrix. Our results represent the first experimental verification of the mutual entanglement shared by virtually all atoms in an ensemble that contains more than a few particles.

5. Discussion and conclusion

In our system, the maximum atom number of ~ 3000 is set by the accuracy of the spin rotation, and can be increased by two orders of magnitude by better magnetic-field control [10]. The achieved state fidelity $\rho_{11} = 0.66 \pm 0.04$ is below the value of $\rho_{11} = 0.73$ required for the Fisher information [14] to exceed that of the CSS. It is limited by a number of factors. At our atom number N_a and the polarization rotation strength ϕ, the polarization impurity $\sim 3 \times 10^{-5}$ of the entire setup gives rise to false heralding events equal to 10% of the quantum spin signal from the CSS. This impurity is also present in the measurement process. In addition, the atoms confined in the optical trap at finite temperature have temporal fluctuations in their coupling strength to the heralding laser pulse and the measurement laser pulse, resulting in $\sim 5\%$ admixture of the CSS to the measured heralded state. The fidelity is further reduced by phase broadening of the atomic state by the heralding laser pulse and by magnetic field fluctuations. Increasing the atom-cavity interaction strength can reduce the false detection events and the atomic state phase broadening. Combined with improved stabilization of the magnetic field, these changes can increase the fidelity and enable more highly-entangled states to be produced.

The above results demonstrate that even with limited resources, i.e. weak atom-photon coupling, heralding schemes can be used to boost the effective interaction strength by a large factor, enabling the production of highly entangled states [21, 17]. Furthermore, by repeated trials and feedback the entanglement generation can be made quasi-deterministic [22, 23]. Our approach is related to other heralded schemes for quantum communication [22, 24, 23, 25] and entangled-state preparation [26, 27, 28], and it would be interesting to generalize the present analysis to infer characteristics of the atomic state from the measured optical signals in those experiments. We note that the same first Dicke state was created in an ensemble of up to 41 atoms with a scheme that uses many heralding photons in a strongly coupled atom-cavity system [13]. The detection of two or more photons prepares Schrödinger cat states [17] of the atomic ensemble with more metrological gain. We expect that heralded methods can generate a variety of nearly pure, complex, strongly entangled states that are not accessible by any other means at the present state of quantum technology.

Journal of Physics: Conference Series **723** (2016) 012054 doi:10.1088/1742-6596/723/1/012054

References

[1] Kitagawa M and Ueda M 1993 *Phys. Rev. A* **47** 5138–5143
[2] Appel J, Windpassinger P J, Oblak D, Hoff U B, Kjærgaard N and Polzik E S 2009 *Proceedings of the National Academy of Sciences* **106** 10960–10965
[3] Takano T, Tanaka S I R, Namiki R and Takahashi Y 2010 *Phys. Rev. Lett.* **104** 013602
[4] Schleier-Smith M H, Leroux I D and Vuletić V 2010 *Phys. Rev. Lett.* **104** 073604
[5] Leroux I D, Schleier-Smith M H and Vuletić V 2010 *Phys. Rev. Lett.* **104**(7) 073602
[6] Gross C, Zibold T, Nicklas E, Estève J and Oberthaler M K 2010 *Nature* **464** 1165–1169
[7] Riedel M F, Böhi P, Li Y, Hänsch T W, Sinatra A and Treutlein P 2010 *Nature* **464** 1170–1173
[8] Hamley C D, Gerving C G, Hoang T M, Bookjans E M and Chapman M S 2012 *Nat Phys* **8** 305 – 308
[9] Sewell R J, Koschorreck M, Napolitano M, Dubost B, Behbood N and Mitchell M W 2012 *Phys. Rev. Lett.* **109** 253605
[10] Bohnet J G, Cox K C, Norcia M A, Weiner J M, Chen Z and Thompson J K 2014 *Nat Photon* **8** 731–736
[11] Leibfried D, Knill E, Seidelin S, Blakestad R B, Chiaverini J, Hume D B, Itano W B, Jost J D, Langer C, Ozeri R, Reichle R and Wineland D J 2006 *Nature* **438** 639–642
[12] Monz T, Schindler P, Barreiro J T, Chwalla M, Nigg D, Coish W A, Harlander M, Hänsel W, Hennrich M and Blatt R 2011 *Phys. Rev. Lett.* **106** 130506
[13] Haas F, Volz J, Gehr R, Reichel J and Estéve J 2014 *Science* **344** 180–183
[14] Strobel H, Muessel W, Linnemann D, Zibold T, Hume D B, Pezz L, Smerzi A and Oberthaler M K 2014 *Science* **345** 424–427
[15] Lücke B, Peise J, Vitagliano G, Arlt J, Santos L, Tóth G and Klempt C 2014 *Phys. Rev. Lett.* **112**(15) 155304
[16] McConnell R, Zhang H, Hu J, Ćuk S and Vuletić V 2015 *Nature* **519** 439–442
[17] McConnell R, Zhang H, Ćuk S, Hu J, Schleier-Smith M H and Vuletić V 2013 *Phys. Rev. A* **88**(6) 063802
[18] Arecchi F T, Courtens E, Gilmore R and Thomas H 1972 *Phys. Rev. A* **6**(6) 2211–2237
[19] Dowling J P, Agarwal G S and Schleich W P 1994 *Phys. Rev. A* **49**(5) 4101–4109
[20] Lvovsky A I, Hansen H, Aichele T, Benson O, Mlynek J and Schiller S 2001 *Phys. Rev. Lett.* **87** 050402
[21] Agarwal G S, Lougovski P and Walther H 2005 *Journal of Modern Optics* **52** 1397–1404
[22] Duan L M, Lukin M D, Cirac J I and Zoller P 2001 *Nature* **414** 413–418
[23] Matsukevich D N, Chaneliere T, Jenkins S D, Lan S Y, Kennedy T A B and Kuzmich A 2006 *Phys. Rev. Lett.* **97** 013601
[24] Kuzmich A, Bowen W P, Boozer A D, Boca A, Chou C W, Duan L M and Kimble H J 2003 *Nature* **423** 731–734
[25] Simon J, Tanji H, Thompson J K and Vuletić V 2007 *Phys. Rev. Lett.* **98** 183601
[26] Choi K S, Goban A, Papp S B, van Enk S J and Kimble H J 2010 *Nature* **468** 412 – 416
[27] Christensen S L, Beguin J B, Sorensen H L, Bookjans E, Oblak D, Müller J H, Appel J and Polzik E S 2013 *New Journal of Physics* **15** 015002
[28] Christensen S L, Béguin J B, Bookjans E, Sørensen H L, Müller J H, Appel J and Polzik E S 2014 *Phys. Rev. A* **89**(3) 033801

Microfabricated Optically-Pumped Magnetometers for Biomagnetic Applications

Svenja Knappe, Orang Alem, Dong Sheng, John Kitching

NIST, 325 Broadway, Boulder CO 80305, USA

Email: knappe@nist.gov

Abstract. We report on the progress in developing microfabricated optically-pumped magnetometer arrays for imaging applications. We have improved our sensitivities by several orders of magnitude in the last ten years. Now, our zero-field magnetometers reach noise values below 15 fT/Hz$^{1/2}$. Recently, we have also developed gradiometers to reject ambient magnetic field noise. We have built several imaging arrays and validated them for biomedical measurements of brain and heart activity.

1. Introduction

Sensors based on laser spectroscopy of atoms have shown impressive precision in the laboratories [1, 2]; atom-based gyroscopes have reached angle-random walks (ARW) of 2×10^{-6} °/√h [3], cold-atom gravimeters have shown sensitivities of $\Delta g/g = 8 \times 10^{-9}$ [4], and atomic magnetometers have reached 160 aT/Hz$^{1/2}$ sensitivity [5]. These levels of performance have opened new opportunities for atom-based sensors in real-world applications and as a result, several companies are now trying to commercialize these advanced sensor approaches. In a commercial setting, especially when high-volume manufacturing is desirable, the challenges of robustness, reliability, and cost must be addressed. The use of microfabrication techniques commonly used in microelectronics and microelectromechanical systems (MEMS) can offer significant advantages for manufacturing and in addition enable improvements in device size, weight and power. Microfabrication has been successfully implemented in commercially available atomic vapor cell clocks [6].

2. Overview of microfabricated optically-pumped magnetometers

Optically-pumped magnetometers have been developed since the 1960s [7, 8], where near-resonant light creates atomic polarization in the ground state of the atoms. The resultant Larmor spin precession can be detected with near-resonant light and serves as a measure of the ambient magnetic field.

Our group developed the first chip-scale optically-pumped magnetometer in 2004 [9]. A thermal vapor of ^{87}Rb atoms was confined in a MEMS vapor cell [10] with an internal volume of 1 mm^3. It utilized coherent population trapping (CPT) to interrogate the atoms on a magnetically-sensitive hyperfine transition. A vertical-cavity surface-emitting laser (VCSEL) was integrated with micro-optics on the sensor head directly, which had a volume of 10 mm^3 without packaging (Figure 1, 2004, green online). The magnetometer reached a sensitivity, i.e., a noise equivalent magnetic field, of 50 pT/Hz$^{1/2}$ at 10 Hz.

By changing to spectroscopy of a Zeeman transition, the spin-destruction collision rate could be reduced and the resonance contrast enhanced, which resulted in a factor of ten improvement in sensitivity [11] (Figure 1, 2006, blue online). The physics package of this device was very similar to the previous one, but it included a pair of field coils to drive the Zeeman resonance magnetically instead of optically. In this so-called Mx configuration, the transmission signal is detected phase-sensitively at the Larmor frequency. Alternatively, we have operated similar magnetometers in an all-optical mode as well as the Mz mode, with similar sensitivities of 5 pT/Hz$^{1/2}$. For microfabricated scalar magnetometers, sensitivity could be increased by implementing a balanced polarimeter [12, 13], a multipass cell [14], or through light narrowing [15].

Higher sensitivities can be achieved when decoherence due to spin-exchange collisions can be suppressed [16] by operating in a regime where the spin-exchange collision rate is faster than the Larmor precession frequency [17]. This limits the dynamic range of the magnetometer, but it could still be useful for biomedical applications, where high sensitivities are required in magnetically-shielded environments. In a tabletop setup, we measured sensitivities around 70 fT/Hz$^{1/2}$ in a MEMS vapor cell of 2 x 3 x 1 mm^3 volume [18] (Figure 1, 2007, grey online). The configuration was simplified by using a single circularly-polarized beam [8] to replace two perpendicular beams of a standard pump-probe configuration [16]. In order to be sensitive to a component of the magnetic field perpendicular to the laser beam, a small oscillating magnetic field is applied in that direction and the transmission signal is detected phase-sensitively. The sensitivity of the magnetometer in this configuration will fundamentally be limited by photon shot noise, which can be replaced by atomic spin projection noise when using separate pump and probe laser beams.

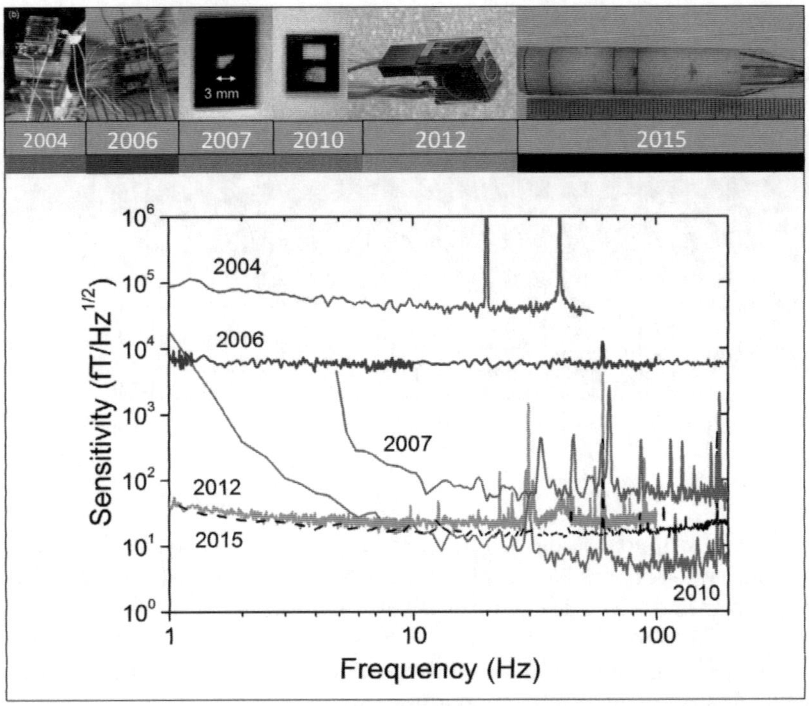

Figure 1: (Bottom) Sensitivity for several generations of our magnetometers. The grey lines are tabletop measurements. (Top) Photographs of the same MEMS magnetometers (colour online).

In a more complicated tabletop setup, a circularly-polarized near-resonant pump laser was split into two arms to optically pump ^{87}Rb atoms in a MEMS vapor cell to create a uniform spin-polarization. A further detuned linearly-polarized probe laser was incident on the cell perpendicular to the pump beam. The polarization rotation of the probe beam was measured with a balanced polarimeter. In a 1 mm^3 volume, we have measured sensitivities as low as 5 fT/Hz$^{1/2}$ (Figure 1, 2010, grey online).

So far, we have built fully packaged sensor heads in single-beam geometries only. For biomedical imaging applications, where large arrays of sensors are needed, we have chosen a design of fiber-coupled sensors heads: A distributed-feedback laser (DFB) is split into up to 16 channels and coupled to the individual sensor heads. We have reached sensitivities between 15 to 20 $fT/Hz^{1/2}$, when averaged between 10 Hz and 50 Hz, in cells with inner volumes of $(1.5\ mm)^3$ (Figure 1, 2012, orange online).

3. Fabrication of microfabricated optically-pumped magnetometers

For biomagnetic imaging applications, the sensor heads are coupled to a control unit through 5 m long optical fibers. That way, the control system can remain outside the shielded room and the sensor heads with footprints of roughly 1 cm^2 are fed to the inside of the room through small holes in the shield. At the heart of the sensor heads are microfabricated alkali vapor cells [10]. Dimensions vary between 1 mm and 4 mm, depending on the application requirements (Figure 2a). These cells have to be heated to roughly 150 °C, while the outside of the package has to remain at room temperature, since it might be directly placed on the skin of a subject.

For this purpose, we have developed low-power MEMS vacuum packages [19] based on anodic bonding of silicon and glass (Figure 2b). The cells are heated optically; glass filters are placed onto the windows of the cell, and light from a second laser at 1550 nm is absorbed by these filters. To reduce conductive heat losses below 5 mW, the cell of volume $(1.5\ mm)^3$ is placed onto a web of polyimide attached to a silicon frame [20]. The radiative heat losses are dominating with roughly 50 mW. When additional gasses are present inside the vacuum package, the required heating power increases due to a convective contribution, which was the case in many of our earlier packages.

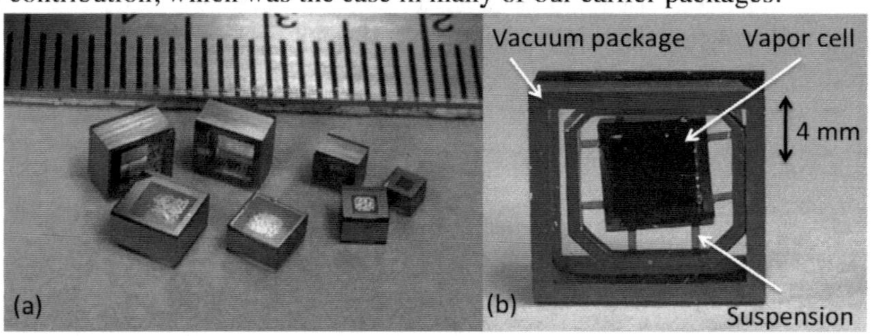

Figure 2: (a) Photograph of several MEMS alkali vapor cells. (b) Photograph of a low-power MEMS vacuum package.

4. Noise contributions

The fundamental performance limits of the single-beam zero-field magnetometers are given by the spin-projection noise and photon shot noise. Under our operating conditions, these noise contributions are usually under 1 $fT/Hz^{1/2}$ and 10 $fT/Hz^{1/2}$, respectively. Besides these fundamental contributions, additional noise sources fall into the categories of laser noise, magnetic noise, and technical noise. As technical noise, we consider contributions from electronics, currents on the magnetic field coils, and vibrations of the sensors in a magnetic gradient environment.

Fluctuations in the intensities, frequencies, and polarizations of the pump and heat lasers can affect the position and shape of the resonance line. By locking the sensor to the peak of the magnetic resonance, rather than running the magnetometer open loop, contributions of the lasers, which do not cause frequency shifts, can be eliminated. This is done by feeding an error signal, generated from the magnetic resonance signal, back to compensating field coils, which are printed onto flexible circuits wrapped around the sensor heads. Field coils must be designed carefully to minimize cross-talk between different sensors in a dense array and this problem still requires more future work.

The largest laser noise contribution of roughly10 $fT/Hz^{1/2}$ stems from laser intensity fluctuations. A major laser contribution that generates fluctuations of the resonance position is the AC Stark shift from the pump laser. The laser is circularly polarized, which will shift the Zeeman levels similar to a magnetic field [21, 22]. Since the laser is locked to the center of the optical resonance, there is no large DC magnetic field offset, but laser frequency fluctuations can be interpreted as magnetic field fluctuations in the direction of the laser beam. In many of our compact designs, the laser is not collimated, which can cause fluctuations in the sensitive direction of the magnetometer. In these cases the laser frequency has to be carefully stabilized in order not to increase the noise equivalent magnetic field at low frequencies.

Finally, magnetic field noise from components with high electrical conductivity or permeability in the proximity of the atoms has to be considered [23]. Calculations of the noise from our mu-metal enclosure suggest noise levels of 10 $fT/Hz^{1/2}$ at 1 Hz and 3 $fT/Hz^{1/2}$ at 100 Hz. Therefore, we added an inner magnetic shield made of a ferrite with an estimated noise of 2 $fT/Hz^{1/2}$ at 1 Hz [24]. We estimate noise from silicon cell walls to be 3 $fT/Hz^{1/2}$, if the standard low-resistivity silicon is used. This can be reduced to 0.01 $fT/Hz^{1/2}$ by use of silicon with a resistivity of 8 kΩ·cm [25]. Our calculations suggest that excess Rb metal on the cell walls could also have significant contribution to magnetic noise; a Rb sphere of diameter 100 μm at a distance of 0.5 mm would create magnetic noise of 7 $fT/Hz^{1/2}$, and a 1 nm thick Rb film on the walls of a 1 mm^3 cell would cause 3 $fT/Hz^{1/2}$ noise in the center of the cell [25]. We have therefore fabricated cells with internal reservoirs to keep the solid Rb metal further away from the interrogation region. While we have identified and mitigated many contributions to our magnetometer noise, the noise budget at frequencies below 10 Hz is still under investigation.

5. Gradiometers

One standard technique to suppress ambient magnetic noise is to operate magnetic field sensors in a gradiometer mode. This is done by placing two magnetometers a certain distance, called the baseline, from each other. When subtracting the outputs of the two magnetometers, readings from far away sources and homogeneous fields will be suppressed by a factor called the common mode rejection ratio (CMRR). In many real-world applications there are large amounts of magnetic background noise and the development of efficient gradiometers is of importance. We have built a gradiometer with two cells separated by 2 cm on the same sensor head. In order to minimize noise contributions from the pump laser, one laser beam is split into two arms on the sensor head and detected with two photodiodes. That way, the photon-shot-noise contributions are increased, but the laser amplitude noise contribution is reduced. Since both of these effects contribute equally to our noise, the noise of one magnetometer is similar to the noise measured when two magnetometer signals are subtracted: 15 $fT/Hz^{1/2}$ between 10 and 50 Hz.

In order to reach a high CMRR, both magnetometers are separately operated under negative feedback, where they are locked to the center of the magnetic resonance. This method eliminates scale factor drifts as a result of changing laser parameters. It also eliminates current noise from a global compensation coil system, which can be used to bring the ambient field into the nanotesla range. The major limitation is currently posed by aligning the sensitive axes of the two magnetometers precisely.

6. Biomagnetic Measurements

Highly sensitive magnetometers are used in a variety of biomedical applications. Often, the magnetic fields on the outside of the body are recorded, which are the result of small currents inside the body. In order to localize the sources, a non-unique inverse problem has to be solved. Currently, low-temperature superconducting quantum interference device (SQUID) magnetometers are the only sensors that are used clinically for these applications, since commercial SQUIDs can reach sensitivities around 3 $fT/Hz^{1/2}$. As a result, these imaging systems have a large Dewar as their defining feature, which houses the magnetometer array. It imposes a rigid sensor geometry, makes these systems expensive, and requires high maintenance. It also limits the system's spatial resolution, since the thermal isolation places the SQUID sensors several centimeters from the Dewar surface.

We have developed several multi-channel imaging systems with microfabricated optically-pumped magnetometer sensor heads and have cross-validated them against SQUIDs for biomagnetic measurements. We have recorded magnetocardiograms (MCG) by placing several sensors over the chest of a healthy subject [26]. In this way magnetic field maps can be generated, with patterns that vary during different sections of a beat [27]. We have also recorded fetal MCG with an array of 25 magnetometers and extracted the maternal and fetal heart signals [28]. In both of these measurements, signal amplitudes of the optically-pumped magnetometers were larger than those measured with SQUIDs, mostly due to the closer proximity to the source.

One biomagnetic imaging application that could benefit from small conformal sensor arrays is magnetoencephalography (MEG), which measures the magnetic fields from neural currents by placing sensors on the scalp of the subject. Due to the curvature of the head, rigid helmets do not allow for close proximity of sensors for every head size. Simulation studies by different groups suggest that the localization error could be reduced substantially with a conformal sensor array where distances between sensitive volume and scalp are less than 5 mm, which is the case for our sensors. We have taken several MEG recordings of spontaneous and evoked brain activity with arrays of several sensors [29]. Recently, we have evaluated a 25-channel array by measuring images of a simple coil, a standard head phantom, and healthy human subjects. The sensors were arranged on a flexible helmet such that each sensor could lightly touch a plastic sphere that housed the phantom dipole or the scalp of the subject. One of the difficulties introduced by a flexible conformal array was to exactly determine the positions of every sensor in the array. Future work includes better determination of sensor locations through a series of externally-applied magnetic fields.

There are several other applications in the biomedical field that could also potentially benefit from uncooled flexible sensor arrays. We have made initial demonstrations of magnetic nanoparticle relaxation [26] and remote detection of nuclear magnetic resonance (NMR) [30] with microfabricated optically-pumped magnetometers.

7. Conclusion

We have developed arrays of microfabricated optically-pumped magnetometers. We have cross-validated our MEG and MCG measurements with those from low-temperature SQUID magnetometers. We have built small gradiometer sensor heads with high common-mode rejection ratios to suppress noise from ambient sources. The complete evaluation of our sensor noise is still ongoing and includes fundamental, technical, laser, and magnetic noise.

We thank Susan Schima for help with the MEMS fabrication, John LeBlanc (Draper Labs.) for building sensor heads for the 25-channel system, and Tilmann Sander and Lutz Trahms (PTB) for the biomagnetic measurements. This is a contribution of NIST, an agency of the U.S. government; and is not subject to copyright.

References

[1] Kitching J, Knappe S, Donley EA 2011 *Sensors Journal, IEEE* **11** 1749-58
[2] Kitching J, Knappe S, Gerginov V, Shah V, Schwindt PDD, Lindseth B, et al. 2008 *Proceedings of the Frequency Standards and Metrology Conference* 445-53
[3] Gustavson TL, Landragin A, Kasevich MA 2000 *Classical and Quantum Gravity* **17** 2385-98
[4] Müller H, Chiow SW, Herrmann S, Chu S, Chung KY 2008 *Physical Review Letters* **100** 031101
[5] Dang HB, Maloof AC, Romalis MV 2010 *Appl Phys Lett* **97** 151110
[6] Knappe S, Shah V, Schwindt PDD, Hollberg L, Kitching J, Liew LA, et al. 2004 *Appl Phys Lett* **85** 1460-2
[7] Bloom A 1962 *Appl Opt* **1** 61-8
[8] Dupont-Roc J, Haroche S, Cohen-Tannoudji C 1969 *Phys Lett A* **28** 628-39
[9] Schwindt PDD, Knappe S, Shah V, Hollberg L, Kitching J, Liew L-A, et al. 2004 *Appl Phys Lett* **85** 6409-11

[10] Liew LA, Knappe S, Moreland J, Robinson H, Hollberg L, Kitching J 2004 *Appl Phys Lett* **84** 2694-6
[11] Schwindt PDD, Lindseth B, Knappe S, Shah V, Kitching J 2007 *Appl Phys Lett* **90** 081102
[12] Shah V, Romalis MV 2009 *Phys Rev A* **80** 013416
[13] Johnson C, Schwindt PDD, Weisend M 2010 *Appl Phys Lett* **97** 243703
[14] Sheng D, Li S, Dural N, Romalis MV 2013 *Phys Rev Lett* **110** 160802
[15] Scholtes T, Schultze V, Ijsselsteijn R, Woetzel S, Meyer HG 2011 *Phys Rev A* **84** 043416
[16] Allred JC, Lyman RN, Kornack TW, Romalis MV 2002 *Phys Rev Lett* **89** 130801
[17] Happer W, Tang H 1973 *Phys Rev Lett* **31** 273
[18] Shah V, Knappe S, Schwindt PDD, Kitching J 2007 *Nature Photonics* **1** 649-52
[19] Mhaskar R, Knappe S, Kitching J 2012 *Appl Phys Lett* **101** 241105
[20] Mescher MJ, Lutwak R, Varghese M 2005 *Solid-State Sensors, Actuators and Microsystems, 2005 Transducers '05 (*311-6 Vol. 1
[21] Happer W, Mathur BS 1967 *Physical Review* **163** 12
[22] Cohen-Tannoudji C, Dupont-Roc J 1972 *Phys Rev A* **5** 968-84
[23] Lee S-K, Romalis MV 2008 *Journal of Applied Physics* **103** 084904
[24] Kornack TW, Smullin SJ, Lee S-K, Romalis MV 2007 *Appl Phys Lett* **90** 223501
[25] Griffith WC, Knappe S, Kitching J 2010 *Optics express* **18** 27167-72
[26] Knappe S, Sander TH, Kosch O, Wiekhorst F, Kitching J, Trahms L 2010 *Appl Phys Lett* **97** 133703
[27] Knappe S, Mhaskar R, Preusser J, Kitching J, Trahms L, Sander T 2012 *5th European Conference of the International Federation for Medical and Biological Engineering (*Springer Berlin Heidelberg) p. 1330-3
[28] Alem O, Sander TH, Mhaskar R, LeBlanc J, Eswaran H, Steinhoff U, et al. 2015 *Phys Med Biol* **60** 4797-811
[29] Sander TH, Preusser J, Mhaskar R, Kitching J, Trahms L, Knappe S 2012 *Biomed Opt Express* **3** 981-90
[30] Ledbetter MP, Savukov IM, Budker D, Shah V, Knappe S, Kitching J, et al. 2008 *Proceedings of the National Academy of Sciences of the USA* **105** 2286-90

NIST on a Chip: Realizing SI units with microfabricated alkali vapour cells

J Kitching[1], E A Donley[1], S Knappe[1], M Hummon[1], A T Dellis[1], J Sherman[1], K Srinivasan[2], V A Aksyuk[2], Q Li[2], D Westly[2], B Roxworthy[2] and A Lal[3]

[1] Time and Frequency Division, National Institute of Standards and Technology, Boulder, Colorado, USA
[2] Center for Nanoscale Science and Technology, National Institute of Standards and Technology, Gaithersburg, USA
[3] Department of Electrical and Computer Engineering, Cornell University, Ithaca, New York, USA

E-mail: john.kitching@nist.gov

Abstract. We describe several ways in which microfabricated alkali atom vapour cells might potentially be used to accurately realize a variety of International System (SI) units, including the second, the meter, the kelvin, the ampere, and the volt, in a compact, low-cost "chip-scale" package. Such instruments may allow inexpensive *in-situ* calibrations at the user's location or widespread integration of accurate references into instrumentation and systems.

1. Introduction

Precision metrology underlies a broad range of modern technology including communications, satellite-based navigation and positioning, manufacturing, and electrical instrumentation. Since the 1990s, there has been a rapid increase in the use of portable, battery-operated electronics and related technologies that include telecommunications (cellular telephones), computing (laptop computers), and even space (cubesats). Such systems require low-cost, low-power and highly miniaturized components that can be mass-produced and integrated easily into new instruments and devices. At NIST, we are currently exploring the possibilities for the development of low-cost, easily manufactured, low-power SI-traceable references suitable for deployment to users for a broad range of *in-situ* calibration purposes. The vision for such instruments is shown in figure 1, although it is unlikely that one chip will incorporate all references. A similar vision has been articulated at the National Physical Laboratory in the United Kingdom [1]. In particular, we are considering possibilities related to microfabricated alkali vapor cells, which were first demonstrated in our group in 2003 [2]. We propose that five SI units (four base units and one derived unit) could potentially be realized using such alkali vapor cells.

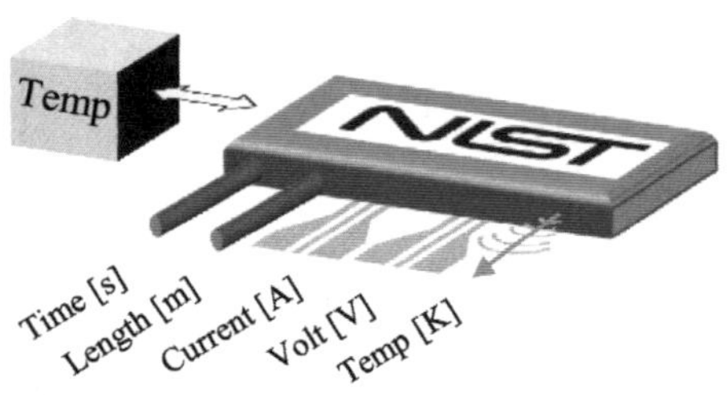

Figure 1. Concept of a possible chip-scale implementation of an SI-traceable set of references (Colour online).

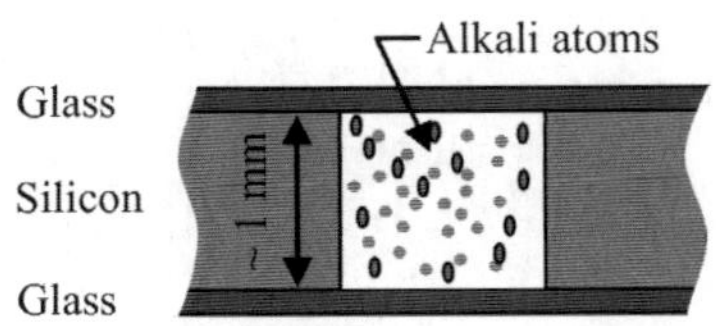

Figure 2 Microfabricated alkali vapour cell design (Colour online).

Microfabricated alkali vapour cells [2] have by now become a fairly well-developed technology, routinely used in commercial atomic clock products [3] and being developed for a number of other technologies such as magnetometry. The basic cell structure is shown in figure 2. A cavity is etched into a polished silicon wafer using standard micromachining processes and glass, bonded to the top and bottom surfaces, confines the alkali atoms to the interior of the cell.

2. Realization of Units

Figure 3 shows a number of ways in which atomic spectroscopy in microfabricated alkali vapour cells can be used to realize a variety of SI units. In all cases, the frequency of some atomic transition is measured in order to realize the unit. The relation between the measured frequency and the specific unit is mediated by a set of fundamental constants, presumed to be known more accurately than the unit will be realized, and in two cases (current and voltage), a measured length.

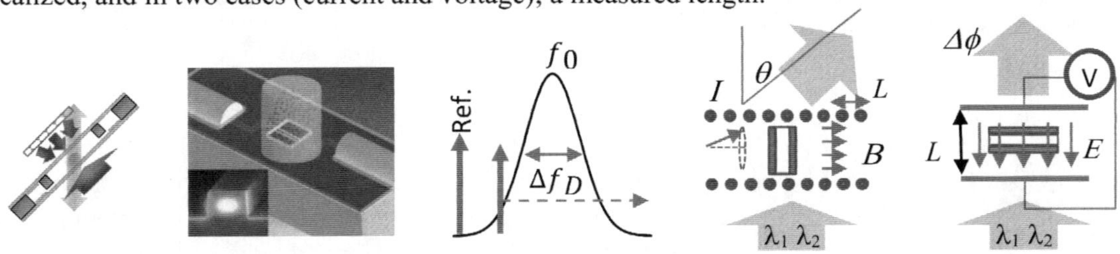

$$t = \frac{1}{f_{Cs}}$$ (a)

$$L = N\frac{c}{f_{opt}}$$ (b)

$$T = \frac{mc^2}{2k_B}\left(\frac{\Delta f_D}{f_0}\right)^2$$ (c)

$$I = \frac{2L}{\mu_0}\frac{\hbar}{g\mu} f_L$$ (d)

$$V = \sqrt{\frac{2hL^2}{\alpha_0}|f_s|}$$ (e)

Figure 3. Measurements of alkali atoms confined in vapour cells, which could lead to realizations of a number of base SI units. (a) A time and frequency standard based on the measurement of the hyperfine frequency of laser-cooled atoms similar to conventional fountain clocks; (b) a length standard based on a laser locked to an optical transition in atoms for which the frequency has been measured; (c) a temperature standard based on Doppler thermometry; (d) a current standard based on the measurement of Zeeman shifts and a diffraction-grating measurement of length; (e) a voltage standard based on DC Stark shifts and an interferometric measurement of length (Colour online).

2.1. Time

Chip-scale atomic clocks [4] have been extensively developed and are beginning to see broad utility in the real world. However, because of the buffer-gas pressure shifts, CSACs require calibration after being fabricated and assembled and the output frequency drifts at a rate of near 10^{-8}/yr [3]. As a result, the SI-traceable uncertainty in the frequency of an uncalibrated commercial CSAC is in the range of 10^{-7}.

We are therefore beginning work to develop a microfabricated platform, shown in figure 2(a) for the creation of laser-cooled atoms in order to realize interaction-free measurements of the Cs hyperfine transition. This platform is based on the standard microfabricated cell fabrication method described above, but with the absence of an intentionally introduced buffer gas and with non-evaporable (or evaporable) getter pumps to maintain the vacuum. A major challenge to such a device is maintaining a vacuum sufficiently low, typically below 10^{-7} Torr, to enable the production of large quantities of cold atoms. Getter pumps can support ultra-high vacuum [5] in combination with other pumps but generally do not pump either He or CH_4. As a result we plan to use aluminosilicate glass as the window material, which is known to have He permeation rates orders of magnitude below [6], for example, borosilicate glass, in order to reduce the He permeation into the cell. We expect lifetimes of months to years by use of glasses with the lowest He permeation rates. A similar proposal was presented in [7].

As with other types of laser-cooled atomic clocks, such an approach enables long interaction times while avoiding collisional broadening and shifts typical of buffer-gas confinement. We anticipate using pulsed coherent population trapping to interrogate the cold atoms [8] in order to avoid the use of a microwave cavity. This approach does introduce additional systematics associated with the first-order Doppler shift and an AC Stark shift, but recent experimental evidence [9, 10] suggests these shifts can be reduced to below 10^{-13} for a Ramsey time of 10 ms, consistent with atoms falling 0.5 mm in free-fall with appropriate design choices. Light needed for this device could be produced with a modulated vertical-cavity surface emitting laser with optical feedback from an external reflector, which both narrows the linewidth and enhances the modulation efficiency at certain modulation frequencies [11].

2.2. Length

It is well-established that because the speed of light is defined as a fixed constant, lasers stabilized to atomic transitions with known optical frequencies become SI-traceable standards of length via interferometry and the relation

$$L = N\frac{c}{f_{\mathrm{opt}}}, \qquad (1)$$

where N is the number of fringes counted as the interferometer mirror separation moves a distance L, f_{opt} is the optical frequency of the interferometer laser (measured with respect to the SI second) and c is the speed of light. Previous work has established a number of suitable references [12], both in the visible and near-infrared and a telecommunications wavelengths [13]. The integration of appropriate atoms with chip-based single-mode photonics may ultimately allow highly miniaturized wavelength references compatible with complementary metal-oxide-semiconductor (CMOS) electronic and photonic fabrication methods.

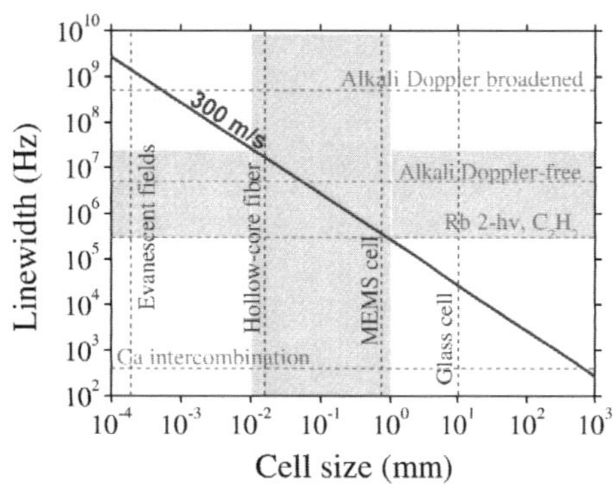

Figure 4. Transit time broadening in a room-temperature thermal vapour of atoms (Colour online).

Transit-time broadening imposes an important limitation on the wavelength reference performance: figure 4 shows this broadening for atoms moving at a typical thermal velocity of 300 m/s as a function of the size of the interaction region. Also drawn in the figure are a number of confinement/interaction platforms that have been used to allow the interaction of the light with atoms (glass cells, microfabricated (MEMS) cells, hollow-core fibers and evanescent fields) as well as a number of atomic transitions suitable as wavelength references. We target the size range between 10 μm and 1 mm, which is compatible with current methods of fabricating silicon-glass alkali vapor cells; the corresponding transit time broadening is 100 kHz to 10 MHz suggesting the use of optical references based on rotational-vibrational transitions in molecular gases, and multi-photon and Doppler-free transitions in alkali atoms.

There appear to be two clear alternatives for integrated atom-photonic systems. The first is to use silicon photonics with light at a wavelength (usually 1.3 μm – 1.5 μm) below the silicon bandgap, combined with molecular transitions in this wavelength range. Both C_2H_2 and HCN have both been investigated [14, 15] as wavelength standards for the telecom band. The strengths of this approach are: a well-developed manufacturing platform for Si photonics and a large number of integrated photonics components already demonstrated [16]; compatibility with the standard CMOS fabrication process; compatibility with optical fiber communications; narrow atomic lines; and many useful transitions. The main disadvantage is that the molecular absorption cross-sections at this wavelength range are quite weak, which results in small signals. An additional difficulty related specifically to miniaturization is transit-time broadening, which reaches the natural linewidth characteristic of molecular transitions of about 300 kHz at a cell length scale of 1 mm; cell sizes smaller than this will not be able to fully realize the advantages of the narrow transition. In addition, the two-photon non-linear absorption in silicon at this wavelength range results in lower Q-factors for optically resonant structures, resulting in difficulty generating frequency combs and limitations on non-linear optical interactions within the material.

The second alternative is to use silicon-nitride photonics [17], transparent at wavelengths down to 300 nm, with alkali vapor cells. There are a number of strengths of this approach including high optical cross-sections for alkali atoms; well-developed microfabricated cell fabrication; CMOS process compatibility; potential for integration with microfabricated frequency combs [18]; and the possibility of optically pumping the atoms. The main weaknesses are broader lines, the lack of free carriers in SiN that enable fast switching, and difficulties incorporating electrostatic actuation for moving components.

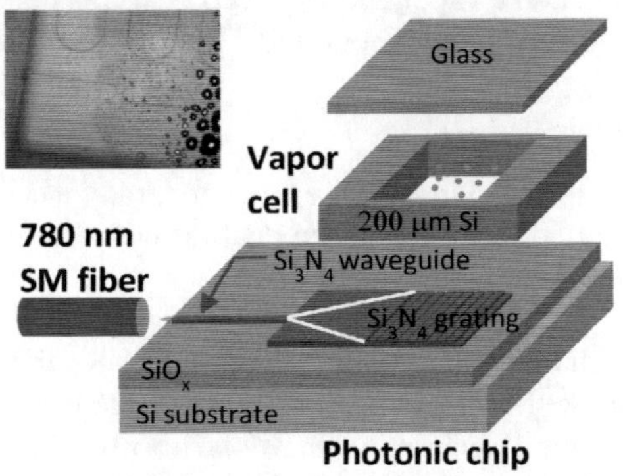

Figure 5. Design for photonically coupled alkali vapor cell. Inset: first such vapor cell fabricated at NIST showing SiN waveguides and alkali atom droplets condensed on the inside wall of the cell (Colour online).

Optical frequency references based on saturated absorption spectroscopy in alkali atoms currently achieve stabilities well below 10^{-12} at one second [19] and near 10^{-12} at one day [20]. Accuracies of this type of reference have been established to about 10^{-10} [21], limited largely by uncontrolled Zeeman shifts. Saturated absorption resonances have also been observed in on-chip hollow waveguides [22] and hollow-core optical fibers [23] and similar Doppler-free resonances are observed in thin glass cells [24]. Saturated absorption spectra have also been observed in microfabricated alkali vapor cells [25, 26]. A final interesting case is the possibility of using SiN photonics to interrogate the Rb two-photon transition at 778 nm for which accuracies 5×10^{-12} [27] have been observed.

In figure 5, we show one possible design, in which light confined in a SiN waveguide is expanded via diffraction in the slab before being coupled into the cell with a diffractive grating output coupler. The conversion from a small optical mode to a larger one allows for the measurement of narrower transition linewidths. The inset in figure 5 shows the first such cell fabricated at NIST. Narrow-linewidth light sources for such a wavelength reference could be a ring-cavity semiconductor laser, microresonator-based Brillouin laser [28], a linewidth-narrowed vertical-cavity surface emitting laser [11] or a vertical-cavity extended cavity laser with macroscopic cavity length.

2.3. *Temperature*

It is anticipated that in the near future, the Kelvin will be redefined by setting Boltzmann's constant to a fixed value and associating the energy $k_B T$ with, for example, the motional energy of atoms in a gas. The measurement of Doppler widths, Δf_D, in molecular and atomic gases can thus be used as an accurate measurement of temperature via the relation [29]

$$T = \frac{mc^2}{2k_B}\left(\frac{\Delta f_D}{f_0}\right)^2, \qquad (2)$$

where mc^2 is the rest energy and f_0 the transition frequency of the atom or molecule being measured. Over the last decade, a number of groups have carried out Doppler thermometry experiments in molecules in order to measure Boltzmann's constant. Uncertainties in k_B of 4×10^{-5} have been achieved [30], with projections of ultimate uncertainties in the 10^{-6} range.

As discussed above, the measurement of molecular transitions in compact vapor cells is challenging because of the small absorption cross-sections of the molecules in the near-infrared wavelength range. We therefore propose to realize an accurate thermometer using electronic transitions in alkali atoms [31]. This approach has the advantages of higher absorption cross-sections and reduced collisional effects but suffers from different systematics including optical pumping, unresolved hyperfine structure and a more limited temperature range of utility due to the change in vapor pressure with temperature.

2.4. Current and Voltage

Atomic vapors can also be used to realize the ampere and the volt via Larmor and DC Stark frequency shifts. Atomic g-factors relate the Larmor precession frequency f_L to the strength of the magnetic field, B, by $f_L = (g\mu/\hbar)B$ where μ is the nuclear or Bohr magneton as appropriate. Many nuclear and atomic electronic g-factors are known, traceable to the SI [32] to about 10^{-7} and hence atomic vapor cells can be used to define accurately a magnetic field. From elementary electrostatics, it is known that an infinite current sheet with a current density per unit length of i creates a magnetic field that is independent of distance from the sheet given by $B = (\mu_0/2)i$. If this current sheet is implemented using lithographically-defined metallic traces spaced by a distance L and a vapor cell is placed near this planar surface (see figure 2(d)), the current I can be determined in terms of the measured Larmor precession frequency by the relation

$$I = \frac{2L}{\mu_0}\frac{\hbar}{g\mu} f_L. \tag{3}$$

The wire spacing, L, which can be on the order of micrometers, can be measured accurately using techniques of grating spectrometry. The major technical challenge appears to be dealing with fringing fields and magnetic shielding. For the latter, AC current measurements may alleviate some of the difficulty.

Finally, in a manner similar to current, measurements of the DC Stark shift in Rydberg or Rydberg-like atoms provides a route to realizing the volt. The DC Stark shift is given as $f_S = -(\alpha_0/2h)E^2$ and the DC polarizabilities, α_0, of some atoms have been measured accurately; the polarizability of the neutral Yb atom, for example, has been measured to near 10^{-5} in optical clock experiments [33] and similar techniques should enable improvements on the 0.1 % uncertainties for shorter-wavelength two-photon transitions in alkali atoms [34]. For an atomic vapor cell placed between two parallel plates separated by a distance L (see figure 2(e)), the voltage on the plates can therefore be determined with respect to the measured Stark shift from

$$V = \sqrt{\frac{2hL^2}{\alpha_0}|f_S|}\,. \tag{4}$$

In this case, the plate separation can be determined (again using a chip-scale wavelength reference) via interferometry. The shift itself is measured using optical spectroscopy of two-photon transitions and a reference cell (with no applied fields) to stabilize the optical frequency. Major technical challenges here are fringing fields and charge build-up on surfaces, similar to those mentioned above for the realization of the ampere. In addition, however, the two-photon transitions often occur at wavelengths for which there are no convenient diode lasers and hence some laser development may be required to ultimately realize a compact, low-power voltage reference in this manner. Similar ideas are being developed for RF electric field measurements [35].

3. Conclusion

We have described a number of ways in which microfabricated alkali vapor cells might be used to realize compact, potentially easily manufactured, SI-traceable references of frequency, wavelength, temperature, current and voltage. All units are realized by some type of frequency-based measurement (RF, optical or microwave) of atomic energy levels. If additional challenges (light sources, packaging, thermal control, wafer-level fabrication, etc.) can be addressed, such references could be used to provide embedded calibration capability across a broad range of instruments and systems.

4. Acknowledgement

This work is a contribution of NIST, an agency of the US government and is not subject to copyright. We acknowledge helpful discussions with M. V. Romalis, comments on the manuscript from S. Kang and W. Loh and thank T. O'Brian for his support of this research program.

References

[1] Hossain K, Milton M and Nimmo B 2013 Metrology for the 2020s. http://www.npl.co.uk/publications/2020vision/
[2] Liew L A, Knappe S, Moreland J, Robinson H, Hollberg L and Kitching J 2004 *Appl. Phys. Lett.* **84** 2694-6.
[3] Lutwak R 2011 The SA.45S Chip-Scale Atomic Clock - Early Production Statistics *Proc. Precise Time and Time Interval (PTTI) Meeting* Long Beach, CA p 207-20.
[4] Knappe S, Shah V, Schwindt P D D, Hollberg L, Kitching J, Liew L A and Moreland J 2004 *Appl. Phys. Lett.* **85** 1460-2.
[5] Benvenuti C, Cazeneuve J M, Chiggiato P, Cicoira F, Santana A E, Johanek V, Ruzinov V and Fraxedas J 1999 *Vacuum* **53** 219-25.
[6] Norton F J 1953 *J. Am. Cer. Soc.* **36** 90-6.
[7] Rushton J A, Aldous M and Himsworth M D 2014 *Rev. Sci. Instr.* **85** 121501.
[8] Chen X, Yang G-Q, Wang J and Zhan M-S 2010 *Chin. Phys. Lett.* **27** 113201.
[9] Esnault F X, Blanshan E, Ivanov E N, Scholten R E, Kitching J and Donley E A 2013 *Phys. Rev. A* **88** 042120.
[10] Blanshan E, Rochester S M, Donley E A and Kitching J 2015 *Phys. Rev. A* **91** 041401(R).
[11] Gavra N, Ruseva V and Rosenbluh M 2008 *Appl. Phys. Lett.* **92** 221113.
[12] Helmcke J 2003 *Meas. Sci. Tech.* **14** 1187-99.
[13] Dennis T, Curtis E A, Oates C W, Hollberg L and Gilbert S L 2002 *J. Lightw. Technol.* **20** 776.
[14] de Labachelerie M, Nakagawa K, Awaji Y and Ohtsu M 1995 *Opt. Lett.* **20** 572-4.
[15] Lee K I, Eah S K and Jhe W 1995 *Opt. Quant. Elec.* **27** 441-5.
[16] Leuthold J, Koos C and Freude W 2010 *Nat. Phot.* **4** 535-44.
[17] Moss D J, Morandotti R, Gaeta A L and Lipson M 2013 *Nat. Phot.* **7** 597-607.
[18] Kippenberg T J, Holzwarth R and Diddams S A 2011 *Science* **332** 555-9.
[19] Qi X H, Chen W L, Yi L, Zhou D W, Zhou T, Xiao Q, Duan J, Zhou X J and Chen X Z 2009 *Chin. Phys. Lett.* **26** 044205.
[20] Affolderbach C and Mileti G 2005 *Opt. Laser Eng.* **43** 291-302.
[21] Udem T, Reichert J, Holzwarth R and Hansch T W 1999 *Phys. Rev. Lett.* **82** 3568-71.
[22] Yang W G, Conkey D B, Wu B, Yin D L, Hawkins A R and Schmidt H 2007 *Nat. Phot.* **1** 331.
[23] Slepkov A D, Bhagwat A R, Venkataraman V, Londero P and Gaeta A L 2010 *Phys. Rev. A* **81** 053825.
[24] Chevrollier M, Fichet M, Oria M, Rahmat G, Bloch D and Ducloy M 1992 *J. Phys. II* **2** 631-57.
[25] Knappe S A, Robinson H G and Hollberg L 2007 *Opt. Exp.* **15** 6293-9.
[26] Gruet F, Vecchio F, Affolderbach C, Petremand Y, de Rooij N F, Maeder T and Mileti G 2013 *Opt. Laser Eng.* **51** 1023-7.
[27] Poulin M, Latrasse C, Touahri D and Tetu M 2002 *Opt. Comm.* **207** 233-42.
[28] Loh W, Green A A S, Baynes F N, Cole D C, Quinlan F J, Lee H, Vahala K J, Papp S B and Diddams S A 2014 *Optica* **2** 225-32.
[29] Bordé C J 2005 *Phil. Trans. R. Soc. A* **363** 2177-201.
[30] Djerroud K, Lemarchand C, Gauguet A, Daussy C, Briaudeau S, Darquie B, Lopez O, Amy-Klein A, Chardonnet C and Borde C J 2009 *Comp. Rend. Phys.* **10** 883-93.
[31] Truong G-W, May E F, Stace T M and Luiten A N 2011 *Phys. Rev. A* **83** 033805.
[32] Phillips W D, Cooke W E and Kleppner D 1975 *Phys. Rev. Lett.* **35** 1619-22.
[33] Sherman J A, Lemke N D, Hinkley N, Pizzocaro M, Fox R W, Ludlow A D and Oates C W 2012 *Phys. Rev. Lett.* **108** 153002.
[34] Antypas D and Elliott D S 2011 *Phys. Rev. A* **83** 062511.
[35] Sedlacek J A, Schwettmann A, Kubler H, Low R, Pfau T and Shaffer J P 2012 *Nat. Phys.* **8**

819-24.

Journal of Physics: Conference Series **723** (2016) 012057 doi:10.1088/1742-6596/723/1/012057

Composite pulses in Hyper-Ramsey spectroscopy for the next generation of atomic clocks

T. Zanon-Willette[1], M. Minissale[2,3], V.I. Yudin[4] and A.V. Taichenachev[4]

[1] LERMA, Observatoire de Paris, PSL Research University, CNRS, Sorbonne Universités, UPMC Univ. Paris 06, F-75005, Paris, France

[2] Aix Marseille Université, CNRS, PIIM UMR 7345, 13397 Marseille, France

[3] Aix-Marseille Université, CNRS, Centrale Marseille, Institut Fresnel UMR 7249, 13013 Marseille, France

[4] Institute of Laser Physics, Siberian Branch of Russian Academy of Sciences, Novosibirsk 630090, Russia, Novosibirsk State University, Novosibirsk 630090, Russia, and Novosibirsk State Technical University, Novosibirsk 630092, Russia

E-mail: `thomas.zanon@upmc.fr`

Abstract. The next generation of atomic frequency standards based on an ensemble of neutral atoms or a single-ion will provide very stringent tests in metrology, applied and fundamental physics requiring a new step in very precise control of external systematic corrections. In the proceedings of the 8th Symposium on Frequency Standards and Metrology, we present a generalization of the recent Hyper-Ramsey spectroscopy with separated oscillating fields using composites pulses in order to suppress field frequency shifts induced by the interrogation laser itself. Sequences of laser pulses including specific selection of phases, frequency detunings and durations are elaborated to generate spectroscopic signals with a strong reduction of the light-shift perturbation by off resonant states. New optical clocks based on weakly allowed or completely forbidden transitions in atoms, ions, molecules and nuclei will benefit from these generalized Ramsey schemes to reach relative accuracies well below the 10^{-18} level.

1. Introduction

The story of very high precision laser pulsed spectroscopy has started in 1930 with the atomic and molecular beam resonance method invented by I.I. Rabi to improve the resolution of frequency measurements [1]. Under a quasi-resonant irradiation, a quantum two-level system can undergo coherent Rabi oscillations. In this case, the initial two-level occupation probabilities can be inverted by a so-called π pulse with an adequate selection of laser field frequency and pulse duration. By scanning the frequency of the irradiation around the exact resonance, a narrow spectroscopic resonance is observed that can be used to realize a very precise discriminator stabilizing a local frequency oscillator to the atomic or molecular transition. To improve even further the frequency resolution, N.F. Ramsey has proposed to replace the single oscillatory field by a double microwave excitation pulse with $\pi/2$ pulses separated by a free evolution time [2] as schematized in Fig. 1. The main advantage of the two-pulse Ramsey spectroscopy is to reduce the probing electromagnetic field perturbation on the atomic transition itself which has drastically impacted the time and frequency metrology with microwave atomic clocks since the 1950s [3].

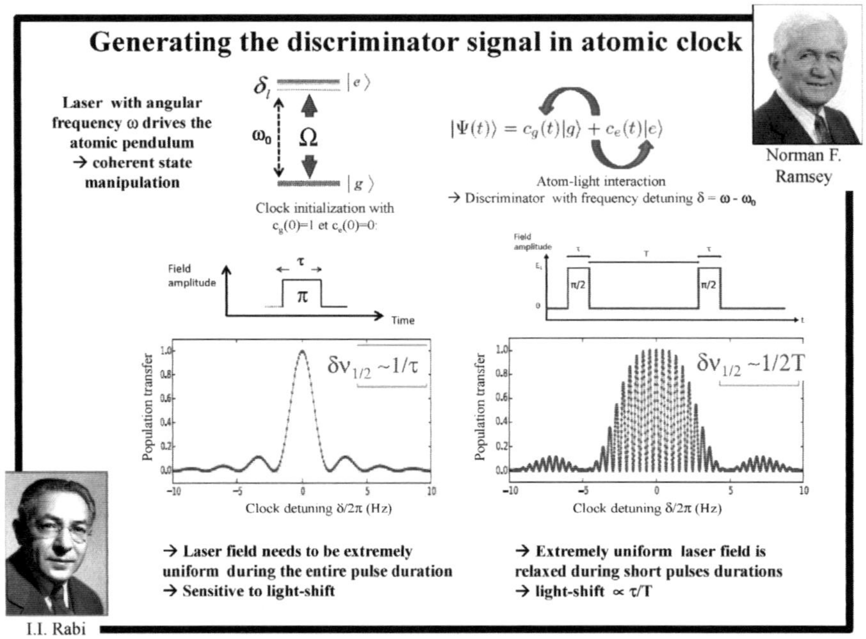

Figure 1. (color online). Quantum system with a narrow $|g\rangle \leftrightarrow |e\rangle$ clock transition probed by a single laser pulse excitation. The unperturbed clock detuning is introduced by δ and the Rabi frequency is Ω. The usual pulse area is defined by the product $\Omega\tau$. Rabi and Ramsey spectroscopy are two laser probing techniques generating a frequency discriminator for an atomic clock. The full width at half maximum is defined by $\delta\nu_{1/2}$.

Optical frequency-standards are today actively developed to replace microwave transitions by very narrow optical resonances in single ion, fermionic and bosonic alkaline-earth atoms [4]. Very long storage time of Doppler and recoil free quantum particles have been achieved using laser cooling techniques and the level of 10^{-18} relative accuracy will be probably almost achieved in the near future [5]. As an example, the lattice fermionic clock requires now a very precise control of atomic interactions to cancel some systematic frequency shifts below the relative 10^{-17} level of accuracy [6, 7]. Light-shifts represent a non negligeable issue for clocks based on bosonic neutral atoms with forbidden dipole transitions activated by mixing static magnetic field with a single laser [8, 9], for clocks with magic-wave induced transition in even isotopes [10] or with a E1-M1 two-photon laser excitation [11, 12, 13]. Because these important issues have to be addressed for bosons, we present in these proceedings a generalization of the Ramsey spectroscopy to reduce at a very low level of correction the light-shift induced by the probing laser itself on the atomic resonance.

2. Clock frequency shift in a generalized Ramsey two-pulse spectroscopy

In a Ramsey laser spectroscopy defined by two pulses with effective area θ_l ($l =$ i,k), the single Rabi's π pulse interaction zone is splitting into two very short $\pi/2$ pulses with duration τ separated by a free evolution time T. The atoms spend a much shorter time in the presence of the external electromagnetic fields minimizing inhomogeneities which may reduce precision in determining the transition frequency. When we consider a light-shift effect during the laser pulses, Ramsey spectroscopy renormalizes the light perturbation by the factor τ/T. If Ramsey fringes are largely shifted by off-resonant states, an additional laser frequency step may be applied to rectify the anticipated shift, thus requiring a redefinition of frequency detunings as

$\delta_l = \delta + \Delta_l$. The uncompensated residual contribution Δ_l present during pulses leads to a net clock frequency-shift of the central fringe expressed as:

$$\delta\nu = -\frac{\Phi(\delta \to 0, \Delta_l)}{2\pi \mathrm{T}} \tag{1}$$

where $\Phi(\delta \to 0, \Delta_l)$ is the zero order phase-shift affecting the central fringe when the unperturbed clock detuning δ between the laser frequency and the atomic transition is closed to zero. If we simply consider the extension of the two-pulse configuration, as in the initial Hyper-Ramsey scheme with different pulse areas [14], we found the global phase-shift expression as [13]:

$$\Phi_{HR} = \arctan\left[\frac{\frac{\delta_\mathrm{i}}{\omega_\mathrm{i}}\tan\theta_\mathrm{i} + \frac{\delta_\mathrm{k}}{\omega_\mathrm{k}}\tan\theta_\mathrm{k}}{1 - \frac{\delta_\mathrm{i}\delta_\mathrm{k}}{\omega_\mathrm{i}\omega_\mathrm{k}}\tan\theta_\mathrm{i}\tan\theta_\mathrm{k}}\right] \tag{2}$$

where the effective area is $\theta_l = \omega_l \tau_l/2$ and $\omega_l = \sqrt{\delta_l^2 + \Omega_l^2}$. The uncompensated part of the residual light-shift correction becomes proportional to the laser pulse area and is reduced to usual Ramsey phase-shift $\Phi_R = 2\arctan\left[\frac{\Delta}{\omega}\tan\theta\right]$ where $\omega = \sqrt{\Delta^2 + \Omega^2}$ in Eq. 2 when $\theta_\mathrm{i} = \theta_\mathrm{k} = \theta$ and $\delta_\mathrm{i} = \delta_\mathrm{k} \equiv \Delta$. Such a linear dependence with $\pi/2$ pulses versus the uncompensated part of the light perturbation is shown by the solid blue line in Fig. 2. The frequency shift response is usually observed in microwave atomic clocks using Ramsey spectroscopy [3]. More importantly, there is another pulse configuration which was never studied by Ramsey to strongly reduce the frequency-shift induced by the laser probe itself. By selecting in Eq. 2 effective pulse areas such

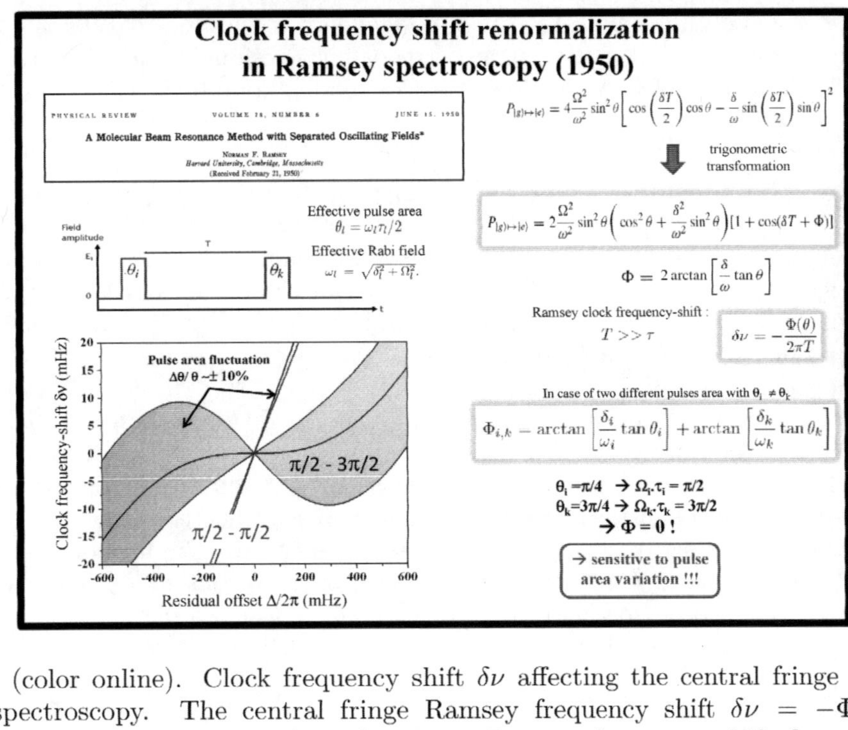

Figure 2. (color online). Clock frequency shift $\delta\nu$ affecting the central fringe in a Ramsey two-pulse spectroscopy. The central fringe Ramsey frequency shift $\delta\nu = -\Phi_R/2\pi T$ with $\pi/2 - \mathrm{T} - \pi/2$ pulses is compared to the Hyper-Ramsey frequency shift $\delta\nu = -\Phi_{HR}/2\pi T$ applying a $\pi/2 - \mathrm{T} - 3\pi/2$ configuration. The shadow area between lines corresponds to an uncompensated $\pm 10\%$ pulse area variation. The Rabi frequency is defined as $\Omega = \pi/2\tau$ where $\tau = 0.1875$ s and T$= 2$ s.

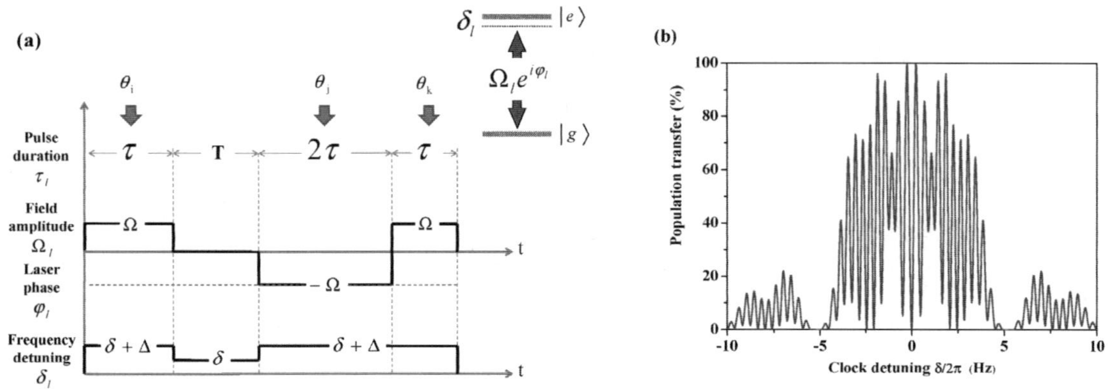

Figure 3. (color online). (a) Sequence of composite pulses $\theta_{\rm i} - {\rm T} - \theta_{\rm j}, \theta_{\rm k}$ with specified laser parameters including detuning δ_l, complex field amplitude $\Omega_l e^{i\varphi_l}$, pulse duration τ_l where $l = $ i,j,k and a free evolution time T between pulses. (b) Generalized Hyper-Ramsey resonance generated by composite pulses.

that $\theta_{\rm i} = \theta$ and $\theta_{\rm k} = 3\theta$ taking for example a pulse duration τ three times longer during the final pulse, we observe a nonlinear clock frequency-shift of the central Ramsey fringe versus the residual correction Δ shown by the solid black line in Fig. 2. The non standard reduction of the residual light-shift perturbation is produced by a combination of pulse areas $\theta \sim \pi/4$ ($\Omega\tau = \pi/2$) and $\theta \sim 3\pi/4$ ($\Omega\tau = 3\pi/2$) in Eq. 2. The linear dependence of the clock frequency shift on small variations of the light-shifts is totally removed except for a residual cubic correction originated by the phase of Eq. 2. However, a relative pulse area modification by $\pm 10\%$ is still affecting the result if it is not well compensated.

3. Generalized Hyper-Ramsey clock frequency shift with composite laser pulses

Generalization of the Ramsey scheme from the original two-pulse configuration [2] to more complex ones including additional intermediate pulses permits to reduce shifts and broadening due to inhomogeneous excitation conditions or shifts that are the result of the excitation itself [15, 16]. Sequence of three excitation pulses θ_l ($l = $ i,j,k) shown in Fig. 3(a) with suitably selected frequency δ_l and phase steps φ_l have been introduced in order to reduce the light-shift and to efficiently suppress the sensitivity of the spectroscopic signal to variations of the probe light intensity [14, 17]. These new schemes, based on the Generalized Hyper-Ramsey (GHR) resonance reported in Fig. 3(b), have been tested experimentally for few clock transitions as the ultra-narrow electric octupole transition in the $^{171}Yb^+$ clock [18]. Composite pulses inspired by NMR techniques are used to compensate simultaneously for noise decoherence, pulse area fluctuation and residual frequency offset due to the applied laser field itself. The first or the second pulse of the usual Ramsey sequence can be separated in two or more contiguous sections which yield to new efficient probing protocols with more degrees of freedom to minimize resonance shifts. The generalized Hyper-Ramsey phase-shift accumulated after a three-pulse interrogation scheme becomes [17]:

$$\Phi_{GHR} = \arctan\left[\frac{\dfrac{\frac{\delta_j}{\omega_j}\tan\theta_j+\frac{\delta_k}{\omega_k}\tan\theta_k}{1-\frac{\delta_j\delta_k+\Omega_j\Omega_k}{\omega_j\omega_k}\tan\theta_j\tan\theta_k}+\dfrac{\frac{\delta_i}{\omega_i}\tan\theta_i+\frac{\delta_{jk}}{\omega_{jk}}\tan\theta_{jk}}{1-\frac{\delta_i\delta_{jk}}{\omega_i\omega_{jk}}\tan\theta_i\tan\theta_{jk}}}{1-\dfrac{\frac{\delta_j}{\omega_j}\tan\theta_j+\frac{\delta_k}{\omega_k}\tan\theta_k}{1-\frac{\delta_j\delta_k+\Omega_j\Omega_k}{\omega_j\omega_k}\tan\theta_j\tan\theta_k}\,\dfrac{\frac{\delta_i}{\omega_i}\tan\theta_i+\frac{\delta_{jk}}{\omega_{jk}}\tan\theta_{jk}}{1-\frac{\delta_i\delta_{jk}}{\omega_i\omega_{jk}}\tan\theta_i\tan\theta_{jk}}}\right] \tag{3}$$

with a reduced variable

$$\frac{\delta_{jk}}{\omega_{jk}}\tan\theta_{jk} = \frac{(\delta_j\Omega_k-\Omega_j\delta_k)\tan\theta_j\tan\theta_k}{\Omega_j\omega_k\tan\theta_j+\Omega_k\omega_j\tan\theta_k} \tag{4}$$

The key point is to establish some efficient quantum control protocols which compensate for frequency-shift. Moreover the quantum control protocols have to be robust to small change in pulse area while achieving a highly contrasted population transfer between the targeted states [14, 17]. By imposing specific pulse areas in Eq. 3 as $\theta_i = \theta_k = \theta$ ($\Omega\tau = \pi/2$) and $\theta_j = 2\theta$ ($\Omega\tau = \pi$) while fixing laser field phases at $\varphi_i = \varphi_k = 0$ and $\varphi_j = \pi$ ($\Omega \to -\Omega$) during the intermediate pulse, a very good compensation of the $\pm 10\%$ relative pulse area variation is demonstrated as reported in Fig. 4.

It seems that combinations of generalized Hyper-Ramsey spectral resonances including some phase-step controls and different pulse durations may demonstrate in the future their capacity to eliminate systematic corrections including potential biases or higher-order light-shift contributions from the probe laser. The next generation of optical clocks based on weakly allowed or completely forbidden transitions in atoms, ions, molecules and nuclei will benefit from these

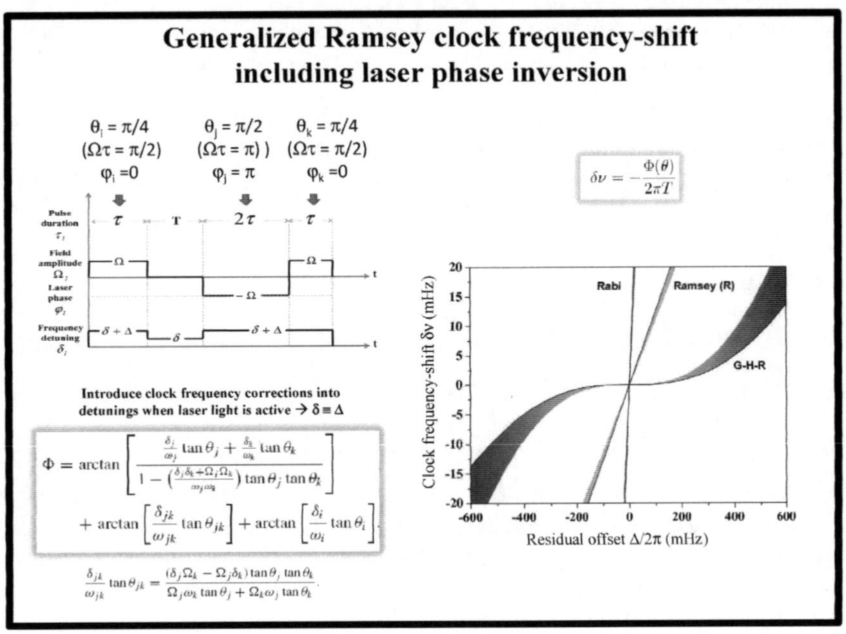

Figure 4. (color online). Generalized Hyper-Ramsey frequency shift $-\Phi_{GHR}/2\pi T$ versus a residual uncompensated light-shift correction. The sequence of composite pulses with $\pi/2 - \mathrm{T} - \pi, \pi/2$ pulses (including $\varphi_j = \pi$ during the intermediate π-pulse) is capable of minimizing the pulse area variation to realize a robust frequency discriminator against a residual light-shift perturbation from the probing laser itself.

generalized Hyper-Ramsey schemes to reach relative accuracies lower than the 10^{-18} level. The Generalized Hyper-Ramsey resonance with separated oscillating fields has also been extended to bosonic three-levels systems using optical stimulated Raman transitions [13] and some recent modified Hyper-Ramsey spectroscopy schemes based on various phase-step modulations of laser pulses have been proposed to completely eliminate the light-shift in optical clocks [19, 20, 21].

Acknowledgments

T. Zanon-Willette deeply acknowledges the organizers of the 8th symposium on frequency standards and metrology. The related proceedings are based on the material presented during the talk delivered on the 16th of october 2015.

V.I.Yu. and A.V.T. were supported by the RFBR (grants 14-02-00712, 14-02-00939, 14-02-00806), by the Russian Academy of Sciences, by Presidium of the Siberian Branch of the Russian Academy of Sciences, by the RF Ministry of Education and Science (state assignment No. 2014/139 project No. 825).

References

[1] Rabi I I, Zacharias J R, Millman S and Kusch P 1938 A New Method of Measuring Nuclear Magnetic Moment *Phys. Rev.* **53** 318

[2] Ramsey N F 1950 A Molecular Beam Resonance Method with Separated Oscillating Fields *Phys. Rev.* **78** 695

[3] Vanier J and Audoin C 1989 *The quantum physics of atomic frequency standards* (Adam Hilger IOP, Bristol)

[4] Ludlow A, D Boyd M M, Ye J, Peik E and Schmidt P O 2015 Optical atomic clocks *Rev. Mod. Phys.* **87** 637

[5] Derevianko A, Katori H 2011 Colloquium: Physics of optical lattice clocks *Rev. Mod. Phys.* **83** 331

[6] Hinkley N, Sherman J A, Phillips N B, Schioppo M, Lemke N D, Beloy K, Pizzocaro M, Oates C W and Ludlow A D (2013) An Atomic Clock with 10^{-18} Instability *Science* **341** 1215

[7] Nicholson T L, Campbell S L, Hutson R B, Marti G E, Bloom B J, McNally R L, Zhang W, Barrett M D, Safronova M S, Strouse G F, Tew W L and Ye J 2015 Systematic evaluation of an atomic clock at 2×10^{-18} total uncertainty *Nature Comm.* **6** 7896

[8] Taichenachev A V, Yudin V I, Oates C W, Hoyt C W, Barber Z W and Hollberg L 2006 Magnetic Field-Induced Spectroscopy of Forbidden Optical Transitions with Application to Lattice-Based Optical Atomic Clocks *Phys. Rev. Lett.* **96** 083001

[9] Barber Z, HoytC, Oates C, Hollberg L, Taichenachev A and Yudin V 2006 Direct Excitation of the Forbidden Clock Transition in Neutral 174Yb Atoms Confined to an Optical Lattice *Phys. Rev. Lett.* **96** 083002

[10] Ovsiannikov V D, Pal'chikov V G, Taichenachev A V, Yudin V I, Katori H, and Takamoto M 2007 Magic-wave-induced $^1S_0$-$^3P_0$ transition in even isotopes of alkaline-earth-metal-like atoms *Phys. Rev.* A **75** 020501(R)

[11] Santra R, Arimondo E, Ido T, Greene C H and Ye J 2005 High-Accuracy Optical Clock via Three-Level Coherence in Neutral Bosonic ^{88}Sr *Phys. Rev. Lett.* **94** 173002

[12] Zanon-Willette T, Ludlow A D, Blatt S, Boyd M M, Arimondo E and Ye J 2006 Cancellation of Stark Shifts in Optical Lattice Clocks by Use of Pulsed Raman and Electromagnetically Induced Transparency Techniques *Phys. Rev. Lett.* **97** 233001

[13] Zanon-Willette T, Almonacil S, de Clercq E, Ludlow A D and Arimondo E 2014 Quantum engineering of atomic phase shifts in optical clocks *Phys. Rev.* A **90** 053427

[14] Yudin V I, Taichenachev A V, Oates C W, Barber Z W, Lemke N D, Ludlow A D, Sterr U, Lisdat Ch and Riehle F 2010 Hyper-Ramsey spectroscopy of optical clock transitions *Phys. Rev.* A **82** 011804(R)

[15] Hahn E L 1950 Spin Echoes *Phys. Rev.* **80** 580

[16] Levitt M H 1986 Composite pulses *Prog. Nucl. Mag. Res. Spect.* **18** 61

[17] Zanon-Willette T, Yudin V I, Taichenachev A V 2015 Generalized hyper-Ramsey resonance with separated oscillating fields *Phys. Rev.* A **92** 023416

[18] Huntemann N, Lipphardt B, Okhapkin M, Tamm Chr, Peik E, Taichenachev A V and Yudin V I 2012 Generalized Ramsey Excitation Scheme with Suppressed Light Shift *Phys. Rev. Lett.* **109** 213002

[19] Hobson R, Bowden W, King S A, Baird P E G, Hill I R, Gill P 2016 Modified hyper-Ramsey methods for the elimination of probe shifts in optical clocks *Phys. Rev.* A **93** 010501(R)

[20] Zanon-Willette T, de Clercq E and Arimondo E 2015 Light-shift elimination in Generalized Hyper-Ramsey quantum clocks *arXiv* **1511.04847** (2015).

[21] Yudin V I, Taichenachev A V, and Basalaev M Yu 2016 Synthetic Frequency Protocol in the Ramsey Spectroscopy of Clock Transitions *arXiv* **1602.00331** (2016).

Journal of Physics: Conference Series **723** (2016) 012058 doi:10.1088/1742-6596/723/1/012058

Compact atom-interferometer gyroscope based on an expanding ball of atoms

S Riedl, G W Hoth, B Pelle, J Kitching and E A Donley

Time and Frequency Division, National Institute of Standards and Technology, Boulder, CO, USA

E-mail: elizabeth.donley@nist.gov

Abstract. We present a compact atom interferometer based on ^{87}Rb atoms that can simultaneously measure rotations and accelerations with a single expanding ball of atoms in a 300 cm^3 vacuum package.

1. Introduction

Inertial sensors based on phase shifts accumulated by matter waves in atom interferometers are a promising technology for high-performance inertial navigation. Acceleration and rotation measurements with these sensors have demonstrated sensitivities of 4.2×10^{-8} m/s^2/$\sqrt{\text{Hz}}$ in a cold-atom gravimeter [1], 80 μdeg/$\sqrt{\text{hr}}$ in a gyroscope based on thermal atomic beams [2, 3], and 400 μdeg/$\sqrt{\text{hr}}$ in a cold-atom gyroscope [4]. These instruments comfortably operate within the specifications for navigation-grade performance (5×10^{-5} m/s^2/$\sqrt{\text{Hz}}$ and 1×10^{-3} deg/$\sqrt{\text{hr}}$) [5]. However, atom interferometers that achieve this performance are many meters in scale. Our goal is to apply these methods in dramatically smaller systems and explore the trade-offs between performance and portability. Here we consider atoms confined in an active volume of 1 cm^3.

While high-performance atom interferometers are becoming more common in laboratory environments, there are only a few published demonstrations performed on moving vehicles [6-8], including an airborne accelerometer demonstration in a vibrationally noisy environment [6, 7]. While portable, these instruments were very complex, and the equipment to run them occupied several cubic meters. Atom-interferometer gyroscopes are even more complex given that they require counter-propagating atom sources with separate atom launch and detection systems. Simplifications to the approach of building atom-interferometer gyroscopes may eventually facilitate their use in portable applications.

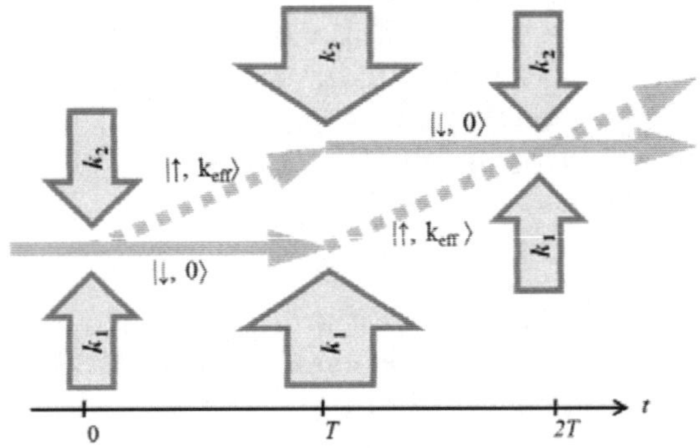

Figure 1. Mach-Zehnder type atom interferometer. The lines represent classical trajectories for the atoms, which undergo transitions between the internal states $|\uparrow\rangle$ (dashed) or $|\downarrow\rangle$ (solid lines). The red arrows represent the Raman pulses. (Colour online).

Most atom-interferometer inertial sensors are based on stimulated Raman transitions [9], where atoms are excited into superposition states with pairs of counterpropagating laser beams tuned into Raman resonance with the atomic hyperfine splitting. The absorption and stimulated emission of photon pairs imparts momentum to the atoms and simultaneously alters the internal hyperfine level [10]. Typically three Raman pulses are used (Red arrows in figure 1). The first $\pi/2$ pulse prepares each atom in a superposition state. The atomic wavepacket is split such that the part that has undergone a stimulated Raman transition also receives a momentum kick of $\hbar\vec{k}_{\text{eff}}$, where $\vec{k}_{\text{eff}} = (\vec{k}_1 - \vec{k}_2)$ is the effective wave vector for the two-photon Raman transition and $\hbar$ is the reduced Plank's constant. After the $\pi/2$ pulse, the phase of each state evolves independently for a period, T, and then a π pulse exchanges the atomic states. The atoms are again allowed to evolve for a period T. Finally, another $\pi/2$ pulse interferes the states and the combined acceleration- and rotation-induced phase shift is determined from a measurement of the final atomic hyperfine population. For atoms moving with an initial velocity $\vec{v}$, this phase shift is

$$\varphi_{\text{int}} = \varphi_{\text{acc}} + \varphi_{\text{rot}} = \vec{k}_{\text{eff}} \cdot \vec{a}T^2 + 2\vec{k}_{\text{eff}} \cdot (\vec{\Omega} \times \vec{v})T^2, \tag{1}$$

where φ_{acc} and φ_{rot} are the acceleration- and rotation-induced phase shifts, $\vec{a}$ is the acceleration, and $\vec{\Omega}$ is the rotation rate. The rotation term can also be expressed in the Sagnac form as $\varphi_{\text{rot}} = \frac{2m}{\hbar} \vec{\Omega} \cdot \vec{A}$, which is proportional to the enclosed area $\vec{A}$ and the rotation rate, and where m is the atomic mass. The interferometer phase is sensitive to accelerations along the direction of the light propagation and rotations perpendicular to this direction (and perpendicular to $\vec{v}$).

The interferometer signal is extracted from the hyperfine population after the third pulse as

$$P(\uparrow) = 1 - P(\downarrow) = \frac{1}{2}(1 - C\cos(\varphi_{\text{int}} + \varphi_0)). \tag{2}$$

$P(\uparrow)$ $[P(\downarrow)]$ is the probability that the atom is in the upper [lower] hyperfine state at the interferometer output, φ_0 is a constant phase offset, and C is the interferometer fringe contrast.

The acceleration and rotation components of φ_{int} cannot typically be distinguished with a single measurement; they are typically extracted by use of two independent counterpropagating atom sources. Because the sign of φ_{rot} depends on the sign of $\vec{v}$, subtracting the $+v$ signal from the $-v$ signal uniquely determines $\vec{\Omega}$. In thermal atomic-beam gyroscopes, this is accomplished with two opposing atom source ovens [2]. For interferometer gyroscopes based on laser-cooled atoms, two separate magneto-optic traps (MOTs) collect and launch atoms in opposite directions [11, 12].

Our system is a cold-atom combination accelerometer/gyroscope based on a single expanding ball of laser-cooled atoms probed using spatially resolved detection of the atomic states. This approach is much simpler than the conventional designs for atom interferometer gyroscopes that require differential measurements from two atom sources. Our work extends the point source interferometry (PSI) technique originally demonstrated in a 10-m tower at Stanford [13] into a new regime for portable applications. Our vacuum system is 100× smaller in height, our atoms are 1000× warmer, and our interaction periods are 200× shorter than those reported in [13]. Our ultimate goal is to determine if such a compact system with a short interrogation period can be competitive in high-performance inertial navigation applications.

2. Experimental Approach

The PSI technique uses a single source of atoms with a broad range of velocities in combination with spatially resolved detection to separate acceleration- and rotation-induced phase shifts. Because the individual atom velocities are correlated with their final positions after the interferometer sequence, rotational phase shifts appear as spatial interference fringes in the atomic states across the atom ball, while the overall phase shift of the central fringe determines the acceleration.

The principles of the sensor are shown in figure 2. An ensemble of ^{87}Rb atoms is collected and cooled in a magneto-optical trap (MOT) and post-cooled with an optical molasses into a volume smaller than a millimeter in diameter. After a brief free fall that allows the Doppler-sensitive and insensitive transitions to be resolved [14] (not shown), the Raman pulses are applied by a pair of laser beams counter-propagating along $+\hat{z}$ and $-\hat{z}$.

Because of the correlation between the atoms' initial velocity and their final position, the final rotation-induced interferometer phase is a function of spatial location. With the $\vec{k}_{\text{eff}}$ along $+\hat{z}$ and for rotation about $\hat{y}$, atoms moving toward $+\hat{x}$ will produce a negative phase shift, while those moving along $-\hat{x}$ will have a positive phase shift. The phase shift will be linear in spatial position $\hat{x}$ at the end of the second interaction period. Similarly, rotation about $\hat{x}$ will cause a linearly varying phase along $\hat{y}$. The spatially-dependent phase is read out using absorption imaging of the cloud with a resonant laser propagating along $\hat{z}$. Under rotation, fringes in the atom population in the absorption image will be observed in the direction $\vec{k}_{\text{eff}} \times \vec{\Omega}$, and can be distinguished around any axis in the $\hat{x} - \hat{y}$ plane.

Physically, this phase shift arises because the expanding cloud contains many atom interferometer trajectories with different enclosed areas. The small size of the initial atom cloud and the radial expansion of the cloud during the interaction period create a correlation between the final atom position and the area enclosed by the atom trajectory. For example, atoms having an initial velocity perpendicular to the Raman beams will have a maximal enclosed area of $\left(\frac{\hbar k_{\text{eff}}}{m}\right) v_0 T^2$, while atoms having an initial velocity v_0 parallel to the Raman beams will have zero enclosed area, because the momentum imparted by the Raman beams is entirely along the direction of the initial velocity. In general, the enclosed area only depends on the projection of the final position of the atom in the $\hat{x} - \hat{y}$ plane perpendicular to $\vec{k}_{\text{eff}}$. Thus, for fixed interaction time, T, all atoms at the same $\hat{x} - \hat{y}$ position (independent of $\hat{z}$ position) undergo the same phase shift and complete the interferometer pulse sequence in the same superposition of hyperfine states.

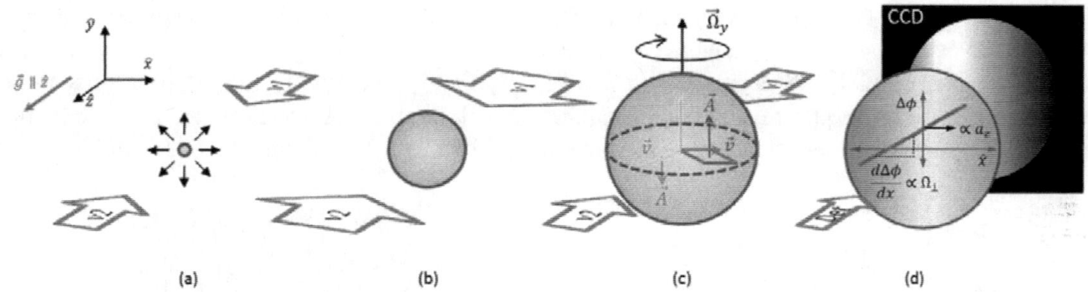

Figure 2. Atom interferometry with an expanding ball of atoms. (a) Initial expanding cloud. (b) Second Raman-Ramsey pulse applied mid-way through expansion. (c) After the full expansion, there is a phase gradient on the cloud. This occurs because the transverse velocity of each atom determines the area enclosed by the trajectory of that atom through the interferometer. This is illustrated here with two sample trajectories corresponding to different transverse velocities. (d) The interferometer phase gradient is imprinted in the atom populations and is read out by a detection beam. The phase gradient is perpendicular to both $\vec{k}_{\text{eff}}$ and $\vec{\Omega}$. The offset of the phase gradient from zero at the cloud's center depends on the acceleration component along $\hat{z}$. In this example as well as in the experiment, the Raman pulses are parallel to gravity. (Colour online).

3. Apparatus

A schematic diagram of the laser system and the physics package is shown in figure 3.

3.1. Vacuum System

The vacuum system is based on a 7.5 cm^3 uncoated spectrometer cell fused to a glass tee that is equipped with two Rb dispensers and sealed onto a 34 mm mini-conflat flange. The cell is mounted to a mini-conflat six-way cross opposite to a glass viewport. A 2 l/s ion pump and a non-evaporable getter pump are also mounted to the six-way cross. The volume of the entire vacuum system is ~300 cm^3. The system is currently not magnetically shielded.

3.2. Laser System

There are currently three distributed Bragg reflector (DBR) lasers used in the interferometer. One laser is used for the MOT cooling light and is locked by saturated absorption spectroscopy and shifted by an AOM to a detuning of two linewidths from the $F = 2 \rightarrow F' = 3$ cycling transition. This laser serves as a reference for the Raman lasers, whose frequency differences from the MOT laser are computer-controlled and set to different values as the lasers play various roles during the experiment [15].

The first of these two frequency-agile lasers is frequency locked to the MOT laser and serves both as the Raman beam for the $F = 2 \rightarrow F'$ transition and the resonant probe beam for detection. The second frequency-agile laser serves as the Raman beam for the $F = 1 \rightarrow F'$ transition, as a repump laser for laser cooling and detection of the $F = 1$ atoms after an interferometer cycle, and for optically pumping the atoms into the $F = 1$, $m_F = 0$ state at the start of each interferometer cycle. The two frequency-agile lasers are phase locked to each other with a frequency offset of 6.834 GHz with an optical phase-locked loop. We typically have about 70 % of the beat-note power in the coherent carrier, which is limited by the frequency modulation bandwidth of the DBR lasers. The difference frequency between the Raman lasers is chirped at a rate of $\vec{k}_{\text{eff}} \cdot \vec{g} = 25.1\ \text{kHz/ms}$ to maintain Raman resonance for all three interferometer pulses. The frequencies of the Raman beams are typically detuned by 200 MHz from the $F' = 0$ resonance. The Raman beams are overlapped with orthogonal polarizations and fiber coupled to the apparatus with a PM fiber. The Raman beams are 7 mm in diameter ($1/e^2$). Typically 4 mW of $F = 1$ light and 10 mW of $F = 2$ light are used for the Raman pulses. The Raman beams go through the cell vertically and the $F = 1 \rightarrow F'$ beam is reflected from a tip/tilt mirror below the cell. The $F = 2 \rightarrow F'$ beam is not retroreflected, and also serves as a probe beam for detection.

The MOT light is fiber coupled to the vacuum chamber via a three-way fiber beamsplitter and each beam is retroreflected. The MOT beams are 7 mm in diameter ($1/e^2$) and 6 mW of laser power is used per beam. Two of the MOT beams enter the cell at 45°. Typically, the cloud contains 10^7 atoms at the start of an interferometer sequence, and the initial Gaussian cloud diameter is ~500 μm (2 σ). The atom ball expands isotropically after release with a typical atom speed of about 3 cm/s ($T \cong 10\ \mu$K).

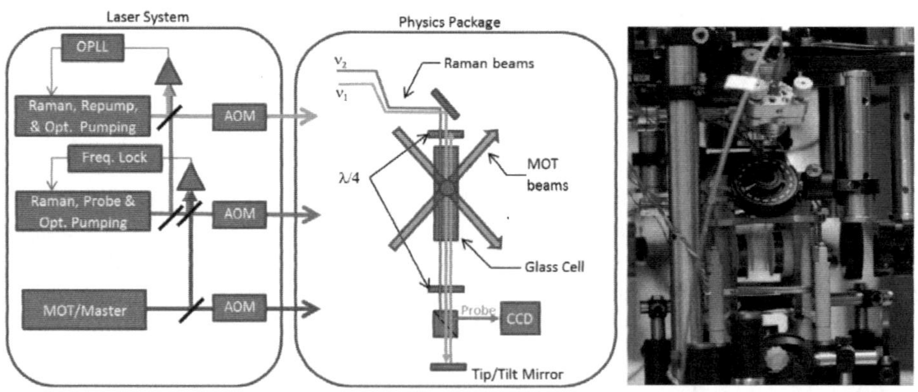

Figure 3. Schematic diagram of the laser system and physics package and an actual photo of the vacuum chamber surrounded by optics and coils. (Colour online).

4. Experiment

4.1. Typical Experimental Sequence

The main steps in a typical interferometer sequence are shown in Table I. The typical cycle time for one measurement is currently 3 s. This is limited by the camera read-out time and by the need to acquire background light and dark images for absorption detection, which does not allow for atom recapture. So far, we have not focused on minimizing the cycle time because we have been characterizing systematics and noise. In the near future, we will minimize the cycle time to optimize the instrument's sensitivity.

Table 1. Main steps in a typical interferometer sequence for observing interferometer fringes. For technical reasons, the acquisition of the light and dark images for absorption imaging is slow and not optimized in the experiment's current implementation. There are many other short trigger and wait periods needed to run the experiment that are not included in this list.

Time	Process
0 to 1.5 s	Load MOT
Next 5 ms	Optical Molasses
Next 1.4 ms	Steps to prepare atoms in $F = 1$, $m_F = 0$
Next 5 ms	Atoms undergo free fall and acquire velocity, which allows the Doppler sensitive and insensitive Raman transitions to be differentiated
Next 3 μs	First Raman $\pi/2$ pulse
Next 10 ms	Wait period T
Next 6 μs	Raman π pulse
Next 10 ms	Second wait period T
Next 3 μs	Second Raman $\pi/2$ pulse
Next 100 μs	Acquire absorption image of atoms in $F = 2$ state
Next 150 μs	Prep and pump $F = 1$ atoms into $F = 2$
Next 100 μs	Acquire absorption image of all atoms ($F = 1$ and $F = 2$)
Next 535 ms	Acquire light and dark reference images for $F = 1$ and $F = 1+2$ imaging conditions.

4.2. Simulating Rotations

We employ a commercial tip/tilt mirror as the Raman-beam retroreflecting optic to simulate and characterize rotations in our system. Compared to rotating the optical table, this approach is a simple way to simulate rotations that is equivalent to rotating the whole apparatus with respect to the local geostationary frame [16]. The mirror begins to rotate a few milliseconds before the first Raman pulse and rotates at a constant rate during the interferometer sequence.

4.3. Initial Results

We have demonstrated sensitivity to acceleration and rotation with our instrument. We have simulated two-axis rotations of the interferometer by tilting the Raman axis between light pulses. The spatial fringes have the expected properties; they are perpendicular to the axis of rotation, have a wavelength that is inversely proportional to the rotation rate Ω, and decrease in contrast with increasing Ω.

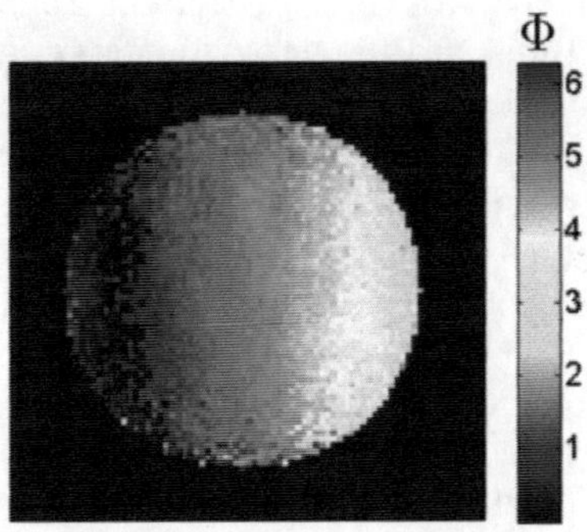

Figure 4. Spatial interference fringes from a rotation rate of Ω = 33 mrad/s (S/N = 40). To construct this image, the Raman chirp rate was scanned over several interferometer fringes (100 images) and Φ was found from a sinusoidal fit to the atom population difference for each pixel. In an actual instrument, 3 images would be required to measure the phase map, and the S/N would be 7. Here, T was 6 ms. The image size is 1.75 mm ×1.75 mm. (Colour online).

5. Outlook

Our current efforts focus on measurements of the gyroscope's scale factor and dynamic range and related systematics. Our medium-term efforts will focus on understanding and reducing the noise, since the current signal-to-noise ratio for the interferometer fringes averaged over the cloud in our interferometer is only ~4. In the long term, we plan to characterize our interferometer as an inertial sensor and to explore the possibility of developing a truly miniaturized system.

Acknowledgments
Special thanks goes to E. Ivanov for technical help and assistance. S. Krzyzewski, S. Papp, M. Lombardi, and W. M. Haynes are gratefully acknowledged for their helpful comments on this manuscript. This work is funded by NIST, which is an agency of the U.S. government, and is not subject to copyright.

References

[1] Hu Z-K, Sun B-L, Duan X-C, Zhou M-K, Chen L-L, Zhan S, Zhang Q-Z and Luo J 2013 Demonstration of an ultrahigh-sensitivity atom-interferometry absolute gravimeter, *Phys. Rev. A* **88** 043610
[2] Gustavson T L, Landragin A and Kasevich M A 2000 Rotation sensing with a dual atom-interferometer sagnac gyroscope, *Classical Quant Grav* **17** 2385-2398
[3] Durfee D S, Shaham Y K and Kasevich M A 2006 Long-term stability of an area-reversible atom-interferometer sagnac gyroscope, *Physical Review Letters* **97** 240801
[4] Berg P, Abend S, Tackmann G, Schubert C, Giese E, Schleich W P, Narducci F A, Ertmer W and Rasel E M 2015 Composite-light-pulse technique for high-precision atom interferometry, *Physical Review Letters* **114** 063002
[5] Jekeli C 2005 Navigation error analysis of atom interferometer inertial sensor, *NAVIGATION: Journal of The Institute of Navigation* **52** 1-14
[6] Stern G, Battelier B, Geiger R, Varoquaux G, Villing A, Moron F, Carraz O, Zahzam N, Bidel Y, Chaibi W, Pereira Dos Santos F, Bresson A, Landragin A and Bouyer P 2009 Light-pulse atom interferometry in microgravity, *The European Physical Journal D* **53** 353-357
[7] Geiger R, Menoret V, Stern G, Zahzam N, Cheinet P, Battelier B, Villing A, Moron F, Lours M, Bidel Y, Bresson A, Landragin A and Bouyer P 2011 Detecting inertial effects with airborne matter-wave interferometry, *Nat Commun* **2:474** doi: 10.1038 / ncomms1479
[8] Schmidt M, Senger A, Hauth M, Freier C, Schkolnik V and Peters A 2011 A mobile high-precision absolute gravimeter based on atom interferometry, *Gyroscopy and Navigation* **2** 170-177
[9] Kasevich M and Chu S 1991 Atomic interferometry using stimulated raman transitions, *Physical Review Letters* **67** 181-184
[10] Borde C J 1989 Atomic interferometry with internal state labelling, *Physics Letters A* **140** 10-12
[11] Canuel B, Leduc F, Holleville D, Gauguet A, Fils J, Virdis A, Clairon A, Dimarcq N, Bordé C, Landragin A and Bouyer P 2006 Six-axis inertial sensor using cold-atom interferometry, *Physical Review Letters* **97** 010402
[12] Müller T, Gilowski M, Zaiser M, Berg P, Schubert C, Wendrich T, Ertmer W and Rasel E M 2009 A compact dual atom interferometer gyroscope based on laser-cooled rubidium, *The European Physical Journal D* **53** 273-281
[13] Dickerson S M, Hogan J M, Sugarbaker A, Johnson D M S and Kasevich M A 2013 Multiaxis inertial sensing with long-time point source atom interferometry, *Physical Review Letters* **111** 083001
[14] Kasevich M, Weiss D S, Riis E, Moler K, Kasapi S and Chu S 1991 Atomic velocity selection using stimulated raman transitions, *Physical Review Letters* **66** 2297-2300
[15] Cheinet P, Pereira Dos Santos F, Petelski T, Le Gouët J, Kim J, Therkildsen K T, Clairon A and Landragin A 2006 Compact laser system for atom interferometry, *Applied Physics B* **84** 643-646
[16] Lan S-Y, Kuan P-C, Estey B, Haslinger P and Müller H 2012 Influence of the coriolis force in atom interferometry, *Physical Review Letters* **108** 090402

Towards a measurement of the nuclear clock transition in ^{229}Th

Simon Stellmer[1], Matthias Schreitl[1], Georgy Kazakov[1], Koji Yoshimura[2] and Thorsten Schumm[1]

[1] Vienna Center for Quantum Science and Technology (VCQ) and Atominstitut, TU Wien, 1020 Vienna, Austria

[2] Research Core for Extreme Quantum World, University of Okayama, Okayama, Japan

E-mail: simon.stellmer@ati.ac.at

Abstract. We investigate a potential candidate for a future optical clock: the nucleus of the isotope ^{229}Th. Over the past 40 years of research, various experiments have found evidence for the existence of an isomeric state at an energy of a few eV. So far, neither the energy nor the lifetime of the isomer have been determined directly. As the scene is not yet prepared for direct laser excitation, other means of populating the isomer need to be explored. We investigate three different approaches, all of which rely on CaF_2 crystals doped with ^{229}Th or ^{233}U. Various kinds of crystal luminescence are discussed in detail.

1. The few-eV isomer in ^{229}Th

Any scientific clock is evaluated along two dimensions: its accuracy to tell the time, and its (in)sensitivity to external perturbations. A promising new platform that could "host" a clock that excels in both aspects is the nucleus of an atom or ion. The system would be intrinsically very small (typical size 5 fm) and could benefit from shielding from environmental effects by the electron cloud. Using the relations $\Delta x \cdot \Delta p > \hbar$ and $E = p^2/2m_p$, the length scale of 5 fm can be related to an energy scale of about 1 MeV. Nuclear decay energies are indeed on the MeV scale, and typical energy differences between adjacent nuclear states are between a few 10 keV and a few 100 keV. Unfortunately, coherent narrow-line X-ray sources, capable of driving such nuclear transitions, are not yet available. To date, the *nuclear clock* remains a vision.

There is, however, a very special case encountered in ^{229}Th (a freak of nature, if you like): this isotope possesses an extremely low-lying state with an excitation energy of only a few eV. According to Fermi's Golden Rule, the lifetime of this state scales as E^{-3} for an M1 transition into the ground state, such that its expected lifetime is comparably large: about an hour. This setting of (i) a transition that can be reached with existing laser technology, (ii) an extremely narrow linewidth, and (iii) a nuclear system well-protected from external perturbation, makes ^{229}Th a promising candidate for a future nuclear optical clock [1]. Being able to control and manipulate a nuclear system with a laser light, even transferring the well-established toolbox of optical spectroscopy to a nuclear system, is an exciting prospect. In addition, nuclear transitions might be highly sensitive to drifts in fundamental constants. Not surprisngly, this laser-driven nuclear two-level system has been proposed for a number of quantum optics applications, among

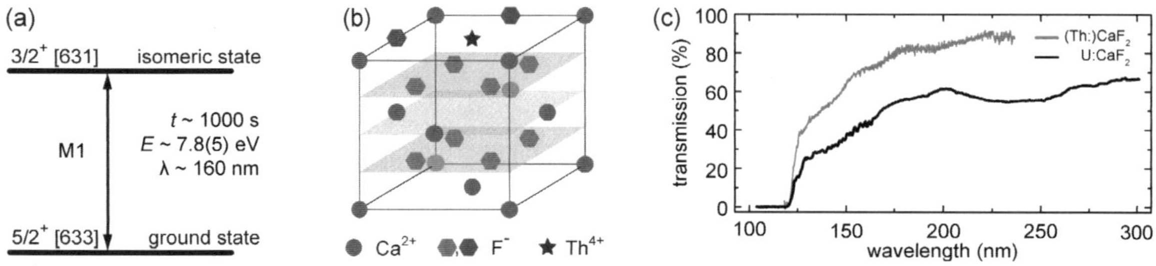

Figure 1. (a) Key parameters of the nuclear clock transition in ^{229}Th. (b) The CaF_2 lattice structure, with a Th^{4+} ion replacing a Ca^{2+} ion. The additional charges are compensated by two fluorine interstitials. (c) Transmission of crystals in the UV range. The grey line shows the transmission of a 5-mm home-grown pure CaF_2 crystal, which is not altered by Th doping concentrations of up to 10^{-5}. The black line is a 4-mm U:CaF_2 crystal at 5×10^{-5} doping concentration.

them a gamma-ray laser. A recent review of E. Peik and M. Okhapkin [2] gives a full account of the current work in this field of research.

Unfortunately, the energy of the isomer is not yet known precisely enough to commence direct laser spectroscopy; the most recent gamma spectroscopy measurement places the energy at 7.8(5) eV, corresponding to a wavelength of 160(10) nm [3]. Key parameters are given in Fig. 1(a). Besides the radiative de-excitation of the isomer, internal conversion (IC) and related processes have been discussed. Indeed, the "demon" of non-radiative decay might preclude fluorescence detection schemes whenever the isomer is given the possibility to decay through coupling to electronic states. The first ionization energy in thorium is below the expected isomer energy, so the ^{229}Th atom needs to be in an ionized state.

There are currently two main platforms envisioned for a future optical clock: crystals and ion traps [1, 2]. Ions traps would offer full control over all quantum degrees of freedom and thus enable high-precision spectroscopy, potentially exploiting the full line-Q of $\Delta E/E \approx 10^{-19}$. As a downside, ion traps can load only a limited number of Th ions (typically between 1 and 10^6). Doped crystals, on the other hand, would enable optical Mössbauer spectroscopy on an ensemble of 10^{15} or more ions. Such crystals form a very secure, robust, room-temperature platform with an obvious potential for small-scale integration. Spectroscopy, however, would suffer from effects related to the crystal structure, such as fast decoherence (of order kHz), line shifts (GHz scale), broadening (kHz widths), and crystal luminescence. One might argue that crystals might be useful for the initial search for the transition and future solid-state applications (e.g. for clocks on satellites), whereas ion traps will provide the ultimate quantum control and most precise measurements.

The Vienna group follows the *solid state* approach and has placed its bets on CaF_2 [4]. Crystals of 17 mm diameter and 40 mm length are routinely produced in our in-house crystal-growing apparatus. We employ the vertical gradient freeze method, which ensures a homogeneous distribution of the dopant over the crystal volume. CaF_2 readily accepts Th and U as dopants, and concentrations of up to 1 % have been achieved. Our ^{229}Th:CaF_2 crystals have a relative concentration of a few 10^{-7}, limited by the available quantity of ^{229}Th. The crystals are cut and polished to discs. Transmission in the UV is found to be slightly smaller compared to commercial crystals, but the transmission through a 1-cm crystal is still well over 50 % at 160 nm; see Fig. 1(c).

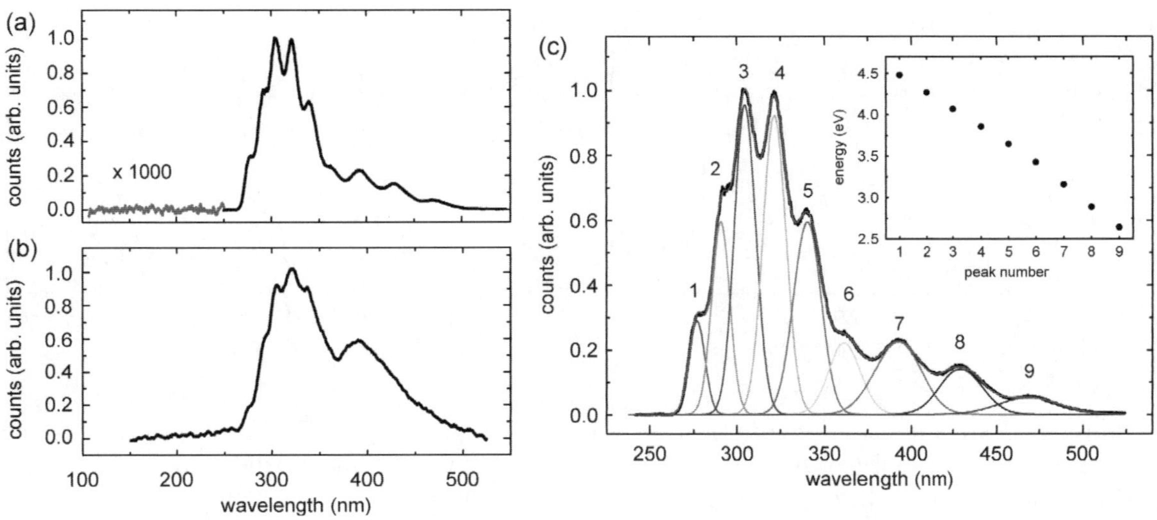

Figure 2. Light-induced luminescence in CaF_2. (a) The famous self-trapped exciton (STE) spectrum of CaF_2, consisting of individual lines. The signal below 240 nm is magnified by a factor of 1000 to show the absence of luminescence. (b) The same lines appear in the thermoluminescence spectrum, however at different relative amplitudes. (c) Zoom into the STE spectrum, with multiple Gaussian peaks fitted to the data. Converting to peak positions into units of energy (inset) reveals an almost equidistant spacing.

2. Luminescence of CaF_2 crystals

The crystal luminescence could potentially mask the the isomer signal if it appears in the same spectral range (around 160 nm) and has a similar decay time (of order 1000 s). Unfortunately, the rare-earth halides (e.g. CaF_2) are well known for their scintillation properties. Luminescence can be induced by various means, such as UV light, X-rays, and dislocations caused by the passage of massive particles following radioactive decay.

2.1. Photoluminescence

To assess the photoluminescence spectrum, we illuminate home-grown crystals with VUV light derived from a deuterium lamp. The excitation is turned off, and the luminescence spectrum emitted by the crystal is measured with a spectrometer; see Fig. 2(a). The dominant spectral feature is the signature of the the well-known self-trapped exciton (STE) structure, consiting of multiple states in a near-harmonic oscillator potential (Fig. 2(c)). The amplitude of the individual lines changes slightly with experimental parameters, but no emission is observed at wavelengths below 260 nm. This is very promising, as the primary search range for the isomer transition is around 160 nm.

Note that STEs are a generic feature of CaF_2: primary crystal defects, as induced by UV light, quickly thermalize into comparably long-lived excitations, such as STEs. These may decay through radiation of a photon. The susceptibility to form short-lived primary defects depends on the "cleanliness" of the crystal. Our home-grown undoped CaF_2 crystals show 10^4 times more photoluminescence compared to commercial crystals. The photoluminescence is not caused, changed, or enhanced by the doping.

Very long-lived defects can be annealed by heating the crystal to a few 100°C, such defects also relax via STEs; see Fig. 2(b). The same emission spectrum is recorded in response to 30-keV X-ray illumination. Detailed studies concerning the time-dependence and temperature dependence of photoluminescence can be found in Ref. [5].

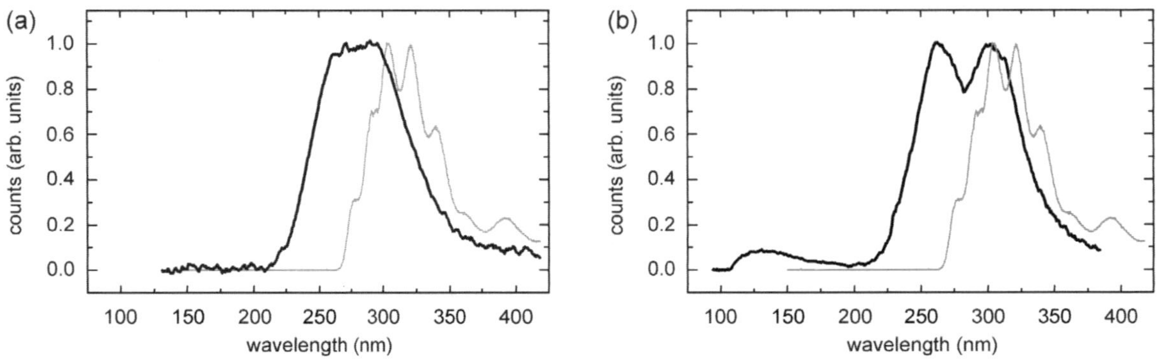

Figure 3. Radioluminescence of CaF_2, (a) induced by 5 kBq of ^{229}Th doped into the crystal (resolution 15 nm), and (b) induced by 7.5 MBq of ^{233}U in proximity to the crystal (resolution 4 nm). The grey line shows the STE-spectrum as a reference. Note the Cherenkov radiation between 120 and 200 nm in panel (b).

2.2. Radioluminescence: alpha decay

We record the emission spectrum of ^{229}Th-doped crystals, where the alpha decay of ^{229}Th (lifetime 7932 years) leads to scintillation. The activity of 5 kBq induces only a small signal (Fig. 3(a)), but we can already see that the emission has an abrupt UV cut-off at 220 nm. This feature is a property of CaF_2 and does not depend on doping, defects, impurities, and the like. The emission spectrum looks very different for other types of crystals, e.g. MgF_2, and it is strikingly different from photoluminescence. Further investigations can be found in Ref. [5].

To increase the signal strength, we bring a 7.5 MBq sheet of ^{233}U in physical contact with a commercial CaF_2 specimen; see Fig. 3(b). The large signal allows us to resolve the substructure of the spectrum. By placing different absorber materials between the source and the crystal, we are able to show that at least 90 % of the emission between 220 and 400 nm is caused by alpha particles.

2.3. Radioluminescence: beta decay

While the lifetime of ^{229}Th is comparably long, all of its daughters along the chain down to the stable ^{209}Bi are short-lived. These decays (5 alpha and 3 beta) contribute to the radioluminescence. The spectral signature of beta decays is very different from the one of alpha decays: electrons above a threshold of 160 keV kinetic energy generate Cherenkov radiation, predominantly at UV wavelengths; see Fig. 3(b). This emission is intrinsically broadband and very "flat", thus is poses a locally homogeneous background for the sought-after narrow-linewidth isomer signal. As all relevant properties of the decay chain are known, the spectrum and absolute amplitude of the Cherenkov radiation can be calculated. High-energy gammas can also generate high-energy electrons through Compton scattering, but this mechanism adds only a minor contribution to the Cherenkov emission.

3. Three experimental approaches to populate the isomer

We will now present three different experiments of our group, ongoing or planned for the near future. All of them rely on doped crystals, but follow very different approaches to populate the isomer. The three schemes are sketched in Fig. 4; different crystals are required for each approach. In all experiments, the VUV gamma emitted during de-excitation of the isomer is measured with a UV spectrometer. Unfortunately, none of these approaches is immune to parasitic non-radiative decay via IC or related processes.

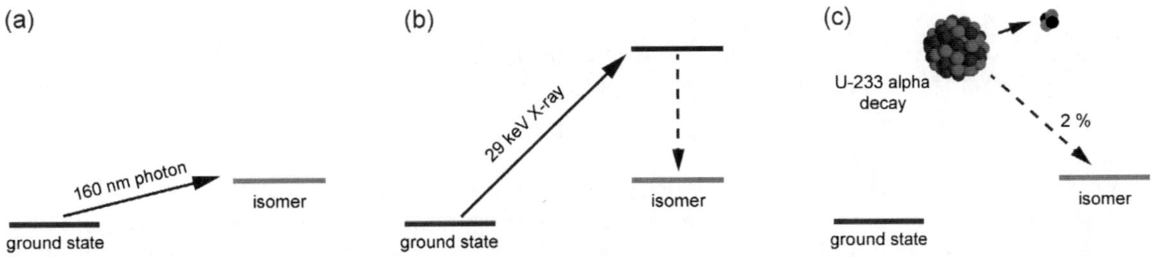

Figure 4. Three approches to populate the isomer: (a) direct excitation on the nuclear clock transition using synchrotron radiation, (b) optical pumping via the 29-keV state, (c) alpha decay of ^{233}U with a 2 % branching ratio into the isomer.

3.1. Direct spectroscopy using synchrotron radiation

Direct spectroscopy of ^{229}Th:CaF_2 crystals with a doping concentration 3×10^{-7} (corresponding to $7 \times 10^{15}\,\mathrm{cm}^{-3}$ in density) will be performed at the MLS synchrotron facility in Berlin. The synchrotron radiation (linewidth about 5 nm, flux about 100 photons/(s × Hz)) can be scanned across the entire search range. Preliminary measurements will be performed with a Cs-Te PMT detector, such that photoluminescence will constitute the main background. Once a signal is detected, either a monochromator in the beamline or a spectrometer can be employed to improve the uncertainty of the wavelength to below 0.1 nm. Two very similar experiments performed recently were unable to find evidence of the isomer emission [6, 7], possibly due to competing non-radiative decay channels.

We expect Cherenkov radiation (induced by radioactive decay of ^{229}Th) to be the main source of background emission below 200 nm. The amplitude and spectrum of Cherenkov radiation can be calculated exactly [8]. We estimate that, for the specific parameters of our experiment and negligible non-radiative decay, the isomer signal should rise above the noise of the Cherenkov background after less than an hour of illumination/detection cycles.

3.2. A nuclear lambda-system using the 29-keV state

Direct excitation with synchrotron radiation in the UV range is complicated by the fact that neither energy nor lifetime of the isomer are known, requiring a two-dimensional search. To circumvent these uncertainties, the isomer can be populated through pumping via the second excited state in the ^{229}Th nucleus; see Fig. 4(b). The energy of this state (29,185.6(2) eV) is known to within the linewidth of the excitation X-ray source, and its lifetime is 200(50) ps. Established detector technologies using fast MCPs allow to detect the re-emitted, delayed 29-keV gamma, proving that the excitation is on resonance and that the isomer is indeed populated. Spectral resolution of the isomer emission is obtained with a spectrometer. Test measurements at SPring-8 have shown that the search range around 160 nm is free of crystal luminescence caused by the excitation X-ray. Mildly doped CaF_2 can stand X-ray illumination for hours.

For experiments using UV excitation, we chose CaF_2 as the crystal material for its optical properties. For the X-ray excitation, choice of the material is guided by its X-ray properties (attenuation and hardness). Given that the mass attenuation increases heavily with mass number, it is desirable to work with crystals composed of lighter atoms. Changing from CaF_2 to MgF_2 already increases the penetration length at 29 keV from 1.3 to 4.7 mm.

Apart from a measurement of the isomer energy and its lifetime, this approach contains all the ingredients of an (admittedly very inefficient) gamma-ray laser: for strong enough pumping, an inversion between ground and isomeric state can be achieved, potentially leading to lasing on the nuclear transition.

This line of expriments is performed in close collaboration with the group of K. Yoshimura at Okayama University, and experiments are performed at the SPring-8 facility.

3.3. Uranium-doped crystals

A large number of past and presently ongoing experiments use the alpha decay of ^{233}U into ^{229}Th as a means to feed the isomer state (branching ratio about 2 %). Most experiments are designed such that a thin layer of ^{233}U is brought near a UV-transparent medium (usually MgF_2). Upon an alpha decay, the ^{229}Th obtains a recoil of up to 84 keV, which allows it to penetrate through the thin ^{233}U layer and deposit onto, or penetrate a few 10 nm into, the crystal. The crystal is then placed into the detector. The main advantage of this approach is the fact that the massive energy released in the alpha decay is separated in both space and time from the isomer detection. We have shown, however, that crystal luminescence caused by alpha decay is spectrally remote from the isomer search region. Equally important, the radioluminescence that does overlap with the expected isomer wavelength (beta radiation from ingrowth of daughters) is fully captured by the detector. Given that the signal rate is intrinsically low (given by the 160 000 years lifetime of ^{233}U and the 15 nm path length through the uranium), it is desirable to increase the signal strength.

We could show that doping ^{233}U directly into suitable crystals has the potential to increase the signal rate by at least two orders of magnitude at the very same S/N as with surface-implantation experiments. The fundamental limitation with uranium-doped crystals stems from the absorption of uranium in the UV range. We have measured the transmission properties of ^{238}U-doped crystals (Fig. 1(c)) and found that the maximum flux of isomer gammas emitted from such a crystal is about $4000/(\mathrm{cm}^2 \times \mathrm{s})$, giving reason for hope that emission on the isomer transition can be detected. The main signal background is caused by beta decay of daughers in the ^{232}U chain, which is an unavoidable contamination in the ^{233}U source. Assuming the absence of IC processes and using parameters of a standard UV spectrometer, we calculate that the isomer signal should rise above the noise of the Cherenkov emission for measurement times of a few hours.

4. Conclusion

We have described three experiments, planned or ongoing, to measure the wavelength of the ^{229}Th isomer transition. The experiments use very different means to excitate the isomer. All of these experiments are sensitive to a background of Cherenkov radiation in the VUV regime [8], and all of them require the radiative de-excitation channel to be of at least a few percent compared to competing non-radiative channels.

Acknowledgments

We greatly appreciate fruitful discussions with E. Peik, L. von der Wense, and P. G. Thirolf. This work was supported by the ERC project 258604-NAC. This project has received funding from the European Union's Horizon 2020 research and innovation programme under grant agreement No 664732 "nuClock".

References

[1] Peik E and Tamm C 2003 *Europhys. Lett.* **61** 181
[2] Peik E and Okhapkin M 2015 *Comptes Rendus Physique* **16** 516
[3] Beck R B *et al.* 2007 *Phys. Rev. Lett.* **98** 142501
[4] Kazakov G A *et al.* 2012 *New J. Phys.* **14** 083019
[5] Stellmer S, Schreitl M and Schumm T 2015 *Sci. Rep.* **5** 15580
[6] Jeet J *et al.* 2015 *Phys. Rev. Lett.* **114** 253001
[7] Yamaguchi A, Kolbe M, Kaser H, Reichel T, Gottwald A and Peik E 2015 *New J. Phys.* **14** 053053
[8] Stellmer S, Schreitl M, Kazakov G A, Sterba J H and Schumm T 2015 *arXiv.1511.07187*

Journal of Physics: Conference Series **723** (2016) 012060 doi:10.1088/1742-6596/723/1/012060

Determination of optimized frequency and frequency ratio values from over-determined sets of clock comparison data

H S Margolis and P Gill

National Physical Laboratory (NPL), Teddington, Middlesex TW11 0LW, UK

E-mail: helen.margolis@npl.co.uk

Abstract. We summarize a recently developed method for analyzing over-determined sets of clock frequency comparison data to derive optimized values for the frequency ratios between each of the contributing standards. This least-squares adjustment procedure, based on the approach used by CODATA to derive a self-consistent set of values for the fundamental physical constants, is used to deduce optimized frequency and frequency ratio values from the international body of clock comparison data available in September 2015.

1. Introduction

The stability and accuracy of the most advanced optical atomic clocks now significantly surpass the performance of the best caesium primary frequency standards [1, 2] and a future redefinition of the SI second by the International Committee for Weights and Measures (CIPM) is anticipated [3, 4]. As a first step in this direction, seven optical clocks can already be used as secondary representations of the second, with recommended frequencies and uncertainties assigned by the Frequency Standards Working Group (WGFS) of the Consultative Committee for Time and Frequency (CCTF) and the Consultative Committee for Length (CCL). These recommended frequency values and uncertainties are periodically reviewed, updated and published on the website of the International Bureau of Weights and Measures (BIPM) [5].

Prior to 2015, almost all the data considered by the WGFS came from absolute frequency measurements made either relative to caesium fountain primary frequency standards or, if such standards were not available in the laboratory concerned, relative to International Atomic Time (TAI). The sole exception was a directly measured optical frequency ratio between the clock transitions in $^{27}Al^+$ and $^{199}Hg^+$ [6], which was used to derive a second absolute frequency value for $^{27}Al^+$ with much lower uncertainty than the directly measured absolute frequency value [7].

At the most recent meeting of the WGFS, held in September 2015, the situation was rather different (figure 1). Not only were a number of new absolute frequency measurements submitted for consideration, five new direct optical frequency ratio measurements were also reported. These included one measurement each of the ratios between the clock transitions in $^{40}Ca^+$ and ^{87}Sr [8], between the E2 and E3 transitions in $^{171}Yb^+$ [9] and between the clock transitions in ^{87}Sr and ^{199}Hg [10], and two independent measurements of the ratio between the clock transitions in ^{171}Yb and ^{87}Sr [11, 12, 13]. The complete body of data was thus over-determined, by which we mean that some frequency ratios can be deduced from the results of more than one experiment.

Journal of Physics: Conference Series **723** (2016) 012060 doi:10.1088/1742-6596/723/1/012060

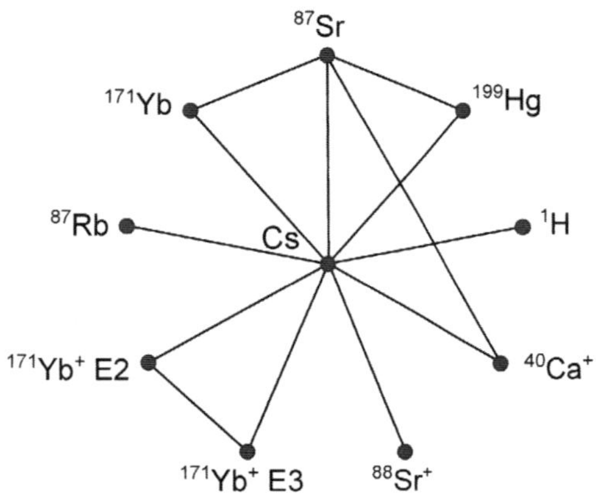

Figure 1. (*Colour online.*) New frequency ratio measurements considered by the WGFS in September 2015. Most were absolute frequency measurements, i.e. frequency ratios involving the caesium primary standard, but four optical frequency ratios had also been measured directly (in one case by two independent groups).

For example, the ratio between the E2 and E3 transitions in $^{171}Yb^+$ can be determined by combining the results of recent absolute frequency measurements of the two transitions made at NPL [9] and PTB [14, 15], but it has also been measured directly at NPL with comparable uncertainty [9]. These multiple routes to deriving this frequency ratio mean that it is no longer possible to treat the two clock transitions in isolation when considering the available data. The same is true of the ^{171}Yb, ^{87}Sr, ^{199}Hg and $^{40}Ca^+$ standards, which are similarly coupled by direct optical frequency ratio measurements.

2. Analysis procedure

We have recently developed a method for handling such over-determined sets of clock frequency comparison data in order to derive optimized values for the frequency ratios between each contributing standard [16]. This is a least-squares adjustment procedure (figure 2), based on the well-established approach used by CODATA to derive a self-consistent set of values for the fundamental physical constants [17].

The input data to the least-squares adjustment are a set of N frequency ratio measurements between clocks based on N_S different reference transitions with frequencies $\nu_1, \ldots \nu_{N_S}$, together with the variances and covariances of these frequency ratio measurements. All input data is treated in a similar way, with absolute frequency measurements simply being a special case of frequency ratios involving the caesium primary standard. The measured frequency ratios are expressed as a function of one or more of a set of $M = N_S - 1$ adjusted frequency ratios z_j, yielding a set of N equations that are, in general, nonlinear. A necessary condition is that no adjusted frequency ratio may be expressed as a function of the others. We choose the set $z_j = \nu_j/\nu_{j+1}$ where $j = 1, \ldots N_S - 1$, and it is the values of these adjusted frequency ratios that are optimized in the least-squares adjustment.

The equations relating the measured frequency ratios to the adjusted frequency ratios are linearized prior to the least-squares adjustment by using a Taylor expansion around initial estimates of the adjusted frequency ratios. This enables linear matrix methods to be applied to yield best estimates for the values of the adjusted frequency ratios, their variances and covariances. Because a linear approximation has been made, this solution will not be exact. However the improved values of the adjusted frequency ratios obtained from the least-squares adjustment can be used as starting values for a new linear approximation and a second least-squares adjustment performed. This process is repeated until the new values of the adjusted frequency ratios obtained from the least-squares adjustment are sufficiently close to the values obtained in the previous iteration. Once this condition has been satisfied, any other frequency ratio of interest (and its uncertainty) can then be calculated from the adjusted frequency ratios and their covariance matrix. Self-consistency checks are also performed on the body of data to

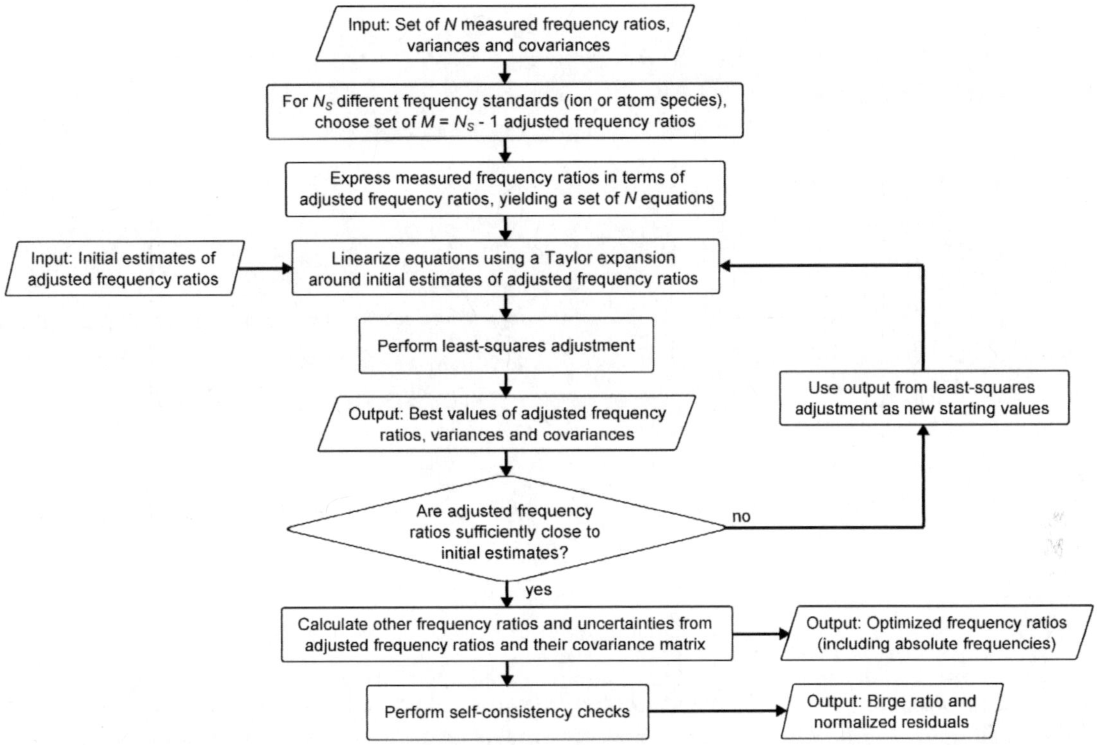

Figure 2. Least-squares analysis procedure used to derive a self-consistent set of optimized frequency and frequency ratio values from an over-determined set of clock comparison data [16].

identify any issues with the uncertainty evaluations for each individual clock comparison.

Our software algorithms, implemented in Matlab, have been shown to reproduce the 2013 CIPM recommended frequency values, when using the same input data employed by the WGFS to derive those values [16]. The uncertainties determined from our least-squares analysis, however, are smaller than those assigned to the CIPM values. The reason for this is that measurements of a particular frequency ratio have usually only been made by a few independent research groups, or in some cases only one. This leads the WGFS to take a conservative approach to uncertainty estimation, typically multiplying the relative standard uncertainty on the weighted mean of a set of frequency values by a factor of two or three.

3. Analysis of data available in September 2015

A number of new frequency comparison results were reported to the WGFS in advance of their September 2015 meeting (figure 1). As well as the five direct optical frequency ratio measurements already discussed, these included 16 new absolute frequency measurements: one each of the 1S–2S transition in ^{1}H [18], the optical clock transitions in ^{171}Yb [19] and ^{88}Sr^{+} [20] and the ground state hyperfine transition in ^{87}Rb [21], two each of the E2 [9, 14] and E3 [9, 15] transitions in ^{171}Yb^{+} and the reference transition in ^{40}Ca^{+} [22], and six of the optical clock transition in ^{87}Sr [23, 24, 25, 26, 27, 28]. In addition, a correction [29] was submitted to a previously-used absolute frquency measurement of the optical clock transition in ^{199}Hg [30] and one earlier absolute frequency measurement of the reference transition in ^{40}Ca^{+} [31] was withdrawn as new data from the same group suggested that the systematic frequency shifts may have been underestimated in this measurement [22].

Journal of Physics: Conference Series **723** (2016) 012060 doi:10.1088/1742-6596/723/1/012060

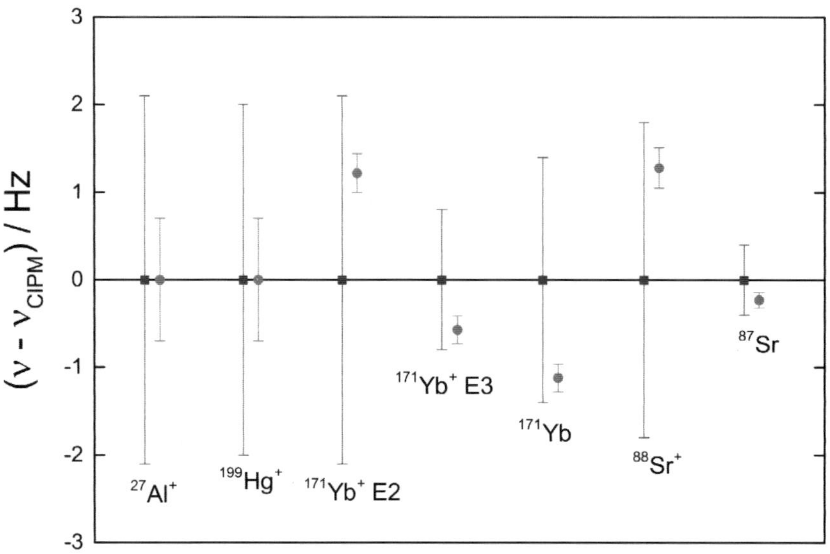

Figure 3. (*Colour online.*) Optimized frequency values obtained for the seven optical secondary representations of the second using all clock comparison data available in September 2015 (red circles) compared to the 2013 CIPM recommended frequency values and uncertainties (blue squares). The uncertainties shown for the new results are those emerging from the least-squares adjustment, not the final uncertainty assigned by the WGFS.

The effects on the optimized frequency values obtained for the optical secondary representations of the second of including this new data in the least-squares analysis are shown in figure 3. Here correlations between the input data are ignored, with the exception of the recent measurements involving the $^{171}Yb^{+}$ standards at NPL [9]. In this case the absolute frequency measurements of the E2 and E3 transitions were performed during the same campaign as the measurement of the direct optical frequency ratio between them, leading to non-negligible correlations between the three values. For example, part of the data from the caesium fountain primary frequency standard is common to the two absolute frequency measurements, leading to correlations from both the statistical fluctuations of the caesium standard and its systematic uncertainty. Similarly, correlations between the E3:E2 ratio measurement and the E2 absolute frequency measurement arise from the statistical and systematic uncertainties associated with the E2 clock transition. The correlation coefficients were thus calculated based on the periods of overlap between the three measurements and the estimated statistical and systematic uncertainties of the three standards ($^{171}Yb^{+}$ E2, $^{171}Yb^{+}$ E3 and Cs fountain).

It is clear from figure 3 that including the new data in the least-squares analysis leads to significant shifts in the optimized frequency values, demonstrating that the conservative approach to uncertainty estimation employed by the WGFS may indeed be a wise one. We emphasize that the uncertainties of the new optimized frequency values shown in figure 3 are the uncertainties determined from the least-squares adjustment, rather than the uncertainties finally assigned by the WGFS. Indeed, much of the discussion during the recent WGFS meeting centred around the magnitude of these uncertainties and whether (and if so, by how much) they should be expanded. For example, the least-squares adjustment yields an uncertainty of

2×10^{-16} for the frequency of the $5s^2\,{}^1S_0$–$5s5p\,{}^3P_0$ transition in ^{87}Sr, which is lower than the uncertainties of the caesium primary frequency standards used as the references for the best absolute frequency measurements of this transition. Since no secondary representation of the second can have a frequency uncertainty lower than that of the best primary frequency standards, the WGFS expanded this uncertainty in updating the list of recommended frequency values. In other cases such as the $^{171}Yb^+$, $^{88}Sr^+$ and ^{199}Hg standards, the optimized frequency values were based on input data from only one or two research groups, and so the WGFS again considered it prudent to expand the uncertainty emerging from the least-squares adjustment.

4. Conclusions and future perspectives

Our analysis procedure allows a self-consistent set of optimized frequency ratio values between high accuracy frequency standards (both optical and microwave) to be derived, based on data from a set of clock comparison experiments, and taking proper account of any correlations among the data. Although the matrix of frequency comparison data available worldwide is currently quite sparsely populated, it is already over-determined. As a consequence the WGFS used our software algorithms in preparing an updated list of recommended frequency values in September 2015, enabling full benefit to be derived from the experimental data available at the time.

The methods developed will become increasingly important as the number of direct optical frequency ratio measurements increases, and will enable valuable information to be derived about the relative performance of different candidates for an optical redefinition of the second. They can also be used to determine optimized values and uncertainties for the absolute frequencies of each optical standard relative to the current definition of the second, maximizing the potential contribution of optical clocks to international timescales prior to a redefinition.

However, in future it is likely that correlations between the input data will become more significant, as increasing numbers of high accuracy clocks are compared in measurement campaigns involving multiple institutions. For example, within the EMRP-funded ITOC project [32] a collection of optical and caesium fountain clocks were simultaneously operated at INRIM, LNE-SYRTE, NPL and PTB for a direct remote comparison via satellite links over a 26-day period. Several local optical frequency ratio measurements were also performed during this campaign. Potentially quite significant correlations can therefore be expected between some of the frequency ratio values resulting from this campaign, even though none of the clocks operated with 100% duty cycle. We therefore expect that it will be increasingly important for the WGFS to gather information about the correlations between the input data to future least-squares adjustments, to avoid biassing the optimized frequency values and underestimating their uncertainties. Such information is not usually readily extracted from publications and so will have to be sought from the research groups that performed the measurements. In addition to this requirement to calculate the correlation coefficients for the input data, it will also be necessary to study in detail the extent to which each input datum contributes to the determination of the optimized frequency ratios as well as to look at the effects of omitting inconsistent or inconsequential data from the least-squares adjustment. The task facing the WGFS is thus likely to increase in complexity over the next few years as the worldwide body of high accuracy clock comparison data continues to grow.

Acknowledgments

This work was performed within the ITOC project as part of the European Metrology Research Programme (EMRP). The EMRP is jointly funded by the EMRP participating countries within EURAMET and the European Union. The authors acknowledge helpful discussions with partners in the ITOC project consortium and with members of the CCL-CCTF WGFS.

References

[1] Poli N, Oates C W, Gill P and Tino G M 2013 *Riivista del Nuovo Cimento* **36** 555–624
[2] Ludlow A D, Boyd M M, Ye J, Peik E and Schmidt P O 2015 *Rev. Mod. Phys.* **87** 637–701
[3] Margolis H 2014 *Nature Physics* **10** 82–3
[4] Riehle F 2015 *C. R. Physique* **16** 506–15
[5] Recommended values of standard frequencies, http://www.bipm.org/en/publications/mises-en-pratique/standard-frequencies.html
[6] Rosenband T *et al* 2008 *Science* **319** 1808–11
[7] Rosenband T *et al* 2007 *Phys. Rev. Lett.* **98** 220801
[8] Matsubara K, Hachisu H, Li Y, Nagano S, Locke C, Nogami A, Kajita M, Hayasaka K, Ido T and Hosokawa M 2012 *Opt. Express* **20** 22034–41
[9] Godun R M, Nisbet P B R, Jones J M, King S A, Johnson L A M, Margolis H S, Szymaniec K, Lea S N, Bongs K and Gill P 2014 *Phys. Rev. Lett.* **113** 210801
[10] Yamanaka K, Ohmae N, Ushijima I, Takamoto M and Katori H 2015 *Phys. Rev. Lett.* **114** 230801
[11] Akamatsu D, Yasuda M, Inaba H, Hosaka K, Tanabe T, Onae A and Hong F-L 2014 *Opt. Express* **22** 7898–905
[12] Akamatsu D, Yasuda M, Inaba H, Hosaka K, Tanabe T, Onae A and Hong F-L 2014 *Opt. Express* **22** 32199
[13] Takamoto M *et al* 2015 *C. R. Physique* **16** 489–98
[14] Tamm C, Huntemann N, Lipphardt B, Gerginov V, Nemitz N, Kazda M, Weyers S and Peik E 2014 *Phys. Rev. A* **89** 023820
[15] Huntemann N, Lipphardt B, Tamm C, Gerginov V, Weyers S and Peik E 2014 *Phys. Rev. Lett.* **113** 210802
[16] Margolis H S and Gill P 2015 *Metrologia* **52** 628–34
[17] Mohr P J and Taylor B N 2000 *Rev. Mod. Phys.* **72** 351–495
[18] Matveev A *et al* 2013 *Phys. Rev. Lett.* **110** 230801
[19] Park C Y *et al Metrologia* **50** 119-28
[20] Barwood G P, Huang G, Klein H A, Johnson L A M, King S A, Margolis H S, Szymaniec K and Gill P 2014 *Phys. Rev. A* **89** 050501
[21] Ovchinnikov Y B, Szymaniec K and Edris S 2015 *Metrologia* **52** 595–9
[22] Huang Y, Guan H, Liu P, Bian W, Ma L, Liang K, Li T and Gao K 2015 *submitted for publication*
[23] Le Targat R *et al* 2013 *Nat. Commun.* **4** 2109
[24] Falke S *et al* 2014 *New J. Phys.* **16** 073023
[25] Akamatsu D, Inaba H, Hosaka K, Yasuda M, Onae A, Suzuyama T, Amemiya M and Hong F-L 2014 *Appl. Phys. Express* **7** 012401
[26] Tanabe T *et al* 2015 *J. Phys. Soc. Japan* **84** 115002
[27] Hachisu H and Ido T 2015 *Jpn. J. Appl. Phys.* **54** 112401
[28] Lin Y-G, Wang Q, Li Y, Meng F, Lin B-K, Zang E-J, Sun Z, Fang F, Li T-C and Fang Z-J 2015 *Chin. Phys. Lett.* **32** 090601
[29] McFerran J J, Yi L, Mejri S, Di Manno S, Zhang W, Guéna J, Le Coq Y and Bize S 2015 *Phys. Rev. Lett.* **115** 219901
[30] McFerran J J, Yi L, Mejri S, Di Manno S, Zhang W, Guéna J, Le Coq Y and Bize S 2012 *Phys. Rev. Lett.* **108** 183004
[31] Huang Y, Cao J, Liu P, Liang K, Ou B, Guan H, Huang X, Li T and Gao K 2012 *Phys. Rev. A* **85** 030503
[32] Margolis H S *et al* 2013 Proc. Joint European Frequency and Time Forum and International Frequency Control Symposium (Prague) (IEEE) p 908

Journal of Physics: Conference Series **723** (2016) 012061 doi:10.1088/1742-6596/723/1/012061

High-Accuracy Ring Laser Gyroscopes: Earth Rotation Rate and Relativistic Effects

N Beverini [1,2,6], A Di Virgilio[2], J Belfi[2], A Ortolan[3], K U Schreiber[4], A Gebauer[4] and T Klügel[5]

[1] Department of Physics, Università di Pisa, Largo Pontecorvo 3, I 56127 Pisa, Italy
[2] INFN, Sezione di Pisa, Largo Pontecorvo 3, I 56127 Pisa, Italy
[3] INFN, Laboratorio Nazionale di Legnaro, I 35020 Legnaro, Italy
[4] Forschungseinrichtung Satellitengeodäsie, Techn. Universität München, Germany
[5] Bundesamt für Kartographie und Geodäsie, GO Wettzell, Germany

e-mail: nicolo.beverini@unipi.it

Abstract. The Gross Ring G is a square ring laser gyroscope, built as a monolithic Zerodur structure with 4 m length on all sides. It has demonstrated that a large ring laser provides a sensitivity high enough to measure the rotational rate of the Earth with a high precision of $\Delta\Omega_E < 10^{-8}$. It is possible to show that further improvement in accuracy could allow the observation of the metric frame dragging, produced by the Earth rotating mass (Lense-Thirring effect), as predicted by General Relativity. Furthermore, it can provide a local measurement of the Earth rotational rate with a sensitivity near to that provided by the international system IERS. The GINGER project is intending to take this level of sensitivity further and to improve the accuracy and the long-term stability. A monolithic structure similar to the G ring laser is not available for GINGER. Therefore the preliminary goal is the demonstration of the feasibility of a larger gyroscope structure, where the mechanical stability is obtained through an active control of the geometry. A prototype moderate size gyroscope (GP-2) has been set up in Pisa in order to test this active control of the ring geometry, while a second structure (GINGERino) has been installed inside the Gran Sasso underground laboratory in order to investigate the properties of a deep underground laboratory in view of an installation of a future GINGER apparatus. The preliminary data on these two latter instruments are presented.

1. Introduction

Ring laser gyroscopes (RLG) have demonstrated to currently be the most sensitive device for testing rotational motion with respect to an inertial frame. A new era for upscaled ring laser gyroscopes started in the 90's of the past century when, thanks to the technological improvements in the production of low loss mirrors, a reflectivity exceeding 99.99% was achieved. Unlocked Earth rotation sensing with a ring laser of about 1 m^2 of area was demonstrated at the University of Canterbury in Christchurch, New Zealand [1]. Subsequent further development led to the construction of the *Gross-Ring* (G), a 16 m^2 area ring laser, located at the Geodetic Observatory in Wettzell (Bavaria) [2]. G has continuously improved over the years, by more and more accurate control of the environmental

[6] To whom any correspondence should be addressed

conditions, finally achieving a rotational sensitivity of the order of 0.25 p-rad/s over 10^4 s, which corresponds to resolving the Earth rotation rate to 3 part in 10^9.

In the VIRGO laboratory in Cascina near Pisa (Italy) a moderate size ring laser of 1.82 m^2 area dubbed G-Pisa was realized by a collaboration between INFN and Department of Physics of the University of Pisa with the support of the expertise of Prof. Schreiber (Technische Universität München (TUM)) in 2008. The effort was inspired by the desire to establish very small ground tilt movements for the VIRGO interferometer. G-Pisa was first operated with the optical cavity orientated in the local horizontal plane and later in the vertical plane [3]. It was operating on March 11th, 2011, when it registered the rotation induced by the seismic Rayleigh waves generated by Fukushima earthquake [4]. Putting together the expertise acquired by the INFN group and that of the other principal groups operating large laser gyros, (TUM and University of Canterbury), a proposal was published in 2011 [5] that suggests a ring laser array with the accuracy needed to observe the Lense-Thirring effect. According to the theory of General Relativity, this effect arises from the metric deformation induced by a massive rotating body. Einstein's equations predicts a 1 ppb correction to the Earth rotational rate, when measured in a laboratory frame bound to the Earth surface with respect to the rate defined by quasar positions measured by the Very Long Baseline Interferometry (VLBI). So far this frame dragging effect has been confirmed in the context of dynamic satellite, experiments only [6, 7]. A "static" measurement in a terrestrial laboratory would give local information as opposed to an averaged measurement as it is provided by the satellite experiments. While the satellite measurements have confirmed the frame dragging effect, they have not yet reached a resolution that would put this theory to a test. In order to provide an improvement, a viable ring laser system would have to improve in sensitivity and above all in the long-term stability. Furthermore it is required to measure the rotation rate of the Earth absolute with sufficient accuracy, since the frame dragging effect appears like a small bias on top of the rate of rotation of the Earth itself.

2. Earth Rotation Rate and Relativistic Effects

2.1. Geophysical Signal Contributions

Large Ring lasers firmly tied to the Earth crust represent the only technology in space geodesy that is directly linked to the instantaneous axis of rotation of the Earth. With a sensor sensitivity of 12 p-rad/s/sqrt(Hz) inferred from the cavity ring-down time, G is sensitive to geophysical signals like polar motion, solid Earth tides, the Annual and the Chandler wobble and local tilts as small as 1 n-rad. A typical measurement sequence is shown in figure 1. Since laser gyros can operate continuously, they provide a continuous set of measurements, which is not yet available for VLBI. However, a good agreement between the VLBI and ring laser technology has been achieved by measuring and comparing the low frequency Chandler- and Annual- Wobble, a free oscillation of the Earth [8]. The remaining discrepancy is due to the fact that local tilts of the single component G ring laser currently cannot be measured with a sufficient long-term stability due to small and slow drift effects in the respective tilt meters. This is work in progress. A moderate improvement in sensitivity and a full 3-dimensional detection of the Earth rotational velocity vector would take the ring lasers into a regime

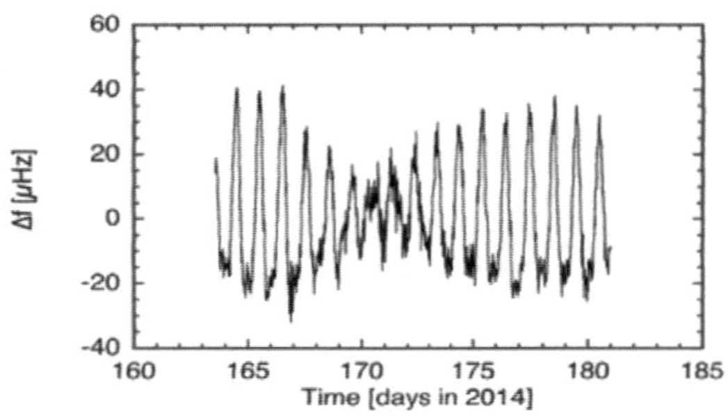

Figure 1: Fluctuation of the Earth rotational rate due to Diurnal polar motion and solid Earth tides, as observed by the laser gyroscope G at the Geodetic Observatory in Wettzell (Bavaria)

Journal of Physics: Conference Series **723** (2016) 012061 doi:10.1088/1742-6596/723/1/012061

where they can be efficiently integrated into the data produced by VLBI observations.

2.2. Relativistic Signal Contributions

In the weak-field approximation of Einstein's equation for General Relativity, the expected rotation signal seen by a ring laser located in a laboratory on the Earth surface, with co-latitude θ and with the axis contained in the meridian plane at an angle ψ with respect to the zenith, can be written as:

$$\nu \cong 4\frac{S}{\lambda p}\Omega_T\left[\cos(\theta+\psi)-2\frac{GM_T}{c^2R_T}\sin\theta\sin\psi+\frac{GI_T}{c^2R_T^3}(2\cos\theta\cos\psi+\sin\theta\sin\psi)\right],$$

where G is Newton's gravitational constant, Ω_T= 7.29·10^{-5} rad/s is the rotational velocity of the Earth, M_T the mass of the Earth, R_T the Earth mean radius, and $I_T \approx 2/5\, M_T\, R_T^2$ the Earth moment of inertia. The first term in the square bracket corresponds to the classical Sagnac signal; the second one, radially oriented, is produced by the coupling of the Earth Newtonian field with the rotation of the laboratory carried by the rotating Earth (*geodetic* or *De Sitter precession*) and the third one, characterized by a dipolar structure, is produced by the rotating mass of the Earth and is proportional to the Earth angular momentum (*Lense-Thirring precession, LT*) [9]. Both these last two relativistic effects are reduced with respect to the classical Sagnac effect by a factor that is of the order of the magnitude of the ratio between R_T and the Schwarzschild radius of the Earth $2GM_T/c^2$. This corresponds to values of around 10^{-9}. These effects can be observed as a difference between the rotation rate observed by a ring laser in the rotating frame of the laboratory and the length of the day determined by the inertial frame of the "fixed stars" by IERS through the technique of VLBI.

3. GINGER project

The performance of G is near the shot noise limit and therefore defined by the properties of the mirrors. An impressive long-term stability of the order of 1 part in 10^8 of the rotation rate of the Earth has been obtained by an accurate modeling of all the environmental effects of geodetic, geophysical, or meteorological origin. However the system lacks absolute accuracy: it is only sensitive to a single component of the angular velocity vector and the absolute orientation of the laser cavity with respect to the fixed stars inertial frame has not been established with the required degree of accuracy. To reach the goal, we need to improve the instrumental setup in several ways:

1. **Reduction of the shot noise threshold:** This can be obtained by enlarging the size of the sensor up to 7 - 9 m (following [8], for equivalent mirrors the sensitivity increases as L^α with $\alpha\approx5$). A larger structure is not useful: this would increase the beam radius and then the mirror losses.
2. **Improvement of the long-term stability:** By means of monitoring and controlling the laser operative parameters and the environment the remaining drift can be reduced. A deep underground sensor location seems necessary to reduce surface tilts due to atmospheric load variation caused by changing weather patterns and local hydrological effects.
3. **Stabilization of the geometrical scale factor:** This requires a strict active stabilization control of the laser beam path geometry and an accurate orientation with respect to the Earth axis.
4. **Correction of the nonlinear laser dynamics.** Intracavity dynamics of the laser gain medium introduce nonreciprocal effects, resulting in a null-shift term in the gyroscope response. To achieve the required accuracy, an estimate of these contributions must be implemented [10].

The long-term stability of G is warranted by its rigid monolithic structure based on the near zero thermal expansion material Zerodur, but this alone is not enough. Small variations of the laboratory temperature and adiabatic compression and expansion with changing atmospheric pressure are causing small changes in the backscatter coupling of the two counter- propagating laser beams and hence a non negligible drift in the measured rotation rate. The present situation does not allow the construction of monolithic structures larger than G, because the necessary ultra-low expansion material is not available in the required size. A viable solution to this problem is the implementation of a closed loop active control system that stabilizes not only the perimeter length, as already done in G, but also the geometrical shape of the ring optical cavity and hence the mechanical scale factor $k=A/p$. In a near-square 4-mirror cavity, the pairs of opposite mirrors define two Fabry-Perot cavities along the diagonals of the ring laser area. When the lengths of these resonators are kept locked to the same value, it is possible to demonstrate that in the square geometry the optical length of the perimeter is a stationary function of the position of the mirrors and consequently any deviation from the exact geometry gives only a second order contribution to k. This property gives also the opportunity to define a procedure in order to approach the correct configuration starting from a non-optimized one [11]. Apart from the scale factor, it must be considered that the RLG signal is also affected by the orientation of the planar ring cavity with respect to the Earth rotational axis. The required accuracy in the sensor orientation is at the level of 1 n-rad, clearly impossible to be achieved by a horizontal cavity in an underground experiment. A set of more (at least 3) equal ring lasers interferometrically referenced to each other, can however allow the reconstruction of the invariant modulus of the rotational velocity of the Earth independently from the knowledge of the absolute orientation. The comparison of this scalar value with VLBI measurements provided by the IERS would then show the LT effect. A 3-dimensional geometry based on a set of six shared mirrors in an octahedral configuration ideally defines three mutually orthogonally nested ring laser cavities and will give a strong internal constraint on the geometry when locking the three Fabry-Perot diagonal cavities to identical values.

4. Toward GINGER

4.1. General Considerations

Work is presently in progress, in order to demonstrate the feasibility of the GINGER concept and to develop and validate the necessary technologies. At first, a detailed model of the laser operation has been realised, which applies a Kalman filter in order to correct the frequency shift induced by the coupling between the counter-propagating beams. The application of the algorithm to the G-Pisa raw data provided a tenfold reduction of the systematic shift error as well as a tenfold reduction of the Allan deviation for times larger than 100 s [10]. Two different installations are presently under test. The first one, GP2, is located at the INFN laboratory of Pisa and is devoted to test the concept of stabilizing the diagonals of a square ring laser and to control of the ring geometry. The second one is an upscaled ring laser structure "GINGERino", placed inside the underground laboratory of Gran Sasso, beneath 1000 m of rock. This ring laser is installed together with tilt meters and more traditional seismic equipment, in order to qualify the site in view of GINGER experiment. As stated before, the localization in a deep underground site will be a critical step to guarantee the necessary long-term stability, reducing the surface tilts due to atmospheric load variations caused by changing weather patterns and local hydrological effects.

4.2. GP2

In order to stabilize the diagonals of the square cavity, the ring laser structure with a length of 1.60 m

Figure 2 – GP-2 gyroscope, located in the Pisa INFN Laboratory (color on line)

Figure 3 – GINGERino apparatus in the underground Gran Sasso INFN laboratories (color on line).

on each side is mounted to a granite slab of 1 ton (figure 2). The plane of the granite is inclined by 47° to the North direction, aligning the cavity orthogonal to the rotational axis of the Earth. In this way, the sensitivity of the interferogram to tilts is minimized. An extension of the vacuum system along the diagonals provides the access to the diagonal Fabry-Pérot cavities. The corner boxes containing the mirrors can be moved by 6 PZT driven actuators to control the geometric parameters in a small range around the ideal shape. The system became operational in the last months. In a previous experiment we had verified the technique to realize an comparison of the absolute lengths and the frequency locking to a stabilized iodine reference He:Ne laser of two Fabry-Perot cavities by a triple-modulated Pound-Drever-Hall circuit [12]. The technique has now been applied to GP-2, and the simultaneous locking of the two diagonal optical cavities to the reference laser has been observed continuously for some hours. At the present stage the stability on the average length of each diagonal resonator in GP-2 reaches the level of 1 nm after hundred seconds of integration of the error signals in closed loop [13].

4.3. GINGERino

The GINGERino structure is the evolution of the previous G-Pisa laser gyroscope to a larger size instrument. It uses the same type of corner boxes of GP-2 that house the mirrors and the same type of mechanical levers for aligning the ring optical cavity. The resonator was mounted on a cross-shaped granite basis of 4 tons of weight (figure 3) and linked up by vacuum pipes in order to obtain a square optical cavity of 3.60 x 3.60 m^2. The granite basis is supported centrally by a monument made of concrete, which is firmly attached to the tunnel floor. A seismic station is located alongside the ring, equipped with two very broad band seismometer (a Guralp-360 and a Trillium-240, the latter linked up to the Italian National seismic array). A two-dimensional tilt meter completes the setup. It was operated from May 2015 onwards. In figure 4 we show a preliminary Allan deviation illustration of the rotational signal, compared with G data. Despite the fact that the mirrors presently mounted in the ring are old and no longer provide the best quality parameters, a level better than $2{\cdot}10^{-6}$ is achieved in

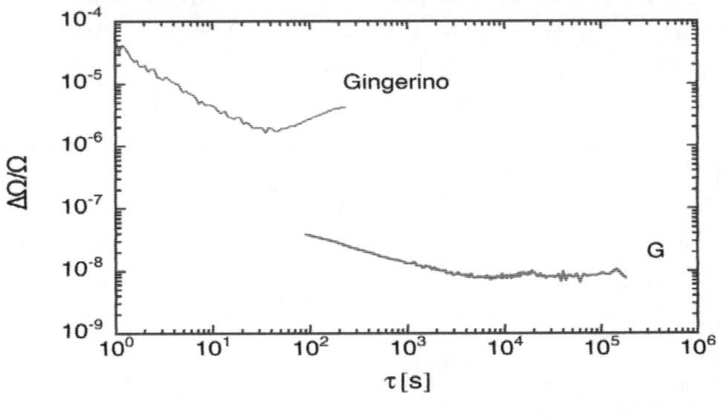

Figure 4 – Relative stability of the rotational signal (Allan deviation) from GINGERino and from G (color on line).

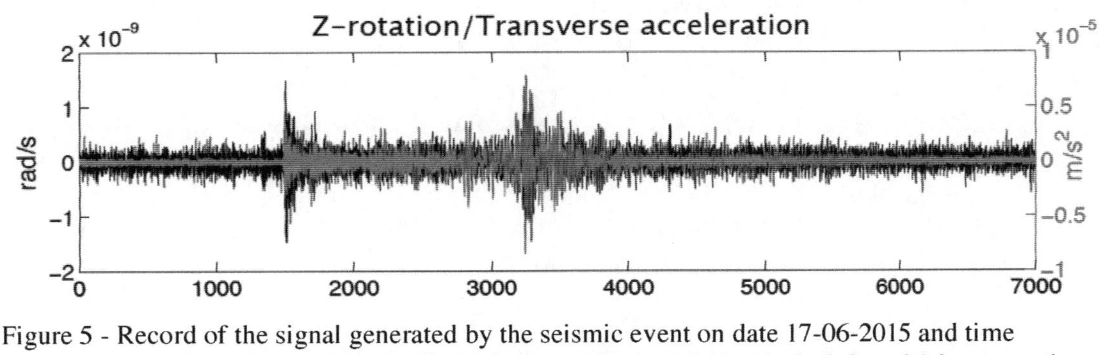

Figure 5 - Record of the signal generated by the seismic event on date 17-06-2015 and time 12:51:38 UTC in Southern Mid Atlantic Ridge (magnitude 6.4). In black (left scale) is reported the rotational signal from GINGERino; in red (right scale) the horizontal seismometer signal [13]. (Color on line)

100 s. Some Earthquakes were registered during these preliminary runs. As an example, Figure 5 shows the record of the rotational Love wave signal generated by the seismic event on 17-06-2015 at 12:51:38 UTC in Southern Mid Atlantic Ridge (magnitude 6.4), together with the seismometer signal. Work on the analysis of the correlation of seismic rotational and translational signals generated by this and by other earthquakes is presently in progress [13].

5. Conclusions

The recent performance of the G ring laser demonstrates the possibility for the ring laser technology to access the resolution required for sensing relativistic effects on rotation. Two main critical points have to be tackled for achieving the required progress. The first is connected to the residual uncontrolled deformation and tilts of the optical cavity of the ring resonator. The second regards the excess of ground rotational motion typically induced by the Earth surface environment. In particular the very low frequency tilts and rotations of the sensor site are critical. Two experiments in Italy are running for the investigation of these two aspects, GP-2 and GINGERino. GP-2 ring laser is designed for the study of active methods for controlling the deformation of the cavity shape, exploiting the stabilization of the diagonal lengths and is currently achieving a resolution of the order of 1 part in 10^{-9}. The performance of such an approach is ultimately limited by the frequency stability of the optical frequency reference and thus compatible with the geometrical stability required by GINGER. Preliminary experiments [14] provide promising results for the stability in the constrained length of the two resonators. The sensitivity of the geometrical scale factor against out of plane angle errors remains to be addressed. The GINGERino ring laser is investigating the local site stability inside the underground laboratory of the INFN in GranSasso. The ring laser stability is currently reaching a level of a few ppm at about 100 s and is limited the ring laser hardware, namely by the fluctuation laser parameters connected to the frequency pulling of a large amount of intra-cavity backscattered light. The present installation demonstrated a level of stability of the environmental parameters that is sufficient to operate the laser for several days with a drift of less than one longitudinal mode spacing. We expect an important reduction of the back-scattering effect by a new set of high quality mirrors that have been commissioned and will be installed in the next months. Improvements in the long term stability are also expected by a better isolation of the laser chamber (pressure stabilization). A full implementation of the already established backscatter correction algorithm is the logical next step. Preparations for this are currently in progress.

References:

[1] Stedman G E 1997 *Rep. Prog. Phys.* **60** 615–688
[2] Schreiber K U and Wells J-P R 2013 *Rev. Sci. Instrum.* **84** 041101
[3] Belfi J, *et al.* 2012 *Appl Phys B* **106**, 271–281
[4] Belfi J, *et al.* 2012 *J. Seismology* **16**, 767-776
[5] Bosi F, *et al.* 2011 *Phys Rev D* **84**, 122002
[6] Ciufolini I, Pavlis C 2004, *Nature* **431** 958-960
[7] Everitt C W F, *et al.*, 2011 *Phys. Rev. Lett.* **106** 221101.
[8] Schreiber K U, Klügel T, Wells J-P R, Hurst R B, Gebauer A 2011 *Phys. Rev. Lett.* **107** 173904
[9] Tartaglia A, Beverini N, Di Virgilio A, Ortolan A, Porzio A, Ruggiero M L 2014 *EPJ Web of Conferences* **74** 03001
[10] Cuccato D, Beghi A, Belfi J, Beverini N, Ortolan A, Di Virgilio A 2014 *Metrologia* **51** 97–107
[11] Santagata R, *et al.* 2015 *Class. Quantum Grav.* **32** 055013
[12] Belfi J, Beverini N, Carelli G, Cuccato D, Di Virgilio A, Maccioni E, Ortolan A, Santagata R 2014 *Class. Quantum Grav.* **31** 225003
[13] Simonelli A, *et al.* 2016 ArXiv 1601.05960 (in press on *Annals of Geophysics*)
[14] Santagata R 2015 *Sub-nanometer length metrology for ultra-stable ring laser gyroscopes* PhD thesis, University of Siena

Journal of Physics: Conference Series **723** (2016) 012062 doi:10.1088/1742-6596/723/1/012062

Paper Laser: a step towards a time scale generation from an ensemble of optical clocks

C A Ortiz[1,2], E de Carlos[1] and J M Lopez[1,2]

[1] Dirección de Tiempo y Frecuencia, Centro Nacional de Metrología, CENAM, km. 4.5 carretera a los Cués, El Marqués, 76246, Querétaro, México.
[2] Centro de Investigación y de Estudios Avanzados del Instituto Politécnico Nacional (IPN), CINVESTAV, Unidad Querétaro, Libramiento Norponiente No. 2000, Fracc. Real de Juriquilla, Santiago de Querétaro, 76230, Querétaro, México.

E-mail: jm.lopez@cinvestav.mx

Abstract. In this paper a simple and innovative technique to combine n optical frequencies with the aim to produce a virtual laser with superior metrological characteristics is introduced. The algorithms to combine a number of clocks to produce a virtual clock, which is also referred as paper clock, are well known. An example of this is the statistical generation of the UTC time scale by the Bureau International des Poids et Mesures (BIPM) using a recursive algorithm (ALGOS). A similar algorithm to combine n optical frequencies, all of them with same nominal value, to produce a "paper laser" whose frequency is known through its difference with respect to the optical frequencies of the ensemble is proposed here. As a demonstration of this, three optical frequencies stabilized to the D_2 Cs-133 line, all of them with similar frequency stability were experimentally combined. A paper laser has been produced during hours whose frequency stability is about $3^{-1/2}$ times with respect to the original optical frequencies. This technique can be applied to combine ultra-stable optical frequencies to produce a paper laser that can be materialized by correcting one of the real optical frequencies of the ensemble. The robustness and stability of a paper laser is very attractive to produce a time scale from its operation.

1. Introduction

In time and frequency metrology, optical frequencies present a lot of advantages in comparison with radio frequencies (RF) based on quartz or sapphire crystals. Some optical oscillators are able to have quality factors (Q) ten thousand times bigger than RF oscillators due to the ratio between the central frequency (10^{14} Hz) and its linewidth (around few Hz). State-of-the-art laser stabilization usually involves phase-locking to a single mode of a passive ultra-stable Fabry-Pérot (FP) cavity. That has allowed a vertiginous advance of the optical clocks during the past three decades and that, in turn, has motivated an international debate about the need to redefine the SI second in terms of an atomic transition in the optical region in order to take advantage of such a grade of stability and accuracy [1]. Also, there are a variety of fundamental physics tests that could be implemented as application of ultra-stable and accurate optical frequencies to increase our understanding of different principles of nature, which remains unexplored.

In addition, using frequency combs, the stability of ultra-stable optical oscillators can be transferred to a wide range of frequencies in the electromagnetic spectrum ranging from RF to the visible [2-5] achieving an outstanding stability. In this paper we propose a simple and innovative technique (similar to those used in the generation of averaged time scales) to improve the stability of an optical oscillator combining n lasers whose frequencies are previously stabilized to an optical reference.

1.1 Motivations

The production of time scales based on the operation of optical clocks is one of the requirements to redefine the SI second in terms of an optical frequency. In computation of time scales, it is desirable to operate simultaneously n different optical clocks. Even more, as in the microwave (MW) clocks case, it is certainly an interesting idea to combine an ensemble of optical clocks to produce a unique virtual optical clock with superior metrological characteristics compared to any member of the ensemble. Based on that, we explore the idea of combining an ensemble of optical frequencies to produce a unique virtual frequency. As a first approach, and to demonstrate the feasibility of this proposal, a simple algorithm has been developed and implemented experimentally to incorporate the frequency differences between three optical oscillators into a mathematical entity, named "paper laser" (PL).

2. The "paper laser" concept

In order to calculate a virtual optical oscillator, an ensemble of n optical oscillators with similar metrological characteristics is required and the frequency differences between them should be determined to feed a recursive algorithm. The PL calculated can be materialized using a member of the ensemble to correct its frequency in almost real time by using an acousto-optical modulator (AOM). In figure 1 a general simplified diagram of a PL generation and materialization is presented. Below the particular algorithm used in this work is described briefly.

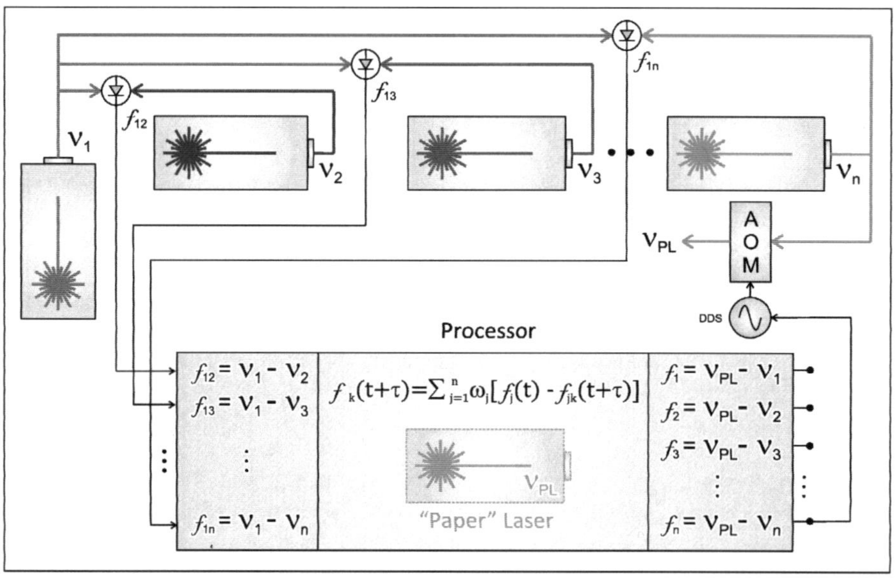

Figure 1. Simplified diagram of generation and materialization of a PL using an ensemble of n independent optical oscillators. AOM: acousto-optical modulator; DDS: direct digital synthesizer.

2.1 Algorithm

Using data obtained from measurements of frequency differences; a simple model, represented by the equation (1), is used to combine the optical frequencies into a single virtual optical frequency ν_{PL}. The frequency difference $f_k(t+\tau)$ at time $t+\tau$ between the PL optical frequency ν_{PL} and the optical frequency ν_k of the k-th member of the ensemble is given by the recursive equation:

$$\nu_{PL}(t+\tau)-\nu_k(t+\tau)=f_k(t+\tau)=\sum_{j=1}^{n}\omega_j\left[f_j(t)-f_{jk}(t+\tau)\right] \tag{1}$$

where f_{jk} is the frequency difference between the optical frequencies of lasers j and k and ω_j represents the weight of the j-th optical frequency of the ensemble. Of course, the condition of normalized weights is given by $\sum\omega_j=1$. For simplicity, and due to characteristics of our experimental setup, the weights are all equal to $1/3$ $(\omega_1=\omega_2=\omega_3=1/3)$. Moreover, the initial conditions on the optical frequency differences are fixed to zero, that is $f_1(t=0)=f_2(t=0)=f_3(t=0)=0$ Hz.

3. Experimental setup

3.1 The optical frequency ensemble

Similarly to the case of MW clocks, an algorithm to combine optical frequencies could produce better results when clocks of the ensemble have similar metrological characteristics. With that in mind, this experiment started with a laser coupled to an ultra-low expansion (ULE) cavity to produce an ultra-stable optical frequency. The master laser is a commercial AlGaAs which is an extended cavity diode laser (ECDL) equipped with a low loss interference filter, 2 W maximum output power, 852 nm wavelength (near to Cs-133's D_2 line) and 20 kHz linewidth. Additionally, a commercial optical FP cavity made of ULE glass is used. The ULE cavity has a free spectral range of 1.49 GHz and a linewidth less than 2.3 kHz. In order to stabilize the ECDL to the optical cavity, the Pound-Drever Hall technique is used [6]. In figure 2, the experimental setup of the ultra-stable laser (USL) is presented.

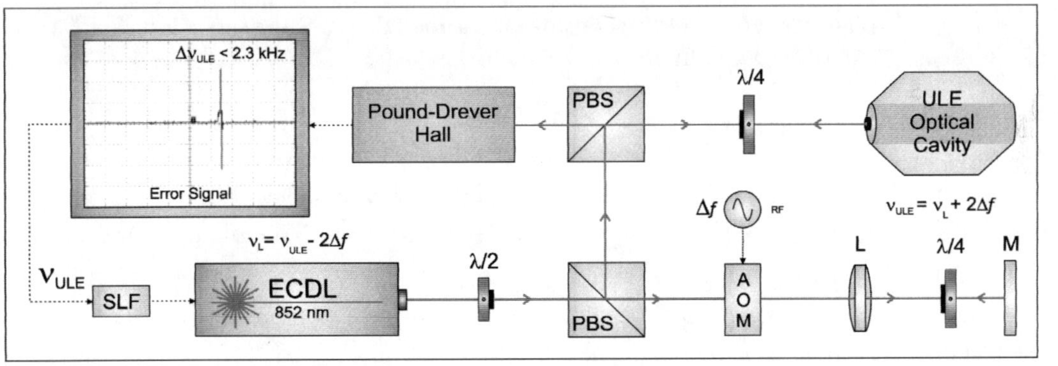

Figure 2. ECDL stabilized to an optical cavity made of ULE glass (USL). Red solid lines represent laser's light and black dashed lines represent electrical connections. PBS: polarizing beamsplitter; M: mirror; λ/2: half-wave plate; λ/4: quarter-wave plate; L: lens; AOM: acousto-optical modulator; SLF: servo loop filter. Colour online.

Then, as shown in figure 3, the USL is used to monitor the most probable D_2 Cs-133 transition $\left[|6^2 s_{1/2}, F=4\rangle \rightarrow |6^2 p_{3/2}, F'=5\rangle\right]$. Using independent AOMs in cat´s eye configuration [7], three independent laser beams (obtained by splitting the USL beam) are "re-locked" to the cesium D_2 line using the polarized spectroscopy technique [8]. To avoid magnetic interferences from external fields, the cesium's cells are located inside a double-layered μ-metal magnetic shield. The process of this frequency locking degrades the original frequency stability of the ultra-stable optical frequency but it results in three independent optical frequencies $(\nu_1,\ \nu_2 \text{ and } \nu_3)$ with similar metrological characteristics.

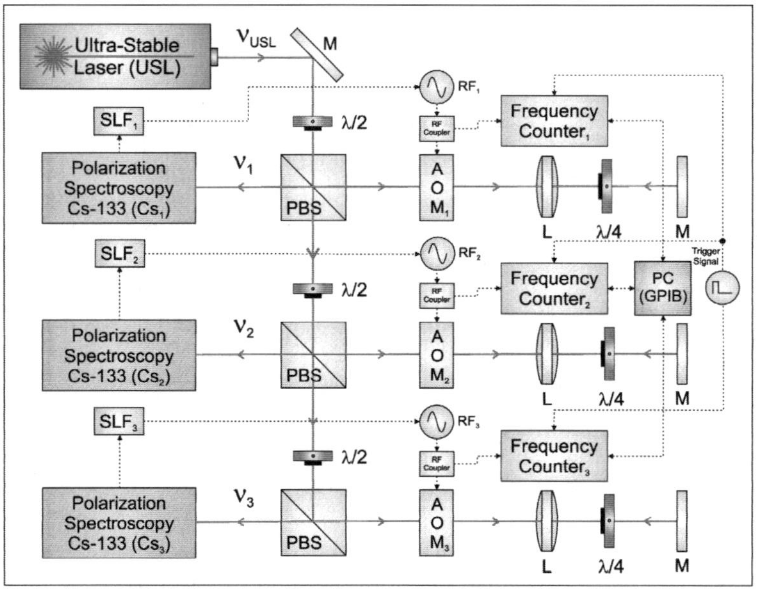

Figure 3. Setup implemented to measure frequency variations of three different and independent spectroscopy experiments simultaneously. Red solid lines represent laser's light and black dashed lines represent electrical connections. PBS: polarizing beamsplitter; M: mirror; λ/2: half-wave plate; λ/4: quarter-wave plate; L: lens; AOM: acousto-optical modulator; SLF: servo loop filter. Colour online.

3.2 Frequency measurements

Variations of frequency RF_i fed into the i-th AOM (i = 1, 2, 3) are measured (figure 3) to monitor the time variations of the frequency difference $\nu_i - \nu_{USL} = 2RF_i$. Due to the fact that ν_{USL} is an ultra-stable frequency, the variations of RF_i allow us to measure the short-term stability of the ν_i optical frequency $(\delta\nu_i \approx 2\delta RF_i)$. The factor of 2 is associated to the double-pass configuration. In order to have simultaneous RF_i measurements, an external trigger signal is sent to frequency counters each second and the readings are sent to a PC before the next trigger pulse.

4. Results and discussion

4.1 Stability analysis

Measurements were taken during 7 days (8 hours per day) to analyze a significant set of data in order to evaluate different kinds of effects affecting the experiment. The long-term drift of data, which is associated to frequency fluctuations of the ULE optical cavity, is suppressed when frequency measurements of each polarization spectroscopy experiment are subtracted given the fact that drift is common. Then, the algorithm uses these data and the PL is mathematically calculated every second. In order to evaluate the frequency stability of $f_k = \nu_{PL} - \nu_k$, the Allan deviation (ADEV) of the PL with respect to members of the ensemble is calculated. Preliminary results in figure 4 are promising because the stabilities of f_{jk} are smaller than the stabilities of the f_k frequencies. That is, the stability of the PL is superior to the stability of any member of the ensemble. Also for completeness, the relative stabilities between USL and PL and the relative stabilities between USL and the optical frequencies

$(\nu_1,\ \nu_2 \text{ and } \nu_3)$ are shown. Therefore, a virtual optical oscillator with superior stability features has been created.

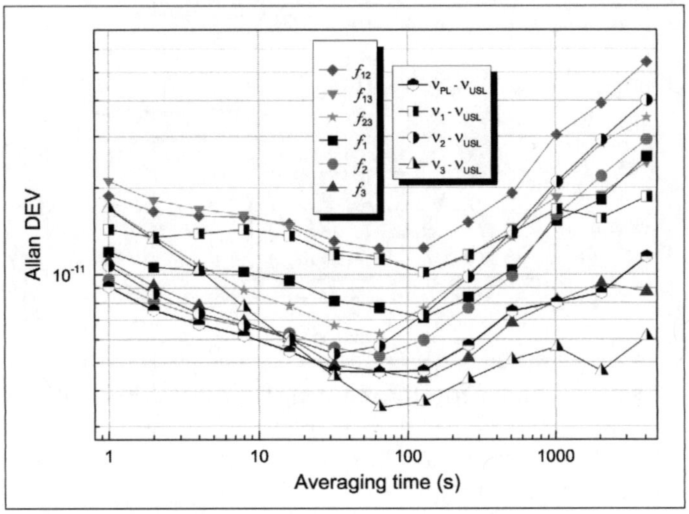

Figure 4. Allan deviation of frequency differences between optical frequencies of the ensemble $(f_{12}, f_{13}, \text{ and} f_{23})$, between optical frequencies and the virtual optical frequency $(f_1, f_2, \text{ and} f_3)$, between PL and USL, and between optical frequencies $(\nu_1,\ \nu_2 \text{ and } \nu_3)$ and the USL.

4.2 Noise analysis

It is well known that the power law noise implies particular slopes μ in the Allan deviation plots. The stability for a generic atomic clock limited by Poisson noise is given by [9]:

$$\sigma(\tau) \approx \frac{\delta\nu}{\nu_0}\frac{1}{\sqrt{N\tau}} \tag{2}$$

where $\delta\nu$ is the linewidth of the transition, ν_0 is the clock frequency, N is the number of detected atoms per second and τ is the averaging time. Analyzing the result of the best fit of the relative stability between the PL and the USL (Figure 5) by using an Allometric function $\sigma(\tau){\sim}A\tau^{-\mu/2}$, it is found that $\mu = 0.37$ and, according to equation (2), the stability of an optical oscillator, in the shot-noise limit, is proportional to $\tau^{-1/2}$. As can be noticed in figure 5, there is an additional contribution to frequency noise that can be attributed to the spectroscopy technique used. In this experimental setup the changes in the offset of the error signal are associated to beam misalignments produced by temperature fluctuations, vibrations during the data acquisition process and alterations in the polarization of the laser beams. Those drifts are erroneously interpreted by servos as frequency changes impacting the values of the relative stability between oscillators. Therefore, μ is smaller than the expected value $(\mu = 1)$. As a perspective to avoid that added noise, modulation transfer (MT) spectroscopy will be used in future experiments. In that case, it is expected that PL continues to have the best metrological features and the stability should be better with respect to results presented in this paper.

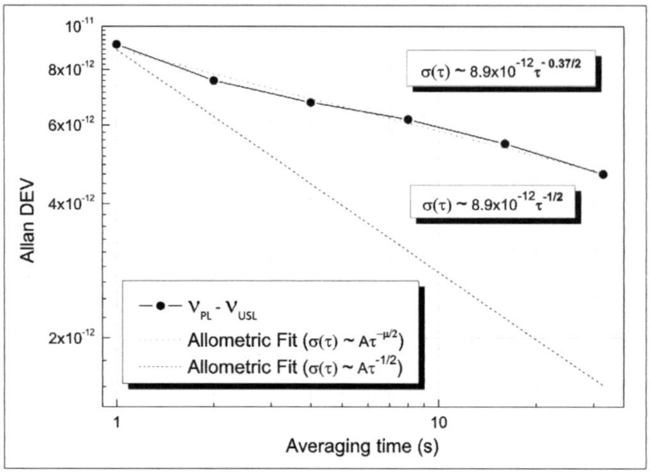

Figure 5. Allometric fit (red dotted) of the stability data corresponding to comparison between the PL and the USL. The dashed blue line corresponds to the Poisson noise limit. Colour online.

Finally, as can be noticed from stability data of figure 4, short-term stability of the virtual optical frequencies is $3^{-1/2}$ times with respect to the optical frequencies of the ensemble as expected, in concordance with equation (2).

5. Conclusions

In this paper a simple and innovative technique to combine optical frequencies into a virtual optical frequency (also called Paper Laser) has been introduced. The power of this technique is that by using a mathematical algorithm, a virtual optical oscillator with superior frequency stability could be generated. In this paper it has been demonstrated that, under particular conditions, the resulting PL has superior frequency stability than those of the optical frequencies of the ensemble, as expected.

Acknowledgments

C A Ortiz thanks CONACYT for his Ph. D scholarship and Cinvestav / CENAM for scientific and technical support.

References

[1] Margolis H 2014 Timekeepers of the future *Nat. Phys.* **10** 82–3
[2] Won R 2014 Frequency comb power *Nat. Phot.* **8** 168
[3] Anon 2011 The next step for metrology *Nat. Phot.* **5** 185
[4] Udem T, Holzwarth R and Hansch T W 2002 Optical frequency metrology *Nature* **416** 233–7
[5] Baynes F N, Quinlan F, Fortier T M, Zhou Q, Beling A, Campbell J C and Diddams S A 2015 Attosecond timing in optical-to-electrical conversion *Optica* **2** 141–6
[6] Black E D 2001 An introduction to Pound–Drever–Hall laser frequency stabilization *Am. J. Phys.* **69** 79–87
[7] Donley E a., Heavner T P, Levi F, Tataw M O and Jefferts S R 2005 Double-pass acousto-optic modulator system *Rev. Sci. Instrum.* **76** 3–8
[8] Wieman C and Hänsch T W 1976 Doppler-Free Laser Polarization Spectroscopy *Phys. Rev. Lett.* **36** 1170–3
[9] Hollberg L, Oates C W, Wilpers G, Hoyt CW, Barber Z W, Diddams S A, Oskay W H and Bergquist 2005 Optical frequency/wavelength references *J. Phys. B At. Mol. Opt. Phys.* **38** S469–S495

Institute of Physics
Dirac House, Temple Back
Bristol BS1 6BE UK

ISSN: 1742-6588
ISBN 978-1-5108-2611-3

9 781510 826113

CURRAN ASSOCIATES INC.
proceedings.com